Kurt Gödel and the Foundations of Mathematics

Horizons of Truth

This volume commemorates the life, work, and foundational views of Kurt Gödel (1906–1978), most famous for his hallmark works on the completeness of first-order logic, the incompleteness of number theory, and the consistency – with the other widely accepted axioms of set theory – of the axiom of choice and of the generalized continuum hypothesis. It explores current research, advances, and ideas for future directions not only in the foundations of mathematics and logic but also in the fields of computer science, artificial intelligence, physics, cosmology, philosophy, theology, and the history of science. The discussion is supplemented by personal reflections from several scholars who knew Gödel personally, providing some interesting insights into his life. By putting his ideas and life's work into the context of current thinking and perceptions, this book will extend the impact of Gödel's fundamental work in mathematics, logic, philosophy, and other disciplines for future generations of researchers.

Matthias Baaz is currently University Professor and Head of the Group for Computational Logic at the Institute of Discrete Mathematics and Geometry at the Vienna University of Technology.

Christos H. Papadimitriou is C. Lester Hogan Professor of Electrical Engineering and Computer Sciences at the University of California, Berkeley, where he has taught since 1996 and where he is a former Miller Fellow.

Hilary W. Putnam is Cogan University Professor Emeritus in the Department of Philosophy at Harvard University.

Dana S. Scott is Hillman University Professor Emeritus of Computer Science, Philosophy, and Mathematical Logic at Carnegie Mellon University in Pittsburgh.

Charles L. Harper, Jr., is Chancellor for International Distance Learning and Senior Vice President of Global Programs at the American University System, as well as President of Vision-Five.com Consulting, in the United States.

Photo courtesy of Notre Dame Archives.

Kurt Gödel and the Foundations of Mathematics

Horizons of Truth

Edited by

Matthias Baaz

Technische Universität Wien

Christos H. Papadimitriou

University of California, Berkeley

Hilary W. Putnam

Department of Philosophy, Harvard University

Dana S. Scott

Computer Science Department, Carnegie Mellon University

Charles L. Harper, Jr.

Vision-Five.com Consulting, United States

CAMBRIDGE
UNIVERSITY PRESS

CAMBRIDGE
UNIVERSITY PRESS

32 Avenue of the Americas, New York NY 10013-2473, USA

Cambridge University Press is part of the University of Cambridge.

It furthers the University's mission by disseminating knowledge in the pursuit of
education, learning and research at the highest international levels of excellence.

www.cambridge.org
Information on this title: www.cambridge.org/9781107677999

First published 2011
First paperback edition 2014

A catalogue record for this publication is available from the British Library

Library of Congress Cataloguing in Publication data
Kurt Gödel and the foundations of mathematics : horizons of truth / edited by
Matthias Baaz ... [et al.].
      p.   cm.
Includes bibliographical references and index.
ISBN 978-0-521-76144-4 (hardback)
1. Gödel's theorem.   2. Mathematics – Philosophy.   3. Gödel, Kurt.   I. Baaz, Matthias.   II. Title.
QA9.65.K87   2011
511.3–dc22   2010048055

ISBN 978-0-521-76144-4 Hardback
ISBN 978-1-107-67799-9 Paperback

To every ω-consistent recursive class κ of *formulae*, there correspond recursive *class signs* r, such that neither (v Gen r) nor Neg(v Gen r) belongs to Flg(κ), where v is the *free variable* of r.[1]

(Any not-too-weak consistent formal theory, in particular any reasonable formalization of number theory, cannot prove everything that is true; i.e., such a theory is necessarily incomplete.)

– Gödel's first incompleteness (undecidability) theorem, 1931

In any not-too-weak formal theory, the formalization of consistency implies the Gödel sentence, which is unprovable if the formal theory is consistent.

(If the formal theory is consistent, then its consistency cannot be proved within the formal theory.)

– Gödel's second incompleteness theorem, 1931

---

[1] See: http://mathworld.wolfram.com/GoedelsIncompletenessTheorem.html.

Contents

Contributors

John D. Barrow
Professor of Mathematical Sciences and
Director of the Millennium Mathematics
Project, Department of Applied
Mathematics and Theoretical Physics,
Centre for Mathematical Sciences,
University of Cambridge,
United Kingdom

Paul J. Cohen
Professor of Mathematics, Emeritus,
Department of Mathematics, Stanford
University, Stanford, California,
United States

B. Jack Copeland
Professor of Philosophy, Department of
Philosophy, University of Canterbury,
New Zealand

Solomon Feferman
Patrick Suppes Family Professor of
Humanities and Sciences, Emeritus,
and Professor of Mathematics and
Philosophy, Emeritus, Departments of
Mathematics and Philosophy, Stanford
University, Stanford, California,
United States

Harvey M. Friedman
Distinguished University Professor of
Mathematics, Philosophy, and Computer
Science, Department of Mathematics,
The Ohio State University, Columbus,
United States

Ivor Grattan-Guinness
Emeritus Professor of the History of
Mathematics and Logic, Middlesex
University Business School, and Centre
for Philosophy of Natural and Social
Science, London School of Economics,
United Kingdom

Petr Hájek
Professor of Mathematical Logic and
Senior Researcher, Institute of Computer
Science, Academy of Sciences of the
Czech Republic, Prague, and President,
Kurt Gödel Society, Vienna, Austria

Juliette Kennedy
Associate Professor, Department of
Mathematics and Statistics, University of
Helsinki, Finland

Ulrich Kohlenbach
Professor of Mathematics, Logic
Research Group, Department of
Mathematics, Technische Universität
Darmstadt, Germany

Georg Kreisel
Professor of Mathematics, Emeritus,
Department of Mathematics, Stanford
University, Stanford, California,
United States

Angus Macintyre
Professor of Mathematics, Department of
Mathematical Sciences, Queen Mary,
University of London, United Kingdom

Piergiorgio Odifreddi
Professor of Mathematical Logic,
Department of Mathematics, University
of Torino, Turin, Italy

Christos H. Papadimitriou
C. Lester Hogan Professor of Electrical
Engineering and Computer Sciences,
Computer Science Division, University
of California, Berkeley, United States

Roger Penrose
Emeritus Rouse Ball Professor of
Mathematics, Mathematical Institute,
University of Oxford, United Kingdom,
and Francis and Helen Pentz
Distinguished Professor of Physics and
Mathematics, Institute for Gravitation
and the Cosmos, Pennsylvania State
University, State College, United States

Hilary W. Putnam
Cogan University Professor, Emeritus,
Department of Philosophy, Harvard
University, Cambridge, Massachusetts,
United States

Wolfgang Rindler
Professor of Physics, Department of
Physics, The University of Texas at
Dallas, Richardson, United States

Karl Sigmund
Professor of Mathematics, Department of
Mathematics, University of Vienna,
Austria

Karl Svozil
Professor of Physics, Institute for
Theoretical Physics, Vienna University
of Technology, Austria

Denys A. Turner
Horace Tracy Pitkin Professor of
Historical Theology, Yale Divinity
School, New Haven, Connecticut,
United States

Avi Wigderson
Herbert Maass Professor of Mathematics,
School of Mathematics, Institute for
Advanced Study, Princeton, New Jersey,
United States

W. Hugh Woodin
Professor of Mathematics, Department of
Mathematics, University of California,
Berkeley, United States

Foreword

While I was writing some words to say about Professor Kurt Gödel's major works for his 2006 centenary celebration at the University of Vienna, it suddenly came to me that for everyone who gathered in his honor, Gödel's extraordinary contributions to and tremendous influence on mathematics would be something of which we were already deeply aware. Thinking that perhaps a repeat of Gödel's results would be unnecessary with this group, I decided to share some of my own personal memories that are recalled when I remember Professor Gödel.

I met Gödel for the first time at the Institute for Advanced Study in Princeton in January 1959, when he was fifty-two years old. At the time, I was a very young thirty-two-year-old whose only interest was my own problem within logic; I knew little of logic as a whole. Throughout my first stay in Princeton, Gödel taught me many new ideas, specifically about nonstandard models and large cardinals. On certain occasions, he would lead me to the library and show me the precise page of a book on which a pertinent theorem was presented, and he advised me on which books I should be reading. He even counseled me that I needed to improve my English to communicate with other mathematicians.

Gödel showed a keen interest in the problem on which I was working then: my fundamental conjecture, that is, the cut elimination theorem on the generalized logic calculus, which is the higher type extension of Gentzen's *logistischer klassischer Kalkül* sequent calculus, as introduced in 1934. At first, Gödel thought that one could find a counterexample using his incompleteness theorem or a nonstandard model. He thought that there must be a counterexample in every impredicative case, that is, a similar situation to my problem in the way that the incompleteness theorem holds. Interestingly, my fundamental conjecture trivially holds in the very impredicative cases. Professor Gödel was surprised to find this and became intrigued with my conjecture. He thought it would help my work if I could meet with Professor Kurt Schutte, whom he immediately invited to the institute. Professor Schutte found the model-theoretic formulation of my fundamental conjecture. Dr. Takahashi's and Dr. Pravitz's later works, which proved my conjecture, were based on Professor Schutte's result. Professor Gödel's insight was correct.

In my later visits with Gödel, we discussed the more philosophical aspects of logic. He seemed to believe that the cardinality of the continuum is small. His theory was that if one assumes the existence of a "beautiful scale" in the real numbers, this conclusion is inevitable. Although I had a hard time understanding this idea, our discussions were stimulating and gave me tremendous pleasure. I wish now that I had pursued his ideas further. My hope is that future generations will take up his many interesting concepts and develop them.

It would be Kurt Gödel's greatest delight to see how his ideas are alive and remain the cornerstones of modern logic more than thirty years after his death. Going further back in time, in a letter to Robert Hooke dated February 5, 1675, Sir Isaac Newton wrote of his own discoveries, "If I have seen further it is by standing on the shoulders of giants."[1] I believe this applies to all who gathered in 2006 and to those who have contributed their work to this volume – for Professor Gödel, although very shy, was truly a giant.

It was in the spirit of acknowledging Professor Gödel's ever-searching imagination and philosophical open-mindedness that the historic 2006 meeting took place and that this book was written. No doubt the legacy of the centenary and this volume will serve as an inspiration for yet other generations of mathematicians (and philosophers). Although I very much regretted that for reasons of health, I could not attend the 2006 celebration, I enthusiastically shared from afar the timeless spirit of Kurt Gödel, which lives on in this book and in our minds.

Gaisi Takeuti
Professor of Mathematics, Emeritus,
Department of Mathematics, University of Illinois,
Urbana-Champaign, United States, and
former President, Kurt Gödel Society,
Vienna, Austria

---

[1] http://www.quotationspage.com/quotes/Isaac_Newton/.

Preface

Kurt Gödel and the Foundations of Mathematics: Horizons of Truth is the culmination of a creative research initiative coorganized by the Kurt Gödel Society, Vienna; the Institute for Experimental Physics; the Kurt Gödel Research Center; the Institute Vienna Circle; the Vienna University of Technology; the Austrian Academy of Sciences; and the Anton Zeilinger Group at the University of Vienna, where the Gödel centenary celebratory symposium "Horizons of Truth: Logics, Foundations of Mathematics, and the Quest for Understanding the Nature of Knowledge" was held from April 27 to April 29, 2006.[1]

More than twenty invited world-renowned researchers in the fields of mathematics, logic, computer science, physics, philosophy, theology, and the history of science attended the symposium, giving the participants the remarkable opportunity to present their ideas about Gödel's work and its influence on various areas of intellectual endeavor. These fascinating interdisciplinary lectures provided new insights into Gödel's life and work and their implications for future generations of researchers.

The interaction among international scholars who only rarely, if ever, have the opportunity to hold discussions in the same room – and some of whom almost never write articles – has produced a book that contains chapters expanded and developed to take advantage of the rich intellectual exchange that took place in Vienna. Written by some of the most renowned figures of the scientific and academic world, the resulting volume is an opus of current research and thinking that is built on the work and inspiration of Gödel.

Several of the contributors were colleagues of or studied with Gödel: Gaisi Takeuti, who contributed the foreword and offers warm remembrances of Gödel's impact on his work; Georg Kreisel, who contributed a detailed chapter on logic and foundations; and Fields Medal winner Paul J. Cohen, who, sadly, died shortly after completing his personal reflections for this volume.

---

[1] See the symposium Web site for further information: http://www.logic.at/goedel2006/.

Also, a number of other chapters in this volume contain extensive biographical details about various aspects of Gödel's life and work to supplement the technical discussions. In addition, we include a short biography of Gödel's life that contains additional biographical resources and a select bibliography of his seminal works, which are frequently cited throughout this book.

The main content of the volume is divided into the following three major parts, broken down further into subparts to highlight the multidimensional impact of Gödel's contributions to academic advancement:

I Historical Context: Gödel's Contributions and Accomplishments
Gödel's Historical, Philosophical, and Scientific Work
Gödel's Legacy: A Historical Perspective
The Past and Future of Computation
Gödelian Cosmology

II A Wider Vision: The Interdisciplinary, Philosophical, and Theological
Implications of Gödel's Work
On the Unknowables
Gödel and the Mathematics of Philosophy
Gödel and Philosophical Theology
Gödel and the Human Mind

III New Frontiers: Beyond Gödel's Work in Mathematics and Symbolic Logic
Extending Gödel's Work
The Realm of Set Theory
Gödel and the Higher Infinite
Gödel and Computer Science

These topics cover not only the technical aspects of Gödel's work and its legacy but also the profoundly reflective aspects of his thinking, augmenting the appeal of the book and ensuring its interest to both a specialized and a multidisciplinary readership.

Because of the stature and diverse research areas of the contributors, we believe this book will appeal not only to mathematicians and logicians but also to computer scientists, physicists, astrophysicists, cosmologists, philosophers, theologians, historians of science, and postdoctoral and graduate students working in these areas as well as to educated and informed others interested in foundational questions.

We hope we have achieved our goal of creating a lasting impact on the academic community by further advancing the legacy of a man without whose stunning contributions to mathematical logic our world's intellectual culture certainly would have been diminished.

Acknowledgments

The editors wish to thank the Kurt Gödel Society, Vienna; the Institute for Experimental Physics; the Kurt Gödel Research Center; the Institute Vienna Circle; the Vienna University of Technology; the Austrian Academy of Sciences; and the Anton Zeilinger Group at the University of Vienna for coorganizing the Gödel centenary celebratory symposium in April 2006.

The symposium was sponsored by the John Templeton Foundation (JTF), United States, which also provided the funding for this book.[1] Additional funding for the centenary symposium was provided in Austria by the Federation of Austrian Industry; the Federal Ministry of Infrastructure; the Federal Ministry of Education, Science, and Culture; the city of Vienna; and the Austrian Mathematical Society. We also thank the Microsoft Corporation for its contribution.

The editors wish to acknowledge the contributions of a number of individuals who contributed to the Gödel research initiative:

Norbert Preining, associate professor at the Research Center for Integrated Science of the Japan Advanced Institute of Science and Technology, former research assistant at the Vienna University of Technology, and Marie Curie Fellow at the University of Siena, as well as publicity chair of the Kurt Gödel Society, codeveloped and cohosted the symposium at the University of Vienna in 2006, in conjunction with Matthias Baaz.

Hyung S. Choi, director of mathematical and physical sciences at JTF, assumed an integral role in developing the academic program for the symposium, in conjunction with Charles L. Harper, Jr. (in his former role as senior vice president and chief strategist of JTF).

Pamela M. Contractor, president and director of Ellipsis Enterprises Inc., working in conjunction with JTF and the volume editors, served as developmental editor of this

---

[1] "Supporting science, investing in the big questions": http://www.templeton.org/.

book along with Robert W. Schluth, senior editor and program director, and Matthew P. Bond, assistant editor and manager, client services, at Ellipsis.

Finally, the editors thank Lauren Cowles, senior editor for mathematics and computer science at Cambridge University Press, New York, for supporting and overseeing this book project.

Short Biography of Kurt Gödel

Gödel seated in the Mathematics–Natural Sciences Library at Fuld Hall, Institute for Advanced Study, Princeton, New Jersey, 1963. Photograph by Alfred Eisenstaedt from the Time and Life Pictures collection. Reproduced with permission from Getty Images.

Gödel's signature. The electronic signature is used with permission from the Institute for Advanced Study.

Kurt Friedrich Gödel is considered one of the most outstanding mathematical logicians of the twentieth century and is thought by many to be the greatest logician since

Aristotle. He was born on April 28, 1906, in what was then Brünn in the Austro-Hungarian Monarchy and today is Brno in the Czech Republic. After attending school in Brünn and graduating with honors, he enrolled at the University of Vienna in 1923 with the original intention of studying physics. He attended lectures on number theory by professor Philipp Furtwängler, who, paralyzed from the neck down, lectured from his wheelchair. Thereafter Gödel became interested in mathematical logic, the field to which he would make his major contributions. As a student, he also attended meetings of what would later become the Vienna Circle (Wiener Kreis), a group composed mainly of philosophers that met to discuss foundational problems, inspired by Ludwig Wittgenstein's *Tractatus Logico-Philosophicus*. The group focused on questions of language and meaning and logical relations such as entailment, originating logical positivism (logical empiricism). Led by Moritz Schlick, who was later murdered by a deranged former student in 1936 at the University of Vienna, its members included Rudolf Carnap, Otto Neurath, Carl Hempel, Hans Reichenbach, Hans Hahn, Karl Menger, and others. A Platonist from an early age, Gödel disagreed with many of his colleagues' views, yet the Vienna Circle had a major influence on his thinking. In his doctoral dissertation, written under the supervision of Hans Hahn, he proved the completeness of first-order predicate logic with identity, which states that any sentence that holds in every model of the logic is derivable in the logic. His dissertation was finished in 1929, and the result was published in 1930.

Also in 1930, at a meeting in Königsberg (David Hilbert's hometown) on September 7 that was attended by, among others, John von Neumann, Gödel, still in his mid-twenties, announced his work demonstrating that systems of mathematics have limits. In particular, he showed that any not-too-weak consistent formal theory (say, any reasonable formalization of number theory) cannot prove everything that is true; that is, such a theory is necessarily incomplete.

Gödel's startling results in formal logic, considered landmarks of twentieth-century logic, were published as the now-famous incompleteness theorems the following year, in 1931, ending many years of attempts to find a set of axioms sufficient for all mathematics and implying that not all mathematical questions are formally solvable in a fixed system. Gödel demonstrated, in effect, that hopes of reducing mathematics to an axiomatic system, as envisioned by mathematicians and philosophers at the turn of the twentieth century, were in vain. His findings put an end to the logicist efforts of Bertrand Russell and Alfred North Whitehead and demonstrated the severe limitations of David Hilbert's program for arithmetic. In the introduction to his 1931 paper, Gödel stated:

> It is well known that the development of mathematics in the direction of greater precision has led to the formalization of extensive mathematical domains, in the sense that proofs can be carried out according to a few mechanical rules. . . . It is reasonable therefore to make the conjecture that these axioms and rules of inference are also sufficient to decide all mathematical questions, which can be formally expressed in the given systems. In what follows it will be shown that this is not the case.[1]

In addition to his proof of the incompleteness of formal number theory, Gödel (1938, 1939, 1953, 1990) published proofs of the relative consistency of the axiom of choice

---

[1] See Gödel (1931) and http://www.ias.edu/people/godel.

Gödel with Einstein at the Institute for Advanced Study, early 1950s. Photograph by Richard F. Arens. From the Shelby White and Leon Levy Archives Center, Institute for Advanced Study, Princeton, New Jersey. Reproduced with permission.

and of the generalized continuum hypothesis. His findings strongly influenced the (later) discovery that a computer can never be programmed to answer all mathematical questions.

After obtaining his *Habilitation*, Gödel joined the faculty of the University of Vienna in 1930, becoming a *Privatdozent* (unsalaried lecturer) in 1933. He would remain there until the Anschluss in 1938, when Austria became part of Nazi Germany. During the 1930s, he made several visits abroad: to the Institute of Advanced Study in Princeton (1933–1934, 1935, and 1938), where he would eventually settle; to the University of Göttingen (1938), where he gave lectures on set theory; and to the University of Notre Dame (1939), where he worked with the newly emigrated Karl Menger. By 1938, Gödel saw that his position as *Privatdozent* would not be continued, and he feared that he would be drafted into the army. He left Europe with his wife via the Trans-Siberian Railway in January 1940; they arrived in San Francisco by ship on March 4. They would never return to Austria.

In Princeton, Gödel joined the Institute for Advanced Study, where he was professor in the School of Mathematics from 1953 until 1976, when he became professor emeritus, holding the mathematics chair until his death from malnutrition on January 14, 1978. There Gödel's interests turned increasingly to philosophy and physics. In the 1940s, he was able to demonstrate the existence of paradoxical solutions to Einstein's field equations in the theory of general relativity, which allowed for the possibility of time travel into the past. Gödel's theorems and other theoretical explorations in physics and philosophy helped usher in the age of computer technology, influencing the innovative work of John von Neumann, Alan Turing, and others in computer science that has so profoundly influenced the world and our attempts to understand and manage it.[2] Gödel's last published paper appeared in 1958.

---

[2] In fact, the most prestigious award for a research contribution in theoretical computer science is called the "Gödel Prize."

During his life, Gödel received several prizes and honorary doctorates and member-ships (and rejected some others). Among them were the Institute for Advanced Study's Einstein Award (1951) as well as the National Medal of Science (in the disciplines of mathematics and computer science) from President Ford (1974) for "laying the foun-dation for today's flourishing study of mathematical logic." Gödel received honorary doctorates from Yale, Harvard, and Rockefeller universities and from Amherst College. He was a member of the National Academy of Sciences of the United States, a foreign member of the Royal Society of London, a corresponding member of the Institute of France, a corresponding Fellow of the British Academy, and an honorary member of the London Mathematical Society. In 2000, *Time* magazine included Gödel among its top one hundred most influential thinkers of the twentieth century.

As noted in the editors' preface, a number of chapters in this volume contain exten-sive biographical details about various aspects of Gödel's life and work to supplement the technical discussions. In the following, we provide additional biographical re-sources and a select bibliography of Gödel's seminal works, which are frequently cited throughout this book.

Additional Biographical Resources

Dawson, John W. Jr. *Logical Dilemmas: The Life and Work of Kurt Gödel*. Wellesley, MA: A K Peters, 1997.

Institute for Advanced Study. http://www.ias.edu/people/godel.

Kurt Gödel Papers. Princeton University Library Manuscripts Division. http://diglib.princeton.edu/ead/getEad?id=ark:/88435/v979v310g#bioghist.

Kurt Gödel Society. http://kgs.logic.at/index.php?id=23.

Sigmund, Karl, and John Dawson. *Gödel's Jahrhundert Ausstellung (Gödel's Centenary Exhibition)*. http://www.goedelexhibition.at/start/.

Sigmund, Karl, John Dawson, and Kurt Mühlberger. *Kurt Gödel: The Album*. Wiesbaden, Germany: Vieweg, 2006. (Available in German and English)

Select Bibliography of Gödel's Seminal Works

Gödel, K. (1930). Die Vollständigkeit der Axiome des logischen Funktionenkalküls. *Monatshefte für Mathematik und Physik*, **37**, 349–60. [Published PhD diss.]

———. (1931). Über formal unentscheidbare Sätze der *Principia Mathematica* und verwandter Systeme I. *Monatshefte für Mathematik und Physik*, **38**, 173–98. [English trans. J. van Heijenoort, ed. (1967). *From Frege to Gödel: A Source Book on Mathematical Logic*. Cambridge, MA: Harvard University Press, pp. 596–616. Repr. with facing English trans. On formally undecidable propositions of *Principia Mathematica* and related systems. I. In *Collected Works*, vol. 1 (1986), pp. 145–95.]

———. (1938). The consistency of the axiom of choice and of the generalized continuum-hypothesis. *Proceedings of the National Academy of Sciences of the United States of America*, **24**, 556–7. [Also in *Collected Works*, vol. 2 (1990), pp. 26–7.]

———. (1939). Consistency-proof for the Generalized Continuum Hypothesis. *Proceedings of the National Academy of Sciences, USA*, **25**, 220–4. [Also in *Collected Works*, vol. 2 (1990), pp. 27–32.]

_____. (1949a). A remark about the relation between relativity theory and idealistic philosophy. In *Albert Einstein: Philosopher-Scientist*, ed. P. A. Schilpp, pp. 557–62. Library of Living Philosophers 7. Evanston, IL: MJF Books. [Also in *Collected Works*, vol. 2 (1990), pp. 202–7.]

_____. (1949b). An example of a new type of cosmological solution of Einstein's field equations of gravitation. *Reviews of Modern Physics*, **21**, 447–50. [Also in *Collected Works*, vol. 2 (1990), pp. 190–8.]

_____. (1952). Rotating universes in general relativity theory. In *Proceedings of the International Congress of Mathematicians*, vol. 1, ed. L. M. Graves et al., pp. 175–81. Cambridge, MA: American Mathematical Society. [Also in *Collected Works*, vol. 2 (1990), pp. 208–16.]

_____. (1953 [1940]). The consistency of the axiom of choice and of the generalized continuum-hypothesis with the axioms of set theory. In *Annals of Mathematics Studies*, vol. 3, rev. ed. Princeton, NJ: Princeton University Press. [Also in *Collected Works*, vol. 2 (1990), pp. 33–101.]

_____. (1958). Über eine bisher noch nicht benüzte Erweiterung des finiten Standpunktes. *Dialectica*, **12**, 280–87. [Repr. English trans. On a hitherto unutilized extension of the finitary standpoint. In *Collected Works*, vol. 2 (1990), pp. 241–51.]

_____. (1964 [1947]). What is Cantor's continuum problem? *American Mathematical Monthly*, **54**, 515–25. [Rev. version in P. Benacerraf and H. Putnam, eds. (1984 [1964]). *Philosophy of Mathematics*. Englewood Cliffs, NJ: Prentice Hall, p. 483. Also in *Collected Works*, vol. 2 (1990), pp. 176–87 (1947 version); pp. 254–70 (1964 version).]

_____. (1986 [1929]). On the completeness of the calculus of logic. PhD diss. In *Collected Works*, vol. 1, pp. 61–101.

_____. (1986–2003). *Collected Works*. 5 vols. Edited by S. Feferman et al. Vols. 1–3, New York: Oxford University Press. Vols. 4 and 5, Oxford: Clarendon Press. [Throughout, referenced as *Collected Works* by vol. number, year, and page.]

_____. (1990 [1939]). The consistency of the generalized continuum hypothesis. In *Collected Works*, vol. 2, p. 27.

_____. (1995a [1949]). Lecture on rotating universes; given at the Institute for Advanced Study, Princeton, May 7, 1949. In *Collected Works*, vol. 3, pp. 269–87.

_____. (1995b [1970]). Ontological proof. In *Collected Works*, vol. 3, pp. 403–4. [Introductory note by R. M. Adams, pp. 388–402. Appendix B: Texts relating to the ontological proof, including Gödel's first version, 1941, pp. 429–37.]

PART ONE

Historical Context: Gödel's Contributions and Accomplishments

Gödel's Historical, Philosophical, and Scientific Work

CHAPTER 1

The Impact of Gödel's Incompleteness Theorems on Mathematics

Angus Macintyre

The Incompleteness Theorems are best known for interpretations put on them beyond – sometimes far beyond – mathematics, as in the Turing Test, or Gödel's belated claims in the Gibbs Lecture, or in Penrose's more recent work. In this chapter, the emphasis is much narrower, in contrast to the title of the 2006 Gödel Centenary Conference in Vienna – "Horizons of Truth: Logics, Foundations of Mathematics, and the Quest for Understanding the Nature of Knowledge" – that preceded the development of this volume. What is discussed here is almost exclusively the impact on pure mathematics.

That the 1931 paper had a broad impact on popular culture is clear. In contrast, the impact on mathematics beyond mathematical logic has been so restricted that it is feasible to survey the areas of mathematics in which ideas coming from Gödel have some relevance. My original purpose in my presentation at the conference was simply to give such a survey, with a view to increasing resistance to the cult of impotence that persists in the literature around Gödel. After the Vienna meeting, Kreisel persuaded me to write an appendix providing some justification for claims I had made concerning formalizing in First-Order Peano Arithmetic (PA) Wiles' proof of Fermat's Last Theorem. Our discussions on this have, in turn, affected Kreisel's (2008) paper, which provides indispensable proof-theoretic background for the appendix.

The Incompleteness Theorems and their proofs are strikingly original mathematics, with something of the charm of Cantor's first work in set theory. The original presentation is free of any distracting reference to earlier mathematics. Unique factorization is used (not proved), but what is used of the Chinese Remainder Theorem is proved. In fact, the paper would have lost little of its drama if the definability of exponentiation from addition and multiplication had been missed. However, this has always seemed to me the prettiest part of the paper, and nearly forty years passed before this definition was refined in the negative solution of Hilbert's 10th problem.

The grand event in Vienna celebrated many aspects of Gödel's work, but I find it hard to imagine that such a popular event would have happened if Gödel had not proved the Incompleteness Theorems of 1931 or had missed such a snappy proof. This is said not to play down the drama and importance of his and Cohen's work on the continuum hypothesis (or Gödel's work in cosmology). For popular scientific culture, the set-theoretic independence, however troubling or challenging, is less dramatic than the general incompleteness phenomenon. In particular, it is hard to imagine the set-theoretic independence alone giving rise to speculations about creativity. (To my knowledge, the independence results for non-Euclidean geometry did not give rise to such speculations, though they did lead to slogans about breakdown of intuitions.) The classic set-theoretic independence results initiated a great, and still ongoing, creative effort, leading to profound mathematics whose future development we can scarcely guess. On the other hand, the technique of diagonalization, so much admired in popular accounts, has had one infusion of new ideas, with the priority methods that first appeared fifty years ago, but it has now gone rather a long time without startling developments.

1.1 His Contemporaries in Logic

I think that one can get a proper perspective on Gödel only by considering the work of his exact contemporaries in logic because there is certainly a case to be made that others had ideas and results that were to prove more fertile than Gödel's. The relevant points are as follows:

- Presburger's work on the ordered abelian group $\mathbb{Z}$, though neither as original nor as difficult as Gödel's work, is in no way dated. Here we have completeness of an intelligible set of axioms, a useful quantifier elimination, and decidability. The result underlies much of p-adic model theory and model-theoretic motivic integration (Denef and Loeser, 2001). There is extensive literature on the computer science side. Lately, Scowcroft (2006) has refined the definability results to give connections to linear geometry and the work of Weyl.
- Tarski's work on real-closed fields (again, a completeness theorem) has inspired the modern theory of o-minimality (see later), which has brought together model theory and real analytic geometry and led to applications in Lie theory. In contrast, Tarski's work on truth, closer to Gödel's work, has had little effect on mathematics beyond logic.
- Skolem had a broad mathematical range and made diverse, suggestive contributions to logic. In the hands of an imaginative mathematician, the Löwenheim-Skolem theorem can be very powerful. In addition, Skolem's work on the ultrapower construction was of exceptional originality and the origin of many fundamental results in pure model theory and the model theory of arithmetic and set theory.
- Herbrand, before Gödel, and Gentzen, after Gödel, did the first (and still among the best) combinatorial work in the transformation of proofs. At this time, Gödel was not doing combinatorial proof theory but rather dramatic unprovability theory. The ideas of Herbrand and Gentzen have been extensively developed in theoretical computer science. On the other hand, functional interpretations, to which Gödel made major contributions, have been very successful in the unwinding of proofs. Very elementary considerations around growth rates of Herbrand terms, brought out by Kreisel, have proved very efficient and memorable in unwinding bounds for the number of solutions in the classical finiteness theorems (Luckhardt, 1996).
- Herbrand and Skolem made important contributions to the thriving number theory of their time, Herbrand to ramification theory and cohomology and Skolem to p-adic

analytic proofs of finiteness theorems for Diophantine equations. Gödel is known to have attended advanced lectures on class field theory but used in his work no more than the Chinese Remainder Theorem, the most primitive of all local-global principles. Julia Robinson (1999) would need much more, around modern formulations of quadratic reciprocity, in her definition of the integers in the rationals. Of course, Gödel's use of the Chinese Remainder Theorem, to code recursions, remained a central idea in the long evolution of his ideas toward the solution of Hilbert's 10th Problem.

- Ramsey's Theorem, proved in connection with what is now regarded as a very peripheral decision problem, was the starting point for combinatorial work that has flourished for many years and is currently very rewarding in connection with harmonic analysis. From the mid-1950s onward, the technology of indiscernibles has been involved in many of the deepest proofs in mathematical logic. That a recursion-theoretic analysis of Ramsey theory could be rewarding was not seen until the late 1960s (Specker, 1971). Curiously, this analysis seems to have had no influence on the the discovery of "Ramsey Incompleteness."

1.2 The Mathematical Evolution of the Ideas

Despite the awe with which the results are still described, the basic ideas are easy and were quickly adapted to obtain a number of striking consequences:

- Even before Gödel, work was under way on recursive definitions. Gödel provided the central technique for negative results, but Turing gave the subject general interest by providing a rigorous notion of machine and computation and by showing the equivalence of his notion of computability to the rather less natural notions based on formal recursion or lambda recursion. Again, Gödel contributed the main technique for proving noncomputability, from which Church readily proved undecidability of the decision problem for first-order predicate calculus. The arithmetization technique was demystified, leading to such notable results as Turing's on universal machines. Somewhat later, one looked at the finer structure of recursively enumerable sets, developed relative computability, and posed the influential Post problem, whose solution led to the introduction of new techniques of diagonalization.
- Rather more slowly, one moved toward the fine structure of definitions in arithmetic. Here Gödel's coding of recursions by the Chinese Remainder Theorem was central to the repertoire, while being recognized as too weak on its own to yield undecidability of Hilbert's 10th problem. At the level of set theory, Tarski adapted Gödel's diagonal argument to get the classic results on undefinability of truth.
- After a while, one had the very polished treatment of undecidable theories in Tarski (1968). The landmark problems, the word problem for groups and Hilbert's 10th, posed well before Gödel, took longer to solve. In the end, both negative solutions revealed some unforeseen structure.

  In the case of Hilbert's 10th problem, there was certainly no reason to think it might be decidable. At the time of Gödel's first work, the only large-scale decidability was in the area of quadratic forms (Siegel), and it took another thirty-five years before Baker's flexible method gave a variety of effective estimates, including a decision procedure for elliptic curves over $\mathbb{Q}$. It still seems to me extremely optimistic to have expected that all recursively enumerable sets should be Diophantine, and miraculous that it turned out so. That said, it remains disappointing that so little has followed from the result. Putnam's neat observation that every Diophantine set is the set of positive values of a polynomial gives the startling result that the set of primes is the set of positive values of a polynomial (and one has been able to spell out such polynomials). But I think it fair to say that the example remains merely striking, and there is no hint of any underlying theory of such representations (e.g., in geometric terms). Again, one would have hoped that the hardcore theory of recursively enumerable sets might by now have told us something quite new and suggestive about Diophantine sets, but this has not happened, and it is notable that no known method for the undecidability of Hilbert's 10th problem gives any information about the corresponding problem for the field of rationals. This makes it all the more encouraging that Poonen (2009) has recently made the first serious progress in that area for many years by showing that the integers have a universal-existential definition in the rational field (improving the old result of Julia Robinson).

- Sarnak's (2006) recent Rademacher lectures on solving equations in primes have an interesting point of contact with the Putnam trick mentioned earlier. It turns out that to get the right formulation of the rather deep theorems of these lectures, he has to allow primes to be positive or negative. Forcing them to be positive would bring the pitfalls of undecidability too close.

1.3 How the Number Theorists React to the Gödel Phenomenon

There has certainly been no cult of impotence. In the last thirty-five years, number theory has made sensational progress, and the Gödel phenomenon has surely seemed irrelevant. The number theorists are keenly aware of issues of effectivity (Hindry and Silverman, 2000) and, indeed, of relative effectivity. Moreover, each of the classical finiteness theorems is currently lacking expected effective information. However, there is not the slightest shred of evidence of some deep-rooted ineffectivity. Some key points are the following:

- The dimensions (of varieties) where unsolvability of Hilbert's 10th problem sets in are far beyond those where current, theory-driven research in number theory takes place or concern varieties with no discernible structure. Lang says somewhere that no undecidability is to be expected for abelian varieties. The sense (and plausibility) of this is pretty clear, even if one can contrive undecidability results about abelian varieties, for example, by considering period lattices with nonrecursive generators.
- The equations whose unsolvability is equivalent (after Gödel decoding) to consistency statements have no visible structure and thus no special interest. Recall Gauss's dismissive remarks about the ad hoc nature of the Fermat equation. What we now appreciate in Wiles' great work is not that a specific family of equations has no nontrivial solution but the link to modularity of elliptic curves over $\mathbb{Q}$ and the profound structure underlying modularity. This has been followed, in an amazingly short time, by fundamental results about Galois representations and the Langlands program (i.e., Sato-Tate conjecture, Serre modularity conjecture).

- The problem of deciding whether curves over $\mathbb{Q}$ have integer points is not yet known to be decidable, but there is a bodyguard of theory, quite independent of logical considerations and by now heavily supported by numerical evidence, that implies that undecidability is not to be expected. Indeed, it is to be dreaded (and this is Shafarevich's "gloomy joke") because of its implications for high theory. There are several proofs of Siegel's theorem, each with at least one region of current ineffectivity (Roth's theorem or the Mordell-Weil theorem). In the case of the Mordell-Weil theorem (needed for abelian varieties in the standard proof of Siegel's theorem), Manin (see Hindry and Silverman, 2000, 463) has shown that two dominant analytic conjectures imply effective bounds for generators of the Mordell-Weil group and thereby remove one of the two regions of ineffectivity. The first conjecture concerns the L-series of abelian varieties A over $\mathbb{Q}$ and says, first, that the series has an analytic continuation to the whole complex plane and, second, that it satisfies a standard functional equation involving the conductor N. By modularity considerations, the conjecture is known to be true for elliptic curves over Q. In general, if one assumes the analytic continuation, one then has the second conjecture, of Birch and Swinnerton-Dyer, relating the order of vanishing at $s = 1$ to the rank of the Mordell-Weil group and giving the leading term of the expansion around $s = 1$. Manin's proof involves showing how these two conjectures give a bound for the regulator and thereby for generators of the Mordell-Weil group. Thus dominant analytic conjectures reduce the ineffectivity to that coming from Roth's theorem.

  Kreisel has made me aware of the curious impact on Weil of issues around incompleteness and undecidability. In 1929, before Gödel, as described in Weil (1980, 526), he confronted issues of effectivity for elliptic curves and made little progress. Much later, in the notes (Weil, 1980) provided for his *Collected Works*, Weil speculates about inherent ineffectivity for the decision problem for curves and expresses the hope that progress in mathematical logic will bring these problems within reach. It seems that he was unaware of how devastating such ineffectivity would be for the edifice of conjectures in modern arithmetic.

- Serre gives another perspective (1989, 99, remark 3). He derives Siegel's theorem from Mordell-Weil and an approximation theorem on abelian varieties. The latter is naturally deduced from Mordell-Weil and Roth. Serre updates some observations of Weil and remarks that from effective Mordell-Weil, one can dispense with Roth in favor of a known abelian analogue of Baker's lower bounds for linear forms in logarithms. Thus it seems that the two analytic conjectures suffice to get the decidability of the decision problem for curves.

- A propos of the decision problem for curves, the natural logical parameters of the problem, such as number of variables and degree of polynomials involved, obscure the geometrical notions (notably genus but also projective nonsingularity and Jacobian varieties) that have proved indispensable to much research since Siegel's work of 1929. (I say "much" because some methods of Diophantine approximation are not naturally described as geometric.) If one's formalism obscures key ideas of the subject, one can hardly expect logic alone to contribute much.

- There is no hint of incompleteness linked to classical analytic number theory. It was, of course, natural to raise the issue of such incompleteness at the time when the prime number theorem did not yet have an "elementary" proof. However, if one is familiar with the standard analytic proofs, using, for example, analytic continuation, Kreisel's sketches (1951, 1952a, 1952b) from the 1950s are totally convincing. Note, too, that *sketch* here is anything but snide. Sketches are the appropriate medium for this kind of mathematics. The whole enterprise is futile for those who do not know the analytic proofs, whereas once one knows them, it typically requires little effort to reorganize the proof into a weaker system. I recommend Kreisel's brief account (2008).

- There is no hint of incompleteness (in the sense that one may need set theory) in the work of Deligne on the Weil conjectures or Faltings on the Mordell conjecture. At the Vienna meeting, I expressed confidence that the Wiles proof of Fermat's last theorem fits into PA. If the discussions on FOM (the "Foundations of Mathematics" moderated e-mail list)[1] are to be taken as representative of opinion among logicians, my claim remains controversial. In the appendix to this chapter, I attempt to open some informed discussion on the matter.

- The appendix is mainly concerned with the role of set theory in the formulation and/or proof of modularity conjectures such as the restricted one used by Wiles in his proof of Fermat's last theorem. One goal is to show that the conjectures themselves are Π^0_1. For this, it is crucial to use the input of Weil to the modularity conjecture, giving precise formulations in terms of *conductors* (see Weil, 1999). Beyond this issue of formulation, one has to survey proofs of modularity and show that each component can be arithmetized. This is not at all a trivial enterprise, even if one aims only at an overview of the proof.

We are fortunate to have the guide (Cornell et al., 1997) to the original proof. This reveals a number of distinct regions of the proof, almost all hitherto unvisited by proof theorists. These include Galois representations, distribution of primes, étale cohomology, modular forms and functions, elliptic curves and their L-series, the analysis behind Langlands's theory (e.g., the proof of Langlands-Tunnell), and advanced commutative algebra of noetherian rings. Even prior to all this, there is basic p-adic and adelic analysis. Some but certainly not all this territory is known to some model theorists who have learned something about definitions and uniformities therein. Moreover, this group of people typically know a fair amount of complex analysis and geometry, and some are accustomed to keeping track of the complexity of inductive arguments.

It is not clear to me what further useful proof-theoretic tool kit one has for the preceding enterprise. I claim that there is no need for a proof of Wiles' theorem (or beyond, e.g., to the modularity theorem) of the use of strong second-order axioms with existential quantifiers involved. The claim leaves open the question of the utility of second-order *formalism* and of metatheorems (such as conservation results) formulated in functional terms. It seems to me likely that these will be useful a little further down the road, if enough logicians begin to pay attention to the structure of large-scale proofs of the type of Wiles (Colin McLarty has recently circulated an interesting paper in this area.) In addition, my own experience suggests that such a global structural analysis will bring to light many local issues of model-theoretic and proof-theoretic interest. For my own purposes, in getting the appendix written, I have relied (perhaps too exclusively) on rules of thumb (and tricks of the trade) from model-theoretic algebra and the model theory of PA and related systems. For a compact account of the basic proof-theoretic

---

[1] See http://www.cs.nyu.edu/mailman/listinfo/fom/.

issues, informed by over sixty years of case studies, one should consult Kreisel (2008). Thus I choose to give arithmetic interpretations of, for example, those parts of real, complex, or p-adic analysis (or topology) used in the proof.

- We are dealing with very specific functions, such as zeta functions, L-series, and modular forms, that are directly connected to arithmetic. There is little difficulty in developing the basics of complex analysis for these functions, on an arithmetic basis, sufficient for classical arithmetical applications (e.g., Dirichlet's theorem or the prime number theorem) (Kreisel, 1952b), and for me, nothing would be gained by working in a second-order formalism in a very weak system. At best, such systems codify elementary arguments of general applicability. About more sophisticated analysis, as involved in Langlands's theory, I see no difficulty in principle, but one must be aware that this analysis is much more delicate than the early analysis of classical analytic number theory, and so one has simply got to go out and do it in arithmetic style. It seems to me very unlikely that one will encounter in this part of the proof any inductions that somehow involve either unlimitedly many first-order quantifier changes or second-order quantifiers, the hallmarks of incompleteness.

- Given the complexity of the inductive arguments used in the foundations of étale cohomology, one would have to be more careful in that area, *if one were using the theory for general schemes*, to get by with inductions of bounded complexity. Fairly extensive experience on the metamathematics of Frobenius leaves me confident of avoiding trouble in any application (via the Weil conjectures) to varieties over finite fields (and in any case, one does not need very general varieties in Wiles' proof).

    Étale cohomology of schemes can be used to prove the basic facts of the coefficients of zeta functions of abelian varieties over finite fields, but there are more elementary ways to get those results in PA. However, étale cohomology provides a huge variety of Galois representations, and so any arithmetization of the latter theory is likely to need a corresponding arithmetization of the former. Though I stress that such an arithmetization is a distinctly nontrivial matter, I hasten to dissociate the problems therein from the discussions on the FOM online forum suggesting that *higher set theory* (beyond ZFC) might be needed for Wiles' proof because of a perceived need for higher set theory in the foundations of étale cohomology.

    Fortunately, there is an FOM entry by Timothy Chow (2007) quoting an anonymous number theorist on the daftness of such claims about higher set theory and étale cohomology. I urge the reader to seek out these remarks, which I expect to be very useful in the enterprise of detaching the innocent from this cult of impotence.

- Again, existing work on model theory of various complete noetherian rings should give confidence that one will find the right setting for the arguments in the deformation-theoretic part of Wiles' proof. This point and ones related to the last few sections are developed in the appendix.

- I want to stress that it is by no means clear to me right now how much induction is needed for a transcription of the mathematics of Cornell et al. (1997) to get a proof of Wiles' theorem in PA. In particular, I think there is little evidence that bounded arithmetic plus the totality of exponentiation would suffice. It is obvious that the latter proves some of the basics around distribution of primes, but one has so little experience of more modern mathematics in the system that one should be cautious.

- In the appendix, I try to open a road to a proof of Fermat's last theorem that does not use set theory at all. Indeed, I seek to show that the modularity conjecture is itself arithmetical and that its proof needs no set theory. It is not claimed, though I see no reason to doubt it, that PA can reproduce the general theory of étale cohomology of varieties over number fields and the resulting Galois representations. That general theory is not needed for Wiles' proof.

1.4 Group Theory

The evolution here is of particular interest. After the original proof by Novikov of the unsolvability of the word problem, one had (in the Adjan-Rabin theorem) confirmation that most classical decision problems about finitely presented groups are unsolvable. Positive results were very rare in those days, mainly around the intricate small cancellation theory, which had a geometrical interpretation (Lyndon and Schupp, 2004). Rather later, in the 1980s, the important isomorphism problem for finitely generated nilpotent groups was shown to be decidable by the essential use of the geometric and number-theoretic ideas of Siegel (Grunewald and Segal, 1980a, 1980b). But for a while, the emphasis was mainly on undecidability, and the tools from Gödel and his followers were appropriate. In 1960, Higman (1961) gave a delightful shift of emphasis. He changed the logical part of undecidability into something positive and versatile. In effect, he took the neat observation from the 1940s on finite axiomatizability by extra predicates and put it into group theory, by essential use of amalgamated products and other basics of the subject to prove that the finitely generated subgroups of finitely presented groups are exactly the finitely generated, recursively presented groups. This became a powerful tool for constructing pathological groups, and it yielded the existence of universal finitely presented groups (in analogy with universal Turing machines). It readily yielded characterizations of groups with solvable word problems, in terms free of recursion theory. Yet Higman's work was virtually the end of the importance of Gödel's ideas in combinatorial group theory. Serre's work on trees gave a new perspective on amalgamation and HNN extensions. By the 1980s, the deep ideas of Gromov around metrical and hyperbolic aspects of group theory had begun to penetrate the subject. These relate to the geometry of small cancellation theory but are much more conceptual. That the class of hyperbolic groups is quite restricted may be a defect from the point of view of logic, but it is not at all from the standpoint of those who connect geometry, topology, and group theory. For masterly accounts of how the subject looks now, one should consult (Bridson, 2002; Bridson and Haefliger, 1999) and Bridson's recent ICM lecture (Bridson, 2007). Machines have not quite disappeared from the scene, inasmuch as finite automata are basic.

My point here is that the ideas from Gödel were very important in establishing the limitations of algorithmic methods for general finitely presented groups and got a beautiful final version in Higman's work but are now of little relevance. They delineate the boundaries beyond which there are monsters (the monster itself is a highly structured entity, connected to many currently central parts of mathematics). But one knows now that the monsters are far away, and one has every confidence in the cordon sanitaire that keeps them away from current research. There is no geometry visible in the pathology.

In contrast, classically posed problems, such as that on the elementary theory of free groups, are currently being analyzed positively through the essential use of geometric or arboreal, as opposed to finite combinatorial, ideas (Sela, 2002).

1.5 Geometry and Dynamical Systems

There are two intriguing recent developments that appear to link a part of mainstream geometry to classical Gödelian ideology. Both involve rather difficult mathematics and give formulations in terms of classical recursion theory.

The first is the work of Nabutovsky and Weinberger (2000) (and see also the related logical results in (Soare, 2004), which interprets surprisingly much of the working of a Turing machine in the variational theory of Riemannian metrics on compact manifolds of dimension at least five. Here, as Soare remarks in his useful account, one is contributing to a very natural geometric classification problem using refined notions of recursion theory. It is worth noting that no fuss is made about unprovability, although one could surely torture the proof into an incompleteness proof. There is, however, a remarkable example of a statement in differential geometry, with no reference to recursion theory, whose only known proof uses recursion theory (Soare, 2004). Soare chooses to stress the computability aspect, whereas it seems to me here that it is really the delicate nature of the recursive construction that is important. (It is notable that an early version of the work used the Sacks density theorem.) It is surely natural to explore the possibilities of the techniques in lower dimensions.

The second is the work of Braverman and Yampolsky (2006), showing, within the ideology of computable analysis, that there are quadratic Julia sets that are noncomputable relative to an oracle for the coefficients of the underlying quadratic polynomial. (There are quadratic polynomials for which one has computability.) It would have been startling had the problem turned out to be computable, but the proof, like most undecidability results for mainstream mathematical structures, is intricate and demands detailed knowledge of central work in the specific subject matter. As of now, there is less evidence of coding as refined as one sees in Nabutovsky and Weinberger (2000). In contrast to the latter, the emphasis here is on noncomputability. In neither case do we yet have a shift of emphasis as decisive as that for Higman's theorem, identifying a natural notion in recursion-theoretic terms.

1.6 Set Theory

Here there is no doubt that Gödel's influence remains, and probably will remain, major. The mathematical facts are quite evident. He gave a beautifully focused development of the inner model L and, by an argument certainly not remote from that for Löwenheim-Skolem, established that GCH and AC hold in L whether they hold in V. As with the work on incompleteness, one is struck by the freshness of the work. ZF settles for L, with ease, the issue of AC and CH (issues not to be clarified for ZF for another twenty-five years). Yet Gödel missed some things about L that are important and were not far away. As usual, I collect a few points:

- His own methods show that GCH and AC have no advantage over ZF as far as provability of arithmetic statements is concerned. This observation of Kreisel from the 1950s is enough to ease certain concerns on the part of working mathematicians, mainly about AC. Examples to which I have alluded elsewhere concern homotopy theory, where Serre was well aware of his apparent appeal to AC via duality considerations, and the Weil conjectures, where Deligne was loathe to use an embedding of the p-adics in the complexes (Macintyre, 2005).

- Jensen's work from the late 1960s shows how much more than CH is true in L. His work is highly original and intricate but, in a real sense, elementary. The pattern is the same as in the previous point. Gödel did not ask what more is true in L.

- We now understand, since Cohen (1966), an enormous range of models of ZFC and profound implications from determinacy and the existence of large cardinals. The new models are used not merely for independence results but sometimes in conjunction with absoluteness results to prove results in ZFC. In fact, we now have a new branch of advanced mathematics, notable for its beauty and coherence. What we do not have, and in my opinion will not have, is any significant influence on other vigorous parts of mathematics.

- Of course, some parts of mathematics have been affected by set-theoretic independence. Many very difficult results have been obtained, but I feel that almost no surprises have emerged (an exception being the link between Lebesgue measurability and consistency of an inaccessible (Shelah, 1984), briefly discussed later). Abelian group theory and general topology had each reached an impasse that was obviously set theoretical, and after Cohen, one was able to show in most cases that there were immovable set-theoretic obstructions to solving problems in the outer reaches of these subjects. My point is simply that one must beware of exaggeration. If one knew these subjects, and set theory, one really ought to have suspected that independence was inevitable in problems of extremely general formulation. It is a mathematical achievement to confirm this, but I do not regard this proliferation of independence results as in any real sense an impact on thriving mathematics. I do not exclude that all this can change, but we have no clue where to look for an example.

  The Shelah result mentioned earlier is of quite a different nature. For the basic facts, a useful reference is Miller's (1989) review, which touches on the work of various authors, for example, Raisonnier. At issue is the strength (but not simply consistency strength) of the three properties Lebesgue measurability, the Baire property, and containing a perfect set. There is an old and seemingly little-known result of Specker showing that if all uncountable sets of reals contain a perfect set, then ω_1 is inaccessible in L. There is Solovay's famous result that from the consistency of an inaccessible, there follows the existence of a model of $ZF + DC$ in which every set of reals is Lebesgue measurable, every set of reals has the Baire property, and every uncountable set of reals contains a perfect set. The startling results of Shelah produce a definable family of sets of reals, indexed by reals, so that in any model of $ZF + DC$ in which each set in the family is Lebesgue measurable, ω_1 is inaccessible in L. One can be much more explicit, as Raissonier shows. In unexpected contrast, the assumption that all sets of reals have the Baire property has no such implications. Indeed, any model of ZF has a generic extension in which every set of reals has the Baire property. The main thing to emphasize here is the very specific effect on L of a measurability assumption about sets of reals and the absence of such an effect for the notion of the Baire property (traditionally regarded as cousin to the measurability notion).

1.7 Logical Form

It remains a popular goal to produce independent sentences (for some of the conventional systems) of low logical complexity. I am not excited by this, mainly because my own mathematical experience has taught me that there is normally little connection between logical form and mathematical significance (compare the earlier remarks on the decision problem for curves). I refer to my remark about the absence of any visible structure in the Gödel sentences. In number theory, it is usually hidden geometric and/or analytic structure that determines relevance, and very little has happened in the seventy-five years since 1931 to connect Gödel's theory to geometry or analysis. It is true that the work of Green and Tao on arithmetic progressions in the primes involves very novel ideas closer to combinatorics, and it is certainly reasonable to go looking for independence in this direction. I think it is fairly clear, though, that Gowers's analytic proof of the finite Szemeredi theorem can be fitted into PA.

1.8 Ramsey Independence

Ramsey's work is arguably more influential than Gödel's (and was done in the same time period). The derived notion of indiscernibles is fundamental in all parts of logic, most prominently in model and set theory (where it has deep links to truth definitions). Thus it was an important event when, in the late 1970s, Paris, Kirby, and Harrington discovered new forms of independent statements with no overt reference to Gödel coding (though, in fact, the most famous new independence results were easily seen to be equivalent to statements from the Gödel family). The new statements were seen as mathematical in a way that the original Gödel sentences were not. There are other differences. The Ramsey sentences are easily proved in set theory and do not iterate in such a natural way as the Gödel ones. My own taste favors the link to hierarchies of provably recursive functions and the point of view that the work gives us an illuminating new perspective on Gentzen's work and a better understanding of what can be done in PA. The independence is the least structured aspect of it. I have never regarded PA as in any way a privileged system.

1.9 "Topologie Moderee"

In central parts of mathematics, one does not go looking for monsters, and one does not encourage them to intrude. One of the most deplorable effects of Gödel's work is the fixation with examining phenomena in utmost generality, in territory far beyond mathematical civilization.

Grothendieck has stressed that the foundational efforts of the analysts and general topologists create structures that are irrelevant or distracting for geometry, and indeed for mathematics in the tradition of Riemann or Poincare, where algebraic topology gives the perspective. We now know (and on this, logicians and geometers interact) that there are extensive non-Gödelian (un-Gödelian?) territories in mathematics, where the category of definable sets is rich enough to support most of the constructions of algebraic and differential topology but moderate enough that Gödelian pathologies are avoided (van den Dries, 1998). Of course, one has to go deeply into specifics to get these universes, whereas Gödel's theory is all purpose in most of its applications.

1.10 Conclusion

As far as incompleteness is concerned, its remote presence has little effect on current mathematics. Some of the techniques that originated in Gödel's early work (and in the work of his contemporaries) remain central in logic and occasionally in work connecting logic and the rest of mathematics. The long-known connections between Diophantine equations, or combinatorics, and consistency statements in set theory seem to have little to do with major structural issues in arithmetic. That PA is entirely natural in the context of finite combinatorics can hardly be denied, but no one has succeeded in crossing the gap between finite combinatorics and arithmetic (especially arithmetic geometry). As far as the geometry of sufficiently general Riemannian manifolds is concerned, techniques descended from Gödel have proved illuminating (Soare, 2004), without the results being sold as "natural independence."

Appendix: Modularity, Fermat's Last Theorem, and PA

At present, all roads to a proof of Fermat's last theorem (FLT) pass through some version of a modularity theorem (generically, MT) about elliptic curves defined over Q. The network of ideas around modularity is extremely intricate, involving (inter alia) analytic number theory, algebraic and arithmetic geometry, Galois theory, and advanced commutative algebra of deformation theory. A casual look at the literature may suggest that in the formulation of MT (or in some of the arguments proving whatever version of MT is required), there is essential appeal to higher-order quantification over one of the following:

- $\mathbb{C}$
- modular forms
- p-adic fields
- the topological group $\mathbf{Gal}(\mathbb{Q})$
- Galois representations
- coefficient rings in the sense of [41]

If one does not look at a proof of MT, one can do little logically except give a superficial formulation of MT as a Π_m^l statement (for l and m not very large and $l \geq 1$). FLT itself is known to be Π_1^0 (and indeed without bounded quantifiers, by Matejasevic). We all know that true higher-order sentences may imply Π_1^0 sentences unprovable in PA. There is, however, no evidence whatsoever that something of this kind is going on here. The original purpose of this appendix was to give good reasons for believing that the current proof(s) of FLT can be modified, without abandoning the grand lines of such proofs, to proofs in PA. There is no possibility of giving a detailed account in a few pages. I hope nevertheless that the present account will convince all except professional skeptics that MT is really Π_1^0. For now, I make a few introductory remarks.

A1 Flavors of Incompleteness

In contrast to the situation in set theory, where one has very powerful and varied techniques for constructing models of ZFC, one has very limited means for obtaining incompleteness results for formal systems of arithmetic. On one hand, there are the original Gödel examples, which can be decided easily using natural set-theoretic arguments involving truth definitions and rather less easily using Gentzen's deeper ideas in proof theory. On the other hand, there are combinatorial principles of Ramsey type, involving indiscernibles. Though these were not formulated until the mid-1970s, one had known earlier (in set theory) the connection between truth definitions and indiscernibles. It was thus very satisfying when Ketonen and Solovay (1981) revealed the fine structure underlying both the Paris-Harrington statement and the Gentzen analysis. As I read this history, as of now, we have met arithmetic incompleteness only via either (1) explicit use of higher-order principles or (2) finite combinatorics of Ramsey type. As regards the latter, there is no trace of it in the literature around MT. As regards the former, there is much overt use of higher-order notions (e.g., from the list given earlier), and the main point of what follows is that there is no need for this higher-order apparatus as far as FLT is concerned. Moreover, MT itself can be faithfully formulated without higher-order quantification, but to see this, one must do some analysis of proofs (but not necessarily a proof-theoretic analysis).

A2 Finite Approximations

$\mathbb{R}$, $\mathbb{C}$, $\mathbb{Q}_p$, **Gal**$(\mathbb{Q})$, and the Tate modules [14] are all completions of finitist arithmetical structures. A substantial amount of their first-order theories admits natural finite approximations that do not quantify over second-order entities. Putting proofs of MT into PA will involve finding finite approximations of MT and of all the other principles that go into its proof.

Nearly sixty years ago, Kreisel (1952b), considered such approximations in the basic theory of complex analytic functions. One use of the ideas is to dispel any illusions that the *basic* techniques of complex analysis, such as are used in the analytic proof of the prime number theorem, have inherently second-order aspects when applied to the classical functions. Kreisel's observations are still useful in the much more complex context of MT.

In that context, one considers modular forms for congruence subgroups Γ (Rohrlich, 1997). The Γ will be parameterized (coded) by integers N, and the set of corresponding cusp forms will form a finite-dimensional space over $\mathbb{C}$, with dimension computable, in terms of N, by Riemann-Roch (Diamond and Schulman, 2005; Rohrlich, 1997). An essential point in the modularity theory, equally essential in any useful logical analysis, is that there is a canonical *basis* (of new forms) whose Fourier coefficients are algebraic integers. Moreover, the array of Fourier coefficients for this basis will be Δ_0 in N. These Fourier series are the important ones in (the formulation of) MT. The general *complex* combinations of them are not needed in the formulation of MT.

Kreisel's observations apply efficiently to the basic theory of modular forms and their associated Dirichlet series, but deeper complex-analytic considerations are involved in the proof of MT via the proof of the Langlands-Tunnell theorem. This concerns two-dimensional *complex* representations ρ of the Weil group of an arbitrary number field F, with the image of ρ in $PGL_2(\mathbb{C})$ solvable. The precise statement, is at first glance, logically very complex (again, an illusion), but the main point here is rather to see and understand the complexity of the proof and, in particular, the use of the trace formula. Gelbart sketches the complex argument in Gelbart (1997), and inspection shows no hint of higher-order maneuvers or associated obstruction to finite approximation. At the same time, I concede that spelling out the finite approximations would be a lengthy enterprise, and here it would be useful to have some metatheorems to apply. Gelbart stresses some delicate aspects of the complex analysis, but these have nothing to do with higher-order logic. There is no evidence at all that base change or the trace formula has any essentially higher-order content. It may be instructive, and is certainly fascinating, to read Langlands on Arthur's work (Langlands, 2001).

I have let ideas get a little out of order to fix attention first on the complex analytic aspects, where logical hygiene was put in place a long time ago (Kreisel, 1952b). Now I turn to nonanalytic infinitistic aspects.

A3 Galois Theory

For any field K, **Gal**(K) is the group of automorphisms of K^{alg} (the algebraic closure of K) over K, with the usual Krull topology. This is a compact, totally disconnected group. First-order quantification over **Gal**(K) is certainly not interpretable in K, but there is a basic apparatus of finite approximation, the Cherlin–van den Dries–Macintyre (CDM) method (Cherlin et al., 1981), which gives natural first-order interpretations of central concepts in the infinite Galois theory. In particular, this can be done in PA for **Gal**($\mathbb{Q}$) and uniformly for **Gal**(F_p) for p prime.

A4 The p-adic Fields

It is routine to develop in PA, uniformly in p and n, the theory of congruences modulo p^n. But as for **Gal**($\mathbb{Q}$), one cannot literally interpret, uniformly or not, the completions $\mathbb{Z}_p$ and the corresponding $\mathbb{Q}_p$. It is, however, quite evident how to establish natural, and extensive, finite approximations. Indeed, for classical algebraic questions, one can do a great deal more by arithmetizing the Ax-Kochen-Ershov theory (Ax and Kochen, 1965a, 1965b, 1966; Eršov, 1965), working, for example, with internal versions of the algebraic p-adics.

Moreover – *and this is crucial for what follows* – one may develop in PA the basics of the Frobenius automorphisms of F_{p^n}, the elements of the theory of decomposition and inertia groups, ramification (for number fields), and Cebotarev's theorem.

Going further, one may interpret even the Krull-Galois theory of the $\mathbb{Q}_p$ uniformly. One may show in PA that $\mathbb{Q}_p$ is elementarily equivalent to its subfield of algebraic elements and that any finite extension of $\mathbb{Q}_p$ is obtained by adjoining an absolutely algebraic number (and that there are only finitely many extensions of each degree). From this, one shows, in PA, the existence of a *unique* conjugacy class of closed subgroup H of $\mathbf{Gal}(Q)$ with $\text{Fix}(H)$ elementary equivalent to $\mathbb{Q}_p$ so that H is naturally isomorphic to $\mathbf{Gal}(\mathbb{Q}_p)$. (This is a model-theoretic addendum to Mazur's (1997) discussion in Cornell (1997). If one has a continuous representation ρ from $\mathbf{Gal}(Q)$ to $M_n(A)$, with A being a topological ring, this gives an equivalence class of representations from $\mathbf{Gal}(\mathbb{Q}_p)$ to $M_n(A)$, with an unambiguous notion of the trace of $\rho(\sigma)$, for each σ in $\mathbf{Gal}(\mathbb{Q}_p)$.

It is standard that $\mathbf{Gal}(\mathbb{Q}_p)$ acts on the residue field (the algebraic closure of $\mathbb{F}_p$) of the algebraic closure of $\mathbb{Q}$ (up to the conjugacies mentioned earlier), yielding an exact diagram

$$1 \to I_p \to \mathbf{Gal}(\mathbb{Q}_p) \to \mathbf{Gal}(\mathbb{F}_p) \to 1,$$

with I_p being the *inertia subgroup* of $\mathbf{Gal}(\mathbb{Q}_p)$. $\mathbf{Gal}(\mathbb{F}_p)$ has the canonical PA-definable Frobenius generator $Frob_p$. This lifts to $\mathbf{Gal}(\mathbb{Q}_p)$ only modulo I_p. If ρ is trivial on I_p (a notion independent of the conjugacy class) – in which case, we say that ρ is *unramified at p* – we can write unambiguously up to conjugacy $Frob_p$ for any lifting of the element $Frob_p$ of $\mathbf{Gal}(\mathbb{F}_p)$, and we have an unambiguous notion of the value of trace $Frob_p$.

In certain basic cases, we have a natural way of coding (in PA) continuous Galois representations. One is when the representations are complex. Because the Krull topology is totally disconnected and the complex topology is connected, any ρ must have finite image and be trivial on some $\mathbf{Gal}(\mathbb{K})$ for $\mathbb{K}$ a finite extension of Q. $\mathbb{K}$ can clearly be coded, as can the finite subset that is the image, and thus (the details are not important at this level of discussion) one codes ρ. A similar argument works for the case in which the image of ρ is given as finite, as in the residual representations to be defined later.

When A is a coefficient ring in the sense of Mazur with residue field $\mathbb{F}_q$, we have an induced *residual representation* $\bar{\rho}_p$ to $M_n(\mathbb{F}_p)$. We say that ρ is a *lifting* of the corresponding residual representation. This notion is of fundamental importance. The basic equivalence for

$$\rho, \ \rho' : \ \mathbf{Gal}(\mathbb{Q}) \to GL_n(A)$$

is that there exists M such that

$$\rho' = M^{-1} \rho M, \quad \text{where } M = 1 \quad \text{mod } m_A.$$

Then a deformation of ρ_0 is an equivalence class of liftings.

A5 Some of the Number-Theoretic Notions and Their Logical Aspects

Stevens (1997) gives a clear, compact account of the large-scale structure of Wiles' proof. I refer to his article for the notions of (1) determinant, (2) cyclotomic character χ_p, (3) odd, (4) flat, and (5) absolutely irreducible. All these, and many more given subsequently, are needed in the statement and proof of Wiles' version of MT.

The fundamental Galois representations attached to an elliptic curve E defined over $\mathbb{Q}$ are those connected with l-torsion for a prime l. The Tate module for l is defined by

$$(T)_l(E) = \overset{\lim}{\leftarrow} E[p_n],$$

and it is isomorphic to $\mathbb{Z}_l{}^2$. $\mathbf{Gal}(\mathbb{Q})$ acts naturally on the Tate module, giving a representation

$$\rho_E : \mathbf{Gal}(\mathbb{Q}) \to GL_2(\mathbb{Z}_p)$$

and residual

$$\bar{\rho} : \mathbf{Gal}(\mathbb{Q}) \to GL_2(\mathbb{F}_p).$$

Again, for the preceding profinite structures, there is a simple apparatus of finite approximation.

The remaining arithmetically definable invariants of E (in fact, they are Δ_0) are discriminant, minimal discriminant Δ_E, and conductor N_E. The only real subtlety, as far as getting these definitions Δ_0, concerns the 2-adic and 3-adic exponents of the conductor, where one has to take account of the work of Ogg and Tate (Tate, 1974). I will give details in a later account (the point being to paraphrase the language of group schemes, adapting the presentation used by Edixhoven (1997)). Then one has the general theorem, involving no fundamental difficulties as far as formulation or proof in PA is concerned.

Theorem A5.1 (Stevens, 1997, 6) *Let ρ_E, p be the p-adic Galois representation associated with an elliptic curve $E_{/\mathbb{Q}}$, and let N_E be the conductor of E. Then*

- *the determinant of ρ_E, p is χ_p and*
- *ρ_E, p is unramified outside of pN_E.*

In particular, ρ_E, p is odd. If E is semistable with minimal discriminant Δ_E, then the residual representation $\bar{\rho}_E$, p has the following local properties:

- *If $l \neq p$, then $\bar{\rho}_E$, p is unramified at $l \Leftrightarrow p | ord_l(\Delta_E)$.*
- *$\bar{\rho}_E$, p is flat at $p \Leftrightarrow p | ord_p(\Delta_E)$.*

A6 L-Series

The notion of L-series of an elliptic curve E, defined over a number field, is defined by Silverman (1997). This is a Δ_0 notion, in the obvious sense, because it is defined in terms of the number of points of E in residue fields. In addition, the Riemann hypothesis for curves can readily be proved in PA so that one has good estimates for the growth rate of the coefficients of the L-series. Going much further, the Faltings Isogeny theorem goes readily into PA.

On the other hand, a modular form has a Fourier expansion (see Rohrlich, 1997) and an associated Dirichlet series. In its simplest form, MT says that if E is defined over $\mathbb{Q}$, and not just over the algebraic numbers (where Belyi's theorem applies), then E has the same L-series as a modular form of level of the conductor of E. The Eichler-Shimura theory attaches to a modular form of weight 2 an odd two-dimensional Galois representation over the ring of integers generated by the coefficients of the form (see Rohrlich, 1997). Motivated by this, one gives a definition of modularity of a Galois representation, say, as in Stevens (1997, section 4.3). There is no present need to spell out the definition as many equivalent definitions will appear later.

A7 Serre Conjecture

Serre (1987) isolated various key invariants of representations ρ over finite fields of characteristic p, namely, N, k, and ϵ. N, *the conductor of* ρ, is not divisible by p but takes into account all other l where ρ is ramified, with a computable exponent (see Edixhoven, 1997). Variable k reflects the ramification at p (see Edixhoven, 1997), and it requires work to show that this is Δ_0 to develop its basic properties in PA. Edixhoven indicates the underlying Galois-theoretic explanation on page 215 of Edixhoven (1997); ϵ is Δ_0-defined on page 213.

Now, Serre went on to formulate a very precise conjecture about the modularity of such representations, linking the preceding invariants to invariants attached to modular forms (Edixhoven, 1997; Serre, 1987). The proof of this conjecture was not available at the time of Wiles' work but has recently become available. As far as I can see, the proof (Khare and Wintenberger, 2004, n.d.-a, n.d.-b) goes into PA. Serre showed that his conjecture implies FLT, using the representation that emerged from the work of Hellegouarch and Frey.

A8 A Special Representation Coming
from a Counterexample to FLT

Let p be a prime, at least 5, and suppose $a, b, c \in \mathbb{Z}$, with $abc \neq 0$ and

$$a^p + b^p + c^p = 0.$$

Without loss of generality for what follows, we can assume b is even and a congruent to 3 modulo 4. Let E be defined by

$$y^2 = x(x - a^p)(x + b^p),$$

the curve of Frey-Hellegouarch.

E is an elliptic curve over $\mathbb{Q}$. Let ρ be the residual representation, modulo p, for p-torsion. Then, easily enough over PA, ρ is absolutely irreducible and odd, and its Serre invariants are 2,2,1. The conductor of E is $rad(abc)$, the product of all primes dividing abc. It is square-free. Serre's conjecture implies the existence of an eigencuspform for $\Gamma(2)$ of weight 2 and determinant 1, but, easily in PA, there is no such form.

The curve E is *semistable*, a Δ_0 condition, and Wiles proved that all such curves are modular, thereby proving FLT. For the rest of this appendix, I restrict myself to a discussion of metamathematical aspects of the ideas in Wiles' proof.

Kreisel pointed out in conversation that one virtue of informal metamathematical analyses is to bring out questions about the scope of implications such as that from modularity to Fermat. One may work over larger number fields, typically totally real, with a more complex notion of modularity. In some cases, the argument from the

Frey curve, via Ribet and Wiles, adapts to give a clear analogue (Jaruis and Meakin, 2004), and in other cases, where the exotic modular form (from Frey) does exist, there is still scope for useful application of the method, for example, by showing that the form is in some other way incompatible with the Frey representation (Bugeaud et al., 2006).

A9 Modular Forms

This is an exceptionally rich subject (Rankin, 1997; Diamond and Schulman, 1995). There is first the nineteenth-century analysis, geometry and group theory, concerning $GL_2^+(\mathbb{R})$ and, in particular the *modular group* $SL_2(\mathbb{Z})$, action of the former on $\mathcal{H}$, *the upper half-plane*, action of the latter on *lattices* and *elliptic curves*, and construction, via series, of *modular functions and modular forms*. In addition, there is the beginning of the theory of the modular curves associated with congruence subgroups. This is a vital part of mathematics about which logic has had little to say.

Now, if one chooses to ignore the mathematical substance and formulate the theory in a general set-theoretic way, for example, by quantifying over any complex numbers, modular forms, or lattices, one can easily contrive to encounter the Gödel phenomenon. This is artificial, ignoring the important aspects of the situation:

- The crucial functions have recursive expansions (as sums over lattices, Fourier series, or Dirichlet series), and the basic complex analysis can be done in the style of Kreisel's finite approximations.
- The hyperbolic geometry (e.g., of cusps) has a corresponding approximation theory, allowing one to get by with arithmetic (and very restricted) quantification.
- The Riemann surface theory of the modular curves, and the correspondence, for weight 2, between $S(N)$ and the space of holomorphic differentials on $(X_0(N))$, as well as the Riemann-Roch theory needed to compute dimension via genus, are entirely arithmetic, with corrresponding finite approximations.
- The basic theory of the Hecke operators and the Peterssen metric is arithmetic (in the logical sense). Crucially, the relevant eigenforms have *algebraic* Fourier coefficients, each form generating a number field. The set of newforms of level N is a recursive (in N) finite set.

The point of the latter is to show that the modularity theorems are in fact *arithmetical* over weak subsystems. The mathematical substance is conveyed by Π_1^0 sentences. Remarks on the FOM online forum suggest that this is not known. However, even a casual glance at Langlands (1990) makes clear that it is very well known to the practitioners. Serre's (1999) obituary of Weil presents more such evidence, when he stresses the advance in arithmetical understanding that came with Weil's formulation of MT.

A10 More on the Conductor

Fundamental to everything is the link between the *conductor* N of E and the *level* of the corresponding newform. Ribet's brilliant proof (level reduction), which reduced the required modularity hypothesis needed to use the Frey curve to prove FLT, goes by a highly sophisticated induction (doable in PA). The following is Ribet's Theorem:

Theorem A10.1 (Ribet's Theorem) *Let f be a weight 2 newform of conductor Nl where l is prime and does not divide N. Suppose $\bar{\rho}_f$ is absolutely irreducible and that one of the following is true:*

- *$\bar{\rho}_f$ is unramified at l*
- *$l = p$ and $\bar{\rho}_f$ is flat at p.*

Then there is a weight-2 newform g of conductor N such that $\bar{\rho}_f \simeq \bar{\rho}_g$.

Note the use of the Δ_0-notion *finite*. The contradiction comes because the induction leads to a nonexistent (even over PA) modular form.

A11 Equivalences

A striking feature of the subject is the variety of equivalent versions of MT, where the equivalences are, in general, highly nontrivial. See Breuil et al. (2001) for the following equivalent versions and Diamond and Schulman (2005) for a detailed treatment of many of the equivalences.

1. The L-function $L(E, s)$ of E equals the L-function $L(f, s)$ for some eigenform f.
2. The L-function $L(E, s)$ of E equals the L-function $L(f, s)$ for some eigenform f of weight 2 and level $N(E)$.
3. For some prime l, the representation $\rho_{E,l}$ is modular.
4. For all primes l, the representation $\rho_{E,l}$ is modular.
5. There is a nonconstant holomorphic map $X_1(N)(\mathbb{C}) \to E(\mathbb{C})$ for some positive integer N.
6. There is a nonconstant morphism $X_1(N_E) \to E$ that is defined over $\mathbb{Q}$.

Statement 5, due to Mazur, is particularly interesting logically because it is so algebraic geometric, as opposed to arithmetic-algebraic geometric. The proof needs a serious descent. This complex version cries out for a bound on N and the degree of the morphism, to be combined in some useful way with quantifier elimination for algebraically closed fields. I am indebted to Richard Taylor and Bjorn Poonen for providing me with a good (Δ_0) bound for the modular degree (which I will not use explicitly until a later publication). With that information, one can translate MT (over a weak subsystem) into an exotic simple formula, as follows. There is an elementary recursive map

$$N \to \Psi_N(a, b),$$

where $\Psi_N(a, b)$ is a quantifier-free formula in two variables a and b so that MT for conductor N is equivalent to the statement that $\Psi_N(a, b)$ holds for all a and b such that the curve

$$y^2 = x^3 + ax + b$$

has conductor N. I will omit details until a future publication.

A12 Wiles' Proof

This involves crucially residual representations and their liftings. On one hand, there is an appeal to Langlands-Tunnell, showing that residual representations with solvable image are modular. On the other hand, there is an elaborate machinery for lifting modular residual representations to actual modular representations.

The Langlands-Tunnell work is certainly much more complex from the analytic standpoint than the kind of work finitized by Kreisel. For a clear account, one can consult Gelbart's (1997) work. The results are about *complex* representations, whereas Wiles has to work over finite fields. As Gelbart points out, the proof goes via *automorphic cuspidal representations* (adelic entities), built from *local* representations (discussed later).

But simply to get mod 3 representations modular, Gelbart gives an explicit argument on pages 160–63. By inspection of this, and Deligne and Serre (1975), this is in PA. The Deligne-Serre proof is patently arithmetical. From one arithmetical modular form, one constructs another by an easily formalized recursion. (That is not to say the proof is easy! To do the whole project in PA, one has to do many such analyses of recursions. For now, I do them on an ad hoc basis.)

For talking about the automorphic representations, one has to develop first in PA the appropriate finitizations of Bargmann's work on $GL_2(\mathbb{R})$ and the more recent p-adic theory. Then one goes to adelic representations. The basic formalization poses no problems. Perhaps the most inspiring part is the dictionary on page 170, and the point to make is that we need it only in an arithmetic version. Finally, the proof of the required base change and trace formula results has no hint of the Gödel phenomenon. The specific algebraic computations needed (p. 202ff.) pose no problems.

A13 Deformation Theory

The construction by de Smit and Lenstra (1997) of the universal deformation ring of an absolutely irreducible representation ρ of a profinite group G over a field is essentially model theoretic. The result is somewhat more amenable to finitization than the other available constructions discussed by Mazur. The construction is a double projective limit and quite explicit, posing no problems of *finite approximation*.

Matters are much more delicate for the *universal modular deformation ring* of a modular representation.

$$\rho_0 : \mathbf{Gal}(\mathbb{Q}) \to GL_2(k),$$

where k is finite of characteristic p, and ρ_0, restricted to $Q(\sqrt{-3})$, is absolutely irreducible.

For both, one imposes, as needed, supplementary conditions on the deformation type. In the case Wiles needed, these conditions are determined by a finite set of unramified primes distinct from p (see Stevens, 1997). In any case, the conditions on the primes are Δ_0.

The modular deformation ring is described by Diamond and Ribet (1997). Because analytic objects are involved, the construction is considerably more elaborate. However, no new infinitistic ideas are needed beyond those used in the discussion of algebraicity of Fourier coefficients earlier. On page 361 of Chapter 12, a completion occurs, but this is readily approximated arithmetically.

The all-important isomorphism

$$R_\Sigma \to T_\Sigma$$

then has to be established. This is very delicate, but not for Gödelian reasons. Conrad (1997) gives an eloquent account of what is involved, including knowledge of Hecke rings and knowledge of deformation rings. The latter involves Galois cohomology and, as usual, will admit finite approximations. The algebraic ideas are very sophisticated but free of any suggestion of Gödel incompleteness.

A14 Conclusion

In the preceding, I have drawn attention to many places in proofs of modularity where the use of higher order notions can be eliminated and where central notions, apparently higher order, are in fact first order and often of Δ_0 or Π_1^0 complexity. It remains to be seen what proof-theoretic analysis can do with this information. Kreisel (2008) points out that the classical cut elimination for PA embedded in a system with (effective) infinitary derivations, combined with what I have sketched, implies that in some sense, the detour via modularity to a proof of FLT (where there is a use of a cut involving the implication, between Π_1^0 sentences, from MT to FLT) can be avoided. Whether this can lead to a more direct proof of FLT remains to be seen.

References

Ax, J., and Kochen, S. (1965a). Diophantine problems over local fields. I. *Amer. J. Math.*, **87**, 605–30.

———. (1965b). Diophantine problems over local fields. II. A complete set of axioms for *p*-adic number theory. *Amer. J. Math.*, **87**, 631–48.

———. (1966). Diophantine problems over local fields. III. Decidable fields. *Ann. Math.*, **83**, 437–56.

Braverman, M., and Yampolsky, M. (2006). Noncomputable Julia sets. *J. Amer. Math. Soc.*, **19**, 551–78.

Breuil, C., Conrad, B., Diamond, F., and Taylor R. (2001). On the modularity of elliptic curves over **Q**: Wild 3-adic exercises. *J. Amer. Math. Soc.*, **14**, 843–939.

Bridson, M. R. (2002). The geometry of the word problem. In *Invitations to Geometry and Topology*, ed. M. R. Bridson and S. M. Salamon, Oxford Graduate Texts Mathematics 7, pp. 29–91. Oxford: Oxford University Press.

———. (2007). Non-positive curvature and complexity for finitely presented groups. In *Proceedings of the International Congress of Mathematicians, Madrid 2006*, vol. 2, ed. M. Sanz-Sole, J. Soria, J. L. Varona, and J. Verdera, pp. 961–87. Madrid, European Mathematical Society.

Bridson, M. R., and Haefliger, A. (1999). *Metric Spaces of Non-positive Curvature*. Grundlehren der Mathematischen Wissenschaften 319. Berlin: Springer.

Bugeaud, Y., Mignotte, M., and Siksek, S. (2006). Classical and modular approaches to exponential Diophantine equations I. Fibonacci and Lucas perfect powers. *Ann. Math.*, **163**, 969–1018.

Cherlin, G., van den Dries, L., and Macintyre, A. (1981). Decidability and undecidability theorems for PAC-fields. *Bull. Amer. Math. Soc. (N.S.)*, **4**, 101–4.

Chow, T. (2007). FLT and ZFC, again. http://www.cs.nyu.edu/pipermail/fom/2007-December/012351.html.

Cohen, P. J. (1966). *Set Theory and the Continuum Hypothesis*. New York: W. A. Benjamin.

Conrad, B. (1997). The flat deformation functor. In [Cornell 1997, pp. 373–420].

Cornell, G., Silverman, J., and Stevens, G. (1997). *Modular Forms and Fermat's Last Theorem.*, New York: Springer.

Deligne, P., and Serre, J.-P. (1975). Formes modulaires de poids 1. *Ann. Sci. École Norm. Sup.*, **7**, 507–30.

Denef, J., and Loeser, F. (2001). Definable sets, motives and p-adic integrals. *J. Amer. Math. Soc.*, **14**, 429–69.

de Smit, B., and Lenstra, H. W., Jr. Explicit construction of universal deformation rings. In [Cornell 1997, pp. 313–26].

Diamond, F., and Ribet, K. A. ℓ-adic modular deformations and Wiles's "main conjecture." In [Cornell 1997, pp. 357–72].

Diamond, F., and Shulman, J. (2005). *A First Course in Modular Forms*. Graduate Texts in Mathematics 228. New York: Springer.

Edixhoven, B. Serre's conjectures. In [Cornell 1997, pp. 209–42].

Eršov, J. L. (1965). On elementary theories of local fields. *Algebra i Logika Sem.*, **4**, 5–30.

Gelbart S. Three lectures on the modularity of $\bar{\rho}_{E,3}$ and the Langlands reciprocity conjecture. In [Cornell 1997, pp. 155–208].

Grunewald F., and Segal, D. (1980a). Some general algorithms. I. Arithmetic groups. *Ann. Math.*, **112**, 531–83.

———. (1980b). Some general algorithms. II. Nilpotent groups. *Ann. Math.*, **112**, 585–617.

Higman G. (1961). Subgroups of finitely presented groups. *Proc. Roy. Soc. A*, **262**, 455–75.

Hindry M., and Silverman, J. H. (2000). *Diophantine Geometry: An Introduction*. Graduate Texts in Mathematics 201. New York: Springer.

Jarvis, F., and Meakin, P. (2004). The Fermat equation over $\mathbb{Q}(\sqrt{2})$. *J. Number Theory*, **109**, 182–96.

Ketonen, J., and Solovay, R. (1981). Rapidly growing Ramsey functions. *Ann. Math.*, **113**, 267–314.

Khare, C., and Wintenberger, J.-P. (2004). On Serre's reciprocity conjecture for 2-dimensional mod p representations of Gal($\bar{\mathbb{Q}}/\mathbb{Q}$). arXiv:math/0412076v1. http://xxx.lanl.gov/abs/math.NT/0412076.

———. (n.d.-b). Serre's modularity conjecture (I). http://www.math.utah.edu/~shekhar/papers.html.

———. (n.d.-b). Serre's modularity conjecture (II). http://www.math.utah.edu/~shekhar/papers.html.

G. Kreisel. On the interpretation of non-finitist proofs. I. *J. Symbolic Logic*, **16** (1951), 241–67.

———. (1952a). On the interpretation of non-finitist proofs. II. Interpretation of number theory. Applications. *J. Symbolic Logic*, **17**, 43–58.

———. (1952b). Some elementary inequalities. *Nederl. Akad. Wetensch. Proc. Ser. A. 55 = Indagationes Math.*, **14**, 334–38.

———. Manuscript in preparation for Darmstadt meeting on Herbrand, September 2008.

Langlands, R. P. (1990). Representation theory: Its rise and its role in number theory. In *Proceedings of the Gibbs Symposium (New Haven, CT, 1989)*, pp. 181–210, Providence, RI: Amer. Math. Soc.

———. (2001). The trace formula and its applications: An introduction to the work of James Arthur. *Canad. Math. Bull.*, **44**, 160–209.

Luckhardt, H. (1996). Bounds extracted by Kreisel from ineffective proofs. In *Kreiseliana: About and around George Kreisel*, ed. P. Odifreddi, pp. 289–300. Wellesley, MA: A A Peters.

Lyndon, R. C., and Schupp, P. E. (2001). Combinatorial group theory. Reprint of the 1977 edition. Berlin: Springer.

Macintyre, A. (2005). The mathematical significance of proof theory. *Philos. Trans. Roy. Soc. A*, **363**, 2419–35.

Mazur, B. (1997). An introduction to the deformation theory of Galois representations. In [Cornell 1997, pp. 243–311].

McLarty, C. (forthcoming). "What Does It Take to Prove Fermat's Last Theorem Grothendiek and the Logic of Number Theory," The Bulletin of Symbolic Logic.

Miller, A. (1989). Review in *J. Symbolic Logic*, **54**, 633–35.

Nabutovsky, A., and Weinberger, S. (2000). Variational problems for Riemannian functionals and arithmetic groups. *Pub. Math, IHES*, **92**, 5–62.

Poonen B. (2009). Characterizing integers among rational numbers by a universal existential formula. *Amer. J. Math.*, **131**, 675–82.

Rankin, R. A. (1977). *Modular Forms and Modular Functions*. Cambridge: Cambridge University Press, 1977.

Robinson, J. (1949). Definability and decision problems in arithmetic. *J. Symbolic Logic*, **14**, 98–114.

D. E. Rohrlich. Modular curves, Hecke correspondences, and *L*-functions. In [Cornell 1997, pp. 41–100].

Sarnak, P. (2006). Transparencies from "Fall 2006 – Hans Rademacher Lectures in Mathematics." http://www.math.princeton.edu/sarnak/.

Scowcroft, P. (2006). Nonnegative solvability of linear equations in certain ordered rings. *Trans. Amer. Math. Soc.*, **358**, 3535–70.

Sela, Z. (2002). Diophantine geometry over groups and the elementary theory of free and hyperbolic groups. In *Proceedings of the International Congress of Mathematicians*, vol. 2 (Beijing, 2002), pp. 87–92, Beijing: Higher Ed. Press.

Serre, J.-P. (1987). Sur les representations modulaires de degr 2 de Gal($\overline{Q}/Q$). *Duke Math. J.*, **54**, 179–230.

———. (1989). *Lectures on the Mordell-Weil theorem*, Aspects of Mathematics 15. Vieweg, Braunschweig.

———. (1999). André Weil. 6 May 1906–6 August 1998. *Biogr. Mem. Fellows Roy. Soc.*, **45**, 520–29.

Shelah, S. (1984). Can you take Solovay's inaccessible away? *Israel J. Math.*, **48**, 1–47.

Silverman, J. H. A survey of the arithmetic theory of elliptic curves. In [Cornell 1997, pp. 17–40].

Soare, R. I. (2004). Computability theory and differential geometry. *Bull. Symbolic Logic*, **10**, 457–86.

Specker, E. (1971). Ramsey's theorem does not hold in recursive set theory. In *Studies in Logic and the Foundations of Mathematics*, vol. 61. ed. R. O. Gandy and C. M. E. Yates, pp. 439–42. Amsterdam, Netherlands: North-Holland.

Stevens, G. An overview of the proof of Fermat's last theorem. In [Cornell 1997, pp. 1–16].

Tarski, A. (1968). Undecidable theories. In *Studies in Logic and the Foundations of Mathematics*, 2nd ed., pp. xi, 98. Amsterdam, Netherland: North-Holland.

Tate, J. T. (1974). The arithmetic of elliptic curves. *Invent. Math.*, **23**, 179–206.

van den Dries, L. (1998). *Tame Topology and o-Minimal Structures*. London Mathematical Society Lecture Note Series 248. Cambridge: Cambridge University Press.

Weil, A. (1980). *Collected Papers*. Vol. 1. Berlin: Springer.

CHAPTER 2

Logical Hygiene, Foundations, and Abstractions: Diversity among Aspects and Options

Georg Kreisel

Preamble *around* (*asymmetric*) *commonplaces* about aspects (of objects in view) and options (for attention to rewarding aspects) in the subtitle related – also, but not only – to the present occasion: the idea(l)s of logical hygiene, foundations, and abstractions have been familiar for decades or more, but different aspects (and different functions of each aspect) of those idea(l)s have rewarded attention, often to points of diminishing returns in prominent directions, as experience around them has changed.

At least as read by some (of us) with logical BOR (background and other resources), many passages in the world's literature ever since the pre-Socratics refer to diversity among aspects, actually of any object including idea(l)s: any relation between the object and other things determines a property of the object, aka aspect (as seen from that other thing).

Remark on other commonplaces, too: *not* knowing them (or rather not remembering such items of common knowledge) when they are useful can be tiresome, but without combining them with more specific and thus more demanding (generally discovered) knowledge, they are rarely enough for progress; cf. below on logical hygiene for more on such asymmetric knowledge.

As to (changing) options – tacitly, with due regard to *suitable* equivalence relations – even for an object in view, as knowledge around it changes, different aspects of the object are liable to reward attention (with that changing BOR). Familiar pieties about first steps toward a focus on rewarding aspects often say more about the (b)rash faithful than the faiths.

Remark tempering the general slogan about *unity in diversity:* time and again, relatively few aspects are discovered that, suitably combined, are rewarding (for BOR available) in relatively many situations to which attention is given; cf. the would-be dramatic news of some relativity of (mere) truth (at least as I have come to understand it). As to – feelings and measures of – reward, it is equally common knowledge that their basins of attraction often evolve: both in consumer interests and in professional traditions in the commerce of ideas; more about this later when the time is ripe.

More specifically – on the present occasion, around the titles of Gödel's most famous logical discoveries in the 1930s[1]* – shifts in attention to diverse aspects of completeness and (relative) consistency will be (re)viewed subsequently under the headings in the title of this piece. Both Hilbert's idea(lization)s of completeness and consistency – together with decidability and logical (in)dependence, an order relation – made familiar by him in the first three decades of the twentieth century, and some broader common meanings (without premature precision, but attention to suitable, usually tacit understandings), will come up in that review.

Remark for professional (pure) mathematicians: they need not be reminded that when such (familiar) words (as completeness, etc.) become part of mathematical trade jargon, they are used as labels for abstractions – roughly speaking, without definite meaning.[2]

A little more will be said in Section 2.3 (and elsewhere) about relations between this practice and – the half truth about – the unreasonable effectiveness of mathematics in natural science, but also of one branch of mathematics in others, for example, geometry in number theory.[3]

2.1 Logical Hygiene: Asymmetric Items in Logical Dress[4]

One aspect of logic most prominent in contemporary mathematics is its simple linguistic structure (here meant in contrast to logical deduction, which seems to have struck MAMNE (many a mind's naked eye) most; more specifically, as in Endnote 1, MAMNE with the BOR of a famous Greek, here Aristotle).[5]

Remark Even such austere texts as Bourbaki's *Éléments* use other literary forms besides logic, for example, in comments in the vernacular in the course of proofs. But in the particular case of – what might be called – mathematical commonplaces across the board, about (a) equivalence and (b) order relations, that aspect has a useful function[6]: the items can be used as parts of proofs; cf. counterparts for branches of mathematics, aka foundational (for the branch in view; in contrast to Section 2.2).

(a) Tacit understandings concerning – the meaning, sometimes aka use, of – equality (familiar since the pre-Socratics)[7] are, as it were, reflected in carping at the axioms for $\equiv$ in logic texts about a century ago, for example, transitivity. (Admittedly, equality as seen by MAMNE or the literal naked eye is often not transitive.) But since then, a more demanding (qualitative) requirement on equivalence relations has become standard, variously known as compatibility with or respect for *suitable* functions: in the case of monadic functions $\forall x \forall y[x \equiv y \Rightarrow f(x) \equiv f(y)]$.

Suitable is underlined inasmuch as the replacement by *some* (or *all*) function(s) is unsuitable except, perhaps, for logical one- or two-finger exercises; cf. Endnote 7. NB

---

* [n] refers to the nth endnote: $1 \leq n \leq 30$. Some readers may prefer the option of reading the text without attention to the endnotes and afterward read them selectively, if at all.

As far as I can see, the material in these notes is of interest only to readers with suitable BOR, occasionally also outside logic or even mathematics. Now one of the less superficial functions of trade jargon is a warning of the BOR assumed (here meant in contrast to literally superficial functions such as brevity). So trade jargon will be used freely.

Outside, but also within mathematics, this has counterparts for functions in a broader (colloquial) sense.

Reminders, especially around logical – or other would-be universal formal – equivalence (where *new* is taken to be nonequivalent and at a later date): pieties about coarseness or any other such extreme of these equivalence relations simply distract from the *discovery* of suitable functions in the situations in view.

(b) Although the (commonplace) reminder of diversity and the (often demanding) matter of discovering relatively few items and their combinations suitable in relatively many situations have come up in (a) with respect to equivalences, they are more familiar around *orderings*.

Reminders about logic a century ago: at one extreme, there were elementary axioms without any compatibility conditions; at another, there was well-foundedness, especially in Cantor's set theory, tacitly, for *arbitrary* descending sequences.[8]

At least in algebra, and more generally in category theory, the emphasis has shifted to suitable *families* of sequences (not always spiced with agony over – the mere legitimacy of the idea of – arbitrary sets). In the simplest cases, the only compatibility required is that the functions in sight preserve order; occasionally this compatibility is satisfied as a corollary to more general considerations, for example, in the lively subject of *o*-minimal structures.[9]

Viewed in the preceding terms, specifically round the acronyms MAMNE and BOR, questions of applied mathematics force themselves on the attention. On one hand, there are – descriptions of – orders that strike MAMNE and the discovery of functions for which the orders are suitable (survival being a favorite in the biological tradition). On the other hand, there is the *discovery* of orderings in a world that strikes MAMNE as diverse: famously, (dis)order (measured) by entropy.[10]

Less generally (and thus more demandingly), there is the order relation familiar as *interpretability*; in particular, definitions of a new symbol (for functions or predicates) are always formally independent but sometimes interpretable.[11]

Remark A moment's thought (or a look back at Endnote 6) shows that with growing BOR, new items of (logical) hygiene may become prominent, while others will be put on the back burner (without being invalidated); after all, this applies to suitable (literal) hygiene, as conditions change (besides having to be remembered).

(c) The following (and last) item in this section fits the present occasion inasmuch as Gödel associated it in the 1940s with a breakdown of our logical intuitions but, by the 1960s, dismissed it without further ado.[12] Here it will be related to a review by Cantor of Frege's *Grundlagen der Arithmetik* (some twenty years before Russell derived a paradox – not so much from "our" but from Frege's intuitions).

Admittedly (actually, at least to me, decidedly apologetically), Cantor used medieval language of "extension" and "intension" in his critique of Frege's thoughtless, sometimes dubbed "naughty," axiom asserting that every predicate determines a set. But some time earlier, Cantor had emphasized – not a formal definition of, but rather – an *aperçu* about sets; to wit, varieties (*Vielheiten*) that can be grasped as unities (*Einheiten*). How else should a variety be described but by a predicate? *Teenage emotion recollected in tranquility*: when encountering this *aperçu*, I – did *not* think of the predicate $x \notin x$, but – did think of $x = x$ as a candidate for a variety that cannot be grasped as a unity. In short, not all predicates P are expected to define sets (whether or not a paradox results from this assumption), but only those such that $\exists z \forall x \ [P(x) \Rightarrow x \in z]$, in other words, those *bounded* by some (set) z. In this case, P *defines* a set, say, y: in symbols $\exists y \forall x \ [P(x) \Leftrightarrow x \in y]$.

Remarks Every predicate of the form $(x \in z) \wedge P(x)$ is bounded in the preceding sense, which is in fact Zermelo's correction of Frege's naughty axiom. Inasmuch as the preceding formulation in logical symbolism is felt to be an improvement on (Cantor's) medieval language, Frege('s formalism) may be said to have contributed to the correction.[13]

Disclaimer By itself this does not establish consistency (and, as mentioned in Endnote 11, it did not for me until the early 1950s[14]) but removes the particular malaise of a paradox; cf. Section 2.2 for more on consistency of principles, aka consistency-in-principle.

Reminders As so often happens when imaginative people thrash about for something to say about a (possibly imagined) problem (or artifact in experimental or observational science), the chances are that something of interest, necessarily elsewhere, will be broached.[15] At least in my – admittedly distinctly limited – knowledge of the scholarly literature, the chances are less rewarding if the "thrashing" is directed at old straws, again admittedly, at least, to me. For example, fitting the present occasion, there is the old straw of the Liar, which seems to have wide appeal, perhaps more than the alternative of relating incompleteness to diagonalizing sequences of partial functions.

2.2 Logical Foundations: A (Here) Suitable Sense[16]

As just mentioned, removing a prima facie case for inconsistency is one thing; establishing consistency, at least, in principle, is another.

Reminder Not only the proverbial man-on-the-Clapham-bus is ill at ease with merely such consistency; so are experienced and distinguished mathematicians.[17] More formally (but without exaggeration), there is demand for consistency with tacit (but, in the situations in view, valid) assumptions, not merely among stated principles and their formal consequences.

Remark Traditional consistency proofs by models approach the demand inasmuch as they establish consistency with other (yet undiscovered, and in any case, unstated) properties of the model in view (especially if it satisfies second-order axioms such as continuity in non-Euclidean geometry); reducing the likelihood of error in practice requires different tactics, including cross-checks.[18]

If formalism, which – like other *-isms* – restricts attention to its own privileged aspects, is not regarded as an ideal, but simply as an idea, Hilbert's kind provides a *definite*, albeit at least to MAMNE without specialized BOR, totally sterile meaning, aka semantics (for closed formulae A): A is a theorem by given rules.

Reminder Gödel's incompleteness (and thus at first sight negative) results have been popularized in terms of some (in arithmetic) true formula or, specifically, of the (true) consistency statement (for the rules considered), not being formally derivable; cf. Endnote 18. Without some usually doctrinaire restriction on proofs – here, of consistency – the result merely ratifies broad experience (also, but) not only in mathematics:

we don't have all the answers. (In the specific case of consistency, at least for the usual rules stated (not too sloppily), proofs by familiar models are known, at least, tacitly.)

In contrast – but fitting the warning in Section 2.1(a) about functions for which logical equivalence is *not* (a) suitable (equivalence relation) – it is by no means expected by MAMNE or even accepted (without checking a proof) that the addition of the *negation of the consistency statement* (which is false in arithmetic) *is* consistent; cf. Endnote 18 for (so-called derivability) conditions (that, equivalently, are properties of canonical representations of the rules in view). Nor has this fact been sterile, at least, in logic itself.[19]

Remark on the fact(s) observed at various times since the early 1930s (and brought up-to-date in Chapter 1 of this volume) that incompleteness properties of the kind in view here are hardly noted in the mathematical literature: besides, they were anticipated in the 1920s, for example, by A. Weil, actually, in the specific case of binary quartics over $\mathbb{Q}$, *L'arithmétique sur les courbes algébriques* (1928); cf. his *Collected Works* with a P. S. Anyhow, familiarity in abstract mathematics (some groups are, others are not commutative), open properties, aka (so far) undecided propositions, are labeled "hypotheses" or conjectures – with a different meaning up-market and down-market[20] – and explicitly mentioned in logical jargon: as premises of implications. NB As usual, if the human touch is wanted with some embodiment of MAMNE, this can be done, too.[21]

A more demanding matter is to discover phenomena outside formal logic such that a suitable combination of – suitable variants of – the (in)completeness results with other knowledge around those phenomena contributes (to their understanding). This has parallels with other uses of mathematics in natural science and requires suitable BOR, and thus belongs to the endnotes.[22] In short, logic is not for mathematics alone.

2.3 Logical Abstractions

prominent in the twentieth century (and not before): memorable from the contrast between the expositions by Euclid and Hilbert (in his *Grundlagen der Geometrie*, which in turn is in contrast to *Anschauliche Geometrie* by Hilbert and Cohn-Vossen[23]).

A metaphor popular in the middle of the twentieth century for the aspect of logic in view dubs it – the grammar of a – language for (tacitly, abstract) mathematics.[24] As announced in Endnote 8 and echoing a wording in Endnote 3, a principal concern of this section is the unreasonable effectiveness (in particular, in mathematics) of logical abstractions in the form of logically independent, but formally interpretable, definitions; cf. (c) later. In preparation for this, (a) mentions some differences between the logical and common or garden varieties of abstractions such as metaphors and other analogies and, in particular, (b) mentions some aspects of more familiar languages that are and some that are not retained in logic (as language).

(a) As a corollary to (the commonplace about) diversity among aspects, already noted, of any object, observations of and thoughts around an object will focus on a *proper* subset of its aspects, with MAMNE's focus being one.

Reminder Depending on BOR, the focus will be arrived at by (conscious) selection from a larger – but, of course, still partial – set of aspects. Roughly, but precisely enough for the preparation in view, such a focus is dubbed abstraction.

Warning about another meaning, especially in higher set theory: here abstraction is not related to a focus (with neglect of many other aspects) but to the ordinal, aka rank or type, of the objects in view here, which are sets: as it were, a measure of the distance from concrete atoms.

At least generally, metaphors leave wide open *which* aspects of a literal meaning are (to be) retained, while logical abstractions specify them: for flexibility, without premature precision, whence the requirement of indefiniteness.

Reminder of Endnote 2: Russell's catchy *aperçu* has been echoed in different words by experienced mathematicians.[25]

To MAMNE (with very little BOR), the possibility of purely logical deductions from logical formulae, the familiar literary form for logical abstractions, is no doubt striking. But even moderate experience of mathematics is enough to realize that this is an undemanding part of mathematical proofs, both for discovering and for understanding proofs. (Today the relation between genetics and epigenetics provides a good metaphor.[26])

(b) Logic, which, as has become standard in contemporary mathematics, includes set theory, is taken to be a counterpart to grammar; with the addition of explicit definitions for so-called nonlogical terms, it becomes a – counterpart to – language. Thus the aspect of growth of familiar languages is retained in the metaphor.

Remark on malaise about expounding obscurum per obscurius (inasmuch as the functions of logic in view may be felt to be obscure and of familiar languages more so): but malaise, like doubts, can be doubtful, too, in particular, here, by forgetting the – commonplace about the – diversity of aspects and of functions; a skill is required to *discover* in those diversities *suitable* elements with counterparts, as in the case of growth of both vocabulary and interpretations discussed earlier.

Reminders, especially around the (much touted) formal character of logical grammar and its rules and the small basic vocabulary: these are in – certainly to MAMNE with little BOR – striking contrast to their counterparts in more familiar languages (and it is of course a principal business of professional logicians to exploit and correspondingly emphasize such differences). As a corollary, a commonplace for idealizations across the board, the discovery of – as always, for situations in view – suitable relations between formal and those familiar languages, aka formalization, requires attention, naturally, more in proof theory than model theory.[27] This much seems widely, even if mostly tacitly, understood and becomes noteworthy only when neglected (with sounds of agony or other impatience).

Another story, actually already broached in Endnote 8, is the extent to which – tacitly, the, by Section 2.1(a) for a suitable equivalence relation, same – capabilities familiar from the use of metaphors and other analogies in familiar languages have counterparts in the use of logical abstractions in pukka mathematics, even if, in this notoriously lazy phrase, they are not needed-in-principle[28]; cf. (c).

(c) To those with here suitable BOR, such idea(l)s as purity of method in the vernacular and consistency strength in trade jargon will be enough for an introduction.[29] Logical hygiene is enough to temper innocent expectations of those idea(l)s generally, and of this ordering (by consistency strength) in particular. But this is not enough to specify contributions by the use of (logical) abstractions to proofs of theorems (either) stated (or seen to be equivalent, e.g., with respect to a current arithmetic interpretation to one) in purely arithmetic form.

Reminder of elementary geometry in abstract style with and without familiar names like "circle" or "ellipse" for abstract (explicit) definitions, with and without pictures in the imagination or on paper – visualization: not only does this help to see relations between such objects in the plane and suitable proofs (that turn out to follow logically from the principles in view) but also three-dimensional figures, of which the objects are (conic) sections; in the jargon of these pages, three-dimensional aspects. Here views of the more or less educated eye are concerned. In other mathematics, it is the mind's (naked) eye; cf. also the reminder in Endnote 25, here adapted to a shift from visualizing the (defining) elements to the (defined) configuration.

Viewing circles and ellipses as conic sections provides an analogy between these conic sections, but at least generally the plane curves do not present themselves with a label announcing that analogy. The literature I know around – the process of – spotting effective analogies varies, often with emphasis on (diverse) *feelings* around it, ranging *from* malaise over the possibility of forgetting the problem in view while pursing such analogies *to* the thrill reminding some of – tacitly, their experience(s) of or dreams about – adultery.[30]

Remark on – the assumption of – a logical essence of mathematical thought: the analogies in contemporary abstract mathematics (once found) take the form of logical abstractions and are formally expressed by explicit definitions, naturally, together with so-called defining properties; cf. Section 2.1(b). For that assumption to do miracles, those defining properties would have to be *noninterpretable* (and, as expected, the faithful – to that assumption – declare piously that interpretability is a conjecture).

Reminder of the disclaimer in Endnote 11(b): some (of us) are grateful for the efforts of the faithful in this area, too. Finally, as it were for reassurance: if – to our surprise – the efforts are successful, it will be an agreeable surprise to all but irreconcilable skeptics.

Depending on BOR, interested readers will not only add to the material of this section on the mathematical effectiveness of logical abstractions and on feelings about their relations to metaphors or other abstractions but will include fine points from their own logical experience. As they will have noticed, in this section, there was no occasion for such fine points, in contrast to, say, Endnote 11c(ii) on determinacy in set theory or Endnote 14 on arbitrary partitions – subsets, not arbitrary predicates; cf. Endnote 13 – in the 1920s. Certainly, to mathematically educated (naked) eyes, the methods of proof for those partition theorems differ strikingly from those in analytic number theory and other Dutch gardens of mathematics, but they were taken in stride. It was a later logical discovery that the partition theorems also have (tacitly, familiar, not just some) nonrecursive aspects and that bounds for finite versions (of van der Waerden's or Ramsey's theorems) grow relatively fast. Readers with logical BOR will recall here such "fine points" as – by independence not only from Peano arithmetic alone, but together with all true Π_1^0-sentences – provably relatively large lower bounds for the Paris-Harrington version of Ramsey's theorem (used by Gowers in his work on Banach spaces) and a mild use of comprehension in the most straightforward proofs of those partition theorems by compactness from their infinite versions. (Though mild, and of course very far from those higher cardinals in Endnote 11(c) and again later, those uses present *logical* differences from other preceding mathematics.) NB "No occasion" for the time being, leaving open the possibility that uses for extending – what Bourbaki called – the grammar will be discovered; uses, that is, "needs" for effectiveness, not necessarily for provability.

Reminder In the early 1920s, Fraenkel, who had experience around ordinal arithmetic, added to the grammar replacement, which, together with the power set, formally implies largish cardinals. This, too, was taken in stride, presumably by some – certainly not all – by superficial analogy with the case of finite sets.

Potentially delicate is the extent to which logical "fine points" contribute to – feelings of – explanation and generally understanding of such phenomena around bounds, as earlier. This is certainly of some consumer interest, even when professionals are traditionally left to fend for themselves; cf. Endnote 25. Naturally, the matter is realistic only for those whose BOR around logic matches that around the other parts of mathematics in view.

As to axioms for higher cardinals, it seems fitting to go back to Hilbert's metaphor of Cantor's paradise (and thus conclude this long section). The aspect most prominent in Hilbert's view of that paradise was its supply of so-called ideal elements, here aka logical abstractions, with (his) special emphasis on their being eliminable like "manners of speech," here aka (a special kind of) interpretable explicit definitions: their potential is a main theme of this section.

By Endnote 11(d), in Cantor's own (and later, of course, others') view of Cantor's paradise, those higher cardinals were especially prominent, and axioms suitable for them were discovered not to be interpretable in ZF (nor even in a second-order variant). As is well known, Gödel emphasized that Π_1^0 (consistency) statements are consequences. In the present chapter – cf. Endnote 3 and elsewhere – the tradition of deriving axioms from the top down is by no means ignored but is rarely found effective in mathematics – naturally, with the exception of a logical language itself. It started off with the airy-fairy project of logical foundations for arithmetic (by Frege), for mathematics (by Russell), and for the world (by Wittgenstein, who went on to specify shortcomings evident for his grand view, but of course with counterparts elsewhere, more down to earth, too). Taken up by mathematicians, this language was discovered to have a most useful function. NB The project itself did not arise within professional mathematical experience, of which those three authors had precious little.

Personal Acknowledgment

As I see such things, my feelings – of obligation and more – about help given to me by K. Derus, not a professional logician, are a private matter. But perhaps this is a place to note that without his help, this chapter would not have been produced.

Endnotes

1. Taken remotely, literally the idea(l) of completeness, here, of knowledge, is at odds with the commonplace about diversity among aspects. Specifically, truth (tacitly, for a particular interpretation) or even validity (for a class) is only one aspect of a proposition or, generally, of logic. (Specializations and generalizations of terms in proof theory or abstractions of relations in abstract mathematics are other aspects prominent in the literature.) Within logic, distinctions

between deductive and functional completeness are familiar, perhaps less so Gödel's own between proof-theoretic and model-theoretic completeness in his dissertation, but unpublished until his *Collected Works* in the 1990s. Attention to certain aspects has evolved in diverse trades in the commerce of ideas, (with luck) rewarding for those with BOR cultivated in the trade.

Remark on Gödel's interest in epistemology, not prominent in his publications in the 1930s, but later; cf. *Synthese,* **114**, 124(c) (1998) and Endnote 29 for more: here, it seems to me, the emphasis is on aspects that strike MAMNE (as in Section 2.1, short for many a mind's naked eye): tacitly bright, but with limited BOR, illustrated by those still famous Greek writers. (MAMNE's view can change with education; cf. training of the literal naked eye's view.) A notorious example is decidability in a finite number of steps in contrast to (another extreme such as) NP or the possibly even narrower class P.

Reminder (Gödel's results being left aside for the moment): at least in the logical literature that I know from before the 1930s, there was nothing like the efficiency of his publications even if by now this has become routine. In fact, in the last three decades, there has been remarkable streamlining in the exposition of logical material from as late as the 1960s.

2. Labels such as *perfect* or *amicable* (natural) numbers have perhaps never been taken literally. Dedekind associated *Körper* with an organic unity (between + and ×), but its translation *field* is, in my experience, associated with contrasts between algebra and agriculture, in particular, warnings against exaggerated expectations of (fancy) algebra in agriculture. Metaphors and similes aside, the emphasis in abstract mathematics is on abstraction (in the sense of indefiniteness); cf. Section 2.3 of the main text, but also Bertrand Russell's oft-quoted quip: "Thus mathematics may be defined as the subject where we never know what we are talking about nor care whether what we are saying is true"; cf. *The International Monthly,* **4** (1901) and p. 366 in vol. 3 of his *Collected Papers.* (The definite article seems dubious inasmuch as some theologians use the same words of their subject; e.g., they use implications, with Latin tags as premises: *etsi deus daretur* and *etsi deus non daretur.*)

Less frivolously, professional mathematicians know striking examples of *proofs* for theorems stated in terms of such labels, which are felt to be "more interesting" than the theorems because of the "novelty" of the ideas. In this case, it can be rewarding to look for other theorems proved by those "new" ideas; cf. Section 2.1(a) on (in)equivalence relations.

Reminders Early completeness theorems, especially by quantifier elimination, gave rise to decomposition theorems, relative consistency theorems to conservation results (in the case of assuming the GCH with a so-far striking dichotomy: in the main body of mathematics, conservation for arithmetic theorems, seen by inspection of Gödel's *L* is enough, and in descriptive set theory, Shoenfield's absoluteness theorem is not). As to completeness of predicate logic, a purely model-theoretic variant has recommended itself; cf. *Lecture Notes in Mathematics,* **453**, 52–71 (1975) for a description of all (partial or total truth valuations of) countable term models with (a complete set of) logical inference rules as a corollary; as it were, a derivation of the rules in contrast to earlier verifications of more ad hoc rules.

Reminder of malaise by MAMNE with Aristotle's BOR in Met Γ 7, 1012a and 22–23: on one hand, it is emphasized that definitions are only sometimes useful; on the other hand, it is felt that for cogent reasoning, each term must have a definite meaning.

3. The "half truth" is the title of E. P. Wigner's article in *Communications in Pure and Applied Mathematics,* **13**, 1–14 (1960). The effectiveness is not in doubt; (one's feeling of) unreasonableness – e.g., of the effectiveness being a free lunch – depends on BOR. I myself am struck by the additional skills besides (pure) mathematical gifts in spotting rewarding combinations between mathematics and other suitable scientific knowledge. (Phenomena do not present themselves with instructions about which of their aspects will fit which mathematics (available) nor vice versa.) Besides, idea(l)s about mathematical methods in the social sciences provide salutary

object lessons, reminiscent of Plato's complaints in the *Republic* about then-popular (cl)aims for geometry of triangles in operations research (for military tactics).

Digression on a topic raised by Plato (and unusual in being quoted by Aristotle with approval; cf. Nich. Ethics, 1095a32–1095b1): *are we on the way to or from general principles?* It is easy to find examples throughout science and mathematics for each direction. In the particular case of Hilbert's famous *Zahlbericht* at the end of the nineteenth century, he went *to* a relatively small arsenal of number fields (from diverse proofs throughout that century). In logic, he went by and large in the opposite direction, perhaps most familiar in the literary form of (im)possibilities-in-principle: tantamount to assuming those principles. Specifically, (the axioms for) those number fields are patently incomplete and, in most cases, undecidable (without even a hint of decidability); consistency was not an issue at all inasmuch as they are patently satisfied by $\mathbb{Q}$ or finite fields (rings of integers *mod p* for prime numbers *p*). While more specific, and correspondingly more demanding, aspects of this topic will come up later when the time is ripe, it will be expected – and is verified in experience – that the way *from* principles, here aka ideal(ization)s, will be most prominent in such areas as cosmology (especially before there was much observational knowledge), by no means typical of natural science.

Disclaimer I do not know the extent to which Plato felt bound by (schoolmasterly?) rules, if any, on the exclusive and inclusive *or* (in Greek corresponding to *aut* and *vel* in Latin); but I feel that only the exclusive *or* in the preceding question would have been enough for – Plato's taste and talent for – drama.

Logical update (for consolidation) around axioms for ordered real closed fields: starting from the top down, with Dedekind's axioms for complete ordered fields (and arbitrary cuts), first-order formalization leads straight to the schema of all first-order cuts, which is of unbounded logical complexity. Starting from the bottom, with experience in real algebra, especially Sturm's theorem results in the schema of Artin/Schreier (of Π_2 complexity).

Peano's axioms for 1 and successor (induction for arbitrary predicates) were tinkered with for familiar reasons, and (second-order definable) $+$ and $\times$ with their recursion equations added; cf. PA.

4. The expression *logical hygiene* was used in the 1940s by A. Weil in *L'avenir des mathématiques* (a kind of companion to Bourbaki's manifesto, *Architecture des mathématiques: la mathématique or les mathématiques?*), but with emphasis on another function: sterility, tacitly with respect to bugs in contrast to cells of the host, which are helped to grow by (proper) hygiene, not always by refinements in, say, cosmetics. This use is asymmetric, too, inasmuch as, by itself (without a healthy organism), hygiene does not "go deep."

5. At least for mathematicians, it is an open secret that logical deductions are not a particularly demanding (or rewarding) part of proofs, in particular, compared with spotting suitable ideas – presented in the form of axioms – including so-called auxiliary definitions. More broadly logical inferences in senses established in logical trades – though occasionally troublesome to some (e.g., the negation of uniform convergence in a text by Titchmarsh widely used in the 1940s) – are certainly easier to simulate (by computers) than many other inferences that are child's play (for five-year-olds).

   Remark on referring to – felt by some as dragging in – classical authors, above Aristotle: far from (cl)aiming thereby to establish an authority, as stated already (and again in the next endnote), the emphasis is on the limited BOR available at that time; sometimes for contrast with later BOR, sometimes as a reminder that, for the (cl)aim in view, such BOR is enough.

6. On the diversity of aspects and the even greater diversity of their (potential) functions, cf. *Quot homines tot sententiae* (Terence) and *Quot capitum vivunt, totidem studiorum milia* (Horace).

   Remark on a difference of emphasis: in effect, these classical authors take Man as the measure of all things (and thus aspects and functions to which attention is paid, tacitly with available BOR); granted that man (or, for that matter, natural number) is *a* measure of all things, it is not always a good measure or even as good as others, again tacitly for the (class of) situation(s) in view. While in mathematics, other numbers such as Dirac's c-numbers and q-numbers are traditional, comparisons – not only by measures (of men and traditions) but also by such categories as up-market and down-market – are felt to be invidious.

7. Heraclitus famously anticipated – in effect, admittedly, not in so many words – Leibniz's definition of identity in terms of equivalence w.r.t *all* properties; recall the business of not stepping into the same river twice. Here, too, carping about an "exact" meaning of all – in these pages seen as a corollary to the commonplace about diversity among aspects – distracts from the more demanding job of discovering *suitable* functions (even among some impeccably (exactly) defined "totality").

   Reminder from elementary logic, now for several function and predicate symbols: the equivalence relations together with the compatibility conditions for those objects imply logical compatibility for their whole logical closure.

8. The shift of emphasis in algebra to suitable families or categories is – of course, in effect, but presumably with diverse intentions (cf. Horace in Endnote 6) – related to general feelings such as malaise about impredicativity, one of several (b)rash reactions to Russell's paradox taken up later in (c). But under the present heading of "hygiene," relations to down-to-earth mathematical experience are more immediately pertinent.

   (a) The property of *arbitrary* map (with given domain and range) is certainly described with less BOR than such sophisticated families as constructible or continuous maps for this or that topology: it is accessible to MAMNE with limited BOR, aka clear (where the feeling of clarity is often modified by BOR in diverse directions to boot).

   (b) In view of considerable BOR, especially in the twentieth century (of abstract mathematics), the marginal utility of more sophisticated families has turned out to be remarkable (at the price of *suitable* sophistication!), even for results about the clearer totality, e.g., about $\mathbb{R}$ by use of results about all real closed fields.

   Remark (for what it is worth): the BOR in view fits such commonplaces as reculer pour mieux sauter, or discovering what is essential, tacitly, for at least one proof of the particular result in view, rarely essential – either in the sense of necessary or sufficient – for a broad range.

   (c) As an example of an exception, which – of course does not prove, but – may underline the rule, here the preceding (b): even where more or less sophisticated lists are available (e.g., enumerations of arbitrary permutations of finite sets), there is no occasion to use them – memorably for proving that no chessboard can be covered by dominoes over two adjacent squares so as to leave a diagonal uncovered.

   Remark on evidence in mathematics; cf. Endnote 2 with Russell's quip on (lack of concern for) mere truth: as expected (from the quip), evidence for, tacitly, other aspects of knowledge is generally more rewarding here, perhaps relatively more so than in natural science. As a corollary, the popular business of confirming or refuting – specifically, the truth of – this or that is liable to be a distraction (from those other aspects, including the potential for abstraction, and thus for flexibility; in particular, even without settling (mere) truth for the originally envisaged interpretation; cf. the case of Riemann's hypothesis as meant by him and one of its abstractions, e.g., to varieties over finite fields).

9. The familiar compatibilities (with orders) – of the group operation and field operations in ordered groups and fields, respectively – were added, as far as I know, without further ado; after all, without such additions the whole (ordered structure) would *not* be more than the sum of its parts. In the case of o-minimal models, there is, at least, the possibility of a theoretical interpretation: compatibilities in the form of theorems (for o-minimal models) not imposed ad hoc.

10. In the tradition (in Endnote 6) of relying on the "human touch," it seems more usual to view this shift to a microworld when looking for unity within (the) diversity (tacitly, among aspects that

strike MAMNE) as a matter of – feelings of – explanations; cf. the introduction to Dirac's classic on quantum mechanics. In terms of Section 2.1(b) of the present text, such discovered orders turn out to be compatible with a relatively great variety of functions.

Remark So (re)viewed, the adjective *deep* seems apt (and does not merely evoke banter about dirt and diamonds at great depth).

11. For reference in Section 2.3. Once a memorable function of – of course, formally independent, but – interpretable definitions is stated, attention to "mere" independence (tacitly, for the situations in view) is formally ratified.

Reminder As long as only *sets* of theorems without the new symbols – not, e.g., their proofs – are in view, aka in-principle-provable theorems, the addition of interpretable definitions is equivalent to the original axioms.

(a) *Digression* around malaise over *relativity* – here, of equivalences relative to functions – apparently felt passionately (and certainly expressed vividly) by Pope Benedict XVI: whatever his feelings, which at least some (of us) regard as his private (or perhaps professional) concern, his words apply to the genuine difficulties of recognizing functions *suitable* for the situations in view, including the possibility of one (or at least relatively few) function(s) or equivalence relation(s) being rewarding in unexpectedly many situations in view; cf. the idea(l) of an essence with the tacit promise of complete knowledge. Bourbaki's basic structures, originally aka *structures-mères* (and appropriate organic combinations) in mathematics or elementary particles (including those associated with "fundamental" forces) in physics are tempered variants of that idea(l). Background to that malaise – by implication, actually, also over which (among the familiar diversity of) aspects reward attention – is provided in chapter 1 of J. Ratzinger, *Einführung in das Christentum* (Munich: Kösel, 1968), explicitly addressed to a lay audience (based on a course at Tübingen 1967 for the *studium generale*). Whatever – those with the BOR of – the theologian Ratzinger may imagine the concerns of that audience to have been, his words evoke vivid association in (the likes of) me: the crisp Latin principles *Verum est ens, verum quia factum* (p. 35) and *verum quia faciendum* (p. 39) – tacitly, with the ideal of truth as a popular peg on which to hang ideas – fit contemporary terminology; of *beable* (used by J. S. Bell for interpreting quantum theory), *observable* (tacitly, with possibly discovered interpretations and means of observation), and *doable* (in the broad sense of suggesting experiments (to do), for which an idea or theory need not to be true for anything remotely like the interpretation(s) in view); cf. also the digression in Endnote 3 on diverse functions of theoretical thought. By a refrain throughout these pages, the skills required for discovering aspects that reward attention (with the BOR available in the situation(s) in view) are not in question. But at least understood as earlier, the words of chapter 1 are not a reassuring preparation for Ratzinger's later chapters (once again, for the likes of me).

(b) The matter of noninterpretable axioms fits the present occasion inasmuch as Gödel emphasized it, partly explicitly, partly tacitly; e.g., so-called axioms of infinity (AI for short) are even second order – and thus robustly – independent. (Recall here also the digression in Endnote 3 on the way from principles, aka "from the top down.")

Disclaimer around the pursuit of far-out idea(l)s in logic or mathematics, also with strong feelings (of interest), to Keats aka mad: in contrast to such pursuits of social, economic, political, or even big scientific ideas, where the costs can be high by ordinary measures (and the feelings strike me as fanatic), I am not aware of any malaise in my own case, especially around pursuit by others – how else should I know about points of diminishing returns? As to critical reviews, it would not occur to me that they might convert the authors; possibly they provide relief for readers left speechless by those far-out idea(l)s. The next paragraph also begins with a kind of confession:

(c) Not having studied any of those AI in the sense of (b), I have no views, but at most impressions, albeit not unrelated to logical experience or broad commonplaces elsewhere in this chapter. (i) Though in principle (cf. footnote 48[a] of Gödel's incompleteness paper) some previously logically independent arithmetic or even other Π_1^0 theorems follow from the AI in view, they also follow from the consistency of (or at least a suitable reflection principle for) the AI.

Remark Presumably the ideas used for proving equiconsistency of diverse systems can be used for more informative (and actually also formally stronger) conclusions about having the same sets of Π_1^0 or even of arithmetic theorems. However, this is no guarantee for relevance to the main body of mathematics, despite the bit around Shoenfield's absoluteness lemma in the penultimate paragraph of Endnote 2.

(ii) Determinacy of (infinite) games in the tradition of descriptive set theory – that lost its organic connection with the main body of mathematics in the 1920s (at least until recently) – is a little too playful, aka too close to recreational mathematics for my taste. But at least this is proportionate to the "distance" of those AI from that main body.

(iii) Subject to correction by – the likes of – Woodin, I have come to relate (not necessarily elementary) embedding axioms (as in his chapter in this volume) to MAMNE's view of the so-called full set-theoretic hierarchy, specifically, MAMNE with my BOR after the early 1950s, but decidedly not before; cf. Endnote 14 for more. The old concern for *many* iterations of a possibly "thin" power set operation is supplemented by such axioms when pursuing a power set operation "fat enough," and thus (if iterated) unstructured enough to admit such nontrivial embeddings.

(d) **Reminders** of Cantor's days: his – contributions to – descriptive set theory flourished; e.g., the Cantor-Bendixon theorem in the subject of closed sets F. In contrast, (higher) cardinal and ordinal arithmetic were above all remarkable by being viable at all; cf. Dr. Johnson on dancing dogs and preaching women (novelists?). E.g., so viewed, the preceding theorem is read as showing that such F satisfy his CH; reminding those (of us) with a little recursion-theoretic BOR that the degrees of the first r.e. sets to be looked at were 0 and 0′. Depending on BOR, this or that option will be exercised in periods of stagnation. At one extreme is Keats' ideal of negative capability (an ideal seen here as a corollary to diversity among aspects of actually any object or, more poetically, to Hamlet's reminder of things not thought of in – tacitly, any – philosophy); cf. top of p. 150 of *Synthese,* **114** (1998) for more thoughts around it. At another extreme are Cantor's excursions into theology with reports in the literary form of set-theoretic metaphors; cf. C. Tapp, *Kardinalität und Kardinäle* (Stuttgart: Steiner, 2005), with a stern rebuke by Cardinal Franzelin (on p. 327, l.25, after noting Cantor's good will: *noch viel notwendiger ist das demütige Gebet um Erleuchtung und Kraft von Oben*).

Remark in terms of the particular theological tradition mentioned parenthetically in Endnote 2: Cantor pursued the premise *etsi deus daretur*; for the alternative premise, the cardinal's words (as I read them) say that he missed in Cantor an awareness of the limitations of (then known) abstract set theory – in the Anglican tradition aka outward and visible signs of an inward and invisible grace (of humility) – with the hope of a (future) bright idea and the power, aka peace of mind, to develop it.

Still continuing the worldly alternative (in the remark), apart from the preoccupation with *higher* cardinals, in modern jargon, Cantor confined himself to enrichments of sets by equivalence and (well-) ordering relations (as in cardinal and ordinal arithmetic), in particular, when contemplating relations between paintings and symphonies, represented by seven- (and somewhere four-) dimensional (product) spaces; cf. p. 412 and pp. 421–22 of Cantor's *Gesammelte Abhandlungen*, edited by Zermelo. (Furthermore, some kind of well ordering is assumed.) All this continues to strike – presumably not only – me as unrealistic, but not altogether sterile when used for contrast: first, within (abstract) mathematics, other enrichments and corresponding ho-momorphisms have replaced Cantor's preoccupations. As to paintings and music, digitalization has made altogether different representations familiar that are effectively combined with suitable hardware (and "wetware"), aesthetically effective for those with suitable gifts, and for the rest (of us) a reminder of additional requisites beyond abstract representations needed for effectiveness.

Preview Other oddities in periods of stagnation will come up subsequently.

12. On pp. 376–82 of vol. 3 of the *Collected Works*, Gödel refers, actually without being specific, to matters of course that correct his earlier oversights and apologizes for mentioning such matters to boot. In contrast, here, where such matters are dubbed "hygiene," it is recognized as an empirical discovery (and, at least to me, by no means as a matter of course) that such matters are often not remembered (even) when this would be useful.

13. The logical background to Section 2.1(c) is familiar, in particular from the formalism of GB, where Cantor's *Einheiten* and *Vielheiten* correspond to sets and classes (i.e., predicates of sets), respectively, but perhaps less so references to the accessible literature: Cantor's review of Frege's *Grundlagen der Arithmetik* is on pp. 440–41 of his *Gesammelte Abhandlungen* (cf. Endnote 11(d)), where Frege's naughty axiom – on every class being equivalent to a set – is dubbed *unglücklicher Gedanke*. In his letter to Dedekind (of 31.VIII.1899 on p. 448 loc. cit.), Cantor shows that the class V, short for $\{x:x=x\}$, is not a set by use of elementary properties of sets. Russell (famously) showed that Frege's axiom specialized to $\{x:x\notin x\}$, say R, is contradictory, albeit without emphasizing that the contradiction is purely logical (without use of set-theoretic properties of $\in$, cf. top of p. 372 of La prédicativité, *Bulletin de la Société Mathématique de France*, **88**, 371–91 (1960)); in contrast, $\exists y \forall x \ (x \in y \Leftrightarrow x \in V)$ is not logically contradictory but is in conflict with the axiom $\forall x \forall C \exists y (y = C \cap x)$, taking V for x and R for C.

Without exaggeration, in the mathematical tradition, Frege's axiom has long been treated with more or less benevolent neglect, aka an oversight or blunder. In the logical tradition (up-market), it is occasionally pointed out that suitable kinds of *partial* predicates or functions satisfy formally – something like – Frege's axiom around the recursion theorem or models of the λ-calculus.

Reminder of the commonplace about diversity, here of interpretations: the mere existence of some variant to which a (classical) logical contradiction is not an obstacle is a matter of course, but not the discovery of an area where such a variant is rewarding (with the risk in terms of Bourbaki's manifesto in Endnote 4 that the relation to Frege's axiom is its *côté le moins intéressant*).

Anecdote and thus without (cl)aims of statistical significance, but by experience suitable for some consumers in the commerce of ideas, e.g., for comparison with their feelings: in his essay on Russell, Gödel asks for an error in Frege's axiom – that, tacitly, feels – like division by 0 (i.e., cancellation without restriction). Whatever he may have expected when asking (or I, when reading) his question, I now feel that (Russell's) locating a purely logical contradiction as (immediate) consequence of the axiom (schema) does feel like (spotting) division by 0. As to evoking feelings of an oversight or a blunder, this depends for me on the individual in view, actually on classical lines: $\alpha\nu\theta\rho\omega\pi o\varsigma$ $\omega\nu$ $\eta\mu\alpha\rho\tau o\nu$ (Menander, 300 B.C.) and *Cuiusvis hominis est errare; nullius nisi insipientis in errore perseverare* (Cicero).

14. **Anecdote** In the 1940s, I was sufficiently dubious about the mere consistency of GB that I called the result – tacitly a certain prenex normal form of GB has no recursive model – a hypothetical contribution on whether all consistent formulae have such models. (Note on arithmetic models for consistent formulae of the predicate calculus, *Fundamenta Mathematicae*, **37**, 265–85 (1950).) In the early 1950s, I went to the trouble of writing this out for GB without the axiom of infinity (for which I knew a model even then).

Remark on T. Skolem, Einige Bemerkungen zu der Abhandlung von E. Zermelo: Über Definitheit in der Axiomatik, *Fundamenta Mathematicae*, **15**, 337–41 (1930) (which I noticed only recently): one ingredient in the proof – of GB not having recursive models – is clear enough at the end of Skolem's note but is there used only to express his malaise about Zermelo's idea, in modern jargon, of all first-order formulae of set theory being definite. (The malaise expressed was around some kind of inconsistency with countable models.) One missing ingredient was – Gödel's at the time unique, and for several decades by no means commonplace – awareness of the potential of (formal) incompleteness and related indefiniteness, specifically, for replacing malaise about contradictions by simple facts in those formal terms.

Horizons of (in)definiteness in MAMNE's and in (such) other views (as those including attention to scientific effectiveness): as observed in *Fundamenta Mathematicae*, 37 (1950) – or by inspection of Skolem's sketch – here a formally undecided Δ_2^0 sentence comes up in contrast to Gödel's Π_1^0 sentences, in general, not obviously decided one way or the other for (MAMNE's) arithmetic. Cf. L. Manevitz and J. Stavi, Δ_2^0-operators and alternating sentences in arithmetic, *Journal of Symbolic Logic,* **45**, 144–54 (1980), where the (obvious) parameter loc. cit. is considered: orders of enumerations of countable sets of formulae (or of submodels).

For reference To MAMNE, this may not feel like a parallel to different models (of sets) in the case of CH inasmuch as no particular enumeration feels (to MAMNE) comparably privileged to the full hierarchy (which, by personal experience above, some (of us) do not at all feel privileged spontaneously, but do so with a vengeance after attention has been drawn to it). But with broader BOR – around the effectiveness of such abstractions as real closed or real fields compared to $\mathbb{R}$ or, for that matter, number fields in the *Zahlbericht* compared with $\mathbb{Q}$ – such feelings have changed (for some of us).

Reminders around the full power set operation (and hierarchies so generated): in logic itself, it serves foundations (cf. Section 2.2, being definite for suitable segments) and – admittedly asymmetrically – in second-order (in)dependence, including categoricity of blessed memory. Such independence is robust, dependence occasionally useful pedagogically, for contrast between generally undemanding *constatations* (of categoricities) and more demanding corresponding formal independence. In MAMNE, the contrast *between* the easily described full hierarchy (compared to L) *and* its relative intractability has evoked malaise, with corresponding barks and meows, for those (of us) with a little BOR around both set theory and number theory. These are reminiscent of obiter dicta here about (easily described) natural and (less easily described) p-adic numbers with rationals in between, tacitly at least around some familiar classes of diophantine equations; cf. (mid) teenage BOR for linear equations over integers and rationals, BOR in Cassels' *Rational Quadratic Forms* also for p-adics (used on the way to rational and integral solutions, comparable to, but more recondite than, absoluteness properties of L, as in Endnotes 2 and 15(a)). Fitting the commonplace of diversity among aspects, sets have other aspects besides those of cardinal arithmetic (like CH; cf. Endnote 11(d)) and numbers other than those of diophantine geometry. (Since Descartes, organic combinations between geometric and algebraic, if not arithmetic, aspects of equations have been familiar.) Another aspect is the focus of prime number theory, the proverbial irregularity in the distribution of primes fitting recent combinations with – knowledge around – pseudo random sets in ergodic theory.

15. As the word (broach) may have suggested, the items in view are not prominent in the magisterial pronouncements of the authors concerned (and thus not emphasized by professional popularizers); cf. Poincaré's skepticism about and around – what was then touted as – logic or Hilbert's finitism or formalism (take your choice; Poincaré had died before Brouwer's preoccupation with the law of the excluded middle).

Remark (in passing here, with lots of suitable literature elsewhere): with additional, often tacit understandings, each of those pronouncements is – in mathematical jargon, trivially, more usually, said to be – easily seen to be, as the case may be, true or false. NB each of those tacit assumptions is satisfied in many (necessarily different) situations actually encountered.

Samples (a) Poincaré was struck by a certain indefiniteness – not only in the truth value of (closed) formulae but also – of predicates or functions in axiomatic (logistic) systems; cf. p. 47 of *Sechs Vorträge über ausgewählte Gegenstände aus der reinen Mathematik und mathematischen Physik* (Leipzig: Teubner, 1910). In Section 2.2, Gödel's effective use of the *distinction* between (formally) definite and possibly indefinite predicates is prominent.

Remark In the 1960s, this aspect of definitions was emphasized in model-theoretic terms, specifically, in trade jargon: (*semi-*) *invariantly definable* or (so) *definable-on* (absolutely defined terms). It also includes – for suitable classes of models what Gödel called – *absoluteness* (for V and L).

(b) For various reasons, I know more of Hilbert's nonmagisterial ideas; probably I am not conscious of most reasons, but one is that I saw a good deal of Bernays, who liked to remember Hilbert. (With such oral history in mind, published and semipublished material strikes the reader vividly that might otherwise be simply skipped. With the present industry of so-called history of logic or mathematics in bulky tomes, which also function as tombs for diverse bon mots, the chances are that hearsay (at least of – the likes of – Bernays) can also be documented). A most useful fact to remember is Hilbert's stroke in the mid-1920s, from which he never recovered. His style of thought was not the same after it. Various items will come up later.

(c) Brouwer's – what has been variously dubbed deviant or alternative – logic is particularly noteworthy by the following difference (in logic) from a kind of natural history, in terms of aspects that strike MAMNE with little BOR (in contrast to natural science, which often relies on discovered aspects; cf. Endnote 11). In particular, instead of (cl)aiming only to eliminate linguistic incongruities (tacitly, with respect to colloquial practice, dubbed "paradoxes" by the faithful), Brouwer described – what he called – choice sequences that (cogently) required a nonclassical logic.

Reminder from general mathematical experience: it is a separate matter to what extent, if any, this is more rewarding than (familiar) paraphrases (here meant in contrast to the ritual of some formal semantics).

16. The expression *logical foundations* has been long familiar (more widely and longer than logical hygiene; cf. Endnote 4), with different – also tacit – (cl)aims to boot, ranging from security to "deep(er)" aspects not accessible to MAMNE. Admittedly, occasionally, there has been emphasis on aspects accessible to MAMNE with little BOR (as in *back to basics*). Readers with (here) suitable BOR will know the (cl)aims of Bertrand Russell in the introduction to (the second edition of) *Principia Mathematica* for logistic foundations to demonstrate a, if not the, unity, as in the name of the journal *Fundamenta Mathematicae* and the opposition in Bourbaki's *la mathématique*. As to security, in the particular case of consistency in the sense of freedom from contradiction (not requiring consistency in the sense of substance or compactness), it became topical around one hundred years ago after Russell's "paradox" in Section 2.1(c); cf. the following endnotes. The aspect – of definite meaning – emphasized here fits a good deal of the logical literature labeled "foundational" but also serves for contrast to the use of logic in Section 2.3.

17. E.g., Bourbaki's article (Foundations of mathematics for the working mathematician, *Journal of Symbolic Logic*, **14**, 1–14 (1950)) (re-) formulates a shift of emphasis going back to Hadamard *from* mere consistency-in-principle *to* consistency-in-mathematical-practice in, let us say, provocative terms: consistency is an empirical fact, not a metaphysical principle (provoking a reminder of many metaphysical principles refuted by empirical facts; some discovered, others striking to MAMNE, at least after being pointed out).

Background to (such) malaise around the idea(l) of consistency: to MAMNE with the BOR of Aristotle – tacitly, truth functional – consistency apparently seems *necessary* without qualification (not if the commonplace about diversity is remembered). Those with the BOR of Niels Bohr are struck by – what he called – "deep truths," say B, when (in his words) both B and its negation are true. (I have come to regard his *aperçu* as follows: in the first place, B must be sufficiently vague to be complemented by discoveries of conditions C_1 and C_2 actually encountered and both rewarding attention in their ways such that $C_1 \Rightarrow B$ and $C_2 \Rightarrow \neg B$. The reading restores consistency, but understood uncompromisingly, Aristotle's ideal would be an obstacle even to looking for such a (re)reading.) Nearer home, the material in Endnote 13 around partial predicates presents *undefined* as a *tertium datum*; cf. the reminder at the end of Endnote 2 on Aristotle's malaise over indefiniteness. At another extreme, Hilbert's equally flashy slogan proclaimed consistency as *sufficient* in mathematics without – let us say, premature – qualifications. Rewarding ones, and also limitations, tacitly, in diverse areas are familiar in logic, ranging from – arithmetic (but, by Endnote 14, not generally recursive) definitions of – models to truth in arithmetic of Π^0_1-theorems in Σ^0_1-complete systems, but not enough for ω-, or better, 1-consistency.

Reminder of a parallel to the effectiveness of mathematics in Endnote 3, by specialization to mathematical logic, when thought of – as throughout these pages – as providing idea(lization)s of especially mathematical experience of proofs. On one hand, additional (formal) knowledge of the ideas contributes, e.g., consequences of different choices of rules (for describing a set of theorems). On the other hand, organic combinations with suitable mathematical aspects of that experience require attention. As expected from experience in the theoretical natural sciences, "hard-nosed" mathematicians will miss counterparts in the logical idealizations of aspects in their experience to which they have become attached, and "starry-eyed" logicians will be disappointed by lack of interest in their pretty exercises; cf. Endnote 18 for more.

Anecdotes about some pioneers: (a) in the preface to – admittedly, only the first edition of – *Was sind und was sollen die Zahlen?* Dedekind noted (his) malaise over some convincing reward for meticulous logical deductions. He picked on reliability (patently related to the (cl)aims for consistency) by use of a metaphor around reading: the only reliable way is to spell out the words (letter by letter; with (my) present proof-theoretic BOR, an alternative is (to me) more satisfactory: emphasis on different specializations and generalizations of terms in theorems derived by diverse logical rules).

(b) According to Bernays – once again, cf. Endnote 15(b) – Hilbert was asked (before his stroke, loc. cit.) if his claims for the ideal of consistency should be taken literally. In his (then) usual style, he laughed and quipped that the claims served only to attract the attention of mathematicians to the potential of proof theory.

Reminders First, Hilbert's audience – knowing his preoccupation (in logic, not elsewhere) with routine proofs, described by him as like numerical calculations of numerical equations; cf. his wording of the tenth problem – would have understood this as potential for the kind of routine consistency proof he envisaged as bypassing then popular (dramatized) foundational problems and getting on with the job of doing mathematics as he had done (reworded in print after his stroke as a "final solution of all foundational problems"). As viewed here (and in Chapter 1 of this volume, as I read it, but without agony over foundational sensibilities), this has occurred anyway, at least in Bourbaki's tradition, without such consistency proofs, and Gödel's incompleteness theorems provided logical hygiene to Hilbert's ideal but no relief from the malaise felt about it by the (experienced) mathematicians mentioned earlier. Second, more generally, while in practical life, it is often salutary to bypass obstacles *after* having paid attention to them, in my experience – here and exemplified later – the ritual of a formal bypass is seldom more rewarding than more or less benevolent neglect, at least, without a shift of emphasis (there is no free lunch). A price paid by those (of us) later pursuing proof theory has been to discover suitable ideas for adapting consistency proofs to aims rewarding for a broader view, naturally, not an unduly high price for those (of us) attracted by such activity; cf. again Endnote 18 for more.

Gödel himself was skeptical about the interest of Hilbert's program in his Königsberg lecture as consistency does not (even) ensure ω- (or in present jargon 1-) consistency; a few years later, in his Zilsel lecture, he referred to some great epistemological value of a realization of the program. (For all I know, a formal ratification of weaknesses of such venerable idea(l)s as (mere) consistency is deemed to have epistemological value.)

18. While (the) ideal(ization of) consistency of principles (in other words, in-principle) lends itself to theory at all levels of sophistication to boot, including Gödel's charming proof of consistency for $S \cup \{\neg Con\ S\}$ from $Con\ S$ in Section 2.2, in practice, it is rarely pertinent to practical concerns, for which an arsenal of precautions has evolved (including cross-checks of numerical computations by suitable, sometimes most imaginative abstract interpretations).

Remark on a logical idealization of – the ubiquitous practice of – cross-checks used in the *second* edition of Hilbert-Bernays II, pp. 298–300 (there is no counterpart in the *first* edition of HB II; cf. pp. 283–89): while no more realistic as a description of natural phenomena of proof, e.g., in speculation about biological data processing than more familiar formal systems, it provides examples of bona fide formal systems that do prove their own consistency (and have the same derivations, not only theorems, as others that don't); only checking that a derivation is a formal proof is different.

Remarks related to commonplaces around theoretical idea(lization)s, aka formalizations in the case of formal theory:

(a) The would-be catchy formulations of the incompleteness theorems in terms of – some tacit order of – strength of systems are, let us say, infelicitous; cf. Section 2.1(b) on functions of orders. The example in HB II loc. cit. illustrates this vividly.

(b) This was brought up to date by R. G. Jeroslow, Redundancies in the Hilbert-Bernays derivability conditions for Gödel's second incompleteness theorem, *Journal of Symbolic Logic*, **38**, 359–67 (1973).

(c) Another aspect, in particular, generality, of Gödel's proofs of incompleteness – actually, plain from both editions of HB II, albeit not stressed there (in keeping with the broad scheme loc. cit.) – was later made familiar by Tarski and Mostowski in terms of *definability*; cf. Gödel's title without – what would then have been – premature precision, about systems "related" to *Principia Mathematica* (or, if preferred, a formally clean variant of it).

Reminder of the preamble around diversity among aspects, here, of PM: one aspect, aka property, of PM is that its set of axioms is r.e., another is that it is *semi-invariantly definable-on* ω among general models (in terminology of the 1960s).

Pedantic technicality concerning both editions of HB II: while the axioms are not restricted (e.g., to being r.e., aka formal), the numbering is required to be recursive: for the later abstraction, this is *invariantly definable-on*, in Gödel's terminology, *entscheidungsdefinit*.

(d) For readers familiar with rational mechanics with its ideal(izations of) solids, fluids, and gases, describing MAMNE's views of those objects in the literary forms of analysis: the contributions range from the level of – the (very few) likes of – Newton, Laplace, or Poincaré to the Tripos examples at Cambridge (United Kingdom) up to the 1940s. Counterparts are expected (and realized) in the logical literature, including the superstition that a principal weakness (of such applied mathematics) is lack of formal rigor, overlooking the oversights in MAMNE's views themselves (and correspondingly of their formal definitions), tacitly, for improving knowledge of the earlier objects.

19. While Gödel – to the amazement of all with any feeling for mathematics – drew a conclusion from mere consistency (or 1-consistency or what have you), without much direct consequence for mathematics, when combined with suitable consistency proofs, the emphasis shifts *from*

Gödel's conclusion *to* (the negation of) principles used in those proofs, most famously (in logic) Gentzen's use of $\in_0$-induction applied to diverse predicates; cf. H. Friedman, Iterated inductive definitions and Σ_2^1-AC, in *Intuitionism and Proof Theory*, ed. J. Myhill et al. (Amsterdam: North-Holland, 1970), pp. 435–42. For a cozy presentation some thirty years ago, with an exposition in appendix II of (part of) Friedman's result suitable here, see, e.g., Wie die Beweistheorie zu ihren Ordinalzahlen kam und kommt, *Jahresbericht der Deutschen Mathematiker-Vereinigung*, **78**, 177–223 (1977), esp. pp. 214–23.

Reminders First, Friedman uses only the fact of unprov*ability* (in Gödel's second theorem for the systems in view), not details of its proof. By general experience, such results are generally seen more vividly by model theory (even though some that were so stated in the stone age of logic were originally established proof theoretically, with proofs providing additional information; cf. Endnote 17). More specifically – at least my expectations of – advantages of a model-theoretic argument are illustrated by A. Bezboruah and J. Shepherdson, Gödel's second theorem for Q, *Journal of Symbolic Logic*, **41**, 503–12 (1976). (Their own reservations in the introduction to – what is in effect their advertisement in the title of – the emphasis on Q do not spoil this use of their paper.) The non-Archimedean, aka nonstandard, models of the systems in view are more or less brutally unfounded and so, in particular, are nonstandard proofs of $\neg Con\ S$ in them. So one expects (or just hopes) to *see* such proofs (or, in fact, of all formulae) and their *oddity*. So (re-) viewed, a principal weakness of Q is the patently exceptional nonstandard proof good enough here: just using the axiom (schema) $A \to A$ and the rules of *modus ponens* and repetition. (Very simple induction would exclude this.)

Disclaimers On one hand, even though exceptional, it is good enough to show that rules in Endnote 18 requiring cross-checks with the (finite) body of previously proved theorems would exclude this kind of proof; but cf. the later remark. On the other hand, I hold no brief for their (stated) ideals of simplicity and, above all, generality without qualification: alternatives are marginal utility and attention to diversity, here, of generalizations. E.g., L. H. and R. H. inverses are equal in abelian groups, but the extension to all groups is only one viable option; another is the extension to suitable commutative structures that are not necessarily groups.

Remark on, what would be for me, a missed opportunity, a blunder by me in the 1950s (probably published somewhere) corrected by Shepherdson: I had (sensibly) reinterpreted Rosser's so-called trick – eliminating Gödel's assumption of 1-consistency for his incompleteness theorem, beyond consistency – as a shift in the rules considered, requiring cross-checking, albeit asymmetrically: only the negation of the formula in view must not have been proved earlier (in contrast to the symmetric variants in Endnote 18). My premature conclusion was that (such) Rosser systems proved their own consistency. They do not; cf. Zbl. 728.03031, referring to a letter (of 5.X.1991) by De Jongh.

20. The (gradual) introduction of the common word *conjecture* as part of mathematical trade jargon – albeit not throughout the trade, where it is more or less synonymous with *guess* or sometimes just *a possibility* – is here viewed as a vivid, albeit perhaps extreme, case of *change* in idea(l)s with BOR; cf. preamble. Roughly speaking, with experience in up-market mathematics, a distinction within the common or garden variety of guesses has come to be recognized as rewarding with suitable BOR, at first, presumably, tentatively. Correspondingly tentative will then be a suitable sense of *evidence* (for a conjecture in view). Not feeling myself to be a suitable spokesman for the preceding trade, here are a couple of impressions:

(a) Like that favorite idea(l) of truth tout court (which leaves tacit some suitable interpretation, e.g., a (class of) model(s)), it can be delicate to specify for which (among many) aspects such evidence is sought; knowledge has many rewarding (and correspondingly demanding) aspects besides truth, let alone, for some familiar interpretation. So much is only cautionary.

(b) More positively, (my impression is that) conjectures are to be related – not only to some special instances, but – to a broad (possibly informal) idea, comparable to the familiar matter of crucial experiments for a general scientific theory.

Reminder The popular slogans of confirmation and refutation (adumbrated in Endnote 8) and their synonyms revive – the (b)rash practice – in (a) without or prematurely specifying areas of validity envisaged; cf. Endnote 22 for more.

21. As to the (popular?) human touch, it is apt to provide *distraction*, well illustrated by history – not so much as practiced in biology – but particularly in the history of ideas; cf. A. Weil's autobiography (p. 119): *l'histoire des mathématiques ou pour mieux dire la lecture des grands textes mathématiques du passé me fascinait* (after having lectured at an ICM on "History of Mathematics: Why and How?"). Depending on their BOR, readers may be distracted in the sense of being entertained or irritatingly deflected (from logical concerns) by the following historical touches:

(a) W.r.t Endnote 18, Tarski emphasized – in effect (I at least do not know if on purpose inasmuch as I do not know whether he ever looked at those pages in HB II) – the *abstract* aspect of the formulation, Turing the particular case of r.e. systems. Gödel himself – both in his course at Princeton in 1934 (before his transatlantic crossing with Bernays) and afterward in his Gibbs lecture – also took up the particular case (that fascinated Turing) for contrast with a metaphysical principle about Reason being able to answer all, let us say, Reasonable questions. Then there is the term *derivability conditions* used in HB. Those committed to the trade dealing in provability logics view these conditions as selections among proof predicates for a given theory, "identified" with its set of theorems; cf. Section 2.1(a) on functions of equivalence relations. For those (of us) who have found the body of proof theory (we know) most rewarding when applied to formal rules, for which there are then familiar canonical representations, those conditions are then – viewed as – properties of the rules; cf. also Endnote 19 on Rosser's "trick."

(b) With respect to Endnote 19, the malaise expressed in erudite jokes, e.g., by Tarski and Weil, overlooks counterparts (to consistency) in ordinary mathematics they knew, e.g., fields in algebra. NB They may have found erudite jokes about fields alluding to agriculture as stale as I find theirs; but cf. Endnote 13 on – one way of – rethinking proofs of a consistency statement.

Anecdotes about my teens at Cambridge (United Kingdom), where analytic number theory was then popular, with proofs frequently dubbed more interesting than their theorems: in my innocence, I asked, "So why do you not use – tacitly, ideas of – the proofs for theorems of interest proportionate to that of the proofs?" I did not get an answer, nor did I have a clue how rewarding (and, as so often, correspondingly demanding) such concern for "rethinking" of proofs, by (suitable!) abstraction, would be in abstract mathematics of the second half of the twentieth century. In fact, this had to wait until the mid-1950s, when I met Serre, a pioneer in the subject (with – as it were, private – logical interests, and a feeling for logic).

(c) With respect to Endnote 20 around conjectures (in contemporary mathematical jargon), readers of the literature will have come across occasional protests by authors whose impressions or just questions have been dubbed (their) conjectures, sometimes the same authors who associate Wiles' (most) famous proof – not with Fermat's marginal remark several centuries ago, but – with Taniyama's conjecture (or Taniyama/Weil if, in effect, usually not in these words, the latter's sharpening to Π_1^0-form is to be emphasized).

Remark Changes in meaning – with expanding BOR – of everyday terms are familiar in physics, of weight and mass when BOR extends to, say, expeditions to the moon, or the neologism "rest mass" when attention includes, say, the motion of electrons around the nucleus of atoms.

For the human touch (including names), here are a couple of anecdotes, at two extremes to boot:

(i) In his autobiography, shortly before Faltings' proof of – what was aka – Mordell's conjecture (of binary diophantine equations over $\mathbb{Q}$, of genus > 1 having at most finitely many solutions), Weil declares his (negative) feeling about that label (presumably, in contrast to the conjectures that the Clay Prizes reward only if proved, while either answer to NP ? P merits a prize).

Personal remark For a meaning of *conjecture* that has grown on me over the years, a formal statement (to be proved, refuted, or shown to be undecided by suitable conditions) is generally not enough; it has to be supplemented by a suitable text (as formal proofs usually are, even in such austere texts as Bourbaki's *Éléments*). Thus it is not excluded (though I do not know how) that *after* Faltings' proof, comments are found that supplement the formal statement into a bona fide conjecture (even without constituting a proof) or, indeed, the opposite (by specifying a previously overlooked obstacle).

(ii) This is a case ad hominem. In the 1960s, I wanted to shift attention – away from what had struck me as thoughtless (cl)aims around "measures" for complexity – to (discovered) logical uses of literally superficial parameters of length of proofs (number of lines or nodes); at least in my view then, reminiscent of the shift (in the 1940s) from consistency to specialization of (existential) terms in proofs of theorems (with existential quantifiers in suitable normal forms); cf. the P.S. to M. Baaz and P. Wojtylak, Generalizing proofs in monadic languages, *Annals of Pure and Applied Logic*, **154**, 71–138 (2008). Briefly, a property $\wp$ of systems (in the notation of Baaz and Wojtylak) with formulae A was considered in terms of provability $\vdash_k$ by proofs of length k: $\exists k \forall n \in \omega \vdash_k A [x/s^n 0]) \Rightarrow \vdash \forall x A$. (Evidently some systems do, others do not have $\wp$.) Some articles that established $\wp$ for this or that class of systems labeled their discoveries as proving so-and-so's conjecture or as being about length, tacitly, with emphasis on the premise in $\wp$, here meant in contrast to the conclusion (about generalization). So (re)viewed, the label does not fit (my) contemporary understanding. More appropriate, albeit very modest, was a conjecture *after* Parikh's proof of $\wp$ for a monadic variant of standard formal arithmetic, PA, related to general experience in mathematics shifting from the existence of *some* solution (of a set of equations) to descriptions of *solution sets*, with specification of a suitable class of systems and suitable terms (instead of the numerals $s^n 0$: $n \in \omega$): the emphasis shifts to generalization – in place of specialization – of terms.

(d) A separate matter is the – feeling, aka sense, of – reward for such additions of the human touch (or, more generally, of biographical material around authors for an understanding of their ideas). By the commonplace of diversity, which, by now, has become a refrain in these pages, a true but boring answer is evident. This does not exclude the discovery of significant categories (of authors and ideas) on which somebody with taste and talent for the human touch may have something rewarding to say.

22. As is well known, at least tacitly, to those with (rewarding) experience around so-called interdisciplinary knowledge, which in the jargon used in these pages combines knowledge from different disciplines, this requires additional skills (to those in any one discipline); cf. Endnote 3 on Wigner's "half truth." Perhaps it should be added that the kind and degree of satisfaction to be expected from an item of interdisciplinary knowledge will be asymmetric, dependent on the relative BOR around the different disciplines available. (Logical BOR, especially around Gödel's proofs, is assumed later.) After earlier reminders about diversity among aspects (and additional diversity among potential combinations with any one aspect; cf. Endnote 6), the incompleteness of the following samples – even with respect to my own knowledge! – is a matter of course.

(a) At one extreme are organic combinations of common(place) knowledge with professional BOR, here viewed as *consolidating* the former, e.g., the commonplace of diversity and its corollary that literal completeness is an oversight. In logic, even when rules are complete with respect to validity, they need not be so with respect to deduction (in the cut-free case), let alone for proofs, even in the case of classical propositional logic.

Less technically, completeness of rules is understood, at least tacitly (by some), with respect to proofs. Now, in a familiar way, for each $N \in \omega$, and given graph G with N nodes, it can be checked whether G is planar, and a propositional formula P_G can be presented that is satisfiable if and only if G is four-colorable. Thus *a* proof of the satisfiability of P_G uses the four-color theorem as an intermediary. With minimal assumptions on an equivalence relation between proofs (cf. Section 2.1(a)), this has no equivalent among formal derivations for propositional logic.

Reminder   Equivalence is trivial if a set of rules is "identified" with its set of (propositional) theorems.

(b) Some dreams around some theory of quantum gravity (rehashed also in this volume) go back, in effect, if not "historically," to an idea in Gödel's Gibbs lecture (there) with special emphasis on r.e., aka formal, theories; cf. Endnote 21(a).

Reminder   At the time (in the early 1950s), Turing's – cult of the black box, in the form of his proposed – test prompted some reservations. Half a century later, robotic engineers are more impressed by other capacities of humans and (some) other primates they (cl)aim to simulate than by superiority in chess and mathematical *deduction*, here meant in contrast also to other more imaginative elements of mathematical thought.

Admittedly, carping at – instead of ignoring; cf. the first reminder in Endnote 17(b) – weak arguments is not everybody's cup of tea. But here – naturally, with suitable BOR – some additions and also disclaimers seem rewarding.

(i) Gödel states boldly – actually, for formal rules, but equally cogently for the broader kind to which his argument for an undecided but true sentence applies – that the knowledge available (to us) may be definable without (our) knowing a definition, and thus without knowing the truth of that sentence. Perhaps more realistically, the enterprise of axiomatizing a – naturally, growing – body of knowledge constitutes itself an *addition* to that knowledge, when the axiomatization so found no longer includes that addition. Only with that addition is the consistency of the axioms, and thus the truth of the undecided sentence (for its arithmetic interpretation), as vivid as this (inclusion) may be – if not to MAMNE then – to some mind's eye with a little logical BOR; cf. notorious optical illusions.

(ii) Independent of those dreams around unknown theories is the broader question of finding some suitable *aspects* of existing theories in natural science that satisfy laws outside some (would-be privileged) familiar kind, e.g., nonrecursive laws in classical mechanics. There are plenty available, e.g., in texts by Pour-El and Richards. What has so far been lacking is an empirically satisfactory interpretation; cf. the review on pp. 53–55 of *Jahresbericht der Deutschen Mathematiker-Vereinigung*, **94** (1992).

Remark   There is no suggestion that nonrecursive aspects are more prominent in (classical) quantum theory (without attention to gravity) than in classical mechanics; in particular, quantum computation does not (cl)aim to compute nonrecursive functions, only to speed up computations.

(c) C. Koch and K. Hepp, Quantum mechanics in the brain, *Nature*, **440**, 611–12 (2006), largely ignore – in effect, perhaps having (tacitly) recognized – the oversights pointed out in (a) and (b) above and so have no occasion to correct them. Instead, they go back to old (cl)aims of relating consciousness to measurements in quantum mechanics in the 1920s, long before any suggestion of quantum computation. Presumably, there is a potential for a review by reference to such computation and also to discovered (neurobiological) aspects of consciousness (while in the 1920s, only aspects visible to MAMNE without neurobiological BOR were used). Whatever the interest of the article to professionals may be, (we) consumers are apt to balk at the sweeping generality of the title and correspondingly facile slogans about the (relatively large) size – in admittedly far-fetched dreams about functions – of microtubules (compared with effective ranges familiar in quantum mechanics, let alone Planck lengths).

Reminder of the question: why is gold yellow and silver gray? No doubt this question may have been asked at a stage with minimal BOR, when MAMNE looking at lumps of gold and of silver sees nothing moving fast, certainly, not their electrons moving around the nuclei of atoms of those metals. Subject to a chance of confirming with competent specialists a report on this in – the German edition of – *Scientific American*, it appears that relativistic effects of such motion are related to those optical properties; cf. Endnote 21(c) for more around a broader (consumer's) view on such matters, enough to be prepared for contrasts with traditional professional choices (by remembering logical hygiene around diversity) among aspects.

Reminder of an unproblematic counterpart of use in all molecular chemistry, not only biology or neurobiology: metric three-dimensional aspects of chemical bonds derived (by Pauling) quantum theoretically: This incidentally fills a gap on p. 35 of Ernst Mayr, *What Makes Biology Unique?* (Cambridge: Cambridge University Press, 2004), at least if Pauling's is counted as one of the great discoveries made by physics in the twentieth century. Warning: the reminder does not fit the equivalence relation (between sciences) apparently – in at least one sense of this word – tacitly understood in the title.

Elementary reminder prompted by Gödel's incompleteness result: as I have come to see it, it is one of those more or less mathematical gems – like the irrationality of $v\sqrt{2}$, uniqueness of the solution ($v = 0$) of $v = cs$ (for a body starting at rest) noticed by Galileo, and (perhaps, even) Russell's contradiction (at least, if viewed as in Section 2.1(c) and Endnote 13) – that, since classical times, have aroused in each generation exaggerated expectations of the "powers of pure reason," exaggerated by forgetting the additional, usually empirical knowledge needed for progress (e.g., for the discovery of suitable measures when rationals are not); cf. also Endnote 11(d) for some additional, equally elementary items around this reminder.

23. Readers interested in Hilbert's states of mind, in particular, before and after his stroke, will no doubt recall Endnotes 15(b) and 22(a). But independently of this, the material (semi-) published during his lifetime demonstrates his flexibility, at least in effect. E.g., soon after his best seller on geometry – and his recommendation in his lectures of the logical exercises in it as a distraction from boring philosophy of space at the time (at the end of the nineteenth century) – he took up Einstein's philosophy of space(-time) with relish, and without the kind of logic loc. cit.

24. The metaphor occurs – actually, in both forms, of logic as grammar, and logic as language – in Bourbaki's *Foundations* in endnote 17; cf. also the understanding in the remark in the first paragraph of Section 2.1. NB This warns at the start that the author was logically self-educated, albeit without even hinting at the usual consequences: not necessarily lack of general understanding but certainly of what is and what is not generally familiar to the professionals (here, in logic).

Remark on the present case: self-taught Bourbaki had an exceptionally gifted teacher, albeit with one-sided logical BOR, given that Weil had translated Hilbert's *Über das Unendliche* in the mid-1920s. For a kind of update, albeit in a very different style, cf. J. Dieudonné, The work of Bourbaki during the last thirty years, *Notices of the American Mathematical Society*, **29**, 618–23 (1982); cf. Endnote 25.

Correction cf. Section 2.1(a) on (in)equivalence relations with and without an order of priority, aka difference and originality, respectively: Dieudonné's (unauthorized) article repeats Bourbaki's own general view of the *Éléments* as not being original and in particular stresses some kind of close relation to Hilbert's earlier abstract number fields in the *Zahlbericht* and abstract geometries in *Grundlagen der Geometrie*. As I have come to read this material, a difference is striking. Hilbert did not and Bourbaki did stress compatibility conditions (e.g., in the manifesto; cf. Endnote 4).

Remark on literary abstractions in metaphors from literal meanings: options for describing the relation in view are Hilbert's earlier ideas as a *source* (for the *grand courant*) of the *Éléments* or as a *seed* fallen on fertile ground (provided by later mathematicians) with fruits presented loc. cit. Depending on BOR around geology and molecular biology, one or the other option is more suitable for drawing attention to the potential of those compatibility conditions (ranging from tributaries of the majestic river to so-called emergent properties coded in DNA, etc.; cf. epigenetics in Endnote 26). To me, both a particular satisfaction derived from handling mathematical abstractions and the skills needed for doing so effectively are reminiscent of those found in experience with metaphors. This parallel might well have been as informative to Simone Weil as her brother's (in March 1940; cf. Endnote 30 for more); cf. also the reminder in the following endnote.

25. On p. 618, at the bottom of the second column, Dieudonné has some pieties (of the bien-pensants) about everybody being free to give their meaning to mathematics, as if not half the world, at least tacitly, felt in need of guidance rather than freedom. There is lip service paid to the (no doubt up-to-date) idea(l) of communication, without a hint of what (feeling?) is to be communicated, except for something in contrast to meaning.

   For reference in (c), following: Whatever Dieudonné's words were to communicate, they certainly evoke (in me) the following:

   Reminder of a shift in emphasis *from* meanings of individual terms (in common or garden varieties of languages) *to* those of compounds, in particular, describing analogies: their effective use would be hampered by harking back to a semantics for the elements, not only by cramping one's pace but by preventing the meaning of the analogy to grow and adapt with new situations encountered. After all, mathematics is not for morons incapable of such growth.

   Disclaimer This is remembered without forgetting the (cl)aims of up-to-date artificial intelligence (in robotics) with its (lucrative cl)aims of automation (and another market).

26. **Anecdote** from the mid-1950s, when I had (exceptional) private lessons in – visions around – molecular biology, long before the slogan of the (family of) DNA molecule(s) being the molecule(s) of life: what I understood of it did not seem unreasonable (in the sense of unrealistic), inasmuch as the molecule presented itself (to me) as a nondeterministic program, like an axiom system, leaving open what was going to be generated – and tacitly given attention – and in which order the rules were to be active, aka applied, a now familiar concern of epigenetics.

27. Roughly speaking, in (classical) model theory, the main (primary) object of attention is a class of theorems, in proof theory also the proofs (and even the manner of checking the correctness of application of the rules; cf. the first paragraph of Endnote 18).

28. As in Endnote 27, recall Endnote 18, now on tacit assumptions in practice around (absence of) needs-in-principle. Russell stressed the (cl)aim of – *Principia Mathematica* as a parenthesis in – a refutation of Kant, specifically of some kind of intuition for mathematics or perhaps visualization in Section 2.3(c). Russell's idea(l) was to use – tacitly, familiar – logic only. So to speak legalistically, (Gödel's) incompleteness refutes Russell's idea, but with the tacit assumption (surely accepted by Russell) that equivalence with respect to truth in arithmetic was enough.

   Remarks Mathematicians, often at a loss for words when wanting to convey very simple feelings simply (not required in the trade), say that they expect from a proof not only truth but understanding.

   In logic, it has become customary in practice to interpret what is lacking in terms – not of (a feeling of) understanding, but – of length; cf. Endnote 21c(ii) on some measures and (my) reservations. No doubt there are some functions of some proofs for which length is a suitable measure, but I'd find it hard to specify such cases, which those with a feeling for mathematics want to understand at all. For one thing, *descriptions* of proofs are often more suitable.

29. By a fluke in my undergraduate days in the 1940s at Cambridge, idea(l)s of elementary and direct proofs, especially in number theory (in contrast to analytic number theory, which used function theory heavily), were (still) a prominent topic. Selberg's (and Erdös') elementary proof(s) of the

prime number theorem caused a stir. By this fluke, I and others have had some time for second thoughts.

(a) Certainly MAMNE – at least with BOR (in this respect) like mine in those days – saw those ideas sufficiently clearly for confirming them formally, as it turned out, fitting some of the literature I learned a little later; cf. Endnote 30.

(b) After an interlude – away from logic engaged in theoretical hydrodynamics, at the time aka applied mathematics, I did not feel that those (clear) idea(l)s were among the "foolish things one had better learn to forget," but my emphasis changed: where, if anywhere, do they contribute?

Remark As it happens, this question, applied to ideal(izations of) liquids, had been exceptionally rewarding in hydrodynamics. By temperament, I paid particular attention to formal rigor in hydrodynamics where this was (pointlessly) neglected, in contrast to logic, where a – for my taste, disproportionate – cult was made of it ever since Gödel had shown that this paid off at all.

(c) By the time I came across Gödel's term *reductive* proof – cf. *Synthese,* **114**, 122 (1998) – I had lived through several changes in the meaning of *elementary* proof; *key words*: first, the division moved from $\mathbb{R}$ or $\mathbb{C}$ to the p-adics, and later from p-adics to l-adics. Furthermore, attention was paid, albeit not in so many words, to – presumably, not only – my malaise over those elementary proofs: their (auxiliary) constructions were (and are) not memorable to me; cf. the middle paragraph on p. 123 loc. cit., citing W. J. Ellison and M. Mendès France, *Les nombres premiers* (Paris: Hermann, 1975).

For all I know, (Gödel's) *reductiveness* may have become a bona fide staple of the epistemological trade; cf. (c) on p. 124 loc. cit. But this leaves open what, if anything, it has to offer the rest of us.

30. The references in the text are to letters of Gauß and A. Weil (to his sister, of 26.III.1940), respectively.

Forewarned by a passage in an earlier letter (of 29.II.1940, reprinted in his autobiography, of which more in the last remark of this endnote), not his own stated aim, but two other aspects of Weil's letter to his sister of 26.III.1940 are given attention subsequently. The first concerns effective uses of "the" analogy between number and function fields.

(a) B. Mazur and A. Wiles, Analogies between function fields and number fields, *American Journal of Mathematics,* **105**, 507–21 (1983) (in an issue dedicated to Weil on his seventy-seventh birthday), give an account of contributions under their heading over the preceding four decades. But also they have – to me very efficient – reminders of a familiar price: *where* is an analogy in view likely to contribute (rather than distract), and in which *direction*? Thus, on p. 507, it is noted parenthetically that function fields are considered to be easier than number fields, and on (top of) p. 508, they cite function fields (for curves over finite fields introduced in the early 1980s as analogs to classical cyclotomic number fields) that "at least at . . . present . . . show every sign of being just as difficult as their classical number-theoretic counterpart."

Remark with the option of consolidating commonplaces about other abstractions by relations to mathematical experience: this kind of *uncertainty* and awareness of *changing* BOR simply go with the territory.

Disclaimer My BOR around the subject is not good enough to "balance the account" over, say, the last twenty-five years, except for mathematical common(place) knowledge about Langlands conjecture(s) coming under the heading in view, and Fields Medals having been awarded for work on some of the number fields and function fields concerned (to Drinfeld and Laf-forgue, respectively); cf. J. W. Cogdell, Langlands conjectures for GL_n, in *An Introduction to the Langlands Program*, ed. J. Bernstein and S. Gelbart (Boston: Birkhaüser, 2004), pp. 222–48.

(b) Those with BOR weighted toward formally real fields will think of Siegel's extension in the 1920s of Lagrange's theorem (on representing nonnegative elements as sums of four squares) to suitable number fields, with equally early straightforward counterexamples for function fields.

(c) In the 1930s, specialists had developed a feeling good enough to let them transfer the ideas of (usually easier) proofs for the fields of formal power series to *p*-adics. (There was then no question of specifying a class of properties for which the transfer was to be expected.) In the 1960s, experience around ultraproducts provided a suitable class – specifically (in the sense of model theory, not Endnote 29) elementary properties – to correct and prove a corrected version of a conjecture of Artin from the 1930s. While this result exceeded the expectations of mathematicians from model theory and other logic, by general experience – e.g., around the stock market – it raised somewhat unrealistic expectations later. At any rate, it remains a bona fide contribution to pure mathematics; cf. Section 2.3.

At another extreme, in foundational, aka hard core, logic, more or less familiar function fields are used as models; cf. Endnote 19 on Q and other so-called weak systems of arithmetic, about which more later. They are at an opposite extreme – to *p*-adics, etc. – when viewed in terms of the digression in Endnote 3 with such examples as Hilbert's *Zahlbericht*: from the bottom up. The logical tradition starts with Peano's (second-) order axioms, and – more or less mechanically, after replacing the successor by $+$ and $\times$ – goes from the top down, restricting induction by quantifier complexity. (This relies on Kleene's arithmetic hierarchy; in effect, though not in these words, he had applied logical hygiene, specifically for ensuring stability by referring to equivalence with respect to definability; this was later sharpened by Matiyasevich by showing polynomial $=$ recursive enumerability.) My impression is that here the trade between mathematics and logic has been mainly one way, with occasionally ingenious mathematics used for (often) logically not particularly realistic problems; cf. ingenious Tripos problems mentioned at the end of Endnote 18 but also the general reminder in Endnote 17.

Remark on logical (partial) orderings by such measures as consistency or interpretability "strength" or simply set inclusion (of provable theorems of this or that complexity; cf. Endnote 11c(i) with Π_1^0 and Δ_∞^0, presumably with attention somewhere in the literature to logical hygiene suitable to provability in "weak" systems parallel to Kleene's for definability): what – at least to me – is particularly lacking is the kind of logical hygiene (for orderings) emphasized in Section 2.1(b): attention to suitable functions, at least tacitly. Without this, the by now well-known fact that those prominent axioms in ordinary mathematics do not differ dramatically in (any of those logical) strength(s) from weak logical systems for arithmetic is neither particularly surprising nor of much realistic consequence. After all, by Section 2.3 the effectiveness of suitable explicit definitions, which are straightforwardly interpretable (and thereby equivalent in strength), is (to me) a good warning of limitations of those measures.

Another aspect of Weil's letter rewards attention, specifically around feelings about mathematical ideas, in his letter, (MAMNE's) equivalence or similarity relations between feelings about such ideas and about personal entanglements, here between feelings about combinations of such ideas with knowledge from two other areas: one is mechanics in Endnote 29(b), the other foundations in Section 2.2. In jargon used earlier, they are mathematical idea(lization)s of experience in those areas; the matter of feelings is complementary to – aspects of – effectiveness adumbrated in (a) to (c) of this endnote. Briefly, the gingerly – but for the likes of me not particularly secretive – approach in the discovery of areas and idea(lization)s suitable for one another has a particular attraction, admittedly (for me) not altogether unrelated to so-called applications of one mathematical discipline in another. NB The temptations of formal ingenuity unsuitable for organic combinations – as in notorious Tripos problems on rational mechanics, but also in foundations – are familiar: spotting and avoiding those pitfalls are necessary skills for moving in those territories rewardingly.

By – what will continue to be – a refrain in these pages, descriptions of feelings are left to others, e.g., to the likes of those other siblings, Henry and William James, if (descriptions of feelings about) subtleties of a particular kind or varieties of mathematical (but also, when I come to think of it, of adulterous) experience are in demand; cf. Endnotes 2 and 3.

Remark on Weil's letters of 29.II and 26.III.1940 (the former on p. 129 of *Souvenirs d'apprentissage*, in particular, expressing his conviction that it was futile to try to convey his mathematical interests to his sister): some contemporary biographers will presumably search in his life during barely a month for events that may have given him the confidence at least to try (by use of his metaphor). Whatever may be thought of the metaphor (of adultery), I myself was disturbed by the ineptness of another metaphor in the earlier letter, comparing the project of describing his interests (to his sister) to describing musical interests to a deaf person. Goethe's self-description suits me better: *So verwandle ich, Ton- und Gehörloser, obgleich Guthörender, jenen großen Genuss [der Musik] in Begriff und Wort* (Letter to Zelter of 2.V.1820).

Gödel's Legacy:
A Historical Perspective

The Reception of Gödel's 1931 Incompletability Theorems by Mathematicians, and Some Logicians, to the Early 1960s

Ivor Grattan-Guinness

Among those concerned with mathematical logic and axiomatics, the impact of Gödel's 1931 theorems on first-order arithmetic was fairly rapid and positive, although a few notable figures were a little slow to recognize it. However, the reception in the mathematical community in general was much slower and more muted. Evidence for this impression is presented from the literature, especially from general books on mathematics. The slowness is itself evidence of the ambiguous attitude that mathematicians have usually shown toward logic.

Around 1930, developments in the foundations of mathematics were becoming dominated by the program of metamathematics led by David Hilbert (1862–1943) as well as by the reasonable expectation that all logical and mathematical theories that could be axiomatized would be shown to be complete and consistent. Gödel's own Dissertation at the University of Vienna was of this kind. In it, he showed that the first-order predicate calculus (as we now call it) with identity passed muster; a lightly revised version was published (Gödel, 1930). However, in that year, he unexpectedly changed the direction of research for ever. In a paper that was to be accepted as his *Habilitation* at the University of Vienna, he showed two properties of the system P of axiomatized first-order arithmetic (Gödel, 1931). In his first theorem, he proved that if P was ω-consistent, then it contained more truths than theorems. Some months later, he proved the second theorem: that any proof of the consistency of P would require a logically richer theory than that of P itself, not a poorer one.

The history of the early reception of Gödel's paper by logicians has been well recorded (see esp. Dawson, 1985) and needs no special review here beyond pointing out some surprising and little-known details. The rest of this chapter deals with the reception by the mathematical community in general, a topic that, to my knowledge, has not previously been studied. Conclusions for both stories are drawn in the last section.

Three tables of bibliographic references related to Gödel's theorems in the primary literature are presented as the first part of the references. To avoid a considerable duplication of referencing, items from these tables are cited in the text using author and date and with the table number in square brackets (e.g., Rosser, 1953 [2]) and are listed with short titles in the tables chronologically by year. The second part of the references lists all other items in the usual way.

3.1 The Significance of Gödel's Theorems: Four Main Aspects

To start, I summarize the quartet of aspects that capture the significance of the theorems.

1. The first theorem sank the program for mathematics that had been developed by Alfred North Whitehead (1861–1947) and Bertrand Russell (1872–1970) in their three-volume *Principia Mathematica* (1st ed. 1910–1913; 2nd ed. 1925–1927; hereafter referred to as *PM*). They held that "all" mathematics (or at least quite a lot of it) could be derived solely from mathematical logic, including set theory (Grattan-Guinness, 2000a), but they had only hoped for consistency and completeness, without having any means at hand to exhibit them. (I shall use *logicism*, a term introduced by Rudolf Carnap (1891–1970) in the late 1920s, to refer to their position.) This theorem also did not do much good for metamathematics.

2. The second theorem sank the conception of metamathematics that Hilbert had been developing in recent years, for, although he was not always explicit about it, he saw the hierarchy of mathematics to metamathematics as an ascent into simplicity. This theorem also did not do much good for logicism.

3. The proof method was rich in consequences in its own right. The method of Gödel numbering (as it became known) played a major role in the growth of recursion theory and the formation of companion impossibility theorems and conjectures in other contexts by Alan Turing (1912–1954), Alonzo Church (1903–1995), and others.

4. The preceding three lines of influence are well recognized. Here I add another one that is rarely mentioned and yet underlies this trio. Gödel succeeded with his proofs because he recognized the central importance of distinguishing theory from metatheory, logic from metalogic, signs from their referents, and so on, in all contexts. The distinction was not new, metamathematics having already partially anticipated this approach, but elsewhere, it was not well grasped or followed. (*PM* was a disaster in this respect, especially with its all-purpose "implication.") Around the same time, Alfred Tarski (1901–1983) was coming to the same conclusion, partly from other considerations (Feferman and Feferman, 2004, esp. interlude 3), and both men deserve the credit for this major change in the conception of logic. To most of their predecessors, logic was some vast, all-embracing body of knowledge relative to which metalogic could not be formed, but from this point on, hierarchies were essential, whether in standard bivalent logic, within which Gödel's theorem lay, or in any other logic to come.

3.2 The Reception by Some Logicians

Despite the unexpected nature of the theorems, their reception among fellow logicians seems on the whole to have been promptly realized. For example, Gödel did not need to write the successor paper that he had expected would be necessary (the title of his paper ends "I"; Gödel, 1931). However, for four figures, the reception was less than rapid.

3.2.1 Hans Hahn

Gödel's paper appeared in the *Monatshefte für Mathematik und Physik*, then under the editorship of Hans Hahn (1879–1934). However, two years later, Hahn threw away a splendid opportunity to publicize the theorems before a general audience. A bust of the physicist Ludwig Boltzmann was unveiled at the University of Vienna, and for the occasion, Hahn prepared a lecture on "Logic, Mathematics and Recognition of Nature" (Hahn, 1933 [3]). Although he included sections on "mathematics and reality" and "science and reality," for some reason, he did not mention Gödel. Initially, he had found the first theorem rather difficult to grasp (Dawson, 1997, 73), and maybe the difficulty continued. Sadly, he was soon to die, with no further chance to redress this oversight.

3.2.2 W. V. Quine

When Gödel's paper appeared, W. V. Quine (1908–2000) was writing his doctoral thesis at Harvard University under the nominal supervision of Whitehead (who himself seemed not to take any notice of the theorems). Quine was outlining a mathematico-logical system potentially of the scope of *PM* but working out from different primitives. His thesis (Quine, 1932) did not mention Gödel, which is not surprising. However, over the next year, he spent some months in Europe, where he met Gödel, Tarski, and Carnap, among others. On his return to Harvard, he wrote his first book, presenting another system of *PM*-like potential (Quine, 1934) – but still without mentioning Gödel. That, under the new circumstances, is rather a surprise. The book consists largely of a workout of another set of primitives in logic and set theory, with a final chapter on the "deducibility of the system of *PM*" that did not advance as far as first-order arithmetic. The account is fairly limited from a mathematical point of view; a more elaborate statement of the definitions of cardinals and ordinals, and maybe rational and irrational numbers, would have made a more satisfactory presentation, and Gödel's theorems would have duly arisen as an aspect of first-order arithmetic.

I wrote to Quine about his silence over Gödel in this book, and he replied as follows (Quine, 1998):

> I was aware and appreciative of Gödel's result by early 1932, probably via Sheffer at Harvard. I recall Sheffer's envy of it, in connection with a kindred but hopeless project of his own.

H. M. Sheffer (1882–1964), a notorious nonpublisher, was influenced in his logical work by the American tradition of postulation theory (Scanlan, 1991). Quine may have been recalling Sheffer's attempt to fashion a graphical system of notation for logic and (among other aims) thereby to establish consistency (Scanlan, 2000, 209–18).

In his letter to me, Quine continued:

> Also I remember that when first I met Gödel in Vienna in the autumn of 1932, I was already aware and appreciative of his grand achievement. I am surprised to find no mention in

my *System of Logistic* (1934). I can only plead that his theorem had no direct bearing on the technical details of that book. He figures repeatedly in my next book, *Mathematical Logic* (1940).

However, Quine discussed incompletability, and Gödel's first theorem as such, only in the last section of *Mathematical Logic*.

3.2.3 Saunders MacLane

A great tradition of American mathematicians writing *Dissertationen* in German universities ended in 1933 with Hitler's arrival in power. Perhaps the last distinguished example was that of the mathematician Saunders MacLane (1909–2005), who studied at Göttingen University under the supervision of Paul Bernays (1888–1977) from 1931 until 1933, when Hermann Weyl (1885–1955) took over. His thesis (MacLane, 1934) treated methods by which proofs of mathematical theorems could be shortened in systematic ways. Though MacLane's study was not the same kind of study as Gödel's theorems, it would have been worth mentioning that MacLane's results did link to Gödel's theorems in that some of the mathematical theories involved were undecidable anyway; however, MacLane did not make such a connection. I asked him why and was astonished by the answer (Mac Lane, 1999; recalling a long discussion we had had in 1996)[1]:

> It is the case that I was not aware of [Gödel's] paper when I was a student in Göttingen 1931 to 1933! Bernays and Gentzen probably must have known of it, but never spoke to me about it!

Thus, apparently, there was some measure of denial about the theorems in Göttingen at that time, doubtless caused by the anger (or disappointment) over them of Field Marshal Hilbert. This is the only case of denial that I have found, and it was somewhat unreal in that Bernays was then publishing the first volume of his account of metamathematics (Hilbert and Bernays, 1934; the theorems were handled in vol. 2, 1939 [1]), and Gerhard Gentzen (1909–1945) was working on post-Gödelian metamathematics.

In his autobiography, which appeared in 2004 and may have been written after his obviously forgotten contacts with me, Mac Lane wrote (2004, 51):

> Gödel's famous incompleteness theorem appeared in 1931, but at that time I did not study it carefully – an amazing omission, since Gödel's work was among the most important in twentieth-century mathematics and a major turning point in logic. I no longer understand why I missed this event. At the time, Gödel was on the faculty of the University of Vienna; I wonder whether his work was suppressed in Göttingen because it was a blow to Hilbert and others or whether I was just so busy that I ignored the excitement. It was only later, back in the United States, that I studied Gödel's work.

3.2.4 Bertrand Russell

The distinction between theory and metatheory can be considered as part of a hierarchy of languages, to use a phrase from Russell, our last figure here. The *Tractatus*

---

[1] I follow this author in calling him "MacLane" when young and "Mac Lane" in old age.

Logico-philosophicus of Ludwig Wittgenstein (1889–1951) (Wittgenstein, 1921) was published, under a different title, only because Russell wrote an introduction to it (Russell, 1921). There Russell objected to the monism that Wittgenstein had imposed on his system and mooted the possibility

> that every language has, as Mr. Wittgenstein says, a structure concerning which, *in the language*, nothing can be said, but that there may be another language dealing with the structure of the first language, and having itself a new structure, and that to this hierarchy of languages there may be no limit.

This is one of Russell's finest philosophical insights, yet he never properly thought of also applying it to his own logic, which he always saw as an all-embracing subject. Soon afterward, in 1923 and 1924, he prepared new material for the second edition of *PM* but made no use of this insight, nor did he draw on it after Gödel's paper was published. He did not mention Gödel in the new introduction that he wrote in 1937 for the reprinting of his book *The Principles of Mathematics* (1903), where he had explained logicism in prosodic form, nor in his philosophical autobiography, *My Philosophical Development* (1959). In *An Enquiry into Meaning and Truth*, he deployed Gödel numbering for a particular purpose but did not use anything else from Gödel's paper (Russell, 1940, 71–72). In two short articles in which he outlined his philosophy of logical positivism, he recalled his theory of hierarchies and related it both to Tarski's semantic theory of truth and to Gödel's "paradoxes" or "puzzles," but he misstated the first theorem as concerning "any formal system" (Russell, 1945, 1950; see also Russell, 1997, 152, 159).

Furthermore, this newly gained link did not stay with him, for in 1963, in his early nineties, he wrote to the logician Leon Henkin, clearly still mystified by the impact that the theorems had made and stating the first one wrongly again (Grattan-Guinness, 2000a, 592–93). Henkin stated the theorem correctly in his answer, but without effect. In 1965, Russell prepared an addendum to the replies, to critics that he had contributed to the Schilpp volume of his philosophy in 1944. Once again, he gave a wrong statement of the theorem, followed by a pathetic response to it (Russell, 1971, xviii).

Russell may have mentioned Gödel in his addendum because in 1944 he had pleaded lack of time to reply to the article on his logic that Gödel had contributed, very late, to the original volume. However, here comes another surprise: in that article (Gödel, 1944 [3]), Gödel had not mentioned his own theorems!

3.3 The Reception among Mathematicians: The Technical Literature

Apart from the cases just described, those closely linked to Gödel's concerns recognized his theorems and their consequences quite soon after they were introduced.[2] The field of formal logic, along with related topics, such as axiomatic set theory and

---

[2] A search of the back files of the reviewing journal *Zentralblatt für Mathematik* yielded thirty-three papers to 1960 on (versions of) the theorems in which Gödel's name was explicitly mentioned. Doubtless some more articles appeared on the topic without "Gödel" in their titles. On the history of logic in general from around 1930 up to the early 1960s, see, for example, Mostowski (1966) and Mangione and Bozzi (1993, chaps. 5–7).

aspects of model theory, progressed steadily, its social status raised by the launch in 1935 of the Association of Symbolic Logic in the United States and its *Journal* the following year. By then, Gödel's theorems formed part of the furniture for logicians.

The rest of this chapter is concerned with the extent and manner of the reception of the theorems among mathematicians in general – a very different story from the reception among logicians. Let us start with the technical works on formal logic that mathematicians could have seen had they wanted.

Some logicians wrote books with a broader audience than fellow specialists, and mathematicians may have read them. I note some principal examples.

Just before World War II, Bernays published the second volume of his survey of metamathematics, including proofs of both theorems in detail (Hilbert and Bernays, 1939 [1]); however, ensuing events must have limited its circulation. During the War and later, various writings of Tarski are relevant, such as the logic textbook (Tarski, 1941 [2]), which appeared in various languages. After the War, such books included the substantial presentation by Stephen Kleene (1909–1994) of metamathematics (Kleene, 1952 [1]) and the technical monograph by Andrzej Mostowski (1913–1975) on Gödelian results for arithmetic (Mostowski, 1952 [1]). Some years later, Martin Davis published a broad account of recursion theory; naturally, the main hero was Turing, but Gödel was duly noted (Davis, 1958 [1]).

However, in *Logic for Mathematicians*, Barkley Rosser (1907–1989) summarized the treatment of mathematics given in *PM* but deliberately avoided details of the theorems and referred to Gödel rarely (Rosser, 1953 [2]). Unfortunately, I did not ask him about this decision, which seems somewhat regrettable, especially as the summary of *PM* itself was, of course, very capably handled.

Another foundational area of mathematics affected by Gödel's theorems was axiomatic set theory. The two main systems were those launched by Ernst Zermelo (1871–1953) in the first decade of the 1900s and by John von Neumann (1903–1957) twenty years later; both were subsequently modified by others, including Gödel. As first-order arithmetic is expressible within each system, they are both incompletable, as Gödel mentioned in the first paragraph of his paper (Gödel, 1931).

In the 1950s, books began to appear on axiomatic or general set theory that were written for a broader audience, and so one would expect that this important feature would be explained. However, several mathematicians were silent, although they noted Gödel's later consistency results on set theory.

An interesting figure here is Abraham Fraenkel (1891–1965), one of the modifiers of Zermelo's system. He was well familiar with Gödel's paper, for he wrote an excellent review of it for the mathematics review journal *Jahrbuch über die Fortschritte der Mathematik* (Fraenkel, 1938 [1]; the tardy publication reflects the declining state of the journal). He had also heard Gödel lecture on his theorems in September 1931 at the annual meeting of the Deutsche Mathematiker-Vereinigung, where he himself had spoken on set theory.[3] Nevertheless, most of his books after 1931 do not mention the theorems: his *Abstract Set Theory*, although Gödel's paper was listed in the superb

---

[3] The third speaker at this special session on foundations was Zermelo, who was trying to prove the completeness of mathematics using a vague prototheory of infinitely long proofs. The aftermath included a superb letter on

bibliography that graced the first edition (Fraenkel, 1953 [3]); also Bernays' book, to which he contributed a lengthy historical introduction (Bernays, 1958 [3]); or even his own short account of "logic and set theory" (Fraenkel, 1959 [3]). The only one of his books in which incompletability received some notice was the one written with Yehoshua Bar-Hillel (1915–1975) (Fraenkel and Bar-Hillel, 1958 [1], esp. 303–8), making the silence elsewhere all the more noticeable.

Some of these books were presented as textbooks (at an advanced level). Several others on or including axiomatic set theory came out in ensuing years (including a 1973 revision of Fraenkel and Bar-Hillel's book by Azriel Levy, with still more details). Discussion of incompletability in connection with sets was patchy: there was no discussion as such in Suppes (1960 [3]); some in Stoll (1961 [1], chap. 9 on first-order theories); and none really in Kneebone (1963 [1], chap. 11), although the theorems themselves were treated in chapter 8. Quine's own account of the relationship between set theory and logic, which naturally gave some space to the set theories embedded in his own logical systems, did note incompletability, although rather sparingly (Quine, 1963 [2], esp. 110, 305, 325).

3.4 The Reception among Mathematicians: The General Literature

The other main sources for mathematicians were popular writings that covered the pure sides and/or foundations and axiomatics of mathematics. In many cases, the authors themselves were not specialists in logic or set theory. I shall not discuss their accounts in detail; in general, they were satisfactory within the limits of the books, although, for example, some authors stated the first theorem in terms of consistency rather than ω-consistency and so in effect described the version from Rosser (1936) without always saying so.

3.4.1 Reactions to 1940

The earliest item of this kind comes not from a popular source, as that just described, but rather from one that was (to be) widely available to mathematicians: a review of Gödel's paper in the new review journal *Zentralblatt für Mathematik*. The author was the Berlin-based philosopher and mathematician Walter Dubislav (1895–1937); by luck, it was the very first review to appear in the journal's second volume (Dubislav, 1932a [1]).

Among popular sources, the first quintet to contain more than passing mentions are in three different languages: a survey of current philosophies of mathematics by Dubislav (1932b [1], esp. 20–22), together with a recognition of the "epoch-making work of K. Gödel" in the review of the book by Heinrich Scholz (1884–1956) (Scholz, 1933 [2]); a popular lecture on the "new logic" given in Vienna by Gödel's friend Karl Menger (1902–1985) in 1932 and published the following year (Menger, 1933 [1]); a

his theorems that Gödel wrote to Zermelo; it was published in Grattan-Guinness (1979) and also in Gödel's (2003b) *Collected Works*, without the annoying typesetting errors that besmirched the first printing.

survey of philosophies of mathematics by Friedrich Waismann (1896–1959), who was, like Gödel, a member of the Vienna Circle (Waismann, 1936 [1], esp. sec. 9A); and a discussion of axiomatics and formalism by the French philosopher Jean Cavaillès (1903–1944) (Cavaillès, 1938 [1], esp. 143–49).

In addition, E. T. Bell (1883–1960), a historian of mathematics whom historians dislike, gave a general account of the development of mathematics – and one reason for the dislike is that he presented the theorems as concerning logic rather than arithmetic (Bell, 1940 [1], 575–78). On the other hand, his account formed part of a substantial survey of various recent foundational studies that would not have been familiar to many of his readers.

Another book of this kind appeared from two New York authors: *Mathematics and the Imagination* (Kasner and Newman, 1940 [3]). As the authors included a chapter on paradoxes and Russell's work (chap. 6), Gödel could have arisen naturally, but nothing about his work was said. However, the situation for the junior author was to change so substantially that he will become a major figure in our story.

3.4.2 The Key Role of Newman

James Roy Newman (1907–1966) trained as a lawyer at Columbia University in New York. Early in his career, he formed a passion for popularizing mathematics.[4] The first manifestation of this passion was the book just noted. (Edward Kasner (1878–1955) was a professor of mathematics at Columbia.) Soon after the book appeared, Newman conceived of editing a large collection of articles on all aspects of mathematics for a general audience, a project that eventually appeared in 1956 as a four-volume set titled *The World of Mathematics*. It contained several articles concerning axiomatics and logic; in particular, with his old philosophy professor at Columbia Ernest Nagel (1901–1985), Newman wrote "Gödel's Proof" (Nagel and Newman, 1956b [1]). A somewhat different version of this article with the same title appeared that year in the popular journal *Scientific American*, of which Newman was a member of the editorial board (Nagel and Newman, 1956a [1]). The differences between the two versions of the article seem to be due only to copyright concerns; there is no substantial variation in content. Two years later, the authors extended the articles into a short book of the same title (Nagel and Newman, 1958 [1]). It was shorter than they had intended because they had wanted to include an English translation of the original paper, but unfortunately, Gödel imposed such unreasonable conditions on the whole project (Gödel, 2003a, 2–8; 2003b, 135–54) that the plan had to be abandoned.

These three publications considerably raised the visibility of the theorems in the mathematical community, and indeed beyond – the 1956 set sold very widely (for example, I own a set that was given to me by an American composer), and the 1958 book is still in print. Newman and the role he played is a remarkable example of the highly effective popularization of mathematics led by someone outside the profession.

---

[4] No historical work seems to have been done on Newman. See his declaration in *Who's Who in America*, vol. 4 (1968), p. 701.

3.4.3 Other American Authors

The importance of the role of the United States in this story is enhanced further by other general publications. Weyl had played a valuable role before Newman. In 1927, he had published in German a survey of the philosophy of mathematics and science, and when it was translated into English, he added several appendixes, including a substantial one on Gödel's theorems and their historical and philosophical context (Weyl, 1949 [1], Appendix A).

Five years later, another distinguished mathematician with a serious interest in foundational issues, R. L. Wilder (1896–1982), published a general book on foundations, including a section on the theorems (Wilder, 1953 [1], 256–61). This work was echoed five years later when the mathematics educators Howard Eves (1911–2004) and Carroll Newsom (1924–1986) published a book with a very similar title but set at a level for undergraduates (Eves and Newsom, 1958 [2]). The influence of Wilder on Eves and Newsom is clear, as the authors state that one of them had been his student (however, neither of the authors took a doctorate under his direction). The authors followed a historical approach to foundations – Eves was a good historian of mathematics – and ended with Gödel's theorems. They stated that the details were above the level of their book, but they had nicely revealed the context in which the theorems arose. Furthermore, in a revised edition of their book published in 1965, they did provide some of the details of the theorems in a new appendix.

Quite different results came from another American textbook of the same period. In *Logic in Elementary Mathematics*, the mathematics educators Robert M. Exner (1914–2001) and Myron F. Rosskopf adopted quite a formal approach, with substantial (although not axiomatic) accounts of the propositional and first-order predicate calculi (Exner and Rosskopf, 1959 [3]). All was ready for Gödel, yet silence reigned in the mathematical chapters of the book as they did not treat arithmetic at all. They gave no explanation for this silence, and I have not been able to construct one; indeed, it seems that a nice opportunity was missed.

3.5 Reception in Some Other Countries

The last section was dominated by Americans, a fact that forms part of my general conclusions. However, a rounder picture is given by considering Gödel's influence in some other countries in which English is not the mother tongue.

3.5.1 France

France is an obvious case: an exceptionally important mathematical country for some centuries but also well known for its long derision of logic. How did the French react to the theorems? As our story starts in the 1930s, to a substantial extent, the question becomes, how did the Bourbaki group react to the theorems?

The tone was set by Jean Dieudonné (1906–1992) in the 1939 volume of the popular journal *Revue Scientifique* in his article "Modern Axiomatic Methods and the Foundations of Mathematics" (Dieudonné, 1939 [3]). The title fits Gödel almost exactly, but Dieudonné ignored him, although he noted Hilbert's ambition to "*prove* the non-contradiction of mathematics"! He just emphasized some aspects of rigor, deduction from (vaguely known) primitives, and set theory. Similarly, four years later, in the same journal, Henri Cartan (b. 1904) contributed "On the Logical Foundation of Mathematics" – surely a Russellian title (Cartan, 1943 [3]). However, he largely set aside "logistic" and focused instead on transfinite arithmetic, axiomatic set theory, and some aspects of logic, including quantification theory and even the *Entscheidungsproblem*, and he mentioned his compatriot logician Jacques Herbrand (1908–1931). Therefore his silence over Gödel is as striking as that of Dieudonné.

After the War, the first fascicles of Bourbaki's *Eléments des Mathématiques* began to appear, with historical notes in some of them. Gödel was absent, of course, although about twenty-five years later, the newer members of the group were aware of him (see the references in the collection of Bourbaki, 1969; one allusion is quoted subsequently).

At that time, the mathematician François le Lionnais edited a substantial volume on current developments in mathematics (le Lionnais, 1948 [3]). Three articles crossed Gödelian territory: Bourbaki on the architecture of mathematics, emphasizing structure and axiomatics; Dieudonné on aspects of the achievements of Hilbert (who had died in 1943); and two other authors on the logical synthesis of results and research. Silence reigned once more. In 1962, the book was reissued, and the editor added to it a reprint of Dieudonné's 1939 article – so *encore rien*.

The Bourbakists were much concerned with rigor and accuracy in mathematics, but they drew little on logic as a discipline in its own right to fulfill their purposes. This attitude was made very clear by their prominent representative André Weil (1906–1998) in a lecture delivered before the Association of Symbolic Logic late in 1948, in which he outlined the "foundations of mathematics for the working mathematician" (Bourbaki, 1949 [3], written by Weil). He noted some features of logic, such as well-formed formulae and quantification, and some set theory, but that was it. Instead, the usual Bourbakist topics were prominent such as structure and abstraction, along with some set theory.

3.5.2 Other Countries

I conclude this survey with some remarks on other countries. Gödel's own part of the world, central Europe, is an interesting study. Most major logicians and several mathematicians had emigrated after the arrival of Hitler, and logic was judged to be a degenerate academic subject. After the War, some logicians still pursued the subject, but our concern is with the mathematical community in general.

In Germany, the second edition of the *Encyclopädie der mathematischen Wissenschaften* included two articles on logic in which the theorems were described (Schmidt, 1950 [1]; Hermes and Scholz, 1952 [1]). (This made two more articles on logic than were to be found in the first edition (1898–1935)!) However, in contrast to that massive achievement, the second edition petered out rather quickly, and so these articles may not have been widely read.

As in the United States, 1956 was quite a good year. A popular book on mathematics by Herbert Meschkowski (1909–1990) contained a short section (Meschkowski, 1956 [1], chap. 11) on the theorems. A passage from Gödel's paper was quoted in the German source book for "formal logic" by Innocentius M. Bochenski (1902–1995) (Bochenski, 1956 [1]; an English edition of that book appeared five years later).

A pioneering philosophical muse on "incompleteness and undecidability" was offered by Wolfgang Stegmüller (1959 [1], esp. chap. B), although I doubt that many mathematicians would have read it. This remark is not intended as a criticism of Stegmüller's book, but mention of it suggests another interesting historical question that, to my knowledge, has not been studied: what is the history of the reception of Gödel's theorems in the philosophical community?

Most of the impact in Germany seems to have been made from the 1960s onward, that is, later than the level of publicity achieved in American writing. For example, the 1958 book of Nagel and Newman was translated into German only in 1964, and the new material in Weyl's philosophy book of 1949, mentioned earlier, did not appear in German until the third edition of 1966.

My impression is that the main reception of the theorems in other countries also dates largely from the 1960s onward. Moreover, for several countries, the writings were in their native languages and so did not export widely. For example, in Poland, Mostowksi produced relevant material in the late 1940s (see, for example, Mostowksi, 1948, chap. 14),[5] but it will have found a much smaller audience than his monograph of 1952 in English. Similarly, the Dutch logician and philosopher Evert Beth (1908–1964) had been mentioning Gödel in various languages from the 1940s (he also suggested to Mostowski that he write that monograph). His most widely read text was surely his wide-ranging book in English *The Foundations of Mathematics*, which contained a chapter on the theorems (Beth, 1959 [1], chap. 21). Unfortunately, however, he forgot to include Gödel's paper in his bibliography.

3.6 Concluding Remarks

One cites the case of an eminent university figure who, in a lecture given at Princeton in the presence of Gödel, would say that nothing new had been done in logic since Aristotle!
                                                                    – Bourbaki (1969, 14)

The evidence presented suggests that the logicians grasped the significance of Gödel's theorems rather quickly, although with a few slightly tardy cases. By contrast, the mathematical community did not become generally aware of the theorems until around twenty-five years after their publication, when American authors, especially the non-mathematician Newman, played a major role in their dissemination. In this section, I discuss these conclusions metatheoretically from various points of view.

In the preceding section, I confessed that this survey of the reception of Gödel's theorems is knowingly incomplete with respect to various countries. It is also incomplete in three other respects: (1) for reasons of space, I have not discussed all the

---

[5] I am indebted to Roman Murawski for this information. The history of the reception of Gödel's theorems in other countries forms part of the history from the 1930s of formal logic there in general, which seems not to be well studied, even for Poland; however, see Wolenski (1989). Pertinent literature includes Cavalieri (1988) and Anellis (1987) on the Soviet Union, Schenk (1995) on the German Democratic Republic, and Sylvan (1992) on Australia.

items listed in the three bibliographic tables, but the other citations corroborate the impression conveyed; (2) literature in the dominant languages must exist that I have not tracked down; as the theorems relate to such a basic part of mathematics, discussions by mathematicians may be anywhere within a wide range of interests, making all of them very hard to locate; and finally, (3) mathematicians specializing in certain areas sometimes become familiar with other areas (e.g., to give lecture courses on them at their institutions) but do not leave a published trail in the form of papers or books. Doubtless, formal logic formed such an alternative for some mathematicians. One example is G. H. Hardy (1877–1947), who on occasion lectured both on logicism and on metamathematics but did not include any foundational topics in his own popular book on mathematics (Hardy, 1940 [3]).

Despite these examples, I think that the conclusions drawn are broadly correct, especially concerning the difference in speeds of reception between logicians and mathematicians. Moreover, I do not find the differences surprising. In my many years of studying the history of mathematics and the history of logic, I have been aware not only of the long history of both disciplines but especially that they usually had little to do with each other. For example, from antiquity, the Greeks developed both a tradition of rigorous mathematics, as exemplified by Euclid's *Elements*, and a tradition of systematic logical deduction with Aristotle (and a different tradition with the Stoics), but they did not bring them together. The same noncontact is evident in later epochs, down even to the twentieth century. For example, few mathematicians have ever read *PM* and so are unaware of the interesting mathematics that it contains, especially in the second and third volumes (e.g., the role and use of the axioms of choice); I recall Dieudonné exhibiting this attitude perfectly at a meeting in Cortona, Italy, in 1983. Russell made the same point in his letter to Henkin mentioned earlier (Grattan-Guinness, 2000a, 593). Again, despite great efforts at publicity, even the leading mathematician Hilbert was unable to convince mathematicians in large numbers from the 1920s onward to follow and take up metamathematics as a field of study or even for teaching.

Among many differences between formal logic and mathematics, logic is the broader discipline, and many of its concerns are and have been profitably pursued without paying any special attention to mathematics; for example, some books on formal logic published in the period covered in this chapter have not been cited precisely for this reason. The main points of contact lie in some aspects of axiomatics and (since around 1900) model theory (Peckhaus, 1999; Grattan-Guinness, 2000b). Otherwise, to a mathematician, logic is certainly important for supplying rules of deduction such as *modus ponens* and proof by contradiction, and maybe to help in the well formation of formulae, but there is no special need to study matters that, to a logician, are central such as why or in what circumstances a deduction is valid, nor is there a need to consider other foundational aspects of mathematical theories. The common attitude of mathematicians is exemplified, in a markedly forceful way, by the Bourbakists, as previously noted; indeed, to some extent, this point shows their influence on the mathematical community.

One aspect of mathematical theories rarely considered by mathematicians but well noted by logicians is the forms and role of definitions (Dubislav, 1931). Gödel's proof of the first theorem rests on a sequence of forty-six definitions that had to be tight because of the need to keep logic and metalogic rigidly apart. If a mathematician were to work his or her way through the sequence, the mathematician would find a level of rigor that is very likely much higher than that which obtains even in so-called rigorous mathematics. Therefore Gödel's theorems would not apply to a mathematician's way of doing mathematics. One area of mathematics where practitioners might aspire to Gödelian standards of proof is abstract algebra, but even there Mac Lane is very unusual in the interest that he took in logic, both in the early 1930s and later (Mac Lane, 2004).

To logicians, mathematicians are rather sloppy and none too familiar with logic, whereas to mathematicians, logicians are fussy and none too familiar with mathematics. The epigraph to this section typifies both sides, not only because of its content but also because of its context, namely, that Aristotle's writings "encourage philosophers in their neglect of the study of mathematics and block the progress of formal logic" (Bourbaki, 1969, 14).

The wide gap between mathematics and formal logic was evident to me even before I began to work on history. Around 1960, I was studying for a degree in mathematics, that is, three years of proving theorems and proving theorems and proving theorems, but in a degree course bereft of significant reflections on proof theory. In particular, Gödel's theorems were not mentioned in any course, although we did study the completeness theorem of 1930 in a course on sets and models (with the logic explained in a few lines). I record these memories because I am sure that my experience was quite typical of its time, in all, or at least in many, countries.

In these contexts, the differences in the speeds of reception of Gödel's theorems are to be expected; indeed, without Newman the gap might have been still greater. This gap has reduced somewhat from the 1960s onward, especially because of the steadily increasing use of computers in general and the companion rise of computer science as an academic discipline, and also (partly for this reason) because of the continual growth of work in formal logic. Gödel (and also Tarski and Turing) grew to be very well known, and Gödel's theorems became quite popular; for example, three English translations of his 1931 paper appeared in 1962, 1965, and 1967 (see under Gödel, 1931). Formal logic became even somewhat trendy, with many textbook publishers wanting a book on it (maybe including some mathematics) on their publication lists.

Finally, where do we place Russell in this division between a logician's logic and a mathematician's logic? He was certainly practicing the former kind, but always within the all-embracing conception of logic mentioned earlier; he never fully realized that his insight about monism could and should apply to his conception itself. For example, around 1940, he gave a talk on logic and the foundations of mathematics at Harvard University, and when MacLane, in the audience, asked about the bearing upon the talk of Gödel's theorem and of metamathematics, "there was a long silence" (Mac Lane, 2005, 138–39).

From the 1930s onward, thanks largely to Gödel and Tarski, there formed a subdivision within the logician's logic between the all-embracers, such as Russell, and those who recognized the necessity for, and generality of, the distinction between logic and metalogic and theory and metatheory. In my view, this distinction is an essential but underrated part of the fundamental importance and influence of Gödel's extraordinary paper of 1931 on logicians and, eventually, on mathematicians also.

Acknowledgments

For permission to quote from letters sent to me by Quine and Mac Lane, I thank Douglas Quine, Gretchen Mac Lane, and Cynthia Hay.

References Related to Gödel's Theorems (in Three Tables)

Works in Which Gödel's Theorems are Reviewed or Discussed

Year	Author, (Short) Title	Comments
1932a	W. Dubislav, Review of Gödel's paper. *Zentralblatt für Mathematik,* **2**, 1.	None
1932b	W. Dubislav, *Philosophie der Mathematik in der Gegenwart.* Berlin: Junker und Dünnhaupt.	None
1933	K. Menger, Die neue Logik [The new logic]. In *Krise und Neuaufbau in den exakten Wissenschaften.* Leipzig, Germany: Deuticke, pp. 93–122.	English trans. in *Selected Papers in Logic* (Dordrecht, Netherlands, 1979), chap. 1.
1936	F. Waismann, *Einführung in das mathematische Denken.* 1st ed. Vienna: Springer.	6 pp., English trans. 1951
1938	J. Cavaillès, *Méthode Axiomatique et Formalisme.* 3 pts. Paris: Hermann.	4 pp.
1938	A. A. Fraenkel, Review of Gödel's paper. *Jahrbuch über die Fortschritte der Mathematik,* **57**, 54.	None
1939	D. Hilbert and P. Bernays, *Grundlagen der Mathematik.* Vol. 2. Berlin: Springer.	Written by Bernays. Available outside Germany?
1940	E. T. Bell, *The Development of Mathematics.* New York: McGraw-Hill.	3 pp. in 30 pp. on foundations; faulty
1949	H. Weyl, *Philosophy of Mathematics and Natural Science.* Princeton, NJ: Princeton University Press.	17 pp. Only in German in 3rd ed. (1966)
1950	A. Schmidt, Mathematische Grundlagen. In *Encyclopädie der mathematischen Wissenschaften.* 2nd ed. Vol. 1, sec. 1, pt. 2. Leipzig: Teubner: pp. 1–48.	Widely read?
1952	H. Hermes and H. Scholz, Mathematische Logik. In *Encyclopädie der mathematischen Wissenschaften.* 2nd ed. Vol. 1, sec. 1, pt. 1. Leipzig: Teubner , pp. 1–82.	Widely read?
1952	S. C. Kleene, *Introduction to Metamathematics.* Amsterdam, Netherlands: North-Holland.	None
1952	A. Mostowski, *Sentences Undecidable in Formalised Arithmetic.* Amsterdam, Netherlands: North-Holland.	Technical monograph
1953	R. L. Wilder, *Introduction to the Foundations of Mathematics.* New York: John Wiley.	6 pp.
1956	I. M. Bochenski, *Formale Logik.* Freiburg, Germany: Karl Alber.	Source book; English trans. 1961
1956	H. Meschkowski, *Wandlungen des mathematischen Denkens.* Braunschweig, Germany: Vieweg.	2nd ed. 1960; English trans. 1965
1956	F. Desua, Consistency and completeness – a review. *American Mathematical Monthly,* **63**, 295–305.	Gödel's theorems prominent
1956a	E. Nagel and J. R. Newman, Gödel's proof. *Scientific American,* **194**, 71–86.	Pioneer popularization

1956b	Version of Nagel and Newman (1956a [1]). In *The World of Mathematics,* vol. _, ed. J. R. Newman, pp. 1668–95. New York: Simon and Schuster.	Parts of book on axiomatics and on logic
1958	E. Nagel and J. R. Newman, *Gödel's Proof.* New York: New York University Press.	Book version of articles: German trans. 1964
1958	M. Davis, *Computability and Unsolvability.* New York: McGraw-Hill.	Book focused around recursion
1958	A. A. Fraenkel and Y. Bar-Hillel, *Foundations of Set Theory.* Amsterdam, Netherlands: North-Holland.	Section in chap. 5; more in 2nd ed. 1973
1959	E. W. Beth, *The Foundations of Mathematics.* Amsterdam, Netherlands: North-Holland.	A chapter; Gödel's paper not cited!
1959	W. Stegmüller, *Unvollständigkeit und Unentscheidbarkeit.* Vienna: Springer.	Philosophical discourse
1961	R. R. Stoll, *Set Theory and Logic.* San Francisco: Freeman.	Textbook; short section
1962	K. Gödel, *On Formally Undecidable Propositions.* Edinburgh, UK: Nelson.	1st English trans.; introduction by R. B. Braithwaite
1962	W. Kneale and M. Kneale, *The Development of Logic.* Oxford: Oxford University Press.	History; 13 pp.
1963	G. T. Kneebone, *Mathematical Logic and the Foundations of Mathematics.* London: van Nostrand.	Textbook; chap. 8 (14 pp.) and other mentions

Works in Which the Theorems Are Mentioned

Year	Author, (Short) Title	Comments
1933	H. Scholz, Review of Dubislav (1932b [1]). *Jahresbericht der Deutschen Mathematiker-Vereinigung,* **43**,88–90.	None
1934	A. Heyting, *Mathematische Grundlagenforschung. Intuitionismus. Beweistheorie.* Berlin: Springer.	None
1939	R. Carnap, *Foundations of Logic and Mathematics.* Chicago: University of Chicago Press.	Part of *International Encyclopedia*
1941	A. Tarski, *Introduction to Logic.* 1st ed. New York: Oxford University Press.	Later eds. and trans.
1941	R. Courant and H. Robbins, *What Is Mathematics?* New York: Oxford University Press.	Small note; to 4th ed. 1947
1946	H. Weyl, Mathematics and logic. *American Mathematical Monthly,* **53**, 2–13.	General review of foundations
1948	L. Chwistek, *The Limits of Science.* London: Kegan, Paul.	Polish original 1935
1951	E. T. Bell, *Mathematics: Queen and Servant of Science.* New York: McGraw-Hill.	Quote from Weyl (1946)
1953	E. R. Stabler, *An Introduction to Mathematical Thought.* Cambridge, MA: Addison-Wesley.	Only corollary properly conveyed
1953	J. B. Rosser, *Logic for Mathematicians.* New York: McGraw-Hill.	Theorems deliberately minimized
1955	E. W. Beth, *Les Fondements Logiques des Mathématiques.* 2nd ed. Paris: Gauthier-Villars.	Also mentions in other of his books
1958	H. Eves and C. V. Newsom, *An Introduction to the Foundations and Fundamental Concepts of Mathematics.* New York: Rinehart.	Some details given in appendix to 2nd ed. (1965)
1963	W. V. Quine, *Set Theory and Its Logic.* Cambridge, MA: Harvard University Press.	2nd ed. 1969

Works in Which the Theorems Are Not Mentioned

Year	Author, (Short) Title	Comments
1933	H. Hahn, *Logik, Mathematik und Naturekennen.* Vienna: Gerold.	French trans. 1935
1935	E. R. Stabler, An interpretation and comparison of three schools of thought in the foundations of mathematics. *Mathematics Teacher,* **28**, 5–35.	Covers logicism, metamathematics, and postulation theory
1938	H. P. Fawcett, *The Nature of Proof: A Description and Evaluation of Certain Procedures Used in a Senior High School to Develop an Understanding of the Nature of Proof.* New York: Teachers College, Columbia University.	Much oriented about the axioms of geometry and its teaching; not a word about any formal logic or proof theory or any of its practitioners; text unaltered in the reprint (2001, Reston, VA: National Council of Teachers of Mathematics)
1939	J. Dieudonné, Les methodes axiomatiques modernes et les fondements des mathématiques. *Revue Scientifique,* **77**, 224–32.	Repr. in Le Lionnais (1948), 1962 repr.
1940	G. H. Hardy, *A Mathematician's Apology.* Cambridge: Cambridge University Press.	Author well familiar with foundations
1940	E. Kasner and J. R. Newman, *Mathematics and the Imagination.* New York: Simon and Schuster.	Chap. on paradoxes and foundations
1943	W. W. Sawyer, *Mathematician's Delight.* Harmondsworth, UK: Penguin.	Foundations not handled
1943	H. Cartan, Sur le fondement logique des mathématiques. *Revue Scientifique,* **81**, 2–11.	"Logistic" deliberately minimized
1944	K. Gödel, Russell's mathematical logic. In *The Philosophy of Bertrand Russell*, ed. P. A. Schilpp. New York: Tudor, pp. 192–226.	Reviews only *Principia Mathematica*; no mention of his theorems
1948	F. Le Lionnais, ed. *Les Grands Courants de la Pensée Mathématique.* Paris: Librairie Scientifique et Technique.	Three related articles, inc. one by Bourbaki; repr. 1962
1940s	N. Bourbaki, *Eléments des Mathématiques.* Paris: Hermann.	A few mentions in later eds.
1949	A. Darbon, *La Philosophie des Mathématiques.* Paris: Presses Universitaires de France.	Posthumous; detailed study of Russell, but no *PM*; done before 1910?
1949	N. Bourbaki, Foundations of mathematics for the working mathematician. *Journal of Symbolic Logic,* **57**, 221–32.	Lecture delivered to the Association of Symbolic Logic by A. Weil
1951	A. Bakst, *Mathematics, Its Magic and Mystery.* Princeton, NJ: van Nostrand.	Mostly on heuristics
1952	L. Couffignal, *Les Machines à Penser.* Paris: Editions de Minuit.	Popular book on computing; no Turing
1953	B. V. Bowden, ed. *Faster Than Thought.* London: Pitman.	Mainly on computers
1953	A. A. Fraenkel, *Abstract Set Theory.* Amsterdam, Netherlands: North-Holland.	Paper only in the (huge) bibliography
1958	K. Menninger, *Zahlwort und Ziffer.* 2nd ed. Göttingen, Germany: Vandenhoek und Ruprecht.	Historical, but not on foundations
1958	P. Bernays, *Axiomatic Set Theory.* Amsterdam, Netherlands: North Holland.	Historical preface by Fraenkel; incompleteness not explicitly mentioned

1959	A. A. Fraenkel, *Mengenlehre und Logik.* Berlin: Duncker und Humblot.	Ditto re incompleteness; English trans. 1966
1959	R. M. Exner and W. F. Rosskopf, *Logic in Elementary Mathematics.* New York: McGraw-Hill.	Textbook level; arithmetic excluded!
1960	P. Suppes, *Axiomatic Set Theory.* Princeton, NJ: van Nostrand.	None
1962	T. Dantzig, *Number: The Language of Science.* 4th ed. New York: Macmillan.	Historical work; includes sets
1964	*The Mathematical Sciences: Essays for COSRIMS.* Cambridge, MA: MIT Press.	No foundations, but article on continuum hypothesis

General References

Anellis, I. (1987). The heritage of S.A. Janovskaja. *History and Philosophy of Logic*, **8**, 45–56.

Bourbaki, N. (1969). *Eléments d'Histoire des Mathématiques.* Paris: Hermann.

Cavalieri, F. (1988). Il debattito sulla logica in Unione Sovietica. *Rivista di Storia della Filosofia*, **3**, 533–69.

Dawson, J. W., Jr. (1985). The reception of Gödel's incompleteness theorems. In *PSA 1984*, vol. 2, pp. 253–71. East Lansing, MI: PSA. [Also in S. G. Shanker, ed. (1988). *Gödel's Theorem.* London: Croom Helm, pp. 74–95. Also in T. Drucker, ed. (1991). *Perspectives on the History of Mathematical Logic.* Boston: Birkhäuser, pp. 84–100.]

Dubislav, W. (1931). *Über die Definition.* 3rd ed. Leipzig, Germany: Meiner. [Repr. 1981.]

Feferman, A. B., and Feferman, S. (2004). *Alfred Tarski: Life and Logic.* Cambridge: Cambridge University Press.

Gödel, K. (1929). On the completeness of the calculus of logic. PhD diss. In *Collected Works*, vol. 1 (1986), pp. 61–101.

———. (1930). Die Vollständigkeit der Axiome des logischen Funktionenkalküls. *Monatshefte für Mathematik und Physik*, **37**, 349–60. [Published PhD diss.]

———. (1931). Über formal unentscheidbare Sätze der Principia Mathematica und verwandter Systeme I. *Monatshefte für Mathematik und Physik*, **38**, 173–98. [English trans. J. Heijenoort, ed. (1967). *From Frege to Gödel: A Source Book on Mathematical Logic.* Cambridge, MA: Harvard University Press, pp. 596–616. Repr. with facing English trans. K. Gödel (1986). On formally undecidable propositions of *Principia Mathematica* and Related Systems. I. In *Collected Works*, vol. 1 (1986), pp. 145–95.]

———. (1986). *Collected Works.* Vol. 1. *Publications 1929*–1936. New York: Oxford University Press.

———. (1990). *Collected Works.* Vol. 2. *Publications 1938*–1974. New York: Oxford University Press.

———. (1995). *Collected Works.* Vol. 3. *Unpublished Essays and Lectures.* New York: Oxford University Press.

———. (2003a). *Collected Works.* Vol. 4. *Correspondence A–G.* Oxford: Clarendon Press.

———. (2003b). *Collected Works.* Vol. 5. *Correspondence H–Z.* Oxford: Clarendon Press.

Grattan-Guinness, I. (1979). In memoriam Kurt Gödel: His 1931 correspondence with Zermelo on his incompletability theorem. *Historia Mathematica*, **6**, 294–304.

———. (2000a). *The Search for Mathematical Roots, 1870–1940: Logics, Set Theories and the Foundations of Mathematics from Cantor through Russell to Gödel.* Princeton, NJ: Princeton University Press.

———. (2000b). Mathematics and symbolic logics: Some notes on an uneasy relationship. *History and Philosophy of Logic*, **20**, 159–67.

Hilbert, D., and Bernays, P. (1934). *Grundlagen der Mathematik*. Vol. 1. Berlin: Springer. [Written by Bernays alone; 2nd ed. 1968.]

Mac Lane, S. (1999). Letter to the author, 2 February 1999.

Mac Lane, S. (2004). *A Mathematical Autobiography*. Wellesley, MA: A K Peters.

Mangione, C., and Bozzi, S. (1993). *Storia della Logica da Boole ai Nostri Giorni*. Milan, Italy: Garzanti.

Mostowksi, A. (1948). *Logika Matematyczna: Kurs Uniwersytecki*. Warsaw: Panstowe Wydawnictwo Naukova.

———. (1966). *Thirty Years of Foundational Studies*. Oxford: Blackwell.

Peckhaus, V. (1999). Nineteenth century logic between philosophy and mathematics. *Bulletin of Symbolic Logic*, **5**, 433–50.

Quine, W. V. O. (1932). The logic of sequences: A generalisation of *Principia Mathematica*. PhD diss. Harvard University, Cambridge, MA. [Publ. New York: Garland, 1989.]

———. (1934). *A System of Logistic*. Cambridge, MA: Harvard University Press.

———. (1940). *Mathematical Logic*. 1st ed. Cambridge, MA: Harvard University Press.

———. (1998). Letter to the author, March 26, 1998.

Rosser, J. B. (1936). Extensions of some theorems of Gödel and Church. *Journal of Symbolic Logic*, **1**, 87–91.

Russell, B. A. W. (1903). *The Principles of Mathematics*. Cambridge: Cambridge University Press. [2nd ed. London: Allen and Unwin, 1937.]

———. (1921). Vorwort (Foreword) [to (Wittgenstein 1921)]. *Annalen der Naturphilosophie*, **14**, 186–98. [Quoted from the English original in *Collected Papers*. Vol. 9. London: Routledge, 1988, pp. 96–112.]

———. (1940). *An Enquiry into Meaning and Truth*. London: Allen and Unwin.

———. (1945). Logical positivism. *Polemic*, **1**, 6–13. [Cited from *Collected Papers*, pp. 147–55.]

———. (1950). Logical positivism. *Revue Internationale de Philosophie*, **4**, 3–19. [Cited from *Collected Papers*, pp. 155–67.]

———. (1959). *My Philosophical Development*. London: Allen and Unwin.

———. (1971). Addendum to my "Reply to criticisms." In *The Philosophy of Bertrand Russell*, 4th ed., ed. P. A. Schilpp, pp. xvii–xx. La Salle, IL: Open Court. [Also in *Collected Papers*, 64–66.]

———. (1997). *Collected Papers*. Vol. 11. London: Routledge.

Scanlan, M. J. (1991). Who were the American postulate theorists? *Journal of Symbolic Logic*, **56**, 981–1002.

———. (2000). The known and unknown H. M. Sheffer. *Transactions of the C.S. Peirce Society*, **36**, 193–224.

Schenk, G. (1995). Zur Logikentwicklung in der DDR. *Modern Logic*, **5**, 248–69.

Sylvan, R. (1992). Significant moments in the development of Australian logic. *Logique et Analyse*, 137–38, 5–44.

Whitehead, A. N., and Russell, B. A. W. (1910–1913). *Principia Mathematica*. 1st ed. 3 vols. Cambridge: Cambridge University Press. [2nd ed. 1925–1927.]

Wittgenstein, L. (1921). Logische-philosophische Abhandlung. *Annalen der Naturphilosophie*, **14**, 198–262.

Wolenski, J. (1989). *Logic and Philosophy in the Lvov-Warsaw School*. Dordrecht, Netherlands: Kluwer.

CHAPTER 4

"Dozent Gödel Will Not Lecture"

Karl Sigmund

Kurt Gödel spent barely fifteen years in Vienna – less than in Brno, where he was born, and less than in Princeton, where he would die. However, the years from 1924 to 1939 constituted his formative period. He was deeply affected by the extraordinary cultural and intellectual flowering of what has been called "Vienna's Golden Autumn," and he may one day be seen as its most prestigious scion.

The University of Vienna, in particular, played a pivotal part in Gödel's life. To the end, he remained grateful for the education it provided him; however, he could never forget the bureaucratic pettiness and the hassles he encountered from his alma mater's administration (Sigmund, 2006; Dawson and Sigmund, 2006; Sigmund et al., 2006; Buldt et al., 2002).

The story of Gödel's Vienna years can be divided into three parts: the first is a tale of brilliant success leading to early recognition, his PhD degree, and his *Habilitation* (professorial qualifications); the second part deals with the confused period between Hitler's rise to power in Germany and the Anschluss, the annexation of Austria in 1938, and is marked by political unrest as well as by long periods of leave due partly to Gödel's visits to Princeton in the 1930s and partly to his recurrent health problems that repeatedly forced him to withdraw into sanatoriums; and finally, the third part is dominated by Gödel's problems after the German takeover. He was politically suspect, plagued by hostile authorities and under the threat of being drafted into the Wehrmacht (Germany's unified armed forces). Thanks to the persistent efforts of the Institute for Advanced Study (IAS) in Princeton, he finally managed to obtain an emigration visa for himself and his wife. However, the Viennese authorities continued to process his case long after he had left the Third Reich in early 1940 and found a safe haven in Princeton. He was granted the title of *Dozent neuer Ordnung* (lecturer of the New Order) in absentia and was enjoined to return to Germany. Gödel never went back, but until 1945, the university kept him on its books with the terse announcement that "Dozent Gödel will not lecture."

This chapter provides a discussion of Gödel's troubled relationship with the University of Vienna based on material from the archives as well as on private letters.

Facsimiles of most of these letters can be found in Sigmund et al. (2006; hereinafter referred to as *GA*).

4.1 Part 1: A Place in History

The curriculum vitae Kurt Gödel wrote in 1929 when he applied for his PhD may serve as an introduction to his life story:

> I was born in 1906 in Brno, where I attended four grades of elementary school and the eight grades of the German *Staatsrealgymnasium* [grammar school] and in 1924 I passed my A-levels with distinction. In the autumn of that year I enrolled as a regular student in the philosophical faculty of the University of Vienna, where I remained until the end of my studies. I initially attended lectures mostly in theoretical physics (with Prof Thirring); later I turned to purely mathematical studies, stimulated by the lectures of Prof Hahn and Hofrat Furtwängler, whose seminars I visited eagerly. In addition, I attended philosophical lectures by Prof Schlick and Prof Gomperz as well as Dozent Carnap. The area at the border between mathematics and philosophy interested me in particular: this has developed in a very lively way over the past few years through the works of Hilbert, Brouwer, Weyl, Russell et al. My dissertation "The completeness of logical calculus" also belongs in this border region. (*Personalakte* Kurt Gödel, Archiv der Universität Wien, hereinafter referred to as PKGAUV; see also *GA*)

Gödel's family belonged to the upper level of the German-speaking middle class in Brno (then called Brünn), a charming provincial town in southern Moravia (in what was then the Austro-Hungarian Monarchy), about two hours north of Vienna by train. Brno had some industry and was somewhat grandiloquently known as the Czech Manchester. Gödel's father was director and later part owner of a textile firm there (Feferman, 1986; Dawson, 1997).

During the last years of the Habsburg Empire, the strife between the Czechs and the Germans dominated political life. It seems that Kurt Gödel and his brother Rudolf were among the few pupils in the German-speaking school never to enroll for a course in Czech (Kreisel, 1980). After Gödel finished school, with distinction, Vienna represented an obvious choice of university for someone with his background. Once in Vienna, Gödel lived with Rudolf, who was his senior by four years and who studied medicine. The two brothers were well off and could afford to be choosy about their apartments – they rented a new one each year (Schimanovich-Galidescu, 2002; see also *GA*).

Gödel's professor in physics, Hans Thirring, was an expert on general relativity who, in 1920, had predicted, together with the Viennese mathematician Josef Lense, that the rotation of a body (such as the earth) should affect its gravitational field. This effect was experimentally verified only a few years ago (Ruffini and Sigismondi, 2003). At an early stage, Gödel must thus have acquired a solid foundation in Einstein's theory, which would prove fruitful when he turned his attention some twenty years later to cosmological work in Princeton.

Gödel's curriculum vitae also mentions the philosopher Heinrich Gomperz and the number theorist Philipp Furtwängler (whose title of *Hofrat*, meaning "court counselor," was highly coveted by Viennese professors). In a questionnaire that Gödel completed many years later (but never sent), he stressed that the introductory lectures of these two luminaries were of particular importance for the development of his own philosophy (the Grandjean questionnaire; see Dawson, 1997; see also *GA*). There is no reason to doubt that Gödel was already a full-fledged Platonist by the age of twenty (Feferman, 1984; Köhler et al., 2002; van Atten and Kennedy, 2003). In contrast, Hahn, Schlick, and Carnap were the mainstays of the Vienna Circle of logical empiricists (Popper, 1995; Stadler, 2001). Gödel, who was the Vienna Circle's youngest and quietest member, disagreed with many of their views aired during the fortnightly sessions of the circle, but he hardly ever voiced his objections (Menger, 1994). It was only twenty years later that he began to put them down in writing, in an essay intended to refute Carnap's view that mathematics is the syntax of language (Carnap, 1963). None of the six versions of the essay were published during Gödel's lifetime.

Gödel's PhD supervisor was Hans Hahn, an eminent mathematician and one of the founders of functional analysis (Sigmund, 1995). Since the early years of the twentieth century, Hahn had been interested in the foundations of mathematics and, in particular, in the works of Frege, Russell, and Hilbert, although he had not yet published anything on the topic (Hahn, 1988).

It seems that Gödel did most of his PhD work on his own. In a postcard dated February 23, 1929, his friend Herbert Feigl (a student of Schlick and another junior member of the Vienna Circle) wrote, "As I was not sure whether you would have agreed, I did not ask Hahn about your paper. Didn't you write to him?" (Gödel's *Nachlass*, Firestone Library, Princeton University; also see *GA*). The paper mentioned was the thesis Gödel had submitted for his PhD (Gödel, 1929).

In his report on the thesis, Hahn wrote, on July 13, 1929, half a year later:

> The paper deals with the so-called restricted functional calculus of logic [first-order logic], where quantifiers range only over individual variables and not over functional variables. It is shown that the system of axioms for first order logic that Whitehead and Russell used in the *Principia Mathematica* is complete, in the sense that every generally valid formula of this calculus can be derived formally from the system of axioms. Moreover, the independence of the axioms is proved. This solves two problems that were explicitly designed as unsolved in Hilbert-Ackermann, *Foundations of Theoretical Logic*. It follows that for systems of axioms that can be formulated entirely within first order logic every conclusion that is valid (i.e. that cannot be disproved by some counterexample) can be formally proved.
>
> The paper is a valuable contribution to logical calculus, fulfils entirely the requirements for a PhD thesis and deserves to be published in its essential parts. (PKGAUV; see also *GA*)

The paper was indeed published, within a few months, in the *Monatshefte für Mathematik and Physik* (Gödel, 1930), whose editor in chief was Hans Hahn. There are a few, highly significant differences between the submitted thesis (1929) and its published version (1930), and it has been argued that Gödel made the changes to comply with the general views of his thesis adviser (Baaz and Zach, 2002).

Gödel was not offered a position at the Mathematical Seminar after receiving his doctorate – no academic position was available. The economic crisis had hit the university with full force. Moreover, Austria had no funding organization able to offer project money to a postdoc. This did not greatly affect Gödel, who had independent means. His father had died recently. His mother, Marianne, left Brno to live with her two sons in two large, adjacent apartments in a stately house in the Josefstädterstrasse, a short walk from the university. Gödel was free to devote himself to his scientific interests. He usually worked late into the night, slept until late the next morning, and spent part of the day at the Mathematical Seminar, helping Hahn with exercise classes, preparing students for seminar talks, and spending much of his time in the library (Taussky-Todd, 1987). The young Viennese professor of geometry Karl Menger, who was barely four years older than Gödel, headed a group of young mathematicians and students who regularly met for discussions and lectures at the newly founded Mathematical Colloquium (Golland and Sigmund, 2000). Gödel was one of its most active participants and helped Menger in the preparation of the proceedings, the *Ergebnisse eines Mathematischen Kolloquiums* (Sigmund, 2002). Together with Menger, he gradually distanced himself from the Vienna Circle, although he retained close contacts with most of its members, in particular with Carnap. They frequently met in one of the numerous coffeehouses in town.

On August 26, 1930, at one such meeting in the Café Reichsrat, Gödel first mentioned his epoch-making discovery: arithmetic is incomplete. (There exist true statements that cannot be derived by purely formal means from the axioms.) A few weeks later, he mentioned his result, almost casually, during a discussion at a conference in Königsberg dedicated to the foundations of mathematics. John von Neumann, who was present, immediately understood its impact and was soon able to deduce that Hilbert's famous program for proving the consistency of arithmetic could not be carried through. He wrote to Gödel about it, but Gödel, in reply, was able to point out that that he had already realized this consequence and included it in his publication, again in the *Monatshefte*: "On formally undecidable propositions of *Principia Mathematica* and related systems. I" (Gödel, 1931). The impact was tremendous. In particular, both von Neumann and Menger outdid themselves to spread the news of this sensational breakthrough. Gödel had originally planned to follow his paper with a second part, presenting a formalized proof, but it soon became obvious that this was not necessary. His argument was blindingly clear.

Gödel's discovery was certainly more than sufficient to ensure his next step on the academic ladder, namely, the *Habilitation*, that is, his promotion to *Privatdozent* (private lecturer). This grade meant, essentially, that he was qualified for a position as professor and had the right to lecture at the university, but it carried with it no salaried job, and the only prospective income was the so-called *Kollegiengeld*, a small fee proportional to the number of students enrolled in the *Privatdozent*'s course.

An unofficial rule of the University of Vienna stated that the *Habilitation* could not be conferred sooner than four years after completion of a doctorate. However, Gödel was clearly a very special case, and so this period was shortened, but he still had to wait until 1933. During the waiting period, he maintained his way of life, making himself useful in many small ways at the Mathematical Seminar and continuing with his scientific work on logic, publishing most of it in the *Ergebnisse eines Mathematischen Kolloquiums*. In June 1932, he finally submitted his *Habilitation* application.

The application once more required a curriculum vitae:

> I was born in 1906 in Brno as the son of parents of German stock. I attended four grades of elementary school and the eight grades of the German *Staatsrealgymnasium* there and

in 1924 I passed my A-levels. In the autumn of that year I moved to Vienna, where I have remained without any lengthy interruptions ever since. I was awarded Austrian nationality in 1929. In the winter semester of 1924 I enrolled as a regular student in the Philosophical Faculty and studied first physics and subsequently mostly mathematics. I also occupied myself with modern developments of epistemology, inspired by Prof Schlick, whose philosophical circle I often visited. My own scientific activity initially dealt with the field of the foundations of mathematics and symbolic logic. In 1929 I submitted a paper in this field, "The completeness of first order logic," as a PhD thesis and obtained my doctorate in February 1930. In the same year I lectured in Königsberg on the above-mentioned paper. Moreover, I lectured on the paper I submitted for my *Habilitation* at the 1931 meeting of the German Mathematical Society in Bad Elster. In Vienna I participated in the colloquium organised by Prof Menger and also contributed to the editing of its annual proceedings. I was also active in the seminar on mathematical logic held by Prof Hahn in the academic year 1931/32, selecting the material and helping the participants prepare their report. In 1931 I was asked by the editorial team of the *Zentralblatt für Mathematik*, to which I contribute regularly, to write jointly with A. Heyting a report on the foundations of mathematics and I am currently engaged in its elaboration. (PKGAUV; see also *GA*)

This report was never finished. It may be noted that in contrast to the former curriculum vitae, Gödel did mention the Vienna Circle (but omitted that he had passed his A-levels with distinction). An *Habilitation* application also required a list of publications, three possible topics for a probationary lecture to be delivered in the presence of the committee of professors evaluating the *Habilitation*, and a list of topics for lecture courses for the coming years. Gödel proposed for the probationary lecture (1) the construction of formally undecidable propositions; (2) the intuitionistic propositional calculus; and (3) the set of values of conditionally convergent series. The committee chose his second topic.

The committee elected Hans Hahn to act as reporter on the scientific achievements of Gödel. Whereas Hahn's report on Gödel's PhD thesis had been rather matter-of-fact, saying that it "fulfils entirely the requirements for a PhD" – hardly adequate for a work whose impact is still strong after eighty years – this time, Hahn pulled no punches. The minutes of the committee meeting from December 1, 1932, state:

Prof Hahn also reported on the scientific suitability of Dr Gödel. The PhD thesis was itself of very high scientific quality ("The completeness of the axioms of the functional calculus of logic"). By showing that every generally valid formula of first order logic is provable, it solved the difficult and important problem posed by Hilbert of whether the axioms of the functional calculus of logic form a complete system. The *Habilitation* thesis "Formally undecidable propositions of *Principia Mathematica* and related systems" is a scientific achievement of the foremost order that has attracted highest attention in all expert circles and that – as can be predicted with certainty – will have its place in the history of mathematics. Dr Gödel has succeeded in showing that there are problems within the logical system of Whitehead and Russell's *Principia Mathematica* that can be stated but are undecidable within the system and that the same holds for every system of formal logic that encompasses the arithmetic of natural numbers; this also shows that the program proposed by Hilbert to prove the consistency of mathematics cannot be carried to its end.

Of several further papers by Gödel that concern the field of symbolic logic, let us highlight the note "The intuitionistic propositional calculus", in which the following theorems are proved: There is no realization of Heyting's axiom system for an intuitionistic propositional calculus with finitely many truth values for which the provable formulas, and only those, yield arbitrarily assigned truth values specified beforehand. Infinitely many systems lie between Heyting's system and the usual system of ordinary propositional calculus.

The papers submitted by Dr Gödel exceed by far the level usually required for a *Habilitation*. An overview of the present-day state of the research on the foundations of mathematics, which Dr Gödel has been asked to provide by the editors of *Zentralblatt für Mathematik*, is due to appear soon.

Dr Gödel is already accepted as the foremost authority in the field of symbolic logic and research on the foundation of mathematics. In close scientific collaboration with the reporter and with Prof Menger he has also brilliantly proven his worth in other fields of mathematics.

The commission has unanimously decided to request the faculty to endorse the scientific competence of Dr Gödel for a *Habilitation* and to admit him to the scientific colloquium. (PKGAUV; see also *GA*)

The scientific colloquium is an hourlong examination during which committee members may ask the candidate any question related to his field. (This requirement has subsequently been dropped.) Gödel's probationary lecture took place within the Mathematical Colloquium. The audience included a guest from abroad, Oswald Veblen, who was currently talent-scouting on behalf of the IAS in Princeton. Menger and von Neumann had drawn his attention to Gödel, whose lecture made a most favorable impression.

The ritual secret ballot in the faculty meeting came up with forty-nine yes votes and the almost usual, isolated nay from one anonymous objector, who apparently delighted in preventing unanimous outcomes. Afterward, Dean Srbik, an eminent historian who would later play a deplorable part after the Anschluss, reported to the minister of education, Kurt Schuschnigg:

The undersigned Dean reports on the progress of Dr Kurt Gödel's *Habilitation* proceedings.

On 25 June 1932 Dr Kurt Gödel submitted to the College of Professors [Professorenkollegium] of the Philosophical Faculty a request for conferring him the *venia legendi* [right to lecture] in the field of "mathematics."

The committee consisted of the professors Furtwängler, Hahn, Himmelbauer, Menger, Prey, Schlick, Tauber, Thirring and Wirtinger and was headed by the Dean. It first met to consider the application on 25 November 1932 and noted first that in accordance with §§2 and 6 of the *Habilitation* norm the field was sufficiently wide and self-sufficient and secondly that the personality of the above mentioned candidate gave no grounds for doubt on his suitability for a university teaching position.

The following meeting of the committee took place the same day. In it, the papers submitted, in particular the *Habilitation* thesis, were examined and found to be adequate, thus allowing the candidate to proceed to the next steps of the *Habilitation*.

Accordingly, the committee proposed to the faculty meeting of 3 December 1932 that the applicant should be permitted to proceed with the further steps, i.e. with the colloquium.

The College of Professors accepted the committee's recommendation and decided to admit the candidate to the colloquium. The result of the ballot on personal suitability was 51 yes and one no with no abstentions.

The colloquium was held on 13 January 1933 and was found to satisfy the requirements of the law and the faculty meeting of 21 January was therefore asked to allow the applicant to proceed to the probationary lecture.

In its meeting on 21 January 1933, the College of Professors accepted this request with the simple majority required by the *Habilitation* norm.

The probationary lecture took place on 3 February 1933 on the topic "The intuitionistic propositional calculus." The committee found this probationary lecture to satisfy the requirements of the law.

Accordingly, the committee proposed at the faculty meeting on 11 February 1933 that the probationary lecture satisfy the requirements of the law. The College of Professors accepted the proposition with the simple majority required by the *Habilitation* norm.

According to §12 of the *Habilitation* norm, the final ballot on conferring the *venia legendi* was held. The result was 42 yes and one no with one abstention.

Based on the College of Professors' decision, the undersigned Dean proposes that the applicant be granted the right to teach in the field of "mathematics" at the Philosophical Faculty of the University of Vienna and asks the Ministry of Education for confirmation. (PKGAUV; see also *GA*)

4.2 Part 2: Psychiatric Clinics and Ivory Towers

Gödel was awarded the title of *Dozent* (lecturer) on March 11, 1933. His proud mother made sure that Brno's German-speaking newspaper, the *Brünner Tagesbote*, announced that Dr. Kurt Gödel, a son of the late Rudolf Gödel, director of an industrial enterprise, had been awarded his *Habilitation* at the University of Vienna. It also announced that Kurt Gödel had been invited to spend a semester as assistant professor at the IAS in Princeton. Most of the newspaper issue dealt with the political turmoil caused by the National Socialists' rise to power in Germany (see *GA*): a few weeks earlier, Hitler had been appointed *Reichskanzler* (Chancellor of the Reich) and had immediately launched a ruthless campaign of *Gleichschaltung* – a euphemistic term that translates as "coordination" – that rapidly transformed the Weimar Republic into the Third Reich.

As *Dozent*, Gödel was required to lecture for two hours per week during at least one semester in four. Gödel held only three lecture courses at the University of Vienna and thus barely fulfilled the requirement to retain his title. His first course, in the summer semester of 1933, dealt with the foundations of arithmetic; his second, two years later, with selected topics in mathematical logic; and his third, two years later again, with axiomatic set theory. Gödel's scientific genius was well known among the students in the Mathematical Seminar, but so was the fact that his lectures were extremely demanding. One of his students was Edmund Hlawka, who later became an eminent number theorist. Hlawka notes that Gödel lectured at a dazzling speed (the book by Hilbert and Ackermann (1928) was covered in one week) and always kept his back to the student audience, whose size dwindled rapidly so that by the end of the semester, only one student was still following the course – none other than Andrzej Mostowski, the well-known Polish logician.

With so few students, Gödel's lecture fees were ridiculously low. In one semester, he earned two schillings and ninety groschen – barely the price of a few beers. It was understood, of course, that Privatdozent Gödel did not need the income as he had independent means. However, during the 1930s, these means shrank inexorably, and by the end of the decade, little was left. Nevertheless, Gödel, who still lived with his mother, his brother, and an aged aunt in two large, adjacent apartments in the Josefstädterstrasse, was not troubled by financial worries. Moreover, the salary he obtained during his visits to the IAS was generous, and he was able to bring a large part of it back to Vienna. Dollars were a prestigious commodity in crisis-ridden Austria.

In June 1934, soon after his return from the United States, Gödel had to be admitted to a sanatorium in Purkersdorf, close to Vienna, for treatment following a nervous breakdown. A terse receipt mentions the deposit of fifty dollars (see *GA*). According to Gödel's brother Rudolf, by this time an MD, the crisis had been caused by stress from overwork. The state of the country to which Kurt Gödel had returned certainly did nothing to reduce his nervous stress. The political and economic outlook in the wake of the Great Depression was bleak. The Austrian chancellor, Dollfuss, a conservative, was struggling fiercely against Left and Right. Parliament was suspended because of an ill-considered move by some deputies and never resumed. Rather than looking for an impossible majority of votes, Dollfuss preferred to establish an authoritarian regime of a clerical-fascist blend. After a brief workers' revolt in February 1934, which claimed over a hundred victims, the Socialist Party was forced to go underground. So were the National Socialists (Nazis), who vehemently agitated in favor of joining the Third Reich.

Political struggles poisoned university life. A large number of the students were sympathizers of the outlawed Nazi Party. Riots frequently forced the university to close, sometimes for weeks. Authorities had to tell their lecturers how to deal with politically motivated disruptions. It is unlikely that Gödel's sparsely attended lectures were ever the theater for National Socialist demonstrations, but Gödel kept the corresponding memorandums to the end of his life, possibly because they say so much about the atmosphere that reigned at Austrian universities in the 1930s.

The minister for education wrote:

> It has come to my attention that at the end of a university lecture a large number of students sang the German national anthem (or, according to others, the *Horst Wessel Lied*), obviously with the intent of a political demonstration, and that some shouted "Heil Hitler." I am therefore obliged to require that strong measures be taken to prevent all forms of acclamation or protest and all singing, shouting and similar forms of political manifestations. (Gödel's *Nachlass*, Firestone Library, Princeton University)

The memorandum then drew attention to the various laws and disciplinary measures to which the police had recourse. Another circular letter was headed "Subversive propaganda at institutes of higher education – duty of all university staff to oppose it" and stated:

> The Ministry of Education has received complaints about activities on university premises that are directed, in a hidden way, against the interests of the country. In various exercise classes, seminars and excursions and on other occasions where students are required to speak, camouflaged political agitators following orders from abroad utter disobliging remarks of a political or personal nature, as well as political jokes etc. without being stopped by the members of the faculty, by scientific staff or by university employees.

The Ministry of Education draws attention to these misdemeanours and expects that they will immediately be stopped, if necessary using radical measures. The Ministry of Education would regret having to use available disciplinary measures against a staff member who should in future – unexpectedly – fail to react dutifully. (Gödel's *Nachlass*, Firestone Library, Princeton University; see also *GA*)

This was signed by Kurt Schuschnigg, then still minister for education, and was dated June 17, 1934, that is, shortly after Gödel's return from the United States. The contrast to the ivory-tower atmosphere at the IAS must have been painful.

Soon afterward, in July 1934, members of the outlawed Nazi Party stormed the chancellery and killed Dollfuss in his office. The putsch was quickly repressed, but the authoritarian regime resorted increasingly to dictatorial measures. Kurt Schuschnigg, who succeeded Dollfuss as chancellor, continued the latter's policies for keeping Hitler at bay, which relied almost exclusively on Mussolini's goodwill.

In the same month, Hans Hahn, Gödel's teacher, died unexpectedly following surgery for cancer. He had been a supporter of "Red Vienna" and one of the very few professors in vocal opposition to the nationalist-conservative majority at the faculty.

Gödel himself was apolitical, but like many others, he was obliged to join the Fatherland Front, a party organization that was imposed from above. It found little support in the population, but eventually, almost every third Austrian became a member. The members had to pledge absolute loyalty and obedience to the Führer (meaning Kurt Schuschnigg) and to wear a badge. In general, the Fatherland Front was considered to be a toothless counterfeit mimicking the Nazi rituals across the border, a relatively harmless rigmarole, but those who failed to adhere risked their jobs. This was the case with Professor Heinrich Gomperz, the charismatic philosopher who had introduced Gödel to Platonism. Gomperz was branded as lacking loyalty to Austria, was forced to retire, and had to emigrate, at age sixty-three, to the United States.

Gödel's financial contributions to the Fatherland Front were minimal and certainly more a sign of cautiousness than of commitment. Karl Menger (1994), in his *Reminiscences*, writes:

> Kurt Gödel kept himself well informed and spoke with me a great deal about politics without showing strong emotional concern about events. His political statements were always noncommittal and usually ended with the words "don't you think?"... Occasionally, however, he made pointed and original remarks about the situation. Once he said, "Hitler's sole difficulty with Austria lies in the fact that he can only appropriate the country as a whole. Were it possible to proceed piecewise, he would certainly have done so long ago – don't you think?"

Gödel's next voyage to Princeton, in autumn 1935, was a disaster. After a few weeks, he had to resign from the IAS and return to Austria. He experienced his worst psychological crisis and had to spend many months in sanatoriums, this time in Rekawinkel and Aflenz. A little handwritten note, kept at the university archives, says, "Unfortunately I am obliged, again, to cancel my lecture course. Dozent Kurt Gödel" (see *GA*).

In summer 1936, the murder of Moritz Schlick on the stairs of the university shook the Austrian public. The murderer, Dr. Nelböck, was only a few years older than Gödel and had also studied mathematics and philosophy. Schlick had judged his PhD thesis to be "fairly weak." Nelböck suffered from psychological crises, pathological jealousy, and persecution mania and had to be committed to psychiatric clinics on several occasions. He developed the idée fixe that Schlick was thwarting his professional and private prospects, but although the motive behind Schlick's murder was clearly personal, it was quickly given a political and ideological spin by Nelböck's defense attorney. The argument (the accused was a simple farmer's son whose upright Christian beliefs had been subverted by an agnostic philosopher) was successful: Nelböck got away with a ten-year sentence and was freed soon after the Nazis annexed Austria.

Schlick's murder became a cause célèbre. Some of the contemporary press articles were so venomous that Karl Menger decided to emigrate, with his young family, to the United States. In 1937, he joined the University of Notre Dame, where he remained until 1946.

By now, Gödel was scientifically isolated in Vienna. After Schlick's death and Menger's emigration, both the Vienna Circle and the Mathematical Colloquium were leaderless and struggled for survival. The philosopher Edgar Zilsel tried valiantly to keep up the regular meetings of the logical empiricists. They could no longer meet in the Mathematical Seminar (there was no professor left who was sufficiently interested to allow them to use the premises) and thus had to convene in Zilsel's apartment. In 1937, Gödel gave a lecture there on mathematical logic. The manuscript was rediscovered fifty years later in his estate by John Dawson. Menger's collaborators, Franz Alt and Abraham Wald, managed to keep the Mathematical Colloquium alive but had no job and no official title and knew that time was running out on them. From America, Menger wrote, on December 31, 1937, to his former student Franz Alt:

> I am deeply sad to be unable to do anything for the dear beautiful circle of Viennese mathematicians. I believe you should get together from time to time and especially see that Gödel takes part in the Colloquium. It would be of the greatest benefit not only to all the other participants but also to Gödel himself, though he may not realise it. Heaven knows what he might become entangled in if he does not talk to you and his other friends in Vienna from time to time. If necessary, be pushy, on my say-so. (from the personal papers of Franz Alt; see *GA*)

However, the new year would see the end of an independent Austria, and Franz Alt, far from being in a position to help Gödel, could consider himself lucky to obtain an exit visa and escape, barely in time.

Gödel's scientific community in Vienna had disappeared. Nevertheless, he was not alone. Unknown to his colleagues and acquaintances, he had been supported for years by the friendship of Adele Nimbursky, née Porkert, a divorcée seven years older than himself. He must have met her in 1929 when he lived with his brother in an apartment just opposite hers in the Lange Gasse. Adele was the daughter of a photographer. The family was of modest means but had an interest in arts. As a child, Adele had studied piano and ballet. It seems that she later worked at the Nachtfalter, a well-known cabaret (listed as a *Vergnügungsetablissement* (pleasure establishment)) in the city center. When Gödel met her, she was thirty and had returned from a failed marriage to live with her parents. She worked for a few years as a masseuse and was listed as such in the telephone book under her parents' address. A couple years later, she was listed, still under the same address, as "private." Gödel kept her under wraps as well as he could. However, when he was in the sanatorium, plagued by persecution mania and a morbid fear of poisoning, she visited him regularly. In a letter written much later, the director of the IAS stated that Gödel, during his crisis, had refused to eat anything except what Adele cooked for him and tasted under his eyes, from his own plate and with his own spoon.

When Gödel recovered, he decided to marry Adele, despite the objections of his family, who feared a misalliance. He rented an apartment in a stately house in the Himmelstrasse, close to the Vienna Woods. Adele, with her practical and energetic touch, took care of the furnishing. In November 1937, the young couple moved in. Gödel's mother, who had shared the apartment in the Josefstädterstrasse for eight years with her two sons, resigned herself to returning to her villa in Brno.

4.3 Part 3: Escape from the Reich

At the time of the Anschluss, Gödel was in Vienna. He was not persecuted for racial or political reasons, but many of his friends and former colleagues were. Gödel witnessed the political upheaval, the euphoric demonstrations greeting Hitler's troops, the anti-Semitic bedlam in the streets, and the ferocious efficiency of the *Gleichschaltung*, the "coordination" that implied the removal of all those disfavored by the new masters (Siegmund-Schultze, 2002). A swiftly appointed commissioner-rector undertook the *Säuberung* (cleanup) of the university, based on the law for the "reestablishment of a professional civil service."

In his inaugural speech, the new rector bemoaned the former coolness displayed by the professors toward National Socialism and declared that "all this has now changed. The material lesson enjoyed by the professorial body during the Führer's presence in Vienna will not fail to have its effect" (Archiv der Universität Wien).

Gödel's former professor Furtwängler, who left because he had reached retirement age, was replaced by one Anton Huber, a dyed-in-the-wool Nazi who had joined the party while in Switzerland. The only other full professor at the Mathematical Seminar, Karl Mayrhofer, turned out to have been an illegal party member all along.

In Gödel's philosophical faculty, fourteen out of forty-five full professors, eleven out of twenty-two associate professors, thirteen out of thirty-two emeritus professors, and fifty-six out of one hundred fifty-nine lecturers "retired." An essential instrument for this upheaval was the oath to the Führer, to whom all salaried employees at the university (Aryans only, of course) had to pledge their allegiance on March 22, 1938. Gödel did not have to take the pledge because as a private lecturer, he did not rank among university employees.

Franz Alt, who had applied for an immigration visa to the United States a few weeks before the Anschluss, was quickly able to emigrate. Others, such as Abraham Wald and Eduard Helly, had a much harder time and had to live through agonizing months of bureaucratic chicanery, but by the end of summer 1938, practically all Gödel's former colleagues were gone.

Gödel himself was in a privileged position thanks to his multiple-entry visa for the United States. On October 1, 1938, he was able to begin a long-scheduled visit (his third), but his wife, Adele, whom he had married two weeks earlier, remained in Vienna. During the subsequent eight months, which Gödel spent partly at the IAS and partly at Karl Menger's University of Notre Dame in Indiana, the sinister nature of the Third Reich became increasingly evident, in particular through the *Reichskristallnacht* (Night of the Broken Glass), a state-led pogrom that claimed dozens of lives, and through the occupation of Czechoslovakia, a flagrant breach of the Munich Treaty. Hitler was clearly heading toward a major European war.

To Menger's dismay, Gödel nevertheless insisted on returning to Vienna in June 1939. He planned to return to Princeton in the autumn, this time with his wife. Apparently, it took him a long time to realize that this time, he was encountering more than just the ordinary bureaucratic hassle with the Devisenstelle (a branch of the Ministry of Finance that pestered him with demands to account for his American dollars) and the Wehrmacht (who required him to show up for a military physical examination). In a letter to Menger written on August 26, 1939, three days before the outbreak of Hitler's war, Gödel vaguely complained of *Laufereien* (pointless errands) but seemed to have no doubts about his forthcoming return to Princeton.

The situation was much more serious than Gödel had anticipated. His multiple-entry visa for the United States was in his previous Austrian passport and therefore no longer valid. Obtaining a new immigration visa seemed to require an inordinate amount of time: the American Consulate in Berlin was submerged with such applications. There were special rules for professors, but Gödel was no longer even a lecturer.

The Germans had abolished the title of *Privatdozent*. Former lecturers could apply to become a *Dozent neuer Ordnung*, which implied a salaried position. Gödel, with his legalistic turn of mind, was outraged by the suspension of his former title. The official reason given for this act was that he had failed to obtain permission to take up an appointment in the United States. Such permission was required not by Austrian law but by German law. Gödel may initially have overlooked this and later tried to amend it by a curt note penned in Princeton:

> This is to inform you that I have accepted an invitation to Princeton, USA, for the winter semester 1938/39. My address there is Fine Hall, Princeton, New Jersey. Respectfully yours, Kurt Gödel. (PKGAUV; see also *GA*)

This was apparently insufficient to mollify the University authorities. On July 12, 1939, the dean informed the ministry that he had no current information on the whereabouts of Gödel (unaware that the latter was back in Vienna at the time). He confirmed that Gödel had written on October 31, 1938, to say that he had accepted an invitation to Princeton but complained that he had given no specific information about his activities or the duration of his stay. Moreover, Gödel had not asked for leave for the summer semester of 1939. On this basis, the dean asked for the cancellation of Gödel's *Habilitation* at the Philosophical Faculty of the University of Vienna.

On August 12, 1939, the ministry replied regretfully that its hands were tied but suggested a neat way to resolve the case of this obnoxious *Privatdozent*:

> According to a past decree the right to teach of the above mentioned has been provisionally suspended. As the former Austrian *Habilitation* norm is no longer valid, I lack the formal means for the requested administrative measure.
>
> However, his right to teach will be finally cancelled, either when he does not apply before 1 October 1939 for promotion to Dozent of the New Order or else when the

Reichsminister for Science and Education does not grant that promotion. It will be your task to accompany an eventual application by Dr Gödel in such a way that the Minister takes the desired decision.

In my opinion it is superfluous, in this case and in other cases in which the right to lecture is suspended . . . to take special steps against a private Dozent of the Old Order whose further teaching is undesired. (PKGAUV; see also *GA*)

To arrive at a decision, the leader of the National Socialist Association of Lecturers of the University of Vienna, Professor Marchet, was asked for his opinion. Marchet, who was no mathematician, had to ask for the opinion of Gödel's former colleague Mayrhofer, who had been a member of the outlawed Nazi Party. This was when Gödel's habit of discretion paid off. Marchet wrote to the dean:

Dr Kurt Gödel (until now Dozent) is scientifically well considered. His *Habilitation* was supported by the Jewish professor Hahn. He is accused of having always moved in liberal-Jewish circles. It must however be mentioned that during the Systemzeit [i.e. the period of the former Austrian regime] mathematics was strongly jewicized [*verjudet*]. I have not been apprised of any direct statements or acts against National Socialism. His professional colleagues have not known him socially, so that further information about him is not obtainable. (PKGAUV; see also *GA*)

Marchet concluded cautiously:

It is therefore not possible for me expressly to support his promotion to Dozent of the New Order but neither do I have any grounds for objecting to it. (PKGAUV; see also *GA*)

On the basis of this report, the dean wrote, on November 27, 1939, to the rector:

I have been requested to report on the personality of Privatdozent Kurt Gödel, PhD, from the points of view of profession, political stance and character.

Gödel has a high reputation in his field, which comprises the boundary region between mathematics and logic and was particularly favoured by Gödel's teacher, the Jewish professor Hahn, as I learn from the judgements of the two full professors of mathematics at our faculty, K. Mayrhofer and A. Huber; Gödel is particularly well-esteemed in the USA, where such questions on the foundations of mathematics are of interest to wide circles. (PKGAUV; see also *GA*)

At that time, it seemed unconceivable that these "questions on the foundations of mathematics" could be of any practical use. However, other scientific communities took a close interest in foundational questions – even negative answers to them – and, by a sequence of unintended consequences, rapidly advanced the development of the computer. It seems, moreover, that little attention was paid in the Third Reich to the type of mathematics needed for cryptoanalysis. It is interesting to speculate on what Gödel could have achieved in a position like that of Turing at Bletchley Park. Fortunately for Gödel, his field played no part in the *Deutsche Mathematik* (Siegmund-Schultze, 2002), a term was used by Vahlen, Teichmüller, and other Nazi mathematicians. They never managed to come up with a proper definition of the term but agreed that set theory, for instance, did not belong to it and that non-Aryan mathematicians tended to compensate for their lack of intuition by an unfortunate bent for bloodless abstraction.

The dean went on to discuss the political stance of Kurt Gödel:

Concerning the political judgement of Gödel I have taken advice from Prof Marchet, the leader of the association of lecturers at the University, whose opinion exactly matches my own personal impression. Gödel, who grew up at a time when the Viennese mathematicians were completely under Jewish influence, has hardly any inner rapport with National Socialism. He gives the impression of a thoroughly apolitical person. Therefore, it seems unlikely that he will be up to the difficult situations bound to arise for a representative of the New Germany in the USA.

As a character, Gödel gives a good impression; in this respect I have never heard any complaint about him. He has good manners and will certainly not commit any social gaffes that could damage the reputation of his home country abroad.

If political reasons should forbid Gödel to travel to America, this raises the question of providing him with means for living. Gödel has no income at all and he wants to accept the invitation to the USA solely to ensure his livelihood. The whole question of leaving Germany would be void if it were possible to find a suitably paid job within the Reich. (PKGAUV; see also *GA*)

For Gödel, the situation was much graver than the dean knew – it was not only a question of finding a livelihood. Gödel had finally been mustered by the Wehrmacht, and to his consternation, he was found fit for garrison duty, as he wrote in November 1939 in a very despondent letter to Osvald Veblen. This meant that he was quite likely to be called up soon to serve in the German army.

The journey to America suddenly seemed out of reach. The American Consul in Berlin was submerged by applications for immigration visas. University professors were eligible for nonquota visas, but Gödel did not fully qualify: he had not lectured enough, and even his position as *Dozent* had become shaky.

However, the IAS proved up to the task. John von Neumann wrote a masterly analysis for Abraham Flexner, the director of the IAS, showing that Gödel ("a mathematician of the first rank – he is in a class by himself") had a valid case after all. In particular, Gödel could take advantage of the fact that although he had not lectured during the last two calendar years, this was because of circumstances beyond his control: "Gödel's teaching," as von Neumann smoothly argued, "was terminated by his being suspended from office by the German Government after the invasion and annexation of Austria in 1938." He concluded, "I think that there cannot be the slightest doubt that he is a professor in the sense of the law."

This letter probably saved Gödel's life, for it allowed Flexner to convince the State Department and, in particular, the American Consul in Berlin. Against all odds, Gödel's visa problems were suddenly resolved. He and his wife obtained German exit permits on December 19, 1939 – it must have seemed like a Christmas miracle. From then on, things moved very quickly. The American immigration visas were issued on January 8, 1940. Gödel had decided not to risk an Atlantic crossing (this was the time that journals spoke of a "phony war," but there was nothing phony about the U-boats). Instead, Kurt and Adele chose to take the eastbound route, via Siberia and the Pacific. The transit visas through the Soviet Union were obtained on January 12, and the journey began a few days later. The entries on Gödel's passport tell of a political tightrope act (Dawson, 2002). The travelers had to cross Poland (which was occupied and divided) and Lithuania (which was soon to be overrun by Stalin's troops). After changing trains in Moscow, the Gödels continued their winter journey via the Trans-Siberian Railway through the recently occupied Manchuria and on to Japan, which at that time had deeply penetrated into China but had not yet signed the pact with Germany or engaged in war against the United States. In Yokohama, the Gödels missed their ship and had to wait several weeks for another steamer to take them across a still Pacific Ocean on to the safety of San Francisco. As Gödel wrote to his brother, "San Francisco is absolutely the most beautiful of all the towns I have ever seen" (letter from Kurt Gödel, *Handschriftensammlung* der Wienbibliothek).

From there on, it was just a transcontinental train ride to the haven of the IAS. Fellow emigrant Oskar Morgenstern notes in his diary (Dawson, 1997) that "Gödel arrived from Vienna. Via Siberia. This time with wife. When asked about Vienna: 'The coffee is wretched'(!)" (diary of Oskar Morgenstern, Morgenstern papers, Duke University; see also *GA*).

Kurt Gödel would never again return to Vienna, but this was something he himself probably did not realize until decades later. The wheels of the bureaucracy that he had left behind continued to grind (Dawson and Sigmund, 2006). On March 7, 1940, the Ministry for Interior and Cultural Affairs sent an ill-tempered letter to the rector of the university. In one single sentence of inordinate length, even by German standards, it announced the return of some documents, asked for details about Gödel's arrival in Vienna, and chided the rector for supporting Gödel's application to become *Dozent neuer Ordnung*. Obviously, the writer was unaware that Gödel had left again in the meantime:

> Attached I return the collateral material of the report mentioned above (with exception of the copies of the Deans' report, the application of Dr Gödel and the memorandum by the head of the association of lecturers, which are intended for my office) and I ask for a detailed statement concerning the fact that Dr Gödel spent some time abroad on his own, and dealing with the precise circumstances of his return (which has obviously taken place). I request clarification of the apparent contradiction in your support for the application of this private lecturer, whose right to teach has been suspended after the re-unification of Austria with the German Reich, despite the fact that the Dean of the Philosophical Faculty has asked for the (final) withdrawal of this right to teach, with the full consent of the head of the association of lecturers, as expressed in the report of July 12 1939, Z.129 from 1938/39, seen by you, a withdrawal which, however, I was unable to effect due to formal reasons (Decree of August 12 1939, Z. IV-2c-232.954) (*Personalakte* Kurt Gödel, Archiv der Universität Wien; see also *GA*)

A second paragraph displays the everyday chicanery of Nazi bureaucracy:

> In addition, I draw your attention on the fact that the proof of descent of Dr Gödel and his wife is incomplete. It consists indeed of 16 documents (namely 12 – not, as mentioned in the Dean's report, 8 – photocopies, 3 additional certified copies and one document concerning a change in religion by the father of Gödel) but the marriage certificates of Gödel's parents and his wife's parents must be submitted (if not, in addition, the certificates of the four pairs of grandparents). (PKGAUV; see also *GA*)

The rector asked the dean of the Philosophical Faculty to clarify the issue raised in the first paragraph. On April 1, 1940, the dean explained his change of mind, somewhat disingenuously, by the fact that Gödel's application for *Dozent neuer Ordnung* constituted a proof of political goodwill. He flatly stated that it was "doubtlessly incorrect" to say that Gödel had traveled to America without proper permission, apparently referring to Gödel's current leave, not the previous one. The ministry official rather surprisingly gave up. On June 28, 1940, "in the name of the Führer," the minister for education signed a heavily embossed diploma that promoted Kurt Gödel to *Dozent neuer Ordnung* and stated, rather pompously:

> I sign this document in the expectation that the promoted will conscientiously obey his official duties according to his oath of service and will justify the trust placed in him. In return, he can be assured of the Führer's special protection. (PKGAUV; see also *GA*)

This was two weeks after the Führer had taken Paris.

The diploma was never collected. It rests today in the university archives, together with a receipt that still awaits the signature of Kurt Gödel. The rectorate of the university wrote on several occasions to Gödel's brother Rudolf, who replied, again and again, that Kurt Gödel would collect the certificate as soon as he was back from the United States.

A letter from Kurt to Rudolf, dated October 6, 1940, contains a passage in which Kurt asks Rudolf to inquire, if feasible, about the salary that he would receive in the event of his return. This may indicate that Gödel did not at the time entirely exclude the possibility of returning to the Third Reich. It more probably shows that Gödel, while still in Vienna, had so little hope that his application would succeed that he had not even bothered to inquire about the rate of pay.

On January 30, 1941, Rudolf Gödel informed the dean that Kurt Gödel was unable to return to Europe as the German Consulate had urgently advised against the journey. The dean noted, "Gödel can only resume his lectures when circumstances permit it. Immediately after his return, Gödel will present himself at the chancellery and the faculty of the University." Slightly later, the ministry granted Kurt Gödel an extension of his leave up to July 31, 1941, but the tone of the letter was somewhat grudging. On March 4, 1941, the German Ministry for Science and Education asked the Foreign Office to inquire about Gödel. In reply:

> As the German General Consulate in New York recently reported, Dr Gödel has asked for an extension of his leave because the Institute in Princeton has offered him a stipend of 4000$ for further research during the coming academic year. Given that the General Consulate explained to him that his acceptance of this offer is undesired but that another suitable occupation cannot be found, Dr Gödel asked to be informed, if a further leave is not granted, whether after an eventual return [*Heimschaffung*] he would be given a salaried position at a German university. However, he added that because of heart trouble he was incapable of any strenuous administrative or teaching duties. (PKGAUV; see also *GA*)

The scene in the German Consulate, with a straight-faced Gödel, would have been worthy of an Ernst Lubitsch film. The term *Heimschaffung* seems very odd, with its connotation of "being bundled home." The waters of the Atlantic had not become safer, and a return journey via Siberia would not be feasible for much longer. In fact, the express letter from the German Foreign Office to the Ministry for Education was dated June 22, 1941, the day after Hitler's troops invaded the Soviet Union. Less than half a year later, Hitler declared war on the United States. However, amid the turmoil, the elusive Dozent Gödel was not forgotten by the Viennese authorities. To their inquiries Rudolf Gödel replied in an increasingly curt fashion. The dean had to write to the ministry:

> With regard to GZ 2226/946/II of 23 September 1942 I report that I have no knowledge relating to the whereabouts of Dozent Dr Kurt Gödel. (PKGAUV; see also *GA*)

For one semester after another, until spring 1945, the *Vorlesungsverzeichnis* (a booklet on the lectures held at the university) contained the terse statement that "Dozent Kurt Gödel will not lecture."

After the war, the institute of mathematics was quickly purged of its Nazi professors. Johann Radon and Edmund Hlawka were appointed – a brilliant recovery – but the emigrants were not asked back. Most of them would probably have declined to return to postwar Vienna and to live among bombed-out houses in a drab and dismal city occupied by the four Allied powers. However, that they were not even asked was a blot on the early years of the new Austrian Republic. As Gödel wrote in 1946 to his mother (who, in 1944, had returned from Brno to live with her son Rudolf in Vienna):

> On principle one should be obliged (since the current Austrian government obviously considers the Hitler regime as an illegal tyranny) to invalidate all the dismissals from the Universities. In most cases those concerned would opt not to return but apparently they are not even being offered the chance. (letter from Kurt Gödel, *Handschriftensammlung der Wienbibliothek*)

Twenty years later, the Austrian foreign minister Bruno Kreisky (later a longtime chancellor of the Austrian Second Republic) wrote to the rector of the university:

> Your Honour,
>   An eminent personality whom I know well has drawn my attention to Professor Dr Kurt Gödel, currently teaching at the Institute for Advanced Study in Princeton. Gödel is a former Austrian.... The decisive result of his research is that all attempts at a formalistic foundation of mathematics must fail and that there exist mathematical systems for which it can be proved that it is impossible, by means of operations within the system, to decide whether sentences that can be derived are provable or not.... I would be grateful, your Honour, if you could consider whether and how Gödel, who soon will celebrate his sixtieth birthday, could get honoured by his former University. A token of recognition from Austria would, I believe, not only honour Gödel but also honour Austria. (*Personalakte* Kurt Gödel, Archiv der Universität Wien; see also *GA*)

The eminent personality was, of course, Oskar Morgenstern. Soon afterward, the dean of the Philosophical Faculty wrote to Gödel and offered him an honorary professorship. Gödel declined, just as he declined an honorary membership of the Austrian Academy of Science, on the most tenuous of grounds. These honors came too late.

References

Baaz, M., and Zach, R. (2002). Das Vollständigkeitsproblem und Gödels Vollständigkeitsbeweis. In *Kurt Gödel: Wahrheit und Beweisbarkeit, Dokumente und historische Analysen*, vol. 2, ed. B.

Buldt, E. Köhler, M. Stöltzner, C. Klein, and W. DePauli-Schimanovich-Göttig et al., pp. 21–27. Vienna: Hölder-Pichler-Tempsky.

Buldt, B., Köhler, E., Weibel, P., Stöltzner, M., Klein, C., DePauli-Schimanovich-Göttig, W, eds. (2002). *Kurt Gödel: Wahrheit und Beweisbarkeit, Dokumente und historische Analysen*. Vienna: Hölder-Pichler-Tempsky.

Carnap, R. (1963). Intellectual autobiography. In *The Philosophy of Rudolf Carnap*, ed. P. A. Schilpp, pp. 3–84. Library of Living Philosophers 11. LaSalle, IL: Open Court.

Dawson, J. (1997). *Logical Dilemmas: The Life and Work of Kurt Gödel*. Wellesley, MA: A K Peters.

———. (2002). Max Dehn, Kurt Gödel, and the trans-Siberian escape route. *Notices of the American Mathematical Society*, **49**, 1068–75.

Dawson, J., and Sigmund, K. (2006). Gödel's Vienna. *Mathematical Intelligencer*, **28**, 44–55.

Feferman, S. (1984). Kurt Gödel: Conviction and caution. *Philosophia Naturalis*, **21**, 546–62.

———. (1986). Gödel's life and work. *Collected Works*, vol. 1, pp. 1–36.

Gödel, K. (1929). On the completeness of the calculus of logic. PhD diss. In *Collected Works*, vol. **1** (1986), pp. 61–101.

———. (1930). Die Vollständigkeit der Axiome des logischen Funktionenkalküls. *Monatshefte für Mathematik und Physik*, 37, 349–60. [Published PhD diss. Also in *Collected Works*, vol. 1 (1986), pp. 102–23.]

———. (1931). Über formal unentscheidbare Sätze der Principia Mathematica und verwandter Systeme I. *Monatshefte für Mathematik und Physik*, **38**, 173–98. [Habilitation thesis. The first incompleteness theorem first appeared as Theorem VI and the second incompleteness theorem as Theorem XI in the same paper. Also repr. with facing English trans. On formally undecidable propositions of *Principia Mathematica* and Related Systems. I. In *Collected Works*, vol. 1 (1986), pp. 145–95. Trans. J. van Heijenoort, ed. (1967). *From Frege to Gödel: A Source Book on Mathematical Logic*. Cambridge, MA: Harvard University Press, pp. 596–616.]

Golland, L., and Sigmund, K. (2000). Exact thought in a demented time: Karl Menger and his Viennese Mathematical Colloquium. *Mathematical Intelligencer*, **22**, 34–45.

Hahn, H. (1988). *Empirismus, Logik, Mathematik*. Frankfurt, Germany: Suhrkamp.

Hilbert, D., and Ackermann, W. (1928). *Grundzüge der theoretischen Logik*. Berlin: Springer. [2nd ed. 1938 Berlin: Springer. English trans. of 2nd ed. *Principles of Mathematical Logic* New York: Chelsea.]

Köhler, E., Weibel, P., Stöltzner, M., Buldt, B., Klein, C., DePauli-Schimanovich-Göttig, W, eds. (2002). *Kurt Gödel: Wahrheit und Beweisbarkeit, Kompendium zum Werk*. Vienna: Hölder-Pichler-Tempsky.

Kreisel, G. (1980). Kurt Gödel. *Biographical Memoirs of Fellows of the Royal Society*, **26**, 148–224.

Menger, K. (1994). *Reminiscences of the Vienna Circle and the Mathematical Colloquium*. Dordrecht, Netherlands: Kluwer.

Popper, K. (1995). Hans Hahn: Erinnerungen eines dankbaren Schülers. In *Hans Hahn, Gesammelte Werke*, ed. L. Schmetterer and K. Sigmund, pp. 1–20. Vienna: Springer.

Ruffini, R., and Sigismondi, C. (2003). *Nonlinear Gravitodynamics: The Lense-Thirring Effect*. Singapore: World Scientific.

Schimanovich-Galidescu, M. E. (2002). Archivmaterial zu Kurt Gödels Wiener Zeit, 1924–1940. In *Kurt Gödel: Wahrheit und Beweisbarkeit, Dokumente und historische Analysen*, vol. 2, ed. B. Buldt, E. Köhler, M. Stöltzner, C. Klein, and W. DePauli-Schimanovich-Göttig., pp. 135–147. Vienna: Hölder-Pichler-Tempsk.

Siegmund-Schultze, R. (2002). *Mathematiker auf der Flucht vor Hitler*. Mannheim, Germany: Vieweg.

Sigmund, K. (1995). A philosopher's mathematician: Hans Hahn and the Vienna Circle. *Mathematical Intelligencer*, **17**, 16–29.

———. (2002). Karl Menger and Vienna's Golden Autumn. In *Karl Menger, Selecta Mathematics I,* ed. B. Schweizer, A. Sklar, and K. Sigmund, pp. 7–21. New York: Springer.

———. (2006). Pictures at an exhibition. *Notices of the American Mathematical Society*, **53**, 426–30.

Sigmund, K., Dawson, J., and

Mühlberger, K. (2006). *Kurt Gödel: The Album*. Mannheim, Germany: Vieweg.

Stadler, F. (2001). *The Vienna Circle*. New York: Springer.

Taussky-Todd, O. (1987). Remembrances of Kurt Gödel. In *Gödel Remembered: Salzburg, 10–12 July 1983*, ed. P. Weingartner and L. Schmetterer, pp. 29–41. Naples, Italy: Bibliopolis.

Van Atten, M., and Kennedy, J. (2003). Gödel's philosophical developments. *The Bulletin of Symbolic Logic*, **9**, 470–92.

CHAPTER 5

Gödel's Thesis: An Appreciation

Juliette Kennedy

Introduction

With his 1929 thesis, Gödel delivers himself to us fully formed. He gives in it a definitive, mathematical treatment of the completeness theorem,[1] but he also declares himself philosophically, unfolding the meaning of that theorem from a wider, rather mature, fully *philosophical* point of view.[2]

Among the rewards of studying particularly its introductory remarks, are the following. First, light is shed on the timing of the first *in*completeness theorem, construed as a response to Carnap as well as its possible genesis as a response to Brouwer; those remarks also add to our understanding of the separation of the completeness and categoricity concepts, which was emerging just then. In a few crucial places, the remarks can strike the modern reader as peculiar. The view taken here is that these peculiarities are interesting and important, and therefore they are treated at length in this chapter. The introductory remarks were never included in the publication based on the thesis, and indeed, Gödel would not publish such unbuttoned philosophical material until 1944, with his *On Russell's Mathematical Logic*.[3]

5.1 The Introduction to Gödel's Thesis: Different Notions of Consistency

The set of remarks we first consider occur in the first paragraph of Gödel's 1929 thesis, and address the issue of whether consistency is a ground for existence:

---

[1] The completeness question was explicitly posed by Hilbert and Ackermann in their 1928 *Grundzüge der theoretischen Logik.*, and by Hilbert in his 1928 address to the International Congress of Mathematics, held in Bologna. An English translation of the latter is printed in (Mancosu 1998).

[2] The author thanks Steve Awodey, John Baldwin, Paolo Mancosu, and Dana Scott for helpful correspondence and Dana Scott, again, for his careful editing. The author is especially grateful to Jouko Väänänen for extensive discussions on many of the issues raised in this paper; to Wilfried Sieg for helpful discussion and references; and finally, to Curtis Franks, whose close look at a previous draft led to many improvements in it.

[3] Gödels contribution to the Schilpp volume on Russell.

L. E. Brouwer, in particular, has emphatically stressed that from the consistency of an axiom system we cannot conclude without further ado that a model can be constructed. But one might perhaps think that the existence of the notions introduced through an axiom system is to be defined outright by the consistency of the axioms and that, therefore, a proof [that a model exists] has to be rejected out of hand. This definition (if only we impose the self-evident requirement that the notion of existence thus introduced obeys the same operation rules as does the elementary one), however, manifestly presupposes the axiom that every mathematical problem is solvable. Or, more precisely, it presupposes that we cannot prove the unsolvability of any problem. For, if the unsolvability of some problem (in the domain of real numbers, say) were proved, then, from the definition above, there would follow the existence of two non-isomorphic realizations of the axiom system for the real numbers, while on the other hand we can prove the isomorphism of any two realizations. We cannot at all exclude out of hand, however, a proof of the unsolvability of a problem if we observe that what is at issue here is only unsolvability by certain *precisely stated formal* means of inference. For, all the notions that are considered here (provable, consistent, and so on) have an exact meaning only when we have precisely delimited the means of inference that are admitted. (Gödel 1986, 61)

Gödel's argument against what we might call Hilbert's principle was aimed at a precept of formalism, as formulated by Hilbert in, for example, his 1899 correspondence[4] with Frege, that consistency is a sufficient ground for existence. Another way of putting this is that one needs only to establish the *consistency* of an axiom system, say, Zermelo-Fraenkel set theory, to infer the existence of those objects to which the axioms refer.[5] Hilbert argued for the idea that consistency entails existence, and Frege argued against it. Following is a sample of their correspondence (Hilbert to Frege, December 29, 1899):

You write I call axioms propositions. From the truth of the axioms it follows that they do not contradict one another. I found it very interesting to read this very sentence in your letter, for as long as I have been thinking, writing and lecturing on these things, I have been saying the exact reverse: if the arbitrarily given axioms do not contradict each other with all their consequences, then they are true and the things defined by the axioms exist. This is for me the criterion of truth and existence.

We do not know the extent to which Gödel might have been aware of the Hilbert-Frege dispute, although he certainly seems to be worrying about the dispute precisely in his remarks here. What is known is that Gödel would have been responding primarily to Carnap, his close colleague and discussion partner, as detailed in Awodey and Reck (2002) and Goldfarb (2005) – but more about that below.

Let us look carefully at Gödel's argument. He begins by defining *consistency* syntactically: an axiom system is said to be consistent if no contradiction is derivable in finitely many steps from the system in question. Gödel then defines completeness:

[4] See the Frege/Hilbert correspondence, (Frege, 1980).

[5] Hilbert would come to abandon this view: "Denn das Problem der Widerspruchsfreiheit gewinnt nunmehr eine ganz bestimmte, greifbare Form: es handelt sich nicht mehr darum, ein System von unendlich vielen Dingen mit gegebenen Verknüpfungs-Eigenschaften als logisch möglich zu erweisen, sondern es kommt nur darauf an, einzusehen, dass es unmöglich ist, aus den in Formeln vorliegenden Axiomen nach den Regeln des logischen Kalküls ein Paar von Formeln wie A und $\neg A$ abzuleiten." See the 1921–1922 lectures of Hilbert and Bernays, as quoted in Sieg (1999). See also Sieg (2002, 389–90).

Here "completeness" is to mean that every valid formula expressible in the restricted functional calculus (a valid *Zählaussage* as Löwenheim would say) can be derived from the axioms by means of a finite sequence of formal inferences.

Gödel goes on to remark about it that as a means of demonstrating consistency, the theorem that the first-order predicate logic is complete represents a "theoretical completion" of the usual method of proving consistency, the usual method being simply to exhibit a model.

Gödel then goes on to identify two properties of consistency, or more precisely, two different senses in which the consistency of an axiom system can guarantee the existence of a model. An axiom system is consistent in the first sense – let us call this "syntactic consistency" – when no contradiction can be syntactically derived and the construction of an associated model can be carried out, something that can be done with the methods outlined in Gödels thesis. An axiom system is consistent in the second sense – let us call this latter notion "semantic consistency" – if the existence of a corresponding model is assumed outright; a construction is not required as

> ... the existence of the notions introduced through an axiom system is to be defined outright by the consistency of the axioms and that, therefore, a proof [that a model exists] has to be rejected out of hand.

Two notions of consistency exist: one starting from a syntactic condition that gives rise to a model via a construction; another that assumes that the syntactic condition automatically defines a model.[6] Note that if Hilbert's principle is taken to mean that "consistency entails existence," then the latter concept of consistency renders this principle true by default. The latter concept also permits a spectrum of variants. We can simply *define* the existence of a model as the consistency of the axioms, which again renders Hilbert's principle true by default. This leads to a discussion whether *existence* in mathematics can be so defined. In fact, Gödel's criticism does not seem to be pointed at this aspect; rather Gödel seems to take issue with Hilbert's principle in the context where consistency is taken in the first sense, that is to say, syntactically.

We note that both concepts of consistency permit the following weakening, namely, that they apply the concept of consistency only to categorical axiomatizations. Thus, in the first sense, one would hold that an axiom system is consistent if one cannot derive a contradiction and, moreover, any two models would be isomorphic. Note that without a completeness theorem, this does not outright imply the existence of a model, and indeed, it is nontrivial to construct a model even when one knows the system is categorical. In the second sense, one would view the axiom system to be consistent if any two models would be isomorphic and such a model does exist (or in a weaker sense, one would hold that in this case, a model exists by *definition*). Again, the latter variant would render Hilbert's principle true by default, although the principle is now weaker.

---

[6] One might wonder if the idea behind semantic consistency is not simply to start from a model and then derive the syntactic condition, which is a way of demonstrating consistency that mathematicians often revert to today.

Before discussing Gödel's argument in detail, we note that the notion of semantic consistency occurs naturally in the (full) second-order framework, for the following reason: given the second-order axioms, consistency from them is no guarantee of having a model in the sense of full second-order logic. This is manifested by some very simple examples. One example is that of second-order arithmetic, denoted P^2, together with the axioms $\{\underline{n} < c : n < \omega\}$, where c is a new constant symbol added to the language of arithmetic. This theory is clearly syntactically consistent, but it has no models in the full second-order sense. This is because the model would have to be nonstandard, whereas P^2 is *categorical*, that is to say, it has only one (standard) model. For full second-order logic, then, the only way to think of consistency, it seems, is to simply define it as a secondary property, that is, in terms of an already-existent model.

We now consider Gödel's argument. The hypothetical situation Gödel considers is to accept that there is a categorical axiomatization A_2 of the real numbers. Gödel also assumes that consistency entails existence, one way or the other. He concludes from this that there cannot be "unsolvable problems about the reals." For assume that there is a proposition ϕ about the reals, such that both ϕ and $\neg\phi$ are consistent with A_2. It then follows from Hilbert's principle (that consistency implies existence) that there must be two nonisomorphic models[7] M and N of the theories $A_2 + \phi$ and $A_2 + \neg\phi$. But Hilbert's (full) second-order axiomatization A_2 of the reals is categorical.[8] We thus obtain a contradiction, and therefore the statement that there are unsolvable problems about the reals must be false.[9]

Note that the existence of M and N would follow also from Gödel's own completeness theorem (as presented here in Gödel's thesis) if one had a completeness theorem for A_2, but then there could not be unsolvable problems in A_2. Gödel seems to be biased in favor of the existence of unsolvable problems, so he is driven toward the conclusion that there cannot be a completeness theorem for A_2. However, if this is so, then Hilbert's principle is only meaningful in the second (semantic) sense of consistency; that is, one knows that the axioms of A_2 are consistent because one already has a model for them. This last point seems to contradict rather strongly what one might think of as the finitist trend of Hilbert's thinking.[10] It would also seem unsatisfactory from the perspective of logic to conclude that there are no unsolvable problems. For example, at the time, the continuum hypothesis was thought to be unsolvable, at least Skolem speculated as much already in 1922.[11]

---

[7] Or "realization," as Gödel called them.

[8] Hilbert states this in "Über den Zahlbegriff," and there are indications of a proof in unpublished lecture notes from the late 1890s. By the late 1920s, it had become folklore (Wilfried Sieg, pers. comm., 2007).

[9] In modern terminology, the standard full second-order axiomatization of the real numbers is syntactically complete. Note that the first-order theory of the arithmetic of the reals without the arithmetic of the natural numbers permits an effective complete axiomatization by Tarski's famous result. Gödel, on the other hand, clearly thought of the theory of the reals as extending the theory of the natural numbers.

[10] I do not address the question in this chapter of whether Gödel understands Hilbert's philosophical position, or the Hilbert program, correctly.

[11] This view of solvability has emerged in the recent debate about second-order logic, where it has been expressed that, for example, the continuum hypothesis is solved by default, as it were, following as it or its negation does

Returning to Gödel's argument, one can give up the categoricity of the axiomatization, but then, without uniqueness, the principle that consistency entails existence borders on incoherence. One might as well simply adopt the syntactic concept of consistency. But then a slight peculiarity emerges, for Gödel assumes that the existence of two nonisomorphic models of Hilbert's axioms follows from the existence of an unsolvable question ϕ about the reals, where unsolvability means that neither ϕ nor $\neg\phi$ is decided by A_2. Thus we do have a proof system, together with derivation rules, relative to which the undecidability of ϕ allows one to add ϕ, and then in turn its negation $\neg\phi$, to Hilbert's axioms to obtain two different consistent sets of axioms. To conclude the argument, in attempting to derive a contradiction, Gödel must now introduce the first notion of consistency (syntactic consistency) to obtain the two nonisomorphic models. Both syntactic and semantic consistency are needed, but in that case, one relies on a version of the completeness theorem for A_2 – or where would these two models come from?[12]

With the two distinct notions of consistency established, we come to the heart of the argument, namely, that in either case, we contradict categoricity; we also come to the heart of the problem with this argument, as Gödel seems to rely on some version of a completeness theorem for A_2. We now know that such a completeness theorem cannot be proved, a consequence of the existence of unsolvable problems. Gödel noted this consequence of the incompleteness theorem explicitly in 1930 during his announcement of the first incompleteness theorem at Königsberg (see below), reasoning, very likely, as follows: a single, second-order sentence ϕ characterizes the standard model for arithmetic up to isomorphism. But this means that if a completeness theorem could be proved for full second-order logic, arithmetic truth would be an r.e. relation; that is, for all first-order arithmetic sentences ψ, $\mathcal{N} \models \psi$ iff $\phi \longrightarrow \psi$ is valid in second-order logic. That is to say, $\mathcal{N} \models \psi$ would be equivalent to the existence of a proof.

That Gödel may have sensed the subtleties and the difficulties in his argument can be seen from his remarks at the end of the paragraph:

> These reflections, incidentally, are intended only to properly illuminate the difficulties that would be connected with such a definition of the notion of existence, without any definitive assertion being made about its possibility or impossibility. (Gödel 1986, 63)

---

from categoricity. For the working set theorist, of course, this idea leads nowhere. To learn that the continuum hypothesis is solved is to have an understanding of which way it is solved (positively or negatively) and what the method of solution is.

[12] Note that Gödel is clearly not equating the two notions of consistency – the distinction between them is made clear and explicit. Might Gödel have had an intermediate system in mind? Such a system might be categorical for the natural number part, but not for the real part. An example of such a system would be a two-sorted system consisting of an ω-rule for the natural numbers, but with comprehension for reals restricted in the sense of Henkin models of second-order logic (Henkin, 1950). The natural number part of models of this system coincides with the standard integers because of the ω-rule. In hindsight, we can say that from the point of view of Hilbert's principle, there is little difference between this intermediate system and A_2. Indeed, we can rule out the possibility that Gödel was thinking of an intermediate system as he specifically mentions categoricity in connection with axiom systems for the reals rather than for the integers. It is also clear that he must have had Hilbert's second-order axiomatization of the reals in mind. This is because, as we have pointed out (and as Gödel seems to be on the verge of realizing given the topic of his thesis), categoricity would fail for a first-order axiomatization of the reals.

In this thesis, Gödel did not delve any further into the matter of second-order axiomatizability – although he would soon enough – but rather wished to emphasize the possible existence of unsolvable questions, going so far as to use the assumption of unsolvable questions about the reals as the basis of an argument by contradiction. By the time he turned in his thesis in 1929, he had formulated the idea that unsolvability is a genuine phenomenon – and that a conception of provability that ignores it is one that is inherently flawed.

5.2 Categoricity and the Separation of First- and Higher-Order Logic

Before taking up the matter of the incompleteness theorems, what could Gödel have said at that point about second-order provability, completeness theorems for second-order logic, and categoricity? That there can be no completeness theorem for (full) second-order logic follows from Gödel's incompleteness theorem, as was mentioned, but Gödel does not seem to have known this *definitively* in 1929, at least to the extent that he relies on it there.

As for categoricity, Gödel did not take note at the time that the failure of categoricity for first-order theories (at least those with infinite models) is a consequence of the completeness theorem. In fact, in connection with the existence of nonstandard models of Peano arithmetic, Gödel would associate this, even in 1934, with the incompleteness theorem rather than with the completeness theorem.

More precisely, it is a simple consequence of the completeness theorem that categoricity fails for first-order arithmetic. This fact was not published until 1936, with the work of Malcev (1971), although the argument Malcev gives for its existence is a simple one: to the language of Peano arithmetic, one adds a new constant symbol, say, c. To the Peano axioms, one adds the axioms $\{\underline{0} < c, \underline{1} < c, \underline{2} < c, \ldots\}$. This new set of axioms is finitely consistent and hence, by compactness, consistent. But in such a model, the interpretation of the new constant symbol c must be a nonstandard integer.[13]

Indirect evidence suggests that Gödel may not have known this argument by 1934. This can be inferred from his review of Skolem's (1933), a landmark paper in which Skolem constructs a model of so-called true arithmetic[14] not isomorphic to the standard model, using what is commonly known as the definable ultrapower construction – and thereby proving directly that categoricity fails for true arithmetic.

In his review, Gödel associates the failure of categoricity of arithmetic with the *in*completeness theorem, from which the existence of those models, as he put it, also easily follows, rather than with his own completeness theorem – an "extraordinary"

---

[13] The compactness theorem is only implicitly stated in Gödel's thesis, but his 1930 paper, based on his thesis, includes it. The theorem as stated by Gödel in his (1930) is as follows: a countably infinite set of quantificational formulas is satisfiable if and only if every finite subset of those formulas is satisfiable.

[14] True arithmetic is the theory whose axioms are all first-order sentences in the language of arithmetic that are true of the standard integers.

omission on Gödel's part, as Robert Vaught put it in his introduction to Gödel's review.[15] As Gödel writes there:

> The author [Skolem] proves that there is a system N^* of entities, with two operations $+$ and $\cdot$, defined on it, and with two operations $>$ and $=$, that is *not* isomorphic to the system N of natural numbers, but for which nevertheless all statements hold that are expressible by means of the symbols mentioned at the outset and hold for the system N. From this it follows that there is no axiom system employing only the the notions mentioned at the outset (and therefore none at all employing only number-theoretic notions) that uniquely determines the structure of the sequence of natural numbers, a result that follows also without difficulty from the investigations of the reviewer in his *1931*.

Gödels last remark is that it follows from the incompleteness theorem that there is no effective axiomatization of the natural numbers that is categorical – a conclusion that also requires the completeness theorem. Whereas Skolem showed something slightly different and much stronger, namely, that categoricity fails even for true arithmetic; that is, even if we admit a noneffective axiomatization of arithmetic, we still do not reach categoricity – although we do reach completeness.[16]

Gödels lack of interest in nonstandard interpretations, if such a charge can be made of Gödel here, may be why what would seem to be a very important observation – that what the completeness theorem precisely does is to separate first- and second-order logic because of categoricity[17] – was not made by Gödel in his thesis.[18]

Apart from the question of what Gödel may have known,[19] what was known about the first- and higher-order logic distinction, and about categoricity, before 1929? Dedekind noted the categoricity of his second-order axiomatization of arithmetic in 1887, but the concept was isolated very likely for the first time, according to Awodey and Reck (2002), by Huntington in 1902,[20] who called it "sufficiency." (Although the notion of isomorphism relative to which Huntington proved his system of postulates to be categorical, an element missing from other formulations from that time that mention categoricity in some form, including Hilbert's, was explicitly formulated by Dedekind in his 1888 "Was Sind und was sollen die Zahlen?")

The distinction between first- and higher-order logics seems also to have become prominent with Weyl's criticism of Zermelo's axiomatization of set theory in 1910, on the grounds that the concept of "definite property," as Zermelo defined it, was

---

[15] See Gödel (1986, 376).

[16] Note that true arithmetic is a complete, nonrecursively axiomatizable theory and hence the incompleteness theorem does not apply to it; that is, one does not obtain a model of true arithmetic not isomorphic to the standard model by means of that theorem. One *can* obtain such a model by means of Malcev's argument, with true arithmetic standing in for Peano.

[17] Of second-order Peano.

[18] That the completeness theorem separates first- and second-order logic because of categoricity is, of course, not completely true; e.g., it is not the case that *all* second-order theories are categorical.

[19] Hilbert-Ackermann, a text that Gödel knew well, gives a second-order axiomatization of number theory. We do not take up the issue of Hilbert's various axiom systems with regard to their being first- or second-order, or with regard to issues of categoricity or to completeness, here. See Awodey and Reck (2002) for a critical assessment.

[20] Huntington (1902)

unclear.[21] Skolem, in his (1923) (independently, not citing Weyl), also suggested that the correct notion of "definite property" should be given in terms of what we would now call first-order definability – although the dispute between Zermelo and Skolem about Zermelo's axiomatization of set theory was not put by them in those terms.

It appears that Skolem, in his 1923 work, was the only logician of the period for whom the failure of categoricity of the Zermelo-Fraenkel axioms for set theory had foundational importance – in fact, it lead to his dismissal, in that paper, of the idea that Zermelo-Fraenkel set theory should play any foundational role in mathematics. As Skolem remarks, the result has been known to him since 1915, but he did not publish it then as

> ... first, I have in the meantime been occupied with other problems; second, I believed that it was so clear that axiomatization in terms of sets was not a satisfactory ultimate foundation of mathematics that mathematicians would, for the most part, not be very much concerned with it. But in recent times I have seen to my surprise that so many mathematicians think that these axioms of set theory provide the ideal foundation for mathematics; therefore it seemed to me that the time had come to publish a critique.[22]

It is not known whether Gödel ever saw Skolem's 1922 paper before he turned in his thesis; in particular, it is not known whether Skolem's foundational views had any influence, or indeed counterinfluence, on Gödel.[23] Returning to what Gödel surely was aware of, as was mentioned earlier, Hilbert-Ackermann, a text which Gödel knew well, gives a second-order axiomatization of number theory.[24]

By 1929, the understanding of various notions of completeness had deepened on the basis of the work already cited, but also because of the work of Fraenkel, and particularly of Carnap, the closest person to Gödel during the crucial years (1926–1929).[25] The work in which Carnap was engaged at the time consisted of building on, and extending the work of, Fraenkel on the various notions of completeness he had isolated. Carnap's manuscript, which Gödel saw, was titled *Untersuchungen zur allgemeinen Axiomatik* and was not published until recently.[26] Awodey and Reck cite Carnap's achievement in that work as follows:

> In [*Untersuchungen*] Carnap extends Fraenkel's considerations in the following three ways: he makes serious attempts to answer Fraenkel's questions about the precise connections between categoricity, deductive completeness, and semantic completeness.

---

[21] See Mancosu et al. (2009).

[22] See Skolem (1923).

[23] Gödel did make a number of attempts to borrow the proceedings volume in which the paper was published from various libraries in Vienna and also in Berlin. The evidence suggests that he did not succeed. See van Atten (2005). For a comparison of Skolem (1923) and Gödel (1929), see van Atten and Kennedy (2009a).

[24] I do not take up the issue of Hilbert's various axiomatization systems with regard to their being first- or second-order, with regard to categoricity, or with regard to completeness any further here. See Awodey and Reck (2002) for a critical assessment.

[25] Carnap and Gödel met frequently for discussion from about 1926, the year Gödel first began attending the Vienna Circle, although the two also met frequently outside the Vienna Circle seminar. It was Carnap from whom Gödel took his only logic course (as late as 1928), Carnap having taken up his *Privatdozentur* in Vienna in 1926. See Dawson (1997) and also Awodey and Carus (2001) for an account of the extensive interaction between the two.

[26] See Carnap (2000); for an analysis of this work see Awodey and Carus (2001).

Unlike Fraenkel, he puts his investigations into a formal, logical framework, namely that of the simple theory of types. And he picks up on Fraenkel's suggestions concerning the relation between his three notions of completeness, on the one hand, and completeness in the sense of Hilbert's "Axiom of Completeness," on the other. Carnap thus addresses, systematically and in detail, what we would now call metatheoretic issues. (Awodey and Reck 2002, 24)

Although Carnap had moved the discussion significantly forward, in virtue of his having carried out the program outlined earlier, the *Gabelbarkeitsatz*, a principle theorem of the *Untersuchungen*, equated categoricity, which he called monomorphicity, with decidability – for Carnap, a form of completeness. This is true in the sense that if a theory is categorical, then it only has one model, and thus any sentence is "decided" – by the model. In the modern, computational sense of the term *completeness*, though, this fails for higher-order logic, for while these models are categorical and hence decidable in the preceding sense, there is no way to transform truth in the model into provability.

Even so, the *Gabelbarkeitsatz* played, and continues to play, an important role in logic. For example, a recent interesting result due to Solovay, gives a proof that what Awodey and Reck refer to as the general case of the *Gabelbarkeitsatz*, namely, that semantic completeness implies categoricity for all finitely axiomatized theories in higher-order logic, is independent of ZFC.

Awodey and Reck argue in their (2002) that

> ... the now standard restriction to first-order logic in connection with notions such as categoricity and completeness conflicts with the way in which those concepts were initially investigated in the works of Hilbert, Carnap, Gödel, Tarski and others. From a historical point of view such a restriction is, thus, unwarranted and misleading. It is also ill-advised from a technical point of view, insofar as some aspects of these issues are more naturally and fruitfully addressed in higher-order logic, ... (Awodey and Reck 2002)[27]

Whatever one's allegiances are in this respect, and Awodey and Carus make a convincing case for their view, I have been arguing that precisely because higher-order logic was the "natural setting," the distinction simply was not clear before.[28]

5.3 The Incompleteness Theorem

As is well known, Gödel's announcement of his first incompleteness theorem was made at the Erkenntnis conference in Königsberg in September 1930, where Gödel was scheduled to speak about his dissertation results. At the end of a brief discussion of these, Gödel remarked:

---

[27] The reader is referred also to Awodey and Carus (2001) and (2006) for a very comprehensive account of Carnap's work on completeness and categoricity, as well as for a comprehensive account of the development of the concepts of completeness and categoricity from Dedekind through Huntington, Veblen, Fraenkel, and Carnap.

[28] W. Goldfarb also builds a convincing case from the Carnapian perspective in his (Goldfarb, 2005) Goldfarb's incisive analysis of Gödel's thesis remarks has a few points in common with the analysis give here, though the emphasis there is on the influence of Carnap's logical work on that of Gödel's.

I would furthermore like to call attention to to a possible application of what has been proved here to the general theory of axiom systems. It concerns the concept of "decidable" and "monomorphic." . . . One would suspect that there is a close connection between these two concepts, yet up to now such a connection has eluded general formulation. . . . In view of the developments presented here it can now be shown that, for a special class of axiom systems, namely those which can be expressed in the restricted functional calculus, decidability always follows from monomorphicity.[29] . . . If the completeness theorem could also be proved for higher parts of logic (the extended functional calculus), then it would be shown in complete generality that decidability follows from monomorphicity. And since we know, for instance, that the Peano axiom system is monomorphic, from that the solvability of every problem of arithmetic and analysis expressible in *Principia Mathematica* would follow.

Such an extension of the completeness theorem is, however, impossible, as I have recently proved; that is, there are mathematical problems which, though they can be expressed in *Principia Mathematica*, cannot be solved by the logical devices of *Principia Mathematica*. (Gödel 1986, 27–29)

The reader will notice the likeness between the penultimate paragraph here and the paragraph from Gödels thesis with which we have been concerned.[30] Both passages take up the issue of the existence of a completeness theorem for second-order logic, the first implicitly and the second explicitly. Thus had Gödel responded to Carnap, while at the same time clearing up any vagueness that may have lingered from his thesis remarks. The incompleteness theorem, in other words, was announced as demonstrating the following negative result: the completeness theorem fails for second-order logic.

We now know from the second incompleteness theorem that consistency cannot be shown for the circumstances outlined by Hilbert. As for semantic consistency, Gödel's remark in his 1931 paper, that consistency can be shown if one passes to higher types, seems to revive the notion in some form:

> . . . it can be shown that the undecidable propositions constructed here become decidable whenever appropriate higher types are added (for example, the type ω to the system P). (Gödel, 1986, 181, n 48a.)

A preferred method of demonstrating the consistency of, say, arithmetic, today is just this. For Peano arithmetic and for the theory of the reals in the sense that Gödel thinks of it (i.e., as extending arithmetic), the only way of proving the consistency of the axioms is by working in a system of higher type – a method that is, in spirit, somewhat close to the semantic consistency assumption. Proving consistency and then constructing a model for it is out of the question otherwise.[31]

---

[29] First-order theories that are categorical are of finite size; otherwise, if there is an infinite model, there are models in every cardinality by the Lowenheim-Skolem theorem.

[30] The reader will also notice a train of thought somewhat reminiscent of Gödels here, in Hilbert's 1928 Bologna lecture, wherein Hilbert also considers incompleteness in the context of the notions of categoricity and completeness. Hilbert's argument also seems to depend on semantic completeness. See Mancosu (1998, 231). A letter from Gödel to Feigl dated September 24, 1928 indicates that Gödel did not attend this lecture but was rather in Brno at the time. See Gödel (2003, 402).

[31] There is, of course, Gentzen's consistency proof for Peano arithmetic, but this touches on issues that lie beyond the scope of this chapter.

5.3.1 The Timing of the Incompleteness Theorem

As we saw, Gödel remarked in his thesis:

> We cannot at all exclude out of hand, however, a proof of the unsolvability of a problem if we observe that what is at issue here is only unsolvability by certain precisely stated formal means of inference.

This suggests that at the time, Gödel must have been considering unsolvability relative to specific formal systems as distinct from unsolvability by any means at all.[32] A second explanation for the appearance of this paragraph in Gödels thesis, therefore, may be that he must have been occupied with some version of the incompleteness theorem at the time.[33] Recall that in his argument, Gödel assumed the existence of a proposition ϕ such that both ϕ and $\neg\phi$ are consistent with the axioms. Should Gödel not abandon this assumption rather than Hilbert's principle? Does not the fact that Hilbert's axiomatization is categorical imply that there will be no unsolvable problems about the reals? As we saw, Gödel's is an argument by contradiction, following from the assumption that there are no unsolvable problems about the reals – his fundamental assumption.

It is, to date, not known precisely when Gödel arrived at the first incompleteness theorem. We do know that he spoke of it, and possibly of the second theorem, to Carnap and Feigl in the Café Reichsrat on August 26, 1930, before announcing it during a discussion session at the Königsberg meeting in September 1930. We know this from Carnap's diary entry from that day: "Gödel's discovery: incompleteness of the system Principia Mathematica; difficulty of the consistency proof."[34] In fact, Gödel must have started thinking about the theorem at least from March 1928, after hearing Brouwer's two lectures in Vienna (on March 10, on intuitionistic foundations, and on March 14, on the intuitionistic continuum).[35]

We know this from an earlier entry in Carnap's diary, dated December 12, 1929, in which he records a conversation he had with Gödel that day about Brouwer's lecture(s) of the previous year.[36] In his entry, Carnap states that Gödel talked "about the inexhaustibility of mathematics (see separate sheet). He was stimulated to this idea by Brouwer's Vienna lecture. Mathematics is not completely formalizable. He appears to be right."[37] The separate sheet reads as follows:

> We admit as legitimate mathematics certain reflections on the grammar of a language that concerns the empirical. If one seeks to formalize such a mathematics, then with each formalization there are problems, which one can understand and express in ordinary language, but cannot express in the given formalized language. It follows (Brouwer) that mathematics is inexhaustible: one must always again draw afresh from the fountain of

---

[32] Gödel would refer to the latter as "absolute unsolvability" in the 1930s. See Kennedy and van Atten (2004) and its revision van Atten and Kennedy (2009a) for a discussion of this notion in Gödel's writings.

[33] A point Warren Goldfarb has made in his (Goldfarb, 2005).

[34] See Dawson (1997).

[35] See van Atten and Kennedy (2009b) for details of the lectures, for Gödel's reactions to them, as well as for the evidence that he attended at least the second of them. See also Wang (1987, 84).

[36] Gödel turned in his thesis in July 1929.

[37] See Wang (1987, 84).

intuition. There is, therefore, no characteristica universalis for the whole mathematics, and no decision procedure for the whole mathematics. In each and every closed language there are only countably many expressions. The continuum appears only in the whole of mathematics . . . If we have only one language, and can only make elucidations about it, then these elucidations are inexhaustible, they always require some new intuition again.

Thus it appears that Brouwer had arrived well before 1931 at an at least informal form of what might be thought of as a diagonal argument, but for languages. In fact, as van Atten and Kennedy point out in (2009a), Brouwer had already remarked in his 1907 thesis that "the totality of all possible mathematical constructions is 'denumerably unfinished'; by this he meant that 'we can never construct in a well-defined way more than a denumerable subset of it, but when we have constructed such a subset, we can immediately deduce from it, following some previously defined mathematical process, new elements which are counted to the original set.'"[38] Brouwer is directly inspired by Cantor's theorem here and in the subsequent lecture, but his observation is nonetheless a step forward from Cantor. Brouwer's generalized version of the undenumerability theorem is a diagonal argument for (a countable) language. One might think that the step taken from this web of ideas to the first incompleteness theorem, is not wholly impossible.

As for other anticipations of the incompleteness theorem, Hilbert conceives of the possibility explicitly in his 1928 Bologna lecture. Post's 1921 conjecture that "a complete symbolic logic is impossible" should also be cited.[39] Of course, what was missing from both Brouwer's and Post's anticipations, what prevented these anticipations from maturing into a full theorem, was any awareness that a completeness theorem was required. Skolem's remark in 1922[40] that the continuum hypothesis will be shown independent has been cited earlier. Finally, Kripke's interpretation[41] of a 1925 remark of Kuratowski to the effect that one cannot prove the existence of inaccessible cardinals in ZFC renders Kuratowski's insight tantalizingly close to the second incompleteness theorem.

Systematically, so to speak, whether Gödel recognized this explicitly or only sensed it, Gödel's 1929 thesis was precisely the place to bring up incompleteness. This is because only once the completeness theorem has been proved, is it meaningful to talk about unsolvability. Why? The completeness theorem shows that first-order logic is maximal, in the sense of being a proof system strong enough to prove any statement that holds in all models of a particular first-order theory. Therefore it becomes meaningful to ask, whether one can *now* decide every proposition in the language of the theory. If we lack a completeness theorem (as in the case of the full second-order logic), then there are consistent theories without a model. Proving an incompleteness theorem would be irrelevant. It is only in the context of a completeness theorem, when we know that every consistent theory has a model, that it becomes relevant.

---

[38] See van Atten and Kennedy (2009b); quoting in turn Brouwer (1907, 82).

[39] See Davis (1965, 416).

[40] See Skolem (1923).

[41] Address to the American Philosophical Association Eastern Division Meeting 2009, Philadelphia (Kripke 2009)

5.4 Back to Gödel's Thesis

We note a second moral Gödel draws from the completeness theorem in the third paragraph of his thesis. He observes[42] that one might raise the following objection: does not the use of the law of excluded middle in its proof "invalidate the entire completeness proof"? The completeness theorem asserts

> 'a kind of decidability," namely every quantificational formula is either provable or a counterexample to it can be given, whereas the principle of the excluded middle seems to express nothing other than the decidability of every problem.

Thus the proof may be circular: one uses the decidability of every question to prove just that assertion.[43] However, Gödel remarks, what he has shown is the provability of a valid formula from "completely *specified, concretely enumerated* inference rules,"[44] not merely from all rules imaginable, whereas the law of excluded middle is used informally (in the metalanguage, as one would say nowadays); the notion of decidability or solvability asserted by the law is left unspecified. As Gödel puts it:

> ... what is affirmed (by the law of excluded middle) is the solvability not at all through specified means but only through all means that are *in any way imaginable* ... [45]

Gödel tells us that the assistance provided by the completeness theorem is that if we assume solvability by all means imaginable, then we have, in the case of a sentence of first-order predicate calculus, a reduction to solvability by very specific means laid out beforehand.

5.5 Conclusion

The paragraphs in which these two observations occur were not included in the article in *Monatshefte* based on Gödel's thesis.[46] Possibly Gödels advisor Hans Hahn, whom Gödel thanks in its acknowledgments, suggested that Gödel leave it out. One could also speculate that Gödel was worried by October 1929, when he submitted the article, that those remarks, dwelling as they do on incompleteness, would draw too much attention to the possibility of a theorem along those lines; but in fact we do not know the reason these remarks were omitted.

On the philosophical (rather than the historical) side of things, one may ask why Gödel felt pressed to argue in his thesis against the idea that consistency implies existence? In other words, what is this argument doing in his thesis in the first place? Apart from the speculation that Gödel may have wanted to mention what may have

---

[42] See Gödel (1986, 63).

[43] This is the content of Brouwer's "Third Insight," as expressed in his "intuitionistische Betrachtungen über den Formalismus." See Mancosu (1998, 40) for an English translation.

[44] emphasis Gödel's

[45] Gödel remarks in a footnote to this passage that the notion of provability by any means imaginable is perhaps "too sweeping." Nevertheless, this does not affect the basic distinction that Gödel wishes to make between the formal and informal notions of provability.

[46] Gödel *1930* in Gödel (1986, 102).

been very much on his mind by then, namely, unsolvability, one could also speculate that Gödel may have worried that what the completeness theorem precisely does is to *verify* the principle that consistency implies existence; if a theory is consistent, Gödel shows, it has a model. However, if this is so, then the presence of mutually incompatible extensions of, say, set theory opens the door to a relativistic notion of mathematical truth, a view Gödel already claimed to have been at odds with in 1923.[47]

What is the continuing significance of the theorem? The debate about whether second-order logic is the natural way to formalize mathematics has, if anything, intensified in recent decades. What Gödel points to in the first paragraph of his thesis is the idea that for logics that lack a completeness theorem, the proof concept is in some sense not finalized. Therefore the concept of consistency is in some sense also not finalized. Of course, one only needs soundness to define consistency in the second-order context; one does not want to say that the proof concept for second-order logic is not completely clear. In fact, logics that do have a completeness theorem, such as first-order logic, might even be thought of as vague, in the sense that the discourse does not fix a unique model, in general, but rather refers to a group of nonstandard models. However this is misleading. For first-order logic, the proof concept is in a complete state of analysis. If a sentence holds in every model, then it is provable. On the other hand, if a sentence is provable, then we know the models in which it holds – all of them.

We refer the reader to the by now massive literature on the subject and conclude with an observation along this line from Andreas Blass[48]: what the completeness theorem precisely shows, Blass asserts, is that if a first-order sentence holds in all models, it must be because of a uniform reason (proof) and not for an accidental reason that happens in each model or in each case for a different reason.

Blass is noticing that the first-order conception of logic eliminates anything arbitrary about the first-order concept of truth – a quintessentially Gödelian thought.

References

Awodey, S., and Carus, A. W. (2001). Carnap, completeness, and categoricity: the Gabelbarkeitssatz of 1928. *Erkenntnis*, **54**, 145–72.

———. (2006). Carnap and Gödel. *Synthese*, DOI 10.1007/s11229-006-9066-4.

Awodey, S., and Reck, E. H. (2002). Completeness and categoricity. I. Nineteenth-century axiomatics to twentieth-century metalogic. *Hist. Philos. Logic*, **23**, 1–30.

Blass, A. (1998). Some semantic aspects of linear logic. *Journal of the Interest Group in Pure and Applied Logic*, **5**, 115–26.

Brouwer, L. E. J. *On the Foundations of Mathematics*. Dissertation, 1907. Reprinted in: Brouwer, L. E. J. Collected works. Vol. 1. Philosophy and foundations of mathematics. Edited by A. Heyting. North-Holland Publishing Co., Amsterdam-Oxford; American Elsevier Publishing Co., Inc., New York, 1975. xv+628 pp.

---

[47] Grandjean questionaire, see Dawson (1987).
[48] See "Some Semantical Aspects of Linear Logic," (Blass, 1998).

Carnap, R. (2000). *Untersuchungen zur allgemeinen Axiomatik*. Edited by T. Bonk and J. Mosterin. Darmstadt, Germany: Wissenschaftliche Buchgemeinschaft.

Davis, M. (1965). *The Undecidable: Basic Papers on Undecidable Propositions, Unsolvable Problems and Computable Functions*. New York: Raven Press.

Dawson, J. W., Jr. (1997). *Logical Dilemmas*. Wellesley, MA: A K Peters.

Frege, G. (1980). *Philosophical and Mathematical Correspondence*. Chicago: University of Chicago Press.

Gödel, K. (1986). *Collected Works. Vol. I: Publications 1929–1936*. Edited by S. Feferman et al. Oxford: Oxford University Press.

———. (2003). *Collected Works. Vol. IV: Correspondence A–G*. Edited by S. Feferman et al. Oxford: Oxford University Press.

Goldfarb, W. (2005). On Gödel's Way in: The Influence of Rudolf Carnap. *The Bulletin of Symbolic Logic*, **11**, 185–193.

Henkin, L. (1950). Completeness in the theory of types. *Journal of Symbolic Logic*, **15**, 81–91.

Huntington, E. V. (1902). A complete set of postulates for the theory of absolute continuous magnitude. *Transactions of the American Mathematical Society*, **3**, 264–79.

Kennedy, J., and van Atten, M. (2004). Gödel's modernism: On set-theoretic incompleteness. *Graduate Faculty Philosophy Journal*, **25**, 289–349.

Malcev, A. I. (1971). *The Metamathematics of Algebraic Systems. Collected Papers: 1936–1967*. Amsterdam, Netherlands: North-Holland.

Mancosu, P. (1998). *From Brouwer to Hilbert*. New York: Oxford University Press.

Mancosu, P., Zach, R., and Badesa, C. (2009). The development of mathematical logic from Russell to Tarski: 1900–1935. In *The Development of Modern Logic*, ed. L. Haaparanta. New York: Oxford University Press, Chapter 9, pp. 318–470.

Sieg, W. (1999). Hilbert's programs 1917–1922. *The Bulletin of Symbolic Logic*, **5**, 1–44.

———. (2002). Beyond Hilbert's Reach. In *Logicism, intuitionism, and formalism*, Synthese Library, **341**, 449–483, Berlin: Springer, Dordrecht, 2009.

Skolem, T. (1933). Eine bemerkung über gewisse ringe mit anwendung auf die produktzerlegung von polynomen. *Norsk Matematisk Forenings Skrifter*, **10**, 73–82.

———. (1923). Einige bemerkungen zur axiomatischen begründung der mengenlehre. *Matematikerkongressen i Helsingfors 4–7 Juli 1922, Den femte skandinaviska matematikerkongressen, Redogörelse*. Printed by Akateeminen Kirjakauppa (Helsinki) 1923, pp. 217–232.

van Atten, M. (2005). On Gödel's awareness of Skolem's Helsinki lecture. *History and Philosophy of Logic*, **26**, 321–26.

van Atten, M., and Kennedy, J. (2009a). Gödel's logic. In *Logic from Russell to Gödel*, ed. D. Gabbay and J. Woods. Chapter 10 *Handbook of the History of Logic*. Amsterdam, Netherlands: Elsevier.

———. (2009b). Gödel's modernism: On set-theoretic incompleteness, revisited. In *Logicism, Intuitionism, and Formalism; What Has Become of Them?* ed. S. Lindström, E. Palmgren, K. Segerberg, and V. Stoltenberg-Hansen. *Synthese Library 341*. Berlin: Springer.

Wang, H. (1987). *Reflections on Kurt Gödel*. Cambridge, MA: MIT Press.

CHAPTER 6

Lieber Herr Bernays! *Lieber Herr Gödel!* Gödel on Finitism, Constructivity, and Hilbert's Program

Solomon Feferman

6.1 Gödel, Bernays, and Hilbert

The correspondence between Paul Bernays and Kurt Gödel is one of the most extensive in the two volumes of Gödel's *Collected Works* (1986–2003) devoted to his letters of (primarily) scientific, philosophical, and historical interest. Ranging from 1930 to 1975, except for one long break, this correspondence engages a rich body of logical and philosophical issues, including the incompleteness theorems, finitism, constructivity, set theory, the philosophy of mathematics, and post-Kantian philosophy. In addition, Gödel's side of the exchange includes his thoughts on many topics that are not expressed elsewhere and testify to the lifelong warm, personal relationship that he shared with Bernays. I have given a detailed synopsis of the Bernays-Gödel correspondence, with explanatory background, in my introductory note to it in *CW IV* (pp. 41–79).[1] My purpose here is to focus on only one group of interrelated topics from these exchanges, namely, the light that this correspondence – together with assorted published and unpublished articles and lectures by Gödel – throws on his perennial preoccupations with the limits of finitism, its relations to constructivity, and the significance of his incompleteness theorems for Hilbert's program.[2] In that connection, this piece has an important subtext, namely, the shadow of Hilbert that loomed over Gödel from the beginning to the end of his career.

Let me explain. Hilbert and Ackermann (1928) posed the fundamental problem of the completeness of the first-order predicate calculus in their logic text; Gödel (1929) settled that question in the affirmative in his dissertation a year later.[3] Also in 1928, Hilbert (1931a) raised the problem of the completeness of arithmetic in his Bologna address; Gödel settled *that* in the negative another year later, in the strongest possible way, by means of his first incompleteness theorem (Gödel 1931): no consistent formal axiomatic extension of a system that contains a sufficient amount of arithmetic is complete. Both of these deserved Hilbert's approbation, but not a word passed from him in public or in writing at the time. In fact, Hilbert and Gödel never met and never communicated. Perhaps the second incompleteness theorem on the unprovability of the consistency of a system (Gödel 1931) took Hilbert by surprise. We don't know exactly what he made of it, but we can appreciate that it might have been quite disturbing, for he had invested a great deal of thought and emotion into his finitary consistency program, which became problematic as a result. He made just one comment about it, of a dismissive character, four years later; I will return to that below.

The primary link between Gödel and Hilbert was Bernays, Hilbert's assistant in Göttingen from 1917 to 1922 and then his junior colleague until 1934, when Bernays was forced to leave Germany because of his Jewish origins. It was in this period that the principal ideas of Hilbert's consistency program and of his *Beweistheorie* to carry it out were developed and later exposited in the two-volume opus *Grundlagen der Mathematik* (Hilbert and Bernays 1934, 1939) whose preparation was carried out entirely by Bernays . It was Bernays who first wrote Gödel in 1930, complimenting him on the completeness theorem for first-order logic and asking about his incompleteness theorems, and it was Bernays who did what apparently Hilbert did not, namely, puzzle out the proofs and significance of the incompleteness theorems through the landmark year of 1931. Then it was Bernays for whom the results were later decisive in the preparation of volume 2 of Hilbert and Bernays (1939).

I will elaborate, but first, some biographical information about Bernays (1976a) is in order, drawn mainly from his short autobiography. He was born in 1888 in London, from where the family soon moved to Paris and, a little later, to Berlin. There, at the university, Bernays began his studies in mathematics with Landau and Schur; he then followed Landau to the University of Göttingen, where he also studied with Hilbert, Weyl, and Klein. Bernays completed a doctorate on the analytic number theory of quadratic forms under Landau's direction in 1912. From Göttingen, he moved later that year to the University of Zürich, where he wrote his *Habilitationschrift* on analytic function theory and became a *Privatdozent* (unsalaried university lecturer). In 1917, Hilbert came to Zürich to deliver his famous lecture, "Axiomatisches Denken." Having resumed his interest in foundational problems, Hilbert invited Bernays to become his assistant in Göttingen to work with him on those questions. In 1918, Bernays began his work in logic with a second *Habilitation* thesis, "On the Completeness of the Propositional Calculus and the Independence of Its Axioms."

As Hilbert's assistant in Göttingen from 1917 to 1922, Bernays was significantly involved in helping Hilbert develop and detail his ideas about mathematical logic and the foundations of mathematics; by the end of that period, these explorations had evolved into Hilbert's program for finitary consistency proofs of formal axiomatic systems for central parts of mathematics. At Hilbert's urging, Bernays was promoted to the (nontenured) position of Professor Extraordinarius at Göttingen in 1922. He held this position until 1933, when he – as a non-Aryan – was forced by the Nazis to give up his post, but Hilbert kept him on at his own expense for an additional six months. In spring 1934, Bernays returned to Zürich, where he held a permanent position at the Eidgenössische Technische Hochschule (or ETH; Swiss Federal Institute of Technology). (Treated as a temporary position for many years, it finally turned into a regular position – albeit only as a Professor Extraordinarius – in 1945.) Following the 1934 move from Göttingen, Bernays had virtually no communication with Hilbert but continued his work on the two volumes of the *Grundlagen der Mathematik* (Hilbert and Bernays 1934, 1939); again, Bernays was completely responsible for their preparation.

This was the first full exposition of Hilbert's finitary consistency program and beyond, presented in a masterful, calm, and unhurried fashion.[4] Alongside that, beginning in 1930, Bernays was developing his axiomatic system of sets and classes as a considerable improvement on the axiomatization of von Neumann; this was eventually published in the *Journal of Symbolic Logic* in seven parts (Bernays, 1937–1954).

Bernays' distinctive voice in the philosophy of mathematics began to emerge early in the 1920s with a pair of articles on the axiomatic method and an early version of Hilbert's program; the long paper Bernays (1930) is on the significance of Hilbert's proof theory for the philosophy of mathematics. However, in a number of pieces from then on, he distanced himself from Hilbert's strictly finitist requirements for the consistency program and expressed a more liberal and nuanced receptiveness to alternative foundational views, including a moderate form of Platonism.[5] This was reinforced through his contact at the ETH in Zürich with Ferdinand Gonseth, who held an open philosophy that rejected the possibility of absolute foundations of mathematics or science. With Gonseth and Gaston Bachelard, he founded the journal *Dialectica* in 1945.[6]

Bernays visited the Institute for Advanced Study (IAS) in Princeton in 1935–1936, and again in 1959–1960, during which period, he had extensive contact with Gödel. As it happens, it was my good fortune to be at the institute that same year and to make Bernays' acquaintance then. I was two years out from a PhD at the University of California, Berkeley, with a dissertation on the arithmetization of metamathematics and – like so many logicians of those days – had been drawn to the IAS by the chance to confer directly with Gödel and benefit from his unique insights. I had by then gone on to establish my main results on transfinite progressions of theories, extending Turing's earlier work on ordinal logics.[7] Among other visitors that same year were Kurt Schütte and Gaisi Takeuti (my office mate at the IAS), to whom – along with Georg Kreisel in his Stanford days – I am indebted for my way into proof theory. However, that's another story.[8]

Bernays was the most senior of the visitors in logic that year; he was then age seventy-one, compared with Gödel's fifty-three. I knew little of his work at the time, other than as the coauthor of the *Grundlagen der Mathematik* and for the development of his elegant theory of sets and classes. A gentle, modest man, he did not advertise the range of his thoughts and accomplishments; it was only later that I began truly to appreciate his place in the field of logic, and it was only much more recently, when working on his correspondence with Gödel for *CW IV*, that I learned of the depth of the personal and intellectual relationship between the two of them.

Here, for example, is a touching letter that Bernays wrote to Gödel on April 24, 1966, testifying to the highest intellectual and personal esteem in which Bernays held him:

Dear Mr. Gödel,
    I feel quite embarrassed, since I had not prepared anything suitable for the *Festschrift* in honor of your 60th birthday – all the more so, as you made such a significant contribution to the *Dialectica* issues for my 70th birthday.
    In any case, I would like now to express my cordial good wishes on your [just] completed decade of life.

In view of the situation in foundational investigations, you can certainly ascertain with much satisfaction that the discoveries and methods you brought to metamathematics are dominant and leading the way in the research of today. May it be granted to you also in the future to influence the direction of this research in a way that is fruitful.

Yet the foundations of mathematics are of course only one of the concerns of your research; and I would also like to wish that your philosophical reflections may turn into such results that you are induced to publish them.

Last [but] not least, I wish in addition that your general state of health in the coming years be as satisfying as possible. Hopefully you can celebrate your 60th birthday quite beautifully and take pleasure in the certainly impressive statement of the general appreciation of your intellectual work.

With very cordial greetings, also to your wife,

Yours truly,

Paul Bernays (*CW IV*, 251)

In addition to his visit at the IAS in 1959–1960, Bernays paid visits to Gödel during three stays when Bernays was a visiting professor at the University of Pennsylvania between 1956 and 1965. Bernays died in Zürich in September 1977, just one month shy of his eighty-ninth birthday and four months before Gödel's death in January 1978.

6.2 The Year 1931: The Incompleteness Theorems and Hilbert's ω-Rule

The correspondence between Bernays and Gödel begins with a letter from Bernays dated Christmas Eve, 1930, complimenting Gödel on the completeness theorem for first-order logic Gödel (1930) and then asking to see his "significant and surprising results" in the foundations of mathematics, namely, the incompleteness theorems, of which he had heard from Courant and Schur. Gödel sent Bernays one set of proof sheets forthwith. The first four items of correspondence between them in 1931 in *CW IV* are largely devoted to Bernays' struggles to understand the incompleteness theorems against the background of ongoing work on the consistency program in the Hilbert school.[9] Earlier, Ackermann (1924), and then (in revised form) von Neumann (1927), had supposedly given a finitary proof of the consistency of a formal system Z for classical first-order arithmetic, nowadays called Peano arithmetic (PA). It was not yet realized that their proof succeeds only in establishing the consistency of the weak subsystem of Z in which induction is restricted to quantifier-free formulas.

Part of Bernays' perplexities with Gödel's incompleteness theorems had to do with the roughly concurrent work of Hilbert (1931a, 1931b), in which a kind of finitary version of the infinitary ω-rule extending Z to a system Z^* was proposed as a means of overcoming the incompleteness of Z. Roughly speaking, the rule allows one to adjoin a sentence $\forall x A(x)$ (*A* being quantifier-free) to the axioms of Z^* for which it has been shown finitarily that each instance $A(n)$ for $n \in \omega$ is already provable in Z^*. Hilbert claimed to have a finitary consistency proof of Z^*, relying mistakenly on the work of Ackermann and von Neumann for Z. No reference to Gödel's incompleteness theorem appears in these articles. What would have led Hilbert to try to overcome the incompleteness of Z if he was not already aware of it? True, he was already lecturing on the proposed extension of Z in December 1930, before he and Bernays even saw the proof sheets of Gödel's paper. On the other hand, Hilbert could well have learned of Gödel's incompleteness theorem before that from von Neumann, who heard about it in September of that year. Others who might have communicated its essence to him were Courant and Schur. This is not a place to go into all the relevant details.[10] Bernays' own two reports on the matter don't quite jibe: in his survey article on Hilbert's foundational contributions for Hilbert's (1935) collected works, Bernays said that even before Gödel's incompleteness theorems, Hilbert had given up the original form of his completeness problem and in its place had taken up the ideas for an extension of Z by a finitary ω-rule. However, in a letter thirty years later to Hilbert's biographer, Constance Reid, Bernays wrote that Hilbert was angry about doubts that he (Bernays) had already expressed about the conjectured completeness of Z and was then angry about Gödel's results.[11]

In his correspondence with Bernays, Gödel points out things that he did not mention in his review (*1931c*) of Hilbert's article on Z*, namely, that even with the proposed ω-rule, the system is incomplete. He went on to write:

> I do not think that one can rest content with the systems Z*, Z** [a variant proposed by Bernays for "aesthetic reasons"] as a satisfactory foundation of number theory ... , and indeed, above all because in them the very complicated and problematical concept of "finitary proof" is assumed (in the statement of the rule for the axioms) without having been made mathematically precise in greater detail. (*CW IV*, 97)

This resonates with Gödel's statement to Carnap in May 1931 that he viewed the move to Z* as a step compromising Hilbert's program.[12] However, he did not make this fundamental criticism in his review of Hilbert (1931a) itself, as he might well have.

Gödel's own concerns with determining the limits of Hilbert's finitism are there from the beginning. In his watershed paper on formally undecidable propositions, after stating and sketching the proof of the second incompleteness theorem on the unprovability of the consistency of systems by their own means, Gödel writes:

> I wish to note expressly that [this theorem does] not contradict Hilbert's formalistic viewpoint. For this viewpoint presupposes only the existence of a consistency proof in which nothing but finitary means of proof is used, and it is conceivable that there exist finitary proofs that *cannot* be expressed in the formalism of [our basic system]. (Gödel, 1931a, *CW I*, 195)

Von Neumann – who was the first to grasp what Gödel had accomplished in his brief announcement of the first incompleteness theorem at the Königsberg conference in September 1930, and who independently realized the second incompleteness theorem – had been urging the opposite view on him. At any rate, Gödel came around to von Neumann's viewpoint by the end of 1933, when he delivered a lecture at a meeting in Cambridge, Massachusetts, during his first visit to the United States. That is the first of a series of three remarkable lectures that Gödel gave between 1933 and 1941, in which his thoughts about finitism, constructivity, and Hilbert's program took more definite form; these are the subjects of our next two sections. (All three lectures appeared in print for the first time in *CW III* as *1933o*, *1938*, and *1941*, respectively.)

In the same period following 1931, there was a long break in the extant correspondence between Gödel and Bernays. It did not resume again until 1939 and was then devoted almost entirely to set theory, especially to Gödel's proofs of the consistency of the axiom of choice and the generalized continuum hypothesis with the axioms of set theory. Initially presented in terms of the system of Zermelo-Fraenkel, for expository purposes, Gödel later adopted the system of sets and classes that Bernays had communicated to him in 1931.

Of incidental personal note is the change in salutations that also took place in 1931: where, in the first few letters, Bernays was addressed as "Sehr geehrter Herr Professor!" and Gödel as "Sehr geehrter Herr Dr. Gödel!" these now became "Lieber Herr Bernays!" and "Lieber Herr Gödel!" respectively, and so remained throughout their correspondence thenceforth.

6.3 The Year 1933: The Cambridge Lecture

On December 30, 1933, Gödel gave a lecture titled "The Present Situation in the Foundations of Mathematics" for a meeting of the Mathematical Association of America in Cambridge, Massachusetts. Gödel's aim in this lecture was clearly announced in the first paragraph:

> The problem of giving a foundation for mathematics (. . . [i.e.,] the totality of methods of proof actually used by mathematicians) can be considered as falling into two different parts. At first these methods of proof have to be reduced to a minimum number of axioms and primitive rules of inference, which have to be stated as precisely as possible, and then secondly a justification in some sense or other has to be sought for these axioms. (*Gödel*, *1933o*, *CW III*, 45)

He went on to assert that the first part of the foundational problem had been solved in a completely satisfactory way by means of formalization in the simple theory of types when all "superfluous restrictions" were removed. As he explained, that was accomplished by axiomatic set theory à la Zermelo, Fraenkel, and von Neumann, with its underlying cumulative type structure admitting a simple passage to transfinite types. Nevertheless, he said that set theory had three weak spots: "The first is connected with the non-constructive notion of existence. . . . The second weak spot, which is still more serious, is . . . the so-called method of impredicative definitions [of classes]. . . . The third weak spot in our axioms is connected with the axiom of choice" (*Gödel*, *1933o*, *CW III*, pp. 49–50). Consideration of these led Gödel directly to the following stunning pronouncement:

> The result of the preceding discussion is that our axioms, if interpreted as meaningful statements, necessarily presuppose a kind of Platonism, which cannot satisfy any critical mind and which does not even produce the conviction that they are consistent. (*Gödel*, *1933o*, *CW III*, 50)

Not that it was likely that the system was inconsistent, Gödel says, because it had been developed in so many different directions without reaching any contradiction. Given that, one might hope to prove the consistency of the system when treated in exact formal terms. However, not *any* proof would do: Gödel said that "it must be conducted by perfectly unobjectionable [constructive] methods; i.e., it must strictly avoid the non-constructive existence proofs, non-predicative definitions and similar things, for it is exactly a justification for these doubtful methods that we are now seeking." Even with this, he said, the nature of such a proof is not uniquely determined because there are different notions of constructivity and, accordingly, "different layers of intuitionistic or constructive mathematics" (*Gödel, *1933o, CW III*, p. 51).

Concerning the use of the word *intuitionistic* in this last quotation, it should be noted that according to Bernays (1967, 502), the prevailing view in the Hilbert school at the beginning of the 1930s equated finitism with intuitionism.[13] Within a few years, finitism was generally distinguished from intuitionism in the sense of the Brouwer school, in part through Heyting's formalization of intuitionistic arithmetic and Gödel's 1933 translation of the classical system of PA into Heyting arithmetic (HA). In this lecture, however, Gödel never used the words *finitary* or *finitistic*. He also did not speak explicitly about the Hilbert consistency program, except for one indirect reference, which follows below.

Gödel delineated the lowest level of constructive mathematics, which he called the system A, in the following terms:

1. The application of the notion of "all" or "any" is to be restricted to those infinite totalities for which we can give a finite procedure for generating all their elements (as we can, e.g., for the totality of integers by the process of forming the next greater integer and as we cannot, e.g., for the totality of all properties of integers).
2. Negation must not be applied to propositions stating that something holds for all elements, because this would give existence propositions. . . . Negatives of general propositions (i.e. existence propositions) are to have a meaning in our system only in the sense that we have found an example but, for the sake of brevity, do not wish to state it explicitly. I.e., they serve merely as an abbreviation and could be entirely dispensed with if we wished. From the fact that we have discarded the notion of existence and the logical rules concerning it, it follows that we are left with essentially only one method for proving general propositions, namely, complete induction applied to the generating process of our elements.
3. And finally, we require that we should introduce only such notions as are decidable for any particular element and only such functions as can be calculated for any particular element. Such notions and functions can always be defined by complete induction, and so we may say that our system [A] is based exclusively on the method of complete induction in its definitions as well as its proofs. (*Gödel *1933o, CW III*, 51)

Gödel did not spell out formally these conditions on the system A, and there has been considerable discussion about exactly how to interpret it. It is pretty clear that the formulas of A should be taken as universal generalizations of quantifier-free formulas built up from decidable atoms by the propositional operations. When dealing with the provable formulas, one can just as well take them to be quantifier-free by disregarding the initial universal quantifiers. Thus, at first sight, it seems that A should be interpreted as a form of primitive recursive arithmetic (PRA), the quantifier-free system that has the usual axioms for zero and successor, the defining equations for each primitive recursive function – in which the step from each function equational axiom to the next is given either explicitly or by a complete induction (ordinary recursion on one numerical variable) – and finally, a rule of induction on the natural numbers. Initially, Wilfried Sieg, Charles Parsons, William Tait, and I all took this interpretation of A for granted. However, both Sieg and Tait subsequently changed their minds, although for different reasons.[14] The main point raised by Sieg of what is at issue has to do with Gödel's (*1933o, 52) statement that the most far-reaching consistency result obtained by methods in accord with the principles of A is that given by the work of Herbrand (1931). Herbrand's theorem is for a (somewhat open-ended) system that goes beyond PRA by including such functions as the one due to those from Ackermann, given by a double-nested recursion. Nevertheless, it seems to me that it is still possible to construe the system A as being PRA if one interprets Gödel's remark as applying to the general form of Herbrand's argument rather than to the specific statement of his theorem.

At any rate, what Gödel (*1933o, CW III, 51–52) has to say about the potential reach of the system A in pursuit of constructive consistency proofs at least puts an upper bound to its strength:

> This method possesses a particularly high degree of evidence, and therefore it would be the most desirable thing if the freedom from contradiction of ordinary non-constructive mathematics could be proved by methods allowable in this system A. Also, as a matter of fact, all the attempts for a proof for freedom from contradiction undertaken by Hilbert and his disciples tried to accomplish exactly that. But unfortunately the hope of succeeding along these lines has vanished entirely in view of some recently discovered facts [namely, the incompleteness theorems].... Now all the intuitionistic proofs complying with the system A which have ever been constructed can easily be expressed in the system of classical analysis and even in the system of classical arithmetic, and there are reasons for believing that this will hold for any proof which one will ever be able to construct.
>
> ... So it seems that not even classical arithmetic can be proved to be non-contradictory by the methods of the system A.

This is the sole reference to Hilbert and his program in the Cambridge lecture, but there is no discussion there of Hilbert's finitist criterion for consistency proofs, let alone of Hilbert's conception of finitism.[15] This also returns us to the question of how to interpret A, for if it were exactly PRA, Gödel would no doubt have recognized that its consistency could be proved in PA, given his 1931 definition of the primitive recursive functions in arithmetic and the arguments of Herbrand (1931). At any rate, whatever the exact reach of the methods provided by A, it evidently appeared hopeless to Gödel by 1933 to carry out that program for the system PA of arithmetic, let alone for analysis and set theory.

Thus, in the final part of *1933o, Gödel took up the question of whether stronger constructive methods than those provided by the system A ought to be admitted to consistency proofs, in particular, "the intuitionistic mathematics as developed by Brouwer and his followers." That would take one at least up to arithmetic, given the reduction of PA to Heyting's intuitionistic system HA. The essential formal difference of A from HA is that in the latter, all first-order formulas are admitted to the language. However, Gödel is not at all satisfied with intuitionistic logic in pursuit of the consistency program because the meaning of its connectives and quantifiers is explained in terms of the (in his view) vague and unrestricted concept of constructive proof. In particular, a proof of an implication $p \rightarrow q$ is supposed to be provided by a construction that converts *any* proof of p into a proof of q, and a proof of $\neg p$ is supposed to be given by a construction that converts *any* proof of p into an absurdity or contradiction. This justifies, for example, $p \rightarrow \neg\neg p$ in intuitionistic logic because p and $\neg p$ constitute a contradiction. Intuitionistic logic does meet the preceding condition 2 to the extent that a proof of $\exists x P(x)$ is supposed to be a construction that provides an instance t for which $P(t)$ holds, but condition 2 is not met in its *stated* form because negation may be applied to universal statements in Heyting's formalism, and because $\neg \forall x P(x)$ is not intuitionistically equivalent to $\exists x \neg P(x)$, it cannot be considered as an abbreviation for the latter. However, Gödel's main objection to intuitionistic logic is that it does not meet condition 1 because "the substrate on which the constructions are carried out are proofs instead of numbers or other enumerable sets of mathematical objects. However, by this very fact they do violate the principle, which I stated before, that the word 'any' can be applied only to those totalities for which we have a finite procedure for generating all their elements.... And this objection applies particularly to the totality of intuitionistic proofs because of the vagueness of the notion of constructivity. Therefore this foundation of classical arithmetic by means of the notion of absurdity is of doubtful value" (*Gödel, *1933o, CW III*, p. 53).

6.4 Transitions: 1934–1941

Hilbert was unaffected by any of the reconsiderations of the possible limits to finitary methods in pursuit of his consistency program that had been stimulated by the second incompleteness theorem. In his preface to volume 1 of the *Grundlagen der Mathematik*, and in his sole reference anywhere to Gödel or his incompleteness theorems, Hilbert writes:

> This situation of the results that have been achieved thus far in proof theory at the same time points the direction for the further research with the end goal to establish as consistent all our usual methods of mathematics.
> With respect to this goal, I would like to emphasize the following: the view, which temporarily arose and which maintained that certain recent results of Gödel show that my proof theory can't be carried out, has been shown to be erroneous. In fact that result shows only that one must exploit the finitary standpoint in a sharper way for the farther reaching consistency proofs. (Hilbert and Bernays, 1934, v)[16]

Nevertheless, the need for a modified form of Hilbert's consistency program began to become generally recognized by others in his school by the mid-1930s, among them his collaborator on the *Grundlagen*. As Bernays (1967, 502) wrote years later, "it became apparent that the '*finite Standpunkt*' is not the only alternative to classical ways of reasoning and is not necessarily implied by the idea of proof theory. An enlarging of the methods of proof theory was therefore suggested: instead of a restriction to finitist methods of reasoning, it was required only that the arguments be of a constructive character, allowing us to deal with more general forms of inference."

One striking new specific way forward was provided by Gerhard Gentzen in his 1936 paper, in which the consistency of arithmetic is proved by transfinite induction up to Cantor's ordinal ε_0, otherwise using only finitary reasoning. In Hilbert's preface to volume 2 of the *Grundlagen der Mathematik* (Hilbert and Bernays, 1939), in which he thanks Bernays for carrying out the exposition and development of his ideas for proof theory and the consistency program, no mention is made of Gödel or Gentzen, even though the volume contains an extended exposition of Gödel's incompleteness theorems and a description of Gentzen's work in a section titled "Überschreitung des bisherigen methodischen Standpunktes der Beweistheorie."

Gödel reported on the issues of a modified consistency program in a remarkable although sketchy presentation that he made to Edgar Zilsel's seminar in Vienna in 1938, the notes for which were reconstructed from the *Gabelsberger* shorthand script as *Gödel* (*1938, *CW III*). He begins by pointing out that one can only deal with the consistency of partial systems of mathematics as represented in formal systems T, and this must be accomplished by other systems S. Moreover,

> [the choice of such a system S] has ... an epistemological side. *After all we want a consistency proof for the purpose of a better foundation of mathematics (laying the foundations more securely)*, and there can be mathematically very interesting proofs that do not accomplish that. A proof is only satisfying if it either
>
> (i) *reduces to a proper part* or
> (ii) *reduces to something which, while not a part, is more evident, reliable, etc., so that one's conviction is thereby strengthened.* (Gödel, *1938, CW III*, 89; emphasis original)

Although a reduction of kind (i) may be preferred because of its objective character, whereas (ii), by comparison, involves subjective judgments, historically, the latter is the route taken through the reduction of nonconstructive to constructive systems. However, because the concept of constructivity is hazy, it is useful to make a "framework definition, which at least gives necessary if not sufficient conditions" (*Gödel, *1938, CW III*, p. 91). That is provided by conditions like those for the system A in the Cambridge lecture, here broken up into four parts, of which the last is that "objects should be surveyable (that is denumerable)." However, now the system at the lowest level of constructivity is specifically referred to as finitary number theory, and Gödel raises the question, "how far do we get, or fail to get with finitary number theory?" (*Gödel, *1938, CW III*, p.93). His conclusion is that "transfinite arithmetic" (presumably PA) is not reducible to A.[17] It should be noted that the disagreements as to how the system A of *1933o should be interpreted have not been raised for Gödel's informal system of finitary number theory in the 1938 lecture; instead, all who have considered it agree that it may be formalized as PRA.

To go beyond what can be treated by finitary number theory, the fourth condition – that the objects be denumerable – is jettisoned in all three of the further constructive approaches considered at Zilsel's seminar. These are (not in the order presented by Gödel) the modal-logical route, the route of induction on transfinite ordinals, and the route of constructive functions of finite type. The modal-logical route means the intuitionistic logic of Brouwer and Heyting, for which a foundation of the notion of constructive proof underlying it is sought in terms of a modal-like operator B (for *Beweisbar*). After considering several possible conditions on B, Gödel's conclusion is that there is no reasonable way to carry this out and that "this [route] is the worst of the three ways" (*Gödel, *1938, CW III*, p. 103). The second route follows Gentzen's proof of consistency of PA by transfinite induction up to ε_0, concerning which Gödel indicates a new and more intuitive way why the proof works; however, few details are given.[18] As to the ordinal route itself, Gödel questions it on several grounds, the main one being its lack of direct evidence for the principle of transfinite induction up to ε_0, let alone for the ordinals that would be needed to establish the consistency of still stronger systems. The final route, by means of constructive functions of finite (and possibly transfinite) type, is barely indicated, but it is that approach that Gödel himself was to explore in depth, as we shall see shortly.

The Zilsel lecture notes conclude with the following quite interesting general assessment, which is worth quoting at length:

> I would like to return to the historical and epistemological side of the question and then ask (1) whether a consistency proof by means of the three extended systems has a value in the sense of laying the foundations more securely; (2) what is closely related, whether the Hilbert program is undermined in an essential respect by the fact that it is necessary to go beyond finitary number theory.
>
> To this we can say two things: (1) If the original Hilbert program could have been carried out, that would have been without any doubt of enormous epistemological value.... (a) Mathematics would have been reduced to a very small part of itself.... (b) Everything would have been reduced to a concrete basis, on which everyone must be able to agree. (2) As to the proofs by means of the extended finitism, the first [i.e., (a)] is no longer the case at all... [whereas] the second [i.e., (b)] (reduction to the concrete basis, which means increase of the degree of evidence) obtains for the different systems to different degrees, thus for the modal-logical route not at all, for the higher function types the most, [and] for the transfinite ordinal numbers... also to a rather high degree. (*Gödel, *1938, CW III*, 113)

In April 1941, Gödel (**1941, CW III*) delivered a lecture titled "In What Sense Is Intuitionistic Logic Constructive?" at Yale University. He had by then become established as a member of the IAS, having fled Austria (and its threatened conscription) with his wife at the last minute at the beginning of 1940.[19] In the Yale lecture, he gave the first public account of what was later to be called the *Dialectica* interpretation of HA in a quantifier-free system of functionals of finite type – the route that he considered to have the greatest degree of evidence (beyond that of finitism) of the three considered at Zilsel's seminar. His stated aim there is that "if one wants to take constructivity in a really strict sense [then] the primitive notions of intuitionistic logic cannot be admitted in their usual sense. This however does not exclude the possibility of defining in some way these notions in terms of strictly constructive ones and then proving the logical axioms which are considered as self-evident by the intuitionists. It turns out that this can actually be done in a certain sense, namely, not for intuitionistic logic as a whole but for its applications in definite mathematical theories (e.g. number theory)" (*Gödel, *1941, CW III*, 191). The functional interpretation proposed as the means to carry this out is the subject of the final body of correspondence between Gödel and Bernays, taken up in the next section.

As already mentioned, set theory was the main subject of the correspondence between Gödel and Bernays in a flurry of letters that began in 1939 and continued for three years (*CW V*, 114–41). The last one from Bernays in this period is dated September 7, 1942. In it, no mention is made of the war, except to remark on its mild effects on the Swiss. Then Bernays – a bachelor – charmingly concludes his letter as follows:

> Hopefully you are now well settled in Princeton, and married life and the domesticity associated with it is quite salubrious for your physical and emotional health, and thereby also for your scientific work.
>
> Would you please convey my respects to your wife, even though I am not personally acquainted with her.
>
> Friendly greetings to you yourself. (*CW V*, 141)

After 1942, the correspondence between the two lapsed for fourteen years; no doubt the difficulties of transmission during the war were the initial reason. However, once resumed in 1956, the correspondence was to continue in a steady stream until the final exchange in 1975. The date 1956 marked the first postwar visit of Bernays to the United States and his first stay at the University of Pennsylvania, during which he was able to come to Princeton to renew his personal contact with Gödel.[20]

6.5 The *Dialectica* Interpretation

In 1958, in honor of Bernays' seventieth birthday, Gödel published in the journal *Dialectica* his interpretation of intuitionistic number theory – and, thereby, classical number theory – in a quantifier-free theory of primitive recursive functions of finite type; this spelled out the notions and results of what had been presented in the Yale lecture in 1941, and it subsequently came to be known as Gödel's *Dialectica* interpretation (Gödel, 1958, *CW II*). The change in title from the Yale lecture, "In What Sense Is Intuitionistic Logic Constructive?" to that of the *Dialectica* paper, "On a Hitherto Unutilized Extension of the Finitist Standpoint" ("Über eine noch nicht benützte Erweiterung des finiten Standpunktes"), indicates both a change of focus and a more precise attention to the bounds for finitism, at least in Hilbert's sense. This is reinforced by the opening paragraph of the *Dialectica* piece:

> P. Bernays has pointed out on several occasions that, since the consistency of a system cannot be proved using means of proof weaker than those of the system itself, it is necessary to go beyond the framework of what is, in Hilbert's sense, finitary mathematics if one wants to prove the consistency of classical mathematics, or even that of classical number theory. Consequently, since finitary mathematics is defined as the mathematics in which evidence rests on what is *intuitive*, certain *abstract* notions are required for the proof of the consistency of number theory. . . . In the absence of a precise notion of what it means to be evident, either in the intuitive or in the abstract realm, we have no strict proof of Bernays' assertion; practically speaking, however, there can be no doubt that it is correct. (Gödel, 1958, *CW II*, 241)

Seven years after the publication of the *Dialectica* paper, Bernays told Gödel of a plan to publish in the same journal an English translation that had been made of it by Leo F. Boron. However, Gödel was not happy with certain aspects of both the original and its translation and set out to revise them. A year later, he changed his mind and decided instead to improve and amplify the original by means of a new series of extensive footnotes. Preparation of these dragged on for four more years, and it was only after much help and encouragement from Bernays and Dana Scott that a revised manuscript was sent to the printer in 1970. However, when the proof sheets were returned, Gödel was again dissatisfied, especially with two of the added notes. Although he apparently worked on rewriting these until 1972, the paper was never returned in final form for publication. The corrected proof sheets found in his *Nachlass* (Firestone Library, Princeton University) were reproduced for the first time in *CW III*, where they appear as *Gödel* (1972). The full story of the vicissitudes of that paper is told by A. S. Troelstra in his introductory note to *Gödel* (*1958, 1972*) in that volume.

The question of the bounds on Hilbert's finitism comes up repeatedly in the correspondence from this period; in letter 61 from Gödel to Bernays of January 1967, he wrote:

> My views have hardly changed since 1958, except that I am now convinced that ε_0 is a bound on Hilbert's finitism, not merely in practice [but] in principle, and that it will also be possible to prove that convincingly. (*CW IV*, 255)

During the same period, Bernays had been working on a second edition of the *Grundlagen der Mathematik*. Its first volume eventually appeared in 1968, and the second appeared in 1970. As it happens, the latter was to contain a new supplement with an exposition of proofs due to Kalmár and Ackermann – subsequent to Gentzen's – of the consistency of the system of Peano arithmetic by means of transfinite induction on the natural ordering of order-type ε_0. In connection with that, Bernays developed a new and (on the face of it) more perspicuous proof of induction up to ε_0 to be included in the supplement, and he sent that proof along with his letter (67) to Gödel of January 1969. As Bernays puts it there, what he established with this proof was "the weak form of induction, which says that every decreasing sequence comes to an end after finitely many steps," that is, that the relation is well founded.[21] Gödel became quite excited about Bernays' proof, and in July 1969, he prepared a draft of a letter (68a) in which he wrote:

> You undoubtedly have given the most convincing proof to date of the ordinal-number character of ε_0. . . . Since functions of the first level [i.e., sequences of ordinals] can be interpreted as free choice sequences and that concept is obviously decidable, a statement of the form 'For all free choice sequences . . .'" contains no intuitionistic implication, and you have consequently completely eliminated the intuitionistic logic. If one reckons choice sequences to be finitary mathematics, your proof is even finitary. . . . I now strongly doubt whether what was said about the boundaries of finitism [at the beginning of 1958] is really right. For it now seems to me, after more careful consideration, that choice sequences are something concretely evident and therefore are finitary in Hilbert's sense, even if Hilbert himself was perhaps of another opinion.[22] (*CW IV*, 269)

Then Gödel included a draft remark to a footnote of the revised version of 1958 to this effect, adding:

> Hilbert did not regard choice sequences . . . as finitary, but this position may be challenged on the basis of Hilbert's own point of view.[23] (*CW IV*, 269)

Indeed, the idea of "[free] choice sequences" used here is due to Brouwer in a form that is quintessentially intuitionistic and generally regarded as nonfinitary in nature. This is in one of Gödel's letters that is marked "*nicht abgeschickt*" (not sent).

In the letter of July 25, 1969, that Gödel *actually* sent (68b), he changed his mind about the significance of Bernays' proof, although he still regarded it as "extraordinarily elegant and simple":

> At first one also has the impression that it comes closer to finitism than the other proofs. But on closer reflection that seems very doubtful to me. The property of being "well-founded" contains two quantifiers after all, and one of them refers to all number sequences (which probably are to be interpreted as choice sequences). In order to eliminate the quantifiers . . . one would use a nested recursion. . . . But nested recursions are not finitary in Hilbert's sense (i.e. not intuitive). . . . Or don't you believe that? (*CW IV*, 271)[24]

What is at issue in all this for Gödel goes back to his letter of 1931 to Bernays in which he said that "the complicated and problematical concept of 'finitary proof'" needs to be made mathematically precise to decide such questions. In 1969, two such characterizations of finitism were on offer, the first due to Georg Kreisel some ten years before that, in terms of an ordinal logic whose limit is exactly ε_0, that is, the strength of PA (Kreisel, 1960). Gödel had discussed this with both Kreisel and Bernays and given it serious consideration but was equivocal about the conclusion. A second proposed characterization that arrived at PRA as the upper bound of finitism was sketched by William Tait (1968) in his article "Constructive Reasoning" (later spelled out in his article "Finitism"; Tait, 1981). As we have seen, that is the formal interpretation of the system A of finitary number theory indicated by Gödel at the Zilsel seminar in 1938, but no reference to Tait's proposal appears in the correspondence with Bernays.

In any case, only the first of these could be considered an explication of finitism in Hilbert's sense, which unquestionably went beyond PRA. In letter 40 of August 1961, Gödel wrote to Bernays that he had had interesting discussions with Kreisel about his work and that "he now really seems to have shown in a mathematically satisfying way that the first ε-number is the precise limit of what is finitary. I find this result very beautiful, even if it will require a phenomenological substructure in order to be completely satisfying" (*CW IV*, 193). The characterization of finitist proof in Kreisel (1960) was given in terms of a transfinite sequence of proof predicates for formal systems $\sum_\alpha$, or ordinal logics in the sense of Turing (1939), under the restriction that the ordinal stages α to which one may ascend are controlled autonomously; that is, there must be, for each such α, a recognition at an earlier stage β that the iteration of the process α times is (finitarily) justified. Kreisel's main results are that the least nonautonomous ordinal is ε_0, and the provably recursive functions of the union of the $\sum_\alpha$ for $\alpha < \varepsilon_0$ are exactly the same as those of PA. Both the description of the ordinal logic for the proposed characterization and the proofs of the main results are very sketchy (as acknowledged by Kreisel though excused by limitations of space); full details, although promised, were never subsequently published.[25] In addition, Kreisel offered little in the way of convincing arguments to motivate his proposed explication of the informal concept of finitist proof; this was perhaps the reason for Gödel's statement that more would be needed to make it "completely satisfying."[26]

Kreisel's (1960) proposal had actually been made at the 1958 meeting of the International Congress of Mathematicians, and Gödel had already been cognizant of it at that time. It is referred to in footnote 4 to the 1958 version of the *Dialectica* paper, where Gödel says that a possible extension of the "original finitary standpoint . . . consists in adjoining to finitary mathematics abstract notions that relate, in a combinatorially finitary way, only to finitary notions and objects, and then iterating this procedure. Among such notions are, for example, those that are involved when we reflect on the content of finitary formalisms that have already been constructed. A formalism embodying this idea was set up by G. Kreisel." Just what Gödel has in mind here by the "original finitary standpoint" is not clear, but whether he intends it to mean a system like PRA or a system for Hilbert's finitism in practice, either would put its strength well below that of PA.

These matters are revisited in *Gödel (1972)*; in its opening paragraph, finitary mathematics in Hilbert's sense is now defined as "the mathematics of *concrete intuition*" instead of as "the mathematics in which evidence rests on what is *intuitive*," as appeared in *Gödel (1958)*. The relevant footnote 4 now reads:

> Note that an adequate proof-theoretic characterization of concrete intuition, in case this faculty is *idealized* by abstracting from the practical limitation, will include induction procedures which *for us* are *not* concretely intuitive and which could very well yield a proof of the inductive inference for ε_0 or larger ordinals. Another possibility of extending the original finitary viewpoint for which the same comment holds consists in considering as finitary any abstract arguments which only reflect . . . on the content of finitary formalisms constructed before, and iterate this reflection transfinitely, using only ordinals constructed in previous stages of this process. A formalism based on this idea was given by G. Kreisel [1960]. (Gödel, 1972, *CW II*, 274)

However, there is now a *further* footnote, f, to this in *1972*, in which Gödel says that "Kreisel wants to conclude from [the fact that the limit of his procedure is ε_0] that ε_0 is the exact limit of idealized concrete intuition. But his arguments would have to be elaborated further to be fully convincing."

6.6 What, Really, Were Gödel's Views on Finitism and the Consistency Program?

In the preceding, I have concentrated primarily on summarizing the available evidence concerning Gödel's views on finitism and Hilbert's program and its constructive extensions without critically examining those views themselves. Just what were those views, and what was their significance for Gödel? There are two main questions, both difficult: first, were Gödel's views on the nature of finitism stable over time, or did they evolve or vacillate in some way? Second, how do Gödel's concerns with the finitist and constructive consistency programs cohere with the Platonistic philosophy of mathematics that he supposedly held from his student days? Our problem is that the evidence is relatively fragmentary and, except for the published version of the *Dialectica* article, for the most part comes from lectures and correspondence that Gödel himself did not commit to print.

With respect to the first of these questions, in my introductory note to the correspondence with Bernays, I spoke of "Gödel's unsettled views over the years as to the exact upper bound of finitary reasoning," but this characterization has been challenged. In Tait's trenchant essay review of *CW IV* and *V*, he takes me and others to task about such judgments, in contrast to his own interpretation of the available evidence:

> In any case, the alternative to this reading of the situation is to attribute to [Gödel] an unreasonable fluctuation in his views about "finitism." I feel that there has been rather too much easy settling for obscurity or inconsistency on Gödel's part in discussions of his works, especially – but not exclusively – his unpublished work, appearing in volumes III–V of the *Collected Works*. Of course, one can reasonably suppose that the wording in those works, by their nature – lecture notes, letters written in a day, *etc.* – did not receive the same care that he devoted to wording in his published papers. Nevertheless, it seems all the more reasonable that, in such cases, one should look for an interpretation of what he wrote, against the background of all of his writings, that has him saying something sensible. (Tait, 2006, 93)

Tait refers to this as the McKeon principle,[27] but as he himself remarks, one can carry it too far: "In McKeon's hands, it degenerated into the less salubrious view that the great philosophers were never wrong: one only needed to discover the right principle of translation." I agree with both the principle and the caveat, so let's proceed with caution.

Another caveat warrants consideration: in his writings on finitism, reprinted as chapters 1 and 2 with an appendix in *The Provenance of Pure Reason*, Tait (2005a) has rightly emphasized the need to distinguish between the proper characterization of finitism and historical views of it, especially Hilbert's.[28] Furthermore, in his essay reviews of *CW III* and of *CW IV* and *V*, this has been extended to Gödel's views of both the proper characterization of finitism and what he took Hilbert's view to be. In the last of these (pp. 92–98), Tait argues that Gödel's conception of finitism *was* stable and is represented by what he has to say about the system of finitary number theory for Zilsel's seminar, which we have seen is interpreted as PRA. He sets aside the problematic interpretation of the system A in the Cambridge lecture as Gödel does not refer to it as being finitist but rather as being at the lowest level of a hierarchy of constructive systems. Tait says that the ascription of unsettled views to Gödel in the correspondence and later articles "is accurate only of his view of *Hilbert's* finitism, and the instability centers around his view of whether or not there is or could be a precise analysis of what is 'intuitive'" (Tait, 2005a, 94). So, if taken with that qualification, my ascription of unsettled views to Gödel is not mistaken. As to Gödel's own conception of finitism, I think the evidence offered by Tait for its stability is quite slim, but that's nothing I'm concerned to make a case about, one way or another.

I would add to all this only a speculation concerning Gödel's vacillating views of Hilbert's finitism: perhaps he wanted it seen as one of the values of his work in 1958 and 1972 that the step to the notions and principles of the system for primitive recursive functionals of finite type would be just what is needed to go beyond finitary reasoning in the sense of Hilbert to capture arithmetic. For that it would be important to tie down Hilbert's conception precisely to have limit ε_0, by means, for example, of a characterization of the sort proposed by Kreisel.

The second question raised earlier is the more difficult one and, to me, the most intriguing of the whole affair: how does Gödel's well-known Platonic realism cohere with his engagement in the constructive consistency program and his claims for its necessity? One way is to deny that there is a genuine incoherence by denying the seriousness of that engagement. That is the position taken by Kreisel, who is utterly dismissive of it, first in his biographical memoir for the Royal Society (Kreisel, 1980) and later in his piece "Gödel's Excursions into Intuitionistic Logic" (Kreisel, 1987). In the former, he writes:

> [Gödel's] last self-contained publication [*1958*]... was presented as a consistency proof. Between 1931 and 1958... he studied other such proofs.... Very much in contrast to the break with traditional aims, advocated throughout this memoir, Gödel continued to use traditional terminology. For example, the original title of Spector (1962), extending Gödel (1958), did not contain the word "consistency"; it was added for posthumous publication at Gödel's insistence. He knew only too well the publicity value of this catchword, which – contrary to his own view of the matter – had made the second incompleteness theorem more spectacular than the first. (Kreisel, 1980, 174)

The reference is to Spector's posthumously published 1962 article, in which a functional interpretation of a system of second-order number theory was obtained by means of the so-called bar-recursive functionals. The "full story" of the title is recounted in Kreisel (1987, 161):

> Spector's simple title was: *Provably recursive functionals of analysis*. Gödel did not find this exciting, and proposed the addition: a consistency proof of analysis.... Of course, I appreciated his flair for attracting attention, but my views about the sham of the consistency business have remained uncompromising. So, to water down his addition, I proposed the further qualification "by an extension of principles formulated in current intuitionistic mathematics," to which Gödel agreed, albeit reluctantly.

Kreisel (1987, 144) also writes more generally that in later years, "Gödel used crude, hackneyed formulations that had proved to have popular appeal (and had put me off . . .), [though] in his very early writing he was more austere." Given the accumulated evidence that has been surveyed here of Gödel's serious engagement with the constructive consistency program from the beginning to the end of his career, it seems to me that this tells us more about Kreisel than about Gödel.[29]

A second way to deal with the prima facie incoherence of Gödel's Platonism with his engagement in the constructive consistency program is to question his retrospective claims of having held the Platonistic views back to his student days, at least if understood in their latter-day form; the case for their development prior to 1944 is made by Charles Parsons (1995). Also, through extensive quotation, Martin Davis (2005) has pointed out that they were by no means uniform. This could account for such things as the Cambridge lecture and the seminar at Zilsel's but doesn't account for the work on the *Dialectica* interpretation well past the point where he had made plain his adherence to full-blown set-theoretical realism.

A third possible way for a confirmed Platonist, such as Gödel, to make an engagement in the constructive consistency program is to recognize the additional epistemological value of success under that program; that is, even if the Platonist is convinced of the truth of his axioms, and hence of their consistency, he could still appreciate the additional evidence of a different nature that a constructive consistency proof would give.[30] In Gödel's case, this is supported by the final section of his Zilsel presentation that was quoted in full in Section 6.4 above, whether or not he was a Platonist at the time. Though that can be argued in principle, it seems to me curious that Gödel's own engagement in the program never went beyond arithmetic, where he can hardly have thought that a constructive proof would make its consistency more evident than what is provided by the intuitive conception. Perhaps, though, he believed that that was only a test case and that a good start with arithmetic, such as that which he essayed with the *Dialectica* interpretation, could take the program much further, to where intuition is no longer so reliable.

Let me venture a psychological explanation, instead, that goes back to what I suggested at the outset: Gödel simply found it galling all through his life that he never received the recognition from Hilbert that he deserved. How could he get satisfaction? Well, just as (in the words of Bernays) "it became Hilbert's goal to do battle with Kronecker with his own weapon of finiteness,"[31] so it became Gödel's goal to do battle with Hilbert with his own weapon of the consistency program. When engaged in that, he would have to do so – as he did – with all seriousness. This explanation resonates with the view of the significance of Hilbert for Gödel advanced in chapter 3 of Takeuti (2003),[32] who concludes that "Hilbert's existence had tremendous meaning for Gödel" and that his "academic career was molded by the goal of exceeding Hilbert." However, Takeuti says that Gödel did not do any work subsequent to 1940 that was comparable to the work on completeness, incompleteness, and the continuum hypothesis, perhaps, among other things, because "there was no longer the challenge to exceed Hilbert." In that we differ: in my view, the challenge remained well into his last decade for Gödel to demonstrate decisively, if possible, why it was necessary to go beyond Hilbert's finitism in order to prosecute the constructive consistency program.

Acknowledgments

I wish to thank Jeremy Avigad, Charles Parsons, Wilfried Sieg, William Tait, and Mark van Atten for their helpful comments on a draft of this chapter.

Endnotes

1. The five volumes of Gödel's *Collected Works* (ed. S. Feferman et al., 1986–2003) are referred to throughout this chapter and in the references as *CW I, II, III, IV*, and *V*. *CW I* consists of the publications 1929–1936, *CW II* of the publications 1938–1974, *CW III* of unpublished essays and lectures, *CW IV* of correspondence A–G, and *CW V* of correspondence H–Z. References to individual items by Gödel follow the system of these volumes, which are either of the form *Gödel 19xx* or of the form **Gödel 19xx*, with possible further addition of a letter in the case of multiple publications within a given year; the former are from *CW I* or *CW II*, whereas the latter are from *CW III*. Thus, e.g., *Gödel 1931* is the famous incompleteness paper, whereas *Gödel 1931c* is a review that Gödel wrote of an article by Hilbert, both in *CW I*; *Gödel *1933o* is notes for a lecture, "The present situation in the foundations of mathematics," to be found in *CW III*. Pagination is by reference to these volumes, e.g., Gödel (1931, *CW I*, 181), or simply Gödel (*CW*

I, 181). In the case of correspondence, reference is by letter number and/or date within a given body of correspondence, as, e.g., Gödel to Bernays (letter 56), or equivalently, Gödel to Bernays (December 2, 1965), under Bernays in *CW IV*. When an item in question was originally written in German, my quotation from it is taken from the facing English translation. Finally, reference will be made to the various introductory notes written by the editors and colleagues that accompany most of the pieces or bodies of correspondence.

2. William Tait (2006), in his forceful and searching essay review of *CW IV* and *V*, covers much the same ground but from a different perspective; see esp. Section 6.6.

3. Hilbert introduced first-order logic and raised the question of completeness much earlier, in his lectures of 1917–1918. According to Awodey and Carus (2001), Gödel learned of this completeness problem in his logic course with Carnap in 1928 (the only logic course that he ever took).

4. For an excellent introduction to Hilbert's program with a guide to the literature, see the entry by Richard Zach (2003) in the *Stanford Encyclopedia of Philosophy*. A more extended exposition is to be found in part III of Mancosu (1998) as an introductory note to a number of major articles by both Hilbert and Bernays in English translation.

5. See Parsons (2008) for an examination of Bernays' later philosophy of mathematics.

6. In recognition of that, Feferman (2008) was published in *Dialectica*.

7. See Turing (1939) and Feferman (1962).

8. See Takeuti's foreword in this volume.

9. The fifth and last item of 1931, from Bernays to Gödel, deals with one final point on this matter; it is mainly devoted to a preliminary presentation of Bernays' theory of sets and classes. See *CW IV* (pp. 104–15).

10. In any case, I have discussed the evidence at length in my introductory note to Gödel's review 1931c of Hilbert's work on Z^*; cf. *CW I*, 208–13.

11. Cf. Reid (1970, 198–99).

12. Cf. *CW I*, 212. Olga Taussky-Todd (1987, 40) writes in her reminiscence that Gödel "lashed out against Hilbert's paper 'Tertium non-datur' [Hilbert, 1931a] saying something like, 'how can he write such a paper after what I have done?' Hilbert in fact did not only write this paper in a style irritating to Gödel, he gave lectures about it in Göttingen in 1932 and other places. It was to prove Hilbert's faith."

13. Gödel himself, in the postscript to his remarks at the Königsberg conference, speaks of "finitary, (that is, intuitionistically unobjectionable) forms of proof" (*1931a, CW I*, 205). Cf. also Bernays (1930, Part II, sec. 2; Sieg, 1990, 272).

14. See Sieg's introductory note to Gödel's correspondence with Herbrand in *CW V*, 9n. s, and Tait (2006, 98–105) in his essay review of *CW IV* and *V*.

15. What Hilbert meant by *finitism* has been the subject of extensive discussion in the literature; the trouble is that he made no specific effort to delineate it and instead relied on informal explanations and examples. A strong case has been made by Zach (1998, 2001, 2003), among others, that Hilbert's conception of finitism definitely goes beyond primitive recursive arithmetic to include Ackermann's function and other functions obtained by multiply nested recursions, together with certain forms of transfinite recursion (in view of Hilbert, 1926).

16. Translated from the German: "*Dieser Ergebnisstand weist zugleich die Richtung für die weitere Forschung in der Beweistheorie auf das Endziel hin, unsere üblichen Methoden der Mathemattick samt und sonders als widersprchsfrei zu erkennen. Im Hinblick auf dieses Ziel möchte ich hervorheben, dass die zeitweilig aufgekommene Meinung, aus gewissen neueren Ergebnissen von Gödel folge die Undurchfürbarkeit meiner Beweistheorie, als irrtümlich erwiesen ist. Jenes Ergebnis zeigt in der Tat auch nur, dass man für die weitergehenden Widerspruchsfreiheitsbeweise den finiten Standpunkt in einer shärferen Weise ausnutzen muss, als dieses bei der Betrachtung der elementaren Formalismen erforderlich ist.*"

17. Hilbert (1926) refers to the axioms for quantifiers in the context of arithmetic as the "transfinite axioms" because the law of excluded middle for quantified formulas requires the assumption of the "completed infinite."

18. The idea is elaborated in Tait (2005b).

19. See Chapter 4 in this volume for more on Gödel's life in and emigration from Vienna.

20. Dana Scott, who was a student in Princeton at the time, recalls a "conversation" with Church, Gödel, Bernays, Kreisel, and a couple of graduate students at Church's home (pers. comm., 2006).

21. As it happens, Bernays' proof as it stands is mistaken, or at best incomplete, as shown by Tait (2006, 89–91); see Endnote 24 for the reason.

22. Following a remark of Kreisel, Mark van Atten (2006, p. 26) suggests that Gödel, *qua* Platonist, considered choice sequences to belong to the ontological realm of "our own 'constructions' or choices" and hence not to pure mathematics; but the theory could well be considered a matter of applied mathematics.

23. The version of the footnote that did appear in the proofs for *Gödel 1972* (ftn. c, *CWII*, p. 272) reads: "A closer approximation to Hilbert's finitism can be achieved by using the concept of free choice sequence rather than 'accessibility.'"

24. Bernays answered the question concerning nested recursion, saying that the *verschränkte rekursion* of vol. 1 of *Grundlagen der Mathematik* appears to him to be finitary in the same sense as the primitive recursions (see *CW IV*, 277); cf. also Endnote 15. Tait (2006, 91–92) critically examines the questions concerning nested recursion and choice sequences; he argues that these go beyond finitism as it ought to be understood (which may well differ from the way Hilbert understood it). In addition (Tait, 2006, 89–91), he usefully spells out Bernays' proof of induction up to ε_0, considered the limit of the ordinals $\omega[n]$, where $\omega[0] = \omega$ and $\omega[n + 1] = \omega^{\omega[n]}$. His analysis reveals that Bernays' prima facie inductive argument to show for each n that there are no infinite sequences descending from $\omega[n]$ actually only reduces that property for $\omega[n + 1]$ to the assumption of nested recursion on $\omega[n]$; therefore the proof as it stands is mistaken or, at best, is incomplete without the additional claim that the no-descending-sequence property on an ordinal justifies nested recursion on that ordinal. The situation here is related to the earlier result of Tait (1961) that reduces, for any ordinal α, ordinary recursion on ω^α to nested recursion on $\omega \times \alpha$, a fact, as it happens, known to Gödel but not invoked in his enthusiastic response to Bernays' supposed proof.

25. A variant formulation of the proposed ordinal logic that is a little more detailed was presented in Kreisel (1965, sec. 3.4, pp. 168ff.).

26. In Kreisel (1965, 169), it is said that the concept being elucidated is "of proofs that one can *see* or *visualize*.... Our primary subject is a *theoretical* notion for the actual visualizing, not that experience itself." The matter was revisited in Kreisel (1970), where the project is to determine "what principles of proof...we recognize as valid once we have understood...certain given concepts" (p. 489), these being, in the case of finitism, "the concepts of ω-sequence and ω-iteration" (p. 490).

27. Tait (2006, p. 93, fn. 15) describes this principle as follows: "Richard McKeon...preached ...this very salubrious view in connection with the interpretation of the great philosophers, such as Plato: If your interpretation of them makes them fools...then it is likely you have more work to do." But one would hardly say that ascribing changing views to Gödel makes him out to be a fool.

28. This obvious need has been emphasized by other commentators as well (cf., e.g., Zach, 2003, and references therein).

29. According to Gödel's communications to Hao Wang (1981, 654), an additional piece of evidence for the seriousness of his engagement with Hilbert's consistency program from the very beginning

is that he was led to the incompleteness theorems through an aborted attempt to carry out the program for analysis (second-order number theory).

30. This possibility has been emphasized by Jeremy Avigad and Mark van Atten in personal communications.
31. Quoted in Reid (1970, 173).
32. This was brought to my attention by Mark van Atten after reading a draft of this article.

References

Ackermann, W. (1924). Begründung des "tertium non datur" mittels Hilbertschen Theorie der Widerspruchsfreiheit. *Mathematischen Annalen*, **93**, 1–36.

Awodey, S., and Carus, A. W. (2001). Carnap, completeness and categoricity: The *Gabelbarkeitssatz* of 1928. *Erkenntnis*, **54**, 145–71.

Bernays, P. (1930). Die Philosophie der Mathematik und die Hilbertsche Beweistheorie. *Blätter für Deutsche Philosophie*, **4**, 326–67. [Repr. P. Bernays (1976). In *Abhandlungen zur Philosophie der Mathematik*. Darmstadt, Germany: Wissenschaftliche Buchgesellschaft, pp. 17–61; English trans. P. Mancosu (1998). In *From Brouwer to Hilbert: The Debate on the Foundations of Mathematics in the 1920s*. New York: Oxford University Press, pp. 234–65.]

———. (1937–1954). A system of axiomatic set theory. *Journal of Symbolic Logic*, Part I (1937), **2**, 65–77; Part II (1941), **6**, 1–17; Part III (1942), **7**, 65–89; Part IV (1942), **7**, 133–45; Part V (1943), **8**, 89–106; Part VI (1948), **13**, 65–79; Part VII (1954), **19**, 81–96.

———. (1967). Hilbert, David. In *Encyclopedia of Philosophy*, vol. 3, ed. P. Edwards, pp. 496–504. New York: Macmillan.

———. (1976a). Kurze Biographie. In *Sets and Classes: On the Work of Paul Bernays*, ed. G. Müller, pp. xiv–xvi. Amsterdam, Netherlands: North-Holland, pp. xiv–xvi. [English trans. pp. xi–xiii.]

———. (1976b). *Abhandlungen zur Philosophie der Mathematik*. Darmstadt, Germany: Wissenschaftliche Buchgesellschaft.

Davis, M. (2005). What did Gödel believe and when did he believe it? *Bulletin of Symbolic Logic*, **11**, 194–206.

Feferman, S. (1962). Transfinite recursive progressions of axiomatic theories. *Journal of Symbolic Logic*, **27**, 259–316.

———. (2008). Lieber Herr Bernays!, Lieber Herr Gödel! Gödel on finitism, constructivity and Hilbert's program. *Dialectica*, **62**(2), 179–203. [A preprint of the current chapter; some differences arose in the editing process. Also see "Gödel's functional interpretation and its use in current mathematics," a preprint of Chapter 17 in this volume, which appeared in the same issue of *Dialectica*.]

Gentzen, G. (1936). Die Widerspruchsfreiheit der reinen Zahlentheorie. *Mathematische Annalen*, **112**, 493–65. [English trans. G. Gentzen (1969). In *The Collected Papers of Gerhard Gentzen*, ed. M. E. Szabo. Amsterdam, Netherlands: North-Holland, pp. 132–213.]

Gödel, K. (1929). On the completeness of the calculus of logic. PhD diss. In *CW I*, pp. 61–101.

———. (1930). Die Vollständigkeit der Axiome des logischen Funktionenkalküls. *Monatshefte für Mathematik und Physik*, **37**, 349–60. [Published PhD diss. Also in *CW I*, pp. 102–23.]

———. (1931a). Über formal unentscheidbare Sätze der *Principia Mathematica* und verwandter Systeme I. *Monatshefte für Mathematik und Physik*, **38**, 173–98. [English trans. J. van Heijenoort, ed. (1967). *From Frege to Gödel: A Source Book on Mathematical Logic*. Cambridge, MA: Harvard University Press, pp. 596–616. Repr. with facing English trans. On formally undecidable propositions of *Principia Mathematica* and Related Systems. I. In *CW I*, pp. 145–95.]

———. (1931b). Diskussion zur Grundlegung der Mathematik, *Erkenntnis*, **2**, 147–51. [Repr. with facing English trans. in *CW I*, pp. 200–5.]

————. (1931c). Review of Hilbert's work on Z*; cf. *CW I*, pp. 212–15.

————. (*1933o). The present situation in the foundations of mathematics. In *CW III*, pp. 45–53.

————. (*1938). Vortrag bei Zilsel. In *CW III* with facing English trans., pp. 86–113.

————. (*1941). In what sense is intuitionistic logic constructive? In *CW III*, pp. 189–200.

————. (1958). Über eine bisher noch nicht benüzte Erweiterung des finiten Standpunktes. *Dialectica*, **12**, 280–87. [Repr. in English trans. On a hitherto unutilized extension of the finitary standpoint. In *CW II*, pp. 241–51.]

————. (1972). On an extension of finitary mathematics which has not yet been used. In *CW II*, pp. 271–80.

Herbrand, J. (1931). Sur la non-contradiction de l'arithmétique. *Journal für die reine und angewandte Mathematik*, **166**, 1–8. [English trans. J. van Heijenoort, ed. (1967). *From Frege to Gödel: A Source Book in Mathematical Logic, 1879–1931*. Cambridge, MA: Harvard University Press, pp. 618–28.]

Hilbert, D. (1926). Über das Unendliche. *Mathematische Annalen*, **95**, 161–90. [English trans. J. van Heijenoort, ed. (1967). *From Frege to Gödel: A Source Book in Mathematical Logic, 1879–1931*. Cambridge, MA: Harvard University Press, pp. English trans. in van Heijenoort, 1967, pp. 367–92.]

————. (1931a). Die Grundlegung der elementaren Zahlenlehre. *Mathematische Annalen*, **104**, 485–94. [English trans. J. van Heijenoort, ed. (1967). *From Frege to Gödel: A Source Book in Mathematical Logic, 1879–1931*. Cambridge, MA: Harvard University Press, 266–73.]

————. (1931b). Beweis des tertium non datur. *Nachrichten von der Gesellschaft der Wissenschaften zu Göttingen, Math.-Physik. Klasse*, 120–25.

————. (1935). *Gesammelte Abhandlungen*. Vol. 3. Berlin: Springer.

Hilbert, D., and Ackermann, W. (1928). *Grundzüge der theoretischen Logik*. Berlin: Springer. [2nd ed. 1938. English trans. of the 2nd ed. *Principles of Mathematical Logic*. New York: Chelsea.]

Hilbert, D., and Bernays, P. (1934). *Grundlagen der Mathematik*. Vol. 1. Berlin: Springer. [Written by Bernays alone; 2nd ed. 1968.]

————. (1939). *Grundlagen der Mathematik*. Vol. 2. Berlin: Springer. [Written by Bernays alone; 2nd ed. 1970.]

Kreisel, G. (1960). Ordinal logics and the characterization of informal concepts of proof. In *Proceedings of the International Congress of Mathematicians, 14–21, August 1958*, ed. J. A. Todd, pp. 289–99. Cambridge: Cambridge University Press.

————. (1965). Mathematical logic. In *Lectures on Modern Mathematics*, vol. 3, ed. T. L. Saaty, pp. 95–195. New York: John Wiley.

————. (1970). Principles of proof and ordinals implicit in given concepts. In *Intuitionism and Proof Theory*, ed. A. Kino, J. Myhill, and R. E. Vesley, pp. 489–516. Amsterdam, Netherlands: North-Holland.

————. (1980). Kurt Gödel, 28 April 1906–14 January 1978. *Biographical Memoirs of Fellows of the Royal Society*, **26**, 148–224. [Corrections **27**, 697; **28**, 718.]

————. (1987). Gödel's excursions into intuitionistic logic. In *Gödel Remembered: Salzburg, 10–12 July 1983*, ed. P. Weingartner and L. Schmetterer, pp. 65–186. Naples, Italy: Bibliopolis.

Mancosu, P., ed. (1998). *From Brouwer to Hilbert: The Debate on the Foundations of Mathematics in the 1920s*. New York: Oxford University Press.

Parsons, C. D. (1995). Platonism and mathematical intuition in Kurt Gödel's thought. *Bulletin of Symbolic Logic*, **1**, 44–74.

————. (2008). Paul Bernays' later philosophy of mathematics. In *Logic Colloquium 2005*, ed. C. Dimitracopoulos, L. Newelski, D. Normann and J. R. Steel, pp. 129–50. Lecture Notes in Logic 28. New York: Association for Symbolic Logic and Cambridge University Press.

Reid, C. (1970). *Hilbert*. New York: Springer.

Sieg, W. (1990). Relative consistency and accessible domains. *Synthese*, **84**, 259–97.

————. (2003). Introductory note to Gödel's correspondence with Herbrand. In *CW V*, pp. 3–13.

Spector, C. (1962). Provably recursive functionals of analysis: A consistency proof of analysis by an extension of principles formulated in current intuitionistic mathematics. In *Recursive Function Theory*, ed. J. C. E. Dekker. Providence: American Mathematical Society, pp. 1–27.

Tait, W. (1961)., Nested recursion. *Mathematische Annalen*, **143**, 236–50.

———. (1968). Constructive reasoning. In *Logic, Methodology and Philosophy of Science III*, ed. B. van Rootselaar and J. F. Staal, pp. 185–99. Amsterdam, Netherlands: North-Holland.

———. (1981). Finitism. *The Journal of Philosophy*, **78**, 524–46. [Repr. as chap. 1 in W. Tait (2005a).]

———. (2001). Gödel's unpublished papers on foundations of mathematics. *Philosophia Mathematica, Series III*, **9**, 87–126. [Repr. as ch. 12 in Tait (2005a).]

———. (2002). Remarks on finitism. In *Reflections on the Foundations of Mathematics: Essays in Honor of Solomon Feferman*, ed. W. Sieg, R. Sommer, C. Talcott., pp. 410–15. Lecture Notes in Logic 15. 2002, Association for Symbolic Logic, A. K. Peters, Natick, MA. [Repr. with appendix as chap. 2 in W. Tait (2005a).]

———. (2005a). *The Provenance of Pure Reason*. New York: Oxford University Press.

———. (2005b). Gödel's reformulation of Gentzen's first consistency proof for arithmetic: The no-counterexample interpretation. *Bulletin of Symbolic Logic*, **11**, 225–38.

———. (2006). Gödel's interpretation of intuitionism. *Philosophia Mathematica, Series III*, **14**, 208–28.

Takeuti, G. (2003). *Memoirs of a Proof Theorist: Gödel and Other Logicians*. Singapore: World Scientific.

Taussky-Todd, O. (1987). Remembrances of Kurt Gödel. In *Gödel Remembered: Salzburg, 10–12 July 1983*, ed. P. Weingartner and L. Schmetterer, pp. 29–48. Naples, Italy: Bibliopolis.

Turing, A. (1939). Systems of logic based on ordinals. *Proceedings of the London Mathematical Society, Series 2*, **45**, 161–228. [Repr. M. Davis, ed. (1965). *The Undecidable: Basic Papers on Undecidable Propositions, Unsolvable Problems and Computable Functions*. Hewlett, NY: Raven Press, 155–222.]

Van Atten, M. (2006). *Brouwer Meets Husserl: On the Phenomenology of Choice Sequences*. Synthese Library 335. Berlin: Springer.

Van Heijenoort, J., ed. (1967), *From Frege to Gödel: A Source Book in Mathematical Logic*. Cambridge, MA: Harvard University Press.

Von Neumann, J. (1927). Zur Hilbertschen Beweistheorie. *Mathematische Zeitschrift*, **26**, 1–46.

Wang, H. (1981). Some facts about Kurt Gödel. *Journal of Symbolic Logic*, **46**, 653–59.

Zach, R. (1998). Numbers and functions in Hilbert's finitism. *Taiwanese Journal for Philosophy and History of Science*, **10**, 33–60.

———. (2001). Hilbert's program: Historical, philosophical and metamathematical perspectives. PhD diss . University of California, Berkeley.

———. (2003). Hilbert's program. *Stanford Encyclopedia of Philosophy*. http://plato.stanford .edu/entries/hilbert-program/.

The Past and Future
of Computation

CHAPTER 7

Computation and Intractability: Echoes of Kurt Gödel

Christos H. Papadimitriou

Kurt Gödel (1906–1978) lived half his adult life during the computer era – in fact, virtually at the epicenter of the early development of the computer. Yet we have no evidence that the great logician took a serious interest in the nature, workings, power, or limitations of the beast.[1] Nevertheless, I shall argue that Gödel's work has had many and crucial connections to computation for a number of reasons:

- The incompleteness theorem (Gödel, 1931) started an intellectual Rube Goldberg that eventually led to the computer.
- Gödel's early work contains the germs of such influential computational ideas as arithmetization and primitive recursion.
- In a 1956 letter to John von Neumann (see Sipser, 1992), which was not widely known for three decades, Gödel proposed a novel quantitative version of Hilbert's program that turned out to be precisely the P versus NP question.
- Negative results, of which the incompleteness theorem is an ideal archetype, constitute an important and distinguishing tradition in computer science research.

Sections 7.1 and 7.2 recount how the incompleteness theorem motivated a number of mathematicians during the 1930s to further sharpen the result's negative verdict and establish that mathematics, besides being impossible to axiomatize, cannot even be mechanized (axiomatization being only one of the possible ways whereby mathematical thought can be mechanized). For this program to be carried out, however, computation had to be somehow defined, and those brilliant definitions – most notably Alan Turing's (1936) – were the mathematical precursors of the computer. When, ten years and one world war later, the time came for the mathematical idea of the universal computer to be incarnated, it was John von Neumann (1945), the first mathematician to grasp the incompleteness theorem's significance and to eagerly explore possible ways of strengthening it, who came up with the influential design ideas that have to a large extent defined computers ever since. Von Neumann is discussed in Section 7.3.

---

[1] But see Sieg (2006) for a very recent insightful analysis of Gödel's views on computation and computability based mainly on his Gibbs lecture (Gödel, 1951).

Sections 7.4 and 7.5 focus on the P versus NP question and on the importance of negative results. In a letter to a dying von Neumann in 1956, Gödel posed a question as a way to explain a novel quantitative version of the foundations quest (see the subsequent discussion): is it possible to construct a machine which, when presented with a sentence in first-order logic, would find a proof for the sentence (provided such a proof exists) after a number of steps that is proportional to the length of the proof (or, say, its square)? As it turned out, this question was equivalent to P = NP, which was to become, decades later, the central open question in computer science and possibly the deepest open problem in mathematics.

In Section 7.6, I recount the influence that Gödel's arithmetization had in theoretical computer science, ending with a recent result establishing that the problems of finding Brouwer fixed points and Nash equilibria are computationally equivalent (Daskalakis et al., 2006); its proof makes crucial use of arithmetization. Finally, in Section 7.7, I provide an epilogue describing the tragic lives of many of the great figures in twentieth-century logic.

7.1 The Two Foundations Quests

The beginning of the twentieth century was a time of crisis for mathematics. The field was flourishing and expanding in all directions, yet its foundations were contemplated with anxiety and insecurity. Non-Euclidean geometries had revealed the dangers of doing mathematics without a thorough understanding of its axiomatic basis, while Georg Cantor's dissection of infinity exposed new complexities that were considered by many as problematic, even ominous. These problems had already claimed one victim: Gottlob Frege had developed an ingenious notation for first-order logic as well as a formalism for axiomatizing all arithmetic (the mathematics of whole numbers; Frege, 1884), a singularly ambitious work in which, however, Bertrand Russell had discovered a fatal flaw – at the precise time that the second volume of Frege's magnum opus was going to the printer.

This was the context of the launching of Hilbert's foundations quest. David Hilbert was not only the most eminent and accomplished mathematician of the time but also a force of nature – a deeply optimistic man convinced that human ingenuity could eventually conquer any challenge. "There are no unsolvable problems," he would proclaim. "In mathematics there is no ignorabimus.... We *must* know, we *shall* know!"[2] He was interested in founding all mathematics on axioms that are consistent (i.e., no contradiction can arise from them) and complete (i.e., every true sentence in the domain follows from the axioms), and he urged mathematicians to find such axiomatic systems for arithmetic, and from those, for all mathematics – he had already succeeded in axiomatizing geometry (Hilbert, 1899). Closer inspection, however, reveals that Hilbert's program consists, in fact, of two goals that are subtly distinct:

---

[2] Most ironically, Hilbert repeated the latter proclamation in a radio broadcast on September 8, 1930, in Königsberg, East Prussia – that is, at the precise place and on the precise day at which an unknown young mathematician named Kurt Gödel was presenting his devastating refutation of the foundations quest.

- **The axiomatic foundations quest (AFQ):** Find a provably consistent set of axioms whose consequences comprise all theorems in mathematics.
- **The computational foundations quest (CFQ):** Find a mechanical procedure for distinguishing between true and false sentences in mathematics.

The AFQ is what is usually referred to in the literature as Hilbert's program, and it is articulated quite unambiguously in his address at the International Congress of Mathematicians (Hilbert, 1900); in his Heidelberg talk (Hilbert, 1905); and in his address, a decade later, to the Swiss Mathematical Society (Hilbert, 1918), for example. In contrast, the CFQ was, at the beginning, only an undercurrent. It is clear that Hilbert was thinking about algorithms early on; already in his 1900 address, he talked about "a process according to which it can be determined by a finite number of operations," albeit in relation to his tenth problem of solving Diophantine polynomial equations. Ironically, the earliest explicit reference to the mechanical nature of Hilbert's program before the late 1920s may have been Henri Poincaré's (1914, 133) well-known polemical remark likening the goal of the foundations quest to "the famous sausage machine at Chicago where the living pigs are fed in at the top and appear as sausage and ham at the bottom."

In 1928, a limited form of the CFQ was enunciated explicitly in Hilbert's work with Ackermann as the *Entscheidungsproblem* (decision problem), the development of an algorithm for deciding whether a sentence in first-order logic is valid (Hilbert and Ackermann, 1928). In other words, the *Entscheidungsproblem* specializes CFQ to the domain of first-order logic.

It is not a priori obvious that the two foundations quests are equivalent (which indeed they are, by virtue of their both being impossible). In fact, that the axiomatic version implies the computational one is a consequence of the completeness theorem, conjectured by Hilbert and Ackermann (1928) and famously proved by Kurt Gödel (1929, 1930) in his doctoral thesis and presented at the September 1930 Königsberg meeting, one day before the announcement of the incompleteness theorem (Gödel, 1931). The completeness theorem states that any valid formula in first-order logic has a proof (in a natural proof system). Therefore, if mathematics were axiomatizable by first-order axioms, given any mathematical sentence, one could search for a proof that the axioms imply the sentence and also for a proof that they imply its negation. Because the axiomatic system is assumed to be complete, one of the two implications must be valid – and hence, by completeness, provable. Thus a machine could look at all possible proofs, say, one after the other in increasing length, testing for each whether it is a proof of one of the two implications. Because one of the two proofs must exist, this algorithm is guaranteed to succeed in finding it after finitely many steps. The situation advanced from the one in Figure 7.1a to the one in Figure 7.1b.[3]

Finally, less than a year after completing his doctoral dissertation, which established the situation shown in Figure 7.1b, Gödel proved the incompleteness theorem, demonstrating that AFQ, the stronger of the two goals, is impossible. With the incompleteness theorem, the situation shown in Figure 7.1c prevails. Figure 7.1d shows the two foundations quests after Turing's paper (ca. 1936).

---

[3] I am not aware of an explicit mention of this straightforward consequence of the completeness theorem in the literature of the time; however, I have no doubt that it was well understood by the research community.

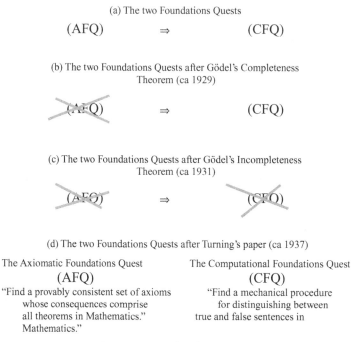

Figure 7.1. A tale of two quests.

7.2 The Dawn of Computation

After this devastating development for the foundations quest, the computational (weaker) version provided the only glimmer of hope. After all, axiomatization is only one of the possible ways to mechanize mathematics, and perhaps some more direct, ad hoc approach might still be possible. But I do not believe that anybody at that time was placing too much faith in this possibility. After the incompleteness disaster, the mood in the foundations camp must have been one of despair and retreat. It is therefore not surprising that several mathematicians worked on refuting the CFQ. Again, on reflection, one realizes that there were three distinct ways in which something could be salvaged from the foundations quest by utilizing computation (in Section 7.4, I discuss a fourth, even more modest goal proposed two decades later by Kurt Gödel himself):

1. Find an algorithm, presumably not based on an axiomatic system, that, given a sentence in some domain of mathematics, would decide in finite time whether it is a theorem.
2. The *Entscheidungsproblem* is the specialization of (1) to first-order logic: find a machine that decides whether a given sentence in first-order logic is valid.[4]
3. Given some (presumably incomplete) axiomatization of mathematics, find a mechanical process that, given a sentence, discovers a proof of the sentence from the axiom system *provided it has one*. This is in fact equivalent to (2), assuming that the axioms are first order, because it entails deciding validity for the first-order logic formula stating that the axioms imply the given sentence.

---

[4] In fact, the proof of the completeness theorem reveals that the *Entscheidungsproblem* is a special case of the decision problem in second-order arithmetic.

But to refute any of these weaker versions of the foundations quest, the intuitive notion of "algorithm" or "machine" or "mechanical process" had to be defined with mathematical precision. This was not necessary previously, when the goal was to carry out the foundations quest: if somebody had indeed come up with a mechanical way of deciding truth in mathematics, the community would probably have little trouble recognizing it as such, once it was presented to them. But now that impossibility was to be established, an exacting discipline of thought was needed.

Because this is the nature of impossibility proofs: to show that something is impossible, one must first define and chart with precision the whole realm of possibilities. And that is the reason why some of the brightest mathematical minds of that era embarked on this momentous task: to define computation.

And define it they did. During the mid-1930s, at least four vastly different and, in most cases, independent definitions of computation were proposed and used to demonstrate the impossibility of CFQ (in fact, of the *Entscheidungsproblem*, the least ambitious version): Alan Turing's (1936) universal machine; Alonzo Church's (1936) λ-calculus; Stephen Kleene's (1936) partial recursive functions; and Emil Post's (1936) machines. Actually, Church and his student Kleene were inspired to embark on this investigation by Gödel himself, who visited Princeton and lectured on the incompleteness theorem as well as on recursive functions in 1934 and 1935 (with Kleene as his note taker).

Post's story is particularly intriguing and touching: already in 1921, he had proved results anticipating the incompleteness theorem, only to keep them in his drawer unpublished, probably out of pathological insecurity. In a letter dated October 31, 1938, to Gödel, whom he admired, Post brushed aside his remarkable achievement and commented, with astonishing modesty, that "after all it is not ideas but the execution of ideas that constitute a mark of greatness." In the mid-1930s, Post (1936) did come up with a computational model that was eventually shown to be equivalent in computational capability to the other three models.

But the most momentous entry into the race for the definition of computation was by Turing (1936). Partly motivated by his interest in the workings of the human mind, he defined what is now known as the Turing machine, an abstract computer model that has deeply influenced both computational thought and computational practice. In particular, Turing's idea of universality (that a machine can perform any computational task if supplied with the right set of instructions) presaged, and arguably brought about, the general-purpose computer and its software.

That these four radically different definitions turned out to be exactly equivalent in computational capability elevated computation to the status of a novel and surprisingly robust – and thus presumably important – mathematical concept. Its real-world incarnation would not be very long in coming.

7.3 Johnny-Come-Lately: Von Neumann, Gödel, and the Computer

Of all the mathematicians involved in the foundations quest, John von Neumann was the first to grasp the significance of Gödel's announcement of the incompleteness theorem. A brilliant and ambitious mathematician who had spent time at Göttingen with Hilbert, von Neumann had worked hard on the quest, and now his labors lay in waste: "It's all over!" he has been reported to have exclaimed at Königsberg. Soon after this encounter, von Neumann realized that Gödel's proof implied another devastating impossibility result, the second incompleteness theorem, stating that any consistent axiomatic system for arithmetic cannot be used to prove itself consistent – consistency proofs being an important part of Hilbert's program. Von Neumann was a man of penetrating mathematical genius who was already well known for his ability to absorb and assimilate rapidly complex work by other mathematicians and to further that work in important directions unanticipated by the original author.

On November 20, 1930, von Neumann wrote to Gödel to announce his discovery of the second incompleteness theorem; as it turns out, on November 17, Gödel had submitted for publication his paper including his second incompleteness theorem, which he had proved independently in the meantime; von Neumann immediately acknowledged Gödel's priority – with how much disappointment we cannot know.

Another interaction between Gödel and von Neumann, this one immediately after the Königsberg presentation, was even more fateful. The result and proof that Gödel sketched at the meeting were quite a bit weaker than the ones submitted for publication two months later. Gödel stumbled on the incompleteness theorem while pursing the foundations quest – he was trying to axiomatize analysis *relative to arithmetic*, that is, assuming an axiomatization of arithmetic. In exploring how truth in arithmetic could be used to establish truth in other domains, he realized with surprise that truth in arithmetic is a little too expressive for its own good, so to speak, an insight that led to the incompleteness theorem. However, his original proof, presented at Königsberg, established the existence of unprovable sentences in the theory of sequences of integers.

According to Gödel's memoirs (see Wang, 1996, 83–84), von Neumann asked him after his talk whether he could extend his result to sentences about integers. Gödel answered that he could see ways of reducing a sequence of integers to a single integer, but these would involve devices such as the "sweeping" functions invented by Cantor and not the ordinary operations of arithmetic. In the next two months, motivated by von Neumann's question, Gödel came up (to his utter surprise, he confessed) with his ingenious encoding scheme of sequences of integers by integers using the Chinese remainder theorem – an encoding that succeeds brilliantly in showing the unprovability of simple sentences about integers.

By all available accounts, after Königsberg, von Neumann lost all interest in logic – in fact, during the next decade, he missed no opportunity to express publicly and unambiguously his total lack of interest in the subject. It is not implausible that this was the result of his traumatic encounter at Königsberg; the mathematical genius could not easily accept that somebody else had irreversibly beaten him to what perhaps was the century's most momentous theorem.

In fact, it appears that von Neumann had also turned a blind eye to the revolution in mathematical computers that followed the incompleteness theorem. In writing a recommendation letter for Alan Turing in 1938 – at the time a graduate student at Princeton, justly famous for having shown the impossibility of the *Entscheidunsproblem* and for having invented what was already being called the Turing machine – von Neumann confined his praise to Turing's work on periodic functions and continuous groups and said not a word about logic and computation.

A chance encounter in 1945 with an army officer named Herman Goldstine at the Aberdeen, Maryland, train station brought an abrupt change to all this. At Goldstine's invitation, von Neumann became involved as a consultant with the design of the ENIAC computer, an effort supported by the military and spearheaded up to that point by University of Pennsylvania engineering professors J. Presper Eckert and John Mauchly. In his work with the ENIAC, and with the succeeding EDVAC project,[5] von Neumann remembered Turing to great effect. His ambitious ideas for a more principled design, based on Turing's concept of universality, divided the ENIAC group, pitching von Neumann, along with several converts, against the more pragmatic engineers on the design team (most ironically, von Neumann's allies in this dispute were dubbed the "logicians"; see Davis, 2000). Von Neumann (1945) proceeded to lay down the principles of the stored program computer architecture in a document with the unassuming title "First Draft of a Report on the EDVAC." Distributed widely by Goldstine under von Neumann's name, the "First Draft" imparts a vision of astonishing clarity and presage and is one of the most influential documents in the history of computation. It is mainly because of "First Draft" that the digital computer is called to this date the "von Neumann machine."

7.4 Hilbert's Last Stand: Gödel's Letter and P versus NP

One decade after the appearance of the "First Draft," as John von Neumann was lying in a hospital bed in Washington, D.C., he received a letter from Kurt Gödel. The great logician inquired, in handwritten German, about the recipient's health with a mixture of human concern, optimism, and a formality oozing with respect. He then continued with a mathematical question. Gödel proposed a less ambitious version of Hilbert and Ackermann's *Entscheidungsproblem* – a version that was still possible, despite the deluge of negative results during the 1930s.

Recall that the *Entscheidungsproblem* sought an algorithm that, given any sentence in first-order logic, would decide whether the sentence has a proof. Instead, Gödel was looking for an algorithm with the following specifications: for any sentence in first-order logic, *if the sentence has a proof,* then the algorithm will succeed in producing it; if the sentence has no proof, the computation may never stop. Had Gödel imposed no further requirement, this feat would be easy to accomplish by a scheme that tries all possible proofs exhaustively, one by one and in increasing length, until the proof of the sentence – assuming it exists – is found and verified. (If the sentence has no proof, then the algorithm would go on computing forever, as it is allowed to do in this case.) Notice that if the sentence indeed has a proof of length L, this scheme will discover this proof after a computation of duration (number of elementary steps) that is exponential in L, simply because there are exponentially many proofs of length L, and the algorithm searches exhaustively through all of them.

Is it possible, asked Gödel, to achieve the same effect within a number of steps that is polynomial in L – as he put it, proportional to L, or perhaps to L^2 or some other power

---

[5] ENIAC = Electronic Numerical Integrator and Computer; EDVAC = Electronic Discrete Variable Automatic Computer. For more information on the development of the computer, see Chapter 8 in this volume.

of L? Can the exhaustive search for proofs be dramatically telescoped? If his question has a positive answer (and we still do not know whether it does), a most intriguing possibility arises: when faced with an interesting mathematical conjecture, we could submit it to this computer algorithm and let it compute for, say, a week. At the end of this period, either we would have obtained a definite proof of the conjecture, or a proof of its negation from a standard set of axioms such as Peano arithmetic. Otherwise, if the algorithm did not stop after a few days, we could conclude the following: if the conjecture – or its refutation – does have a proof, then the proof will have to be so long that it could not be expected to fit within a million or more pages. In other words, for all practical purposes, the conjecture cannot be settled. The existence of such an algorithm would trivialize the practice of mathematics, the collapse of the foundations quest in Gödel's hands a quarter century earlier notwithstanding.

By 1956, as computers were being used more and more often in the solution of practical problems, the dichotomy between polynomial and exponential computation had begun to replace Turing's original contrast between computability and uncomputability as the more meaningful boundary. In many real-world problems of a combinatorial nature, a solution is sought in an exponential population of possible solutions. Some of these problems can be solved by an efficient, polynomial method, whereas others seem to require exponential exhaustive search. Von Neumann (1953) was the first to hint at this dichotomy in a paper in which he succeeded in devising a polynomial algorithm for a particular combinatorial problem (what is now known as the assignment problem); this paper is probably one of the reasons why he was the recipient of Gödel's letter.

The dichotomy between polynomial and exponential search problems was identified explicitly almost a decade later in the work of Alan Cobham (1965) and Jack Edmonds (1965). In today's terminology and notation, the class of all search problems is called "NP," whereas the search problems fathomable in polynomial time constitute the subclass "P." Whether P = NP – that is, whether this dichotomy between exponential and polynomial search is real or fictional, an artifact of our limited ability to devise fast algorithms – has since been recognized as the most important and deep problem in computer science, if not in all mathematics.

Fifteen years later, Stephen Cook (1971), who was not aware of Gödel's 1956 letter, showed that a quantitative form of the *Entscheidungsproblem* called "satisfiability" is universal for NP; that is to say, it is "NP-complete." Leonid Levin (1973) had a similar insight independently, if a little later. The proofs can be considered a quantitative analogue of Turing's (1936) simulation of computation by first-order logic. Thus Gödel's question of whether proofs can be discovered in time proportional to L^c for some constant c can be seen to coincide with asking whether P = NP. The great man had anticipated not only this momentous question but also the central role logic would play in its treatment.

7.5 On Negative Results and Gödel's Influence

Few computer scientists today agree with the optimism regarding the P vs. NP question apparent in Gödel's 1956 letter. NP-completeness of a problem is now considered concrete evidence of intractability. The holy grail of computer science research today is a proof that $P \neq NP$.

This would be a negative result because it states that certain problems cannot be solved in polynomial time. As computer science is about the power and limitations of computation, negative results have been a major part of the field's research agenda and tradition since the very beginning. Indeed, the field's birth certificate, so to speak, is a negative result: Alan Turing (1936) proved that the halting problem (the problem of telling whether a given program will eventually terminate) is undecidable (it cannot be decided by a computer program). Over subsequent decades, as the field faced computational challenges of increasing variety and ambition, which had to be met with limited computational resources, negative results became an important tool that helped focus the research effort on what could actually be achieved. Even though negative results are present in many other fields (e.g., recall the impossibility of squaring the circle or solving quintic equations by radicals, Heisenberg's uncertainty principle, and Arrow's theorem), it is also true that in no other field (besides logic, that is) are negative results such a systematic and integral part of the field's research methodology and culture, and of course, the mother of all negative results is precisely Gödel's incompleteness theorem (of which Turing's halting problem can be seen as a sharpening). In this sense, Gödel has been a major source of inspiration to theoretical computer scientists.[6]

But Gödel's (1931) early work also contains several specific technical ideas that can be ex post recognized as computational. In his proof of the incompleteness theorem, Gödel uses primitive recursive functions, a class of functions from natural numbers to natural numbers (including the familiar functions of addition, multiplication, etc.), each of which can be defined, in an unmistakably algorithmic way, on the basis of certain very elementary functions; in terms of other previously defined functions; or even in terms of the function itself, albeit on smaller arguments. The primitive recursive functions, which had already been known to Ackermann and others, were extended in the hands of Kleene (1936), who arrived at one of the important equivalent ways of defining computation discovered during the 1930s (as recounted in Section 7.2). Gödel's (1940) proof of the consistency of the continuum hypothesis employs what is essentially an extension of the primitive recursive functions to sets.

Gödel's proof of the incompleteness theorem also relies on arithmetization, that is, on representing syntactic elements, such as logical terms, formulas, and proofs, as numbers. As it turns out, arithmetization has also been used crucially in theoretical computer science research. To mention only two celebrated examples, Adi Shamir's (1992) "IP = PSPACE" proof is an interactive algorithm for establishing that a particular quantified Boolean formula is satisfiable that works by first translating the formula into a complex arithmetic expression involving products and sums and then establishing that the expression evaluates to something different from zero. The PCP theorem of Arora et al. (1998) encodes a satisfying truth assignment to a Boolean formula in an ingenious and sophisticated way that has the counterintuitive property that errors (deficiencies in either the satisfaction part or the encoding) are either completely absent or are ubiquitous and so can be detected with high probability by rote sampling.

---

[6] Indeed, the most prestigious award for a research contribution in theoretical computer science is called the "Gödel Prize."

It is hard to discern the extent to which Gödel's proof of the incompleteness theorem was in any way an inspiration for these proofs. Certainly they are both very different in style from Gödel's arithmetization, which is, in some sense, tighter than both those proofs just outlined in that Gödel then turns around and closes the loop by looking at (properties of) the numbers resulting from his arithmetization as syntactic elements, something that does not occur clearly in either of the other cases. The next section recounts a more recent, and very different, use of arithmetization in the investigation of the complexity of Nash equilibria (Daskalakis et al., 2006), for which I can positively attest that Gödel's proof was, to a certain extent, an inspiration.

7.6 Latter-Day Arithmetization: Equilibria in Games

Two years before he met Gödel at Königsberg, von Neumann (1928) – building, as was his habit, on much less conclusive earlier work, this time by Emile Borel (1921) – had proved his famous min-max theorem: every two-player zero-sum game has an equilibrium.

We pause here to remind the reader of certain basic concepts from game theory: a *game* is a situation in which two or more *players* interact, each with a finite set of options (*strategies*). Once all players choose one strategy each, the *payoff* to each player, a real number, is determined. A game is *zero-sum* if, for any combination of strategies, the sum of the payoffs is zero, that is, no net payment ever enters or leaves the system. When the players choose, independently, a *randomized* strategy each (i.e., a distribution over their strategy set), then it is routine to calculate the *expected payoff* for each player. A *best response* of a player, given a strategy choice (randomized or pure) for all other players, is a strategy (again, pure or randomized) that maximizes the player's expected payoff. An *equilibrium* is a set of strategies (randomized or not), one for each player, each of which is a best response to all the rest taken together.

The beginning of game theory coincides with the publication of the book by John von Neumann and Oscar Morgenstern (1944); with this it became apparent that the case treated by von Neumann in his 1928 paper (namely, that of two-player, zero-sum games) is far too specialized. The question arose, are there more general contexts in which equilibria are guaranteed to exist?

Five years later, an ingenious graduate student at Princeton named John F. Nash (1951) established a sweeping result stating that "all games have a randomized equilibrium" – an existence theorem so momentous that the concept whose existence was established has since been called the "Nash equilibrium."

Games are used to model complex and consequential situations, and the development of game theory coincided with that of the computer era. Therefore it is no surprise that ever since Nash's proof, researchers have strived to develop algorithms for finding the equilibria guaranteed to exist by Nash's theorem (given data representing the players' utilities for each strategy combination); however, none of the algorithms developed for this problem is efficient; that is to say, none required only polynomial time. Whether the problem of computing Nash equilibria is in P became a persistent open question (for the special case of two-person, zero-sum games, the answer is positive via the min-max theorem and linear programming).

In search of an algorithm for finding Nash equilibria, it is natural to examine Nash's existence proof and determine whether it can be turned into an algorithm, that is, to see to what extent the proof is constructive. Unfortunately, this leads nowhere: Nash proved the existence of equilibria by a clever reduction to Brouwer's (1910) fixed-point theorem (stating that any continuous function f from any compact set, such as the unit cube, to itself has a fixed point: a point x such that $f(x) = x$), another important existence theorem whose proof is notoriously nonconstructive. Therefore computer scientists focused, predictably, on the complexity of finding a Brouwer fixed point, and it soon became apparent that this is, in many ways, a computationally difficult – "intractable" – task (Hirsch et al., 1989; Papadimitriou, 1994).[7] In other words, it is unlikely that Brouwer fixed points can be found in polynomial time. The question then became, is Nash's reliance on Brouwer's theorem inherent, or is there a different, more constructive proof of Nash's theorem?

This question was recently resolved (Daskalakis et al., 2006): there is a (polynomial) reduction from Brouwer's fixed points to Nash equilibria, and hence the two problems are computationally equivalent. Finding Nash equilibria is an intractable problem, and to return to our concerns in this chapter, this reduction relies crucially on a particular kind of arithmetization. Some highlights of the construction follow (see Daskalakis et al. (2006) for the complete proof and Papadimitriou (2007) for a more extensive overview as well as for a review of related work).

We start from an (appropriately represented) function f from the three-dimensional unit cube to itself (in three dimensions, the Brouwer fixed-point problem has already attained its full computational generality). Our plan is to create a game G_f such that there is a natural correspondence between the Nash equilibria of G_f and the fixed points of f. The game G_f has many players (the number of players is later reduced to four by a different maneuver; actually, it was ultimately lowered to two in subsequent work (Chen and Deng, 2006)).

Each player in G_f has two strategies, and thus the mixed strategy of each player can be considered a number in [0,1] – setting the basis for our arithmetization. Three distinguished players represent the three coordinates (x, y, z) of a point in the cube. The plan is to create a game in which, at any Nash equilibrium, these three players must represent a fixed point of f.

To achieve this, we create games that add, multiply, compare, and so on; that is, small games are defined, typically with two "input" players, one "output" player, and one or two other "internal" players, such that at any Nash equilibrium of the game of the output player (considered as a number, as explained earlier) is the product (or the sum, or the maximum, etc.) of the input players (also considered as numbers). These games are then combined to synthesize a complex game G_f that effectively computes the function f at the point (x, y, z) and (after overcoming a few further complications)

---

[7] The intractability of the problem of finding a Brouwer fixed point, as well as a Nash equilibrium, differs from NP-completeness (the usual concept of intractability in computer science) in important ways, having to do with the fact that in contrast with all search problems known to be NP-complete, fixed points and equilibria are guaranteed to exist. As a result, these problems are intractable in a sense that is both weaker and more sophisticated than NP-completeness; see Papadimitriou (1994, 2007) and Daskalakis et al. (2006) for extensive discussions of this point.

forces the Nash equilibria of G_f to coincide with the fixed points of f, completing the proof.

7.7 Epilogue

Kurt Gödel, the greatest logician since Aristotle, starved himself to death in a crisis of paranoia.[8] Emil Post died of complications from electroshock treatment for his depression. Georg Cantor was in and out of mental hospitals for much of his life, suffering from bipolar disorder. According to Gian-Carlo Rota (1997, 4):

> It cannot be a complete coincidence that several outstanding logicians of the twentieth century found shelter in asylums at some time in their lives: Cantor, Zermelo, Gödel, Peano, and Post are some.

Although I am aware of no justification for the presence of Ernst Zermelo or Giuseppe Peano in Rota's list, the lineup is devastating indeed. Furthermore, Luitzen Egbertus Jan Brouwer, the founder of intuitionistic logic (a point of view that scoffed at the foundations quest) and who, incidentally, introduced Gödel to logic during a 1927 visit to Vienna, died paranoid.[9] If one also recalls that David Hilbert and Bertrand Russell both had schizophrenic sons, then the connection between the foundations quest and insanity becomes a mystery that cannot be easily dismissed.

But the story of the foundations quest and its aftermath leading to the computer seethes with human suffering that goes beyond the tortured minds of many of its protagonists (the book by Martin Davis (2000) is a wonderful account of all this).[10] It unfolded through two world wars and periods of peace that were not much better. Many of its participants fled the rise of the Third Reich, whereas some, like Hilbert, did not – perhaps more tragically. Even though Gottlob Frege's reputation survived the destruction of his magnum opus in Russell's hands, it has been marred forever by his later anti-Semitic writings. John von Neumann died, in his early fifties and a few months after receiving Gödel's letter (which he is not known to have answered), of cancer – probably the result of radiation exposure that he suffered in his capacity as the architect not of the computer but of the U.S. hydrogen bomb. Finally, Alan Turing, the man who contributed as much as anybody to the advent of the computer – and possibly to the war effort – was hunted to his death, on account of his homosexuality, by the British establishment's unimaginable cruelty, bigotry, and ingratitude.

---

[8] Apostolos Doxiadis's play *Seventeenth Night* recounts Gödel's tragic end in a manner that is both touching and mathematically meaningful.

[9] John Nash, though certainly not a logician, did seminal work in game theory, the study of rational strategic behavior – a field that is arguably the rough analogue of logic for the second half of the twentieth century, the time of the rise of capitalism and the advent of the cold war, during which foundations were no longer contemplated or sought. He, too, suffered a mental breakdown, as famously described in *A Beautiful Mind: The Life of Mathematical Genius and Nobel Laureate John Nash*, by Sylvia Nasar (Simon and Schuster, 1998).

[10] The graphic novel *LOGICOMIX* (Bloomsbury, 2009), written by Apostolos Doxiadis and me, with art by Alecos Papadatos and Annie di Donna, tells the story of logic in the twentieth century, focusing on the foundations quest and the advent of the computer from Bertrand Russell's point of view and featuring Kurt Gödel.

Acknowledgments

I am grateful to Phokion Kolaitis, Martha Sideri, and especially Apostolos Doxiadis – my source on all things Gödel, and fellow student of the story of the foundations quest – for valuable discussions and comments on this chapter.

References

Arora, S., Lund, C., Motwani, R., Sudan, M., and Szegedy, M. (1998). Proof verification and the hardness of approximation problems. *Journal of the Association of Computing Machinery*, **45**, 501–55.

Borel, E. (1921). The theory of play and integral equations with skew symmetric kernels. *Comptes Rendus Hebdomadaires des Seances de l'Academie des Sciences*. [English trans. 1953 *Econometrica*, **21**, 97–100.]

Brouwer, L. E. J. (1910). Über eineindeutige, stetige Transformationen von Flächen in sich. *Mathematische Annalen*, **69**, 176–80.

Chen, X., and Deng, X. (2006). Settling the complexity of two-player Nash-equilibrium. In *Proceedings of the Forty-seventh Annual IEEE Symposium on Foundations of Computer Science*, pp. 261–72. Los Alamitos, CA: Institute of Electrical and Electronics Engineers.

Church, A. (1936). An unsolvable problem of elementary number theory. *American Journal of Mathematics*, **58**, 345–63.

Cobham, A. (1965). The intrinsic computational difficulty of functions. In *Proceedings of Logic, Methodology, and Philosophy of Science II*. Amsterdam, Netherlands: North-Holland.

Cook, S. A. (1971). The complexity of theorem-proving procedures. In *Proceedings of the Third Annual Association of Computing Machinery Symposium on Theory of Computing*, pp. 151–58. New York: Association of Computing Machinery.

Daskalakis, C., Goldberg, P. W., and Papadimitriou, C. H. (2006). The complexity of computing a Nash equilibrium. In *Proceedings of the Thirty-eighth Annual Association of Computing Machinery Symposium on Theory of Computing*, pp. 71–78. New York: Association of Computing Machinery.

Davis, M. (2000). *The Universal Computer: The Road from Leibniz to Turing*. New York: W. W. Norton.

Edmonds, J. (1965). Paths, trees, and flowers. *Canadian Journal of Mathematics*, **17**, 449–67.

Frege, G. (1884). *Die Grundlagen der Arithmetik: eine logisch-mathematische Untersuchung über den Begriff der Zahl*. Breslau, Germany: W. Koebner. [Trans. J. L. Austin (1974). *The Foundations of Arithmetic: A Logico-Mathematical Enquiry into the Concept of Number*. Oxford: Blackwell. 2nd rev. ed. (1893). *Grundgesetze der Arithmetik*. Vol. 1. Jena, Germany: Verlag Hermann Pohle. Partial trans. M. Furth (1964). *The Basic Laws of Arithmetic*. Berkeley: University of California Press.]

Gödel, K. (1929). On the completeness of the calculus of logic. PhD diss. In *Collected Works*, vol. 1 (1986), pp. 61–101.

———. (1930). Die Vollständigkeit der Axiome des logischen Funktionenkalküls. *Monatshefte für Mathematik und Physik*, **37**, 349–60. [Published PhD diss. Also in *Collected Works*, vol. 1 (1986), pp. 102–23.]

———. (1931). Über formal unentscheidbare Sätze der *Principia Mathematica* und verwandter Systeme I. *Monatshefte für Mathematik und Physik*, **38**, 173–98. [English trans. J. van Heijenoort, ed. (1967). *From Frege to Gödel: A Source Book on Mathematical Logic*. Cambridge, MA: Harvard University Press, pp. 596–616. Repr. with facing English trans. On formally undecidable propositions of *Principia Mathematica* and Related Systems. I. In *Collected Works*, vol. 1 (1986), pp. 145–95.]

———. (1940). *The Consistency of the Axiom of Choice and of the Generalized Continuum-Hypothesis with the Axioms of Set Theory.* Annals of Mathematics Studies 3. Princeton, NJ: Princeton University Press. [Rev. ed. 1953. Also in *Collected Works*, vol. 2 (1990), pp. 33–101.]

———. (1951). Some basic theorems on the foundations of mathematics and their implications. [Also in *Collected Works*, vol. 3 (1995), pp. 304–23 (Gibbs lecture).]

Hilbert, D. (1899). Grundlagen der Geometrie. In *Festschrift zur Feier der Enthüllung des Gauss-Weber-Denkmals in Göttingen*, pp. 1–92. Leipzig, Germany: Teubner.

———. (1900). Mathematische Probleme. Lecture given at the International Congress of Mathematicians, Paris.

———. (1905). Über die Grundlagen der Logik und der Arithmetik. In *Verhandlungen des dritten Internationalen Mathematiker-Kongresses in Heidelberg*, pp. 174–85. Leipzig, Germany: Teubner.

———. (1918). Axiomatisches Denken. *Mathematische Annalen*, **78**, 405–15.

Hilbert, D., and Ackermann, W. (1928). *Grundzüge der theoretischen Logik.* Berlin: Springer. [2nd ed. 1938. English trans. of 2nd ed. *Principles of Mathematical Logic.* New York: Chelsea.]

Hirsch, M. D., Papadimitriou, C. H., and Vavasis, S. A. (1989). Exponential lower bounds for finding Brouwer fix points. *Journal of Complexity*, **5**, 379–416.

Kleene, S. C. (1936). General recursive functions of natural numbers. *Mathematische Annalen*, **112**, 727–42.

Levin, L. A. (1973). Universal search problems. *Problemi Peredachy Informatsii*, **9**, 265–66.

Nash, J. (1951). Noncooperative games. *Annals of Mathematics*, **54**, 289–95.

Papadimitriou, C. H. (1994). On the complexity of the parity argument and other inefficient proofs of existence. *Journal of Computer and System Sciences*, **48**, 498–532.

———. (2007). The complexity of Nash equilibria. In *Algorithmic Game Theory*, ed. N. Nisan, T. Roughgarden, E. Tardos, V. Vazirani, et al., pp. 29–51, New York: Cambridge University Press.

Poincaré, H. (1914). *Wissenschaft und Methode.* Leipzig, Germany: Teubner.

Post, E. (1936). Finite combinatory processes: Formulation 1. *Journal of Symbolic Logic*, **1**, 103–5.

Rota, G. (1997). *Indiscrete Thoughts.* Boston: Birkhäuser.

Shamir, A. (1992). IP = PSPACE. *Journal of the Association of Computing Machinery*, **39**, 869–77.

Sieg, W. (2006). Gödel on computability. *Philosophia Mathematica, Series III*, **14**, 189–207.

Sipser, M. (1992). The history and status of the P versus NP question. In *Proceedings of the Twenty-fourth Annual Association of Computing Machinery Symposium on Theory of Computing*, pp. 71–78. New York: Association of Computing Machinery.

Turing, A. M. (1936). On computable numbers, with an application to the *Entscheidungsproblem*. *Proceedings of the London Mathematical Society, Series 2*, **42**, 230–65. [Erratum 1937 *Series 2*, **43**, 544–46.]

Von Neumann, J. (1928). Zur Theorie der Gessellshaftsspiele. *Mathematische Annalen*, **100**, 295–320.

———. (1945). First draft of a report on the EDVAC. Philadelphia: Moore School of Electrical Engineering, University of Pennsylvania. [Repr. 1993 *Institute of Electrical and Electronics Engineers (IEEE) Annals of the History of Computing*, **15**, 27–75.]

———. (1953). Certain zero-sum two-person game equivalent to the optimal assignment problem. *Contributions to the Theory of Games, II, Annals of Mathematics Studies*, **28**, 5–12.

Von Neumann, J., and Morgenstern, O. (1944). *Theory of Games and Economic Behavior.* Princeton, NJ: Princeton University Press.

Wang, H. (1996). *A Logical Journey: From Gödel to Philosophy.* Cambridge, MA: MIT Press.

CHAPTER 8

From the *Entscheidungsproblem* to the Personal Computer – and Beyond

B. Jack Copeland

Gödel's classic paper of 1931 did not settle Hilbert's *Entscheidungsproblem*. The young logician Alan Turing took up the problem and, in the course of his attack, conceived the basic principle of the modern computer – the idea of controlling the machine's operations by means of a program of coded instructions stored in the computer's memory. The transition from mathematical logic to electronic hardware took twelve years (1936–1948), and Turing played a central role in this transformation.

In a reference to Turing's famous paper of 1936, "On Computable Numbers, with an Application to the *Entscheidungsproblem*,"[1] one of the pioneers of the electronic computer, John Womersley, aptly described the new machines as "Turing in hardware" (Womersley, 2005 [1946], 39). With Gödel's (1931) introduction of the idea of representing logical and arithmetical statements as numbers, together with his foundational contributions to recursion theory, the new machines might also be described, in a more distant sense, as "Gödel in hardware." Gödel himself, however, took little interest in the development of the electronic computer (Wang, 1987, 171).

By what steps did the logicomathematical results of the 1930s lead to the modern stored-program digital computer? The answer offered by computing folklore runs "von Neumann . . . ENIAC . . . EDVAC. . . . Princeton"; the true story is rather different, however. This chapter charts the development of the electronic stored-program digital computer, from the extraordinary Colossus computers built for code breaking during World War II to the first successful run of a stored program in 1948. Turing's own Automatic Computing Engine (ACE), the fastest of the pioneer machines, became a foundation stone of the developing British computer industry and was the inspiration for the earliest single-user desk-side computers – the first personal computers. The epilogue compares Gödel's and Turing's views on the relationship between computation and mathematics.

---

[1] Turing's paper is incorrectly stated to have been published in 1937 in Gödel's *Collected Works* (see, e.g., vol. 1 (1986), 456) as well as in Andrew Hodges' (1992) biography of Turing. The circumstances of publication of Turing's paper are described in Copeland (2004, 5).

8.1 Turing and the *Entscheidungsproblem*

Gödel's 1931 paper did not settle Hilbert's decision problem for the first-order predicate calculus. In fact, given Church's thesis, Gödel's proof of theorem X of the paper does entail the unsolvability of the decision problem, but this was not noticed until much later. (It seems that the point was first noted by Martin Davis (1965; see also see Kleene, 1986, 136).) It is probably fortunate that Gödel did not settle the decision problem in 1931, for if he had, then the young logician Alan Turing would not have taken it up, and the history of the computer might have unfolded quite differently (and perhaps much less satisfactorily, in the absence of Turing's fundamental concept of universality).

Hilbert's *Entscheidungsproblem* (decision problem) for the *engere Functionenkalkül* **K** (first-order predicate calculus), as stated by Turing, is the following: is there a general (mechanical) process for determining whether a given formula A of the functional calculus **K** is provable (Turing, 1936, 259; also in Copeland, 2004, 84). Turing heard of the *Entscheidungsproblem* in lectures given at Cambridge by Max Newman (a Fellow of St. John's College). Turing, who went up to King's College in October 1931 to read mathematics and was elected a Fellow of King's in spring 1935 at the age of only twenty-two, attended Newman's advanced lectures on the "Foundations of Mathematics" from January to March 1935.[2] Yorick Smythies attended this course of lectures in 1934 and took detailed notes: Newman covered the Hilbert program, the propositional and predicate calculus, cardinals, the theory of types and the axiom of reducibility, Peano arithmetic, Hilbert on proving consistency, and Gödel's first and second incompleteness theorems.[3] Newman mentioned that the *Entscheidungsproblem* had been settled only in the special case of monadic expressions. As everyone knows, Turing took on the *Entscheidungsproblem* and showed it to be unsolvable, along the way inventing the universal Turing machine. In a single brilliant paper, Turing ushered in both the modern computer and the study of unsolvability (the area of mathematical logic concerning problems "too hard" to be solvable by the universal Turing machine).

Newman's contribution was not limited to bringing the *Entscheidungsproblem* to Turing's attention: in the lectures, Newman defined a constructive process as one that a *machine* can carry out. He explained in an interview:

> I believe it all started because [Turing] attended a lecture of mine on foundations of mathematics and logic. . . . I think I said in the course of this lecture that what is meant by saying that [a] process is constructive is that it's a purely mechanical machine – and I may even have said, a machine can do it. . . .
>
> And this of course led [Turing] to the next challenge, what sort of machine, and this inspired him to try and say what one would mean by a perfectly general computing machine.[4]

"On Computable Numbers" is the birthplace of the fundamental principle of the modern computer, the idea of controlling the machine's operations by means of a

---

[2] *Cambridge University Reporter*, April 18, 1935, p. 826.

[3] Notes taken by Yorick Smythies during Newman's "Foundations of Mathematics" lectures in 1934. St. John's College Library, Cambridge.

[4] Max Newman in a tape-recorded interview with Christopher Evans ("The Pioneers of Computing: An Oral History of Computing," London: Science Museum (a set of tape recordings) also quoted in Copeland (2004, 206; transcription by Copeland).

program stored in the computer's memory. Place a different program on the memory tape of the universal Turing machine, and the machine will carry out a different computation. It was a fabulous idea. It led to the now ubiquitous "one-stop-shop" computer: the single slab of machinery of fixed structure that makes use of symbolic instructions stored in rewritable memory to become a word processor, or desk calculator, or chess opponent, or photo editor – or any other machine we have the skill to create in the form of a program. Turing's universal machine has changed the world.

After Newman learned of Turing's universal computing machine early in 1936, he developed an interest in computing machinery, which he described as being, at that time, "rather theoretical."[5] Turing himself was interested right from the start in building a universal computing machine (Newman, 1954) but knew of no suitable technology. It was not until his and Newman's wartime days as code breakers at Bletchley Park that the dream of building a miraculously fast, all-purpose, *electronic* computer took hold of them. However, before turning to that side of the story, let us note what Gödel had to say about Turing's paper of 1936. (The final section of this chapter contains more on this topic.)

Gödel described Turing's analysis of computability as "most satisfactory" and "correct ... beyond any doubt."[6] He also remarked, "We had not perceived the sharp concept of mechanical procedures sharply before Turing, who brought us to the right perspective" (Wang, 1974, 85). Gödel was not persuaded by Church's thesis until he saw Turing's formulation. Kleene (1981, 59–61) wrote, "According to a November 29, 1935, letter from Church to me, Gödel 'regarded as thoroughly unsatisfactory' Church's proposal to use λ-definability as a definition of effective calculability.... It seems that only after Turing's formulation appeared did Gödel accept Church's thesis."

It was Turing's work, Gödel emphasized, that enabled him to generalize his incompleteness result of 1931 (originally directed specifically at the formal system set out by Whitehead and Russell in their *Principia Mathematica* (1910–1913)) to "*every* consistent formal system containing a certain amount of finitary number theory" (Gödel, 1965, 71).

Turing, on the other hand – always a lone worker – seems to have paid little attention to Gödel's work in the course of his attack on the *Entscheidungsproblem*. R. B. Braithwaite, another Fellow of King's, wrote (in a letter to Margaret Boden, October 21, 1982) of "Turing's complete ignorance of Gödel's work when he wrote his 'Computable Numbers' paper," adding, "I consider I played some part in drawing Turing's attention to the relation of his work to Gödel's."

8.2 Early Days

In the Victorian era, the farsighted Charles Babbage dreamed of building a huge, general-purpose calculating machine, a single "engine" (as he called it) that could take over the work of hundreds of human calculators. It was the railway age, and Babbage proposed to build computers from types of mechanical components then found

---

[5] Newman interviewed by Evans, "The Pioneers of Computing: An Oral History of Computing."
[6] "Some Basic Theorems on the Foundations of Mathematics and Their Implications," *Collected Works*, vol. 3 (1995), p. 304, and "Undecidable Diophantine Propositions," *Collected Works*, vol. 3 (1995), p. 168.

in railway engines and other Victorian industrial machinery. His first project was a special-purpose machine, the Difference Engine, for the production of mathematical tables (e.g., logarithm tables).[7] In 1822, Babbage unveiled a working model of the Difference Engine, but he never completed the full-scale machine. A still more ambitious proposal was Babbage's Analytical Engine, an all-purpose mechanical digital computer.[8] Babbage emphasized that "the conditions which enable a finite machine to make calculations of unlimited extent are fulfilled in the Analytical Engine" (Babbage, 1989 [1864], 97). The Analytical Engine was never built, but a small model was constructed just before Babbage's death.

Despite being nonelectrical, Babbage's Analytical Engine had many similarities with modern computers, including the ability to branch "conditionally" according to the result of a previous calculation. It was to have an arithmetic unit, which, in the idiom of the time, Babbage called the "Mill," and an internal memory, which Babbage called the "Store." The Engine was to be controlled by a program of instructions contained on punched cards. Turing (1950, 439, 450) described the Analytical Engine as a "universal digital computer" and said generously that "Babbage had all the essential ideas" of the digital computer (see also Copeland, 2004, 446, 455).

Nevertheless, Babbage's machine lacked a key feature of the universal Turing machine: in Turing's machine, there is no fundamental distinction between program and data. Arguably, it is the absence of such a distinction that demarcates a stored-program computer from a merely program-controlled computer. As Robin Gandy (once Turing's student) put the point, Turing's "universal machine is a stored-program machine [in that], unlike Babbage's all-purpose machine, the mechanisms used in reading a program are *of the same kind* as those used in executing it" (1998, 90; emphasis added).

Post-Babbage, a number of purely mechanical computing machines were constructed, including a modified version of the Difference Engine, built and successfully marketed by the Swedes Georg and Edvard Scheutz. Babbage's idea of an all-purpose calculating engine was never forgotten. In conversation with me in 1996, T. H. Flowers, who designed and built Colossus, the first large-scale electronic computer, recalled Babbage's ideas being discussed at wartime Bletchley Park (the British code-breaking headquarters where Colossus was used). Babbage was an influence on the American pioneer Howard Aiken (Aiken, 1982 [1937]), whose giant electromechanical Automatic Sequence Controlled Calculator at Harvard is sometimes said to have been the first computer. Aiken's machine did not embody the stored-program concept, however. Instructions were fed in on punched tape (echoing Babbage's scheme for programming the Analytical Engine, where instructions were to be fed in on punched cards connected together with ribbon to form a continuous strip). Control was via the punched tape rather than via modification of internally stored instructions (as with modern computers). For example, if a program loop were required, this would be achieved by feeding the same instructions repeatedly through the tape reader.

---

[7] For further information on the Difference Engine, see Campbell-Kelly (1989), Randell (1982), and Swade (2001).

[8] For further information on the Analytical Engine, see Bromley (1982), Lovelace and Menabrea (1953 [1843]), and Swade (2001).

Figure 8.1. Part of Babbage's Difference Engine. This working fragment of the planned machine – one-seventh of the calculating section – has been described as the "first successful physical realisation of automatic calculation" (Swade, 2010). Assembled in 1832, it was the only part of the Engine that Babbage ever completed. Science and Society Picture Library, Science Museum, London. Reproduced with permission.

Aiken's computer was not electronic. Electromechanical equipment, such as the Automatic Sequence Controlled Calculator, was based on the *relay*, a small switch consisting of a mechanical circuit breaker (a metal rod) that is operated by a magnetic field created by electric current passing through a coil of wire. Relays are clunky and slow in comparison with electronic equipment. Others besides Aiken – notably Turing himself, George Stibitz at Bell Telephone Laboratories, and Konrad Zuse in Berlin – produced electromechanical computing machines in the preelectronic era. None of the electromechanical computers embodied the stored-program concept, however. Programming was done by rerouting wires (e.g., by means of plug boards and switches).

To Zuse belongs the honor of having built the first general-purpose programmable digital computer. His electromechanical Z3 (as it was later called) was working in 1941 (Zuse, 1980). Turing's Bombe, designed during the last months of 1939, was a special-purpose electromechanical computer. The Bombe's function was to search through the possible settings of the German Enigma machine until settings were found that would decrypt the enemy's message.[9] The prototype Bombe (named simply "Victory") was installed at Bletchley Park in spring 1940 and was used exclusively by Turing and his unit in their efforts to break German U-boat Enigma messages. Like the other electromechanical computers of the era, the Bombe offered superhuman speed: it could carry out in hours a search that would take a human clerk weeks or months.

Superhuman–but feeble compared with the speed of the electronic machines that would soon be developed. Electronic valves (vacuum tubes) operate very many times faster than relays because the valve's only moving part is a beam of electrons. It was the development of high-speed digital techniques using valves that made the modern computer possible. Valves were used originally for purposes such as amplifying radio signals. The output would vary continuously in proportion to a continuously varying input, for example, a radio signal representing speech. It was the novel idea of using the valve as a very fast switch, producing pulses of current – pulse for 1, no pulse for 0 – that was the route to high-speed digital computation.

8.3 Colossus: The First Large-Scale Electronic Computer

Newman spent the early years of the war lecturing in Cambridge. He and Turing kept in close touch after Turing left Cambridge for Bletchley in 1939. Turing's first wartime letter to Newman starts, "Dear Newman, Very glad to get your letter, as I needed some stimulus to make me start thinking about logic."[10] Despite the pressures of code breaking, Turing found time to collaborate with Newman on a paper on Church's theory of types (Newman and Turing, 1942). If he had a night off-duty from Enigma, Turing would be seen "coming in as usual . . . doing his own mathematical research at night, in the warmth and light of the office, without interrupting the routine of daytime sleep," recalled fellow code breaker Joan Murray, née Clarke (Murray, 1993, 117). In 1942, Newman was also beckoned by Bletchley Park. He wrote to the Master of St. John's College to request leave of absence, and at the end of August, he joined Colonel John Tiltman's Research Section (Newman, 2006). Tiltman, an old hand, had been breaking Russian, Japanese, and German ciphers since 1920. When Newman arrived at Bletchley, Tiltman's group was attempting to break the German cipher machine that the British named "Tunny." Tunny was the raison d'être of Colossus.

Colossus was designed and built by telephone engineer Tommy Flowers, a neglected pioneer of computing.[11] Before the war, he made groundbreaking use of electronic valves. In his laboratory at the Post Office Research Station at Dollis Hill (in North

---

[9] See "Enigma," "History of Hut 8 to December 1941," and "Bombe and Spider" in Copeland (2004).

[10] A selection of their letters is in Copeland (2004, chap. 4); an extract from one is in Section 8.5 below.

[11] Material relating directly to Flowers derives from (1) Flowers in interviews with Copeland, 1996–1998, and (2) Flowers in an interview with Christopher Evans in 1977 (see Endnote 4).

Figure 8.2. Operators at the controls of Flowers' Colossus. The long loop of punched paper tape that is wound around the pulley wheels contained the encrypted German message. Colossus broke its first message at Bletchley Park in January 1944. British National Archives Image Library, Kew. Crown copyright. Reproduced with permission.

London), Flowers designed telecommunications equipment involving as many as three thousand to four thousand valves per installation. Outside Flowers' small group, few knew that electronic valves could be used reliably in such large numbers. According to conventional wisdom, valves were too unreliable to permit large-scale use. Each valve contained a hot, glowing filament, rather like a lightbulb, and the delicacy of this filament meant that valves would suddenly cease operating. The greater the number of valves in the equipment, the shorter the mean time to failure. Large installations were therefore not practical, or so it was believed. This belief was largely based on experience with radio receivers and the like, which were switched on and off frequently. However, Flowers discovered that as long as valves were left on continuously, they were in fact more reliable than relays – and, of course, immensely faster.

The earliest known use of valves for digital calculation was by John Atanasoff at Iowa State College, although his machine was on a much smaller scale than the equipment built by Flowers. During 1939–1942, Atanasoff constructed a small, special-purpose electronic digital computing machine for solving a certain type of mathematical equation (Atanasoff, 1982 [1940]). Unfortunately, his computer never worked properly because errors were introduced into the calculations by a persistently buggy card reader; but the electronic parts of this three-hundred-valve machine did function satisfactorily. Atanasoff left the computer incomplete when he became involved in war work in 1942, and he never returned to it. In 1941, John Mauchly paid a visit to Atanasoff's Iowa laboratory and was greatly interested in his nascent electronic digital computer (Burks, 2003, 52–53). With engineer Presper Eckert, Mauchly went on to design and build the

ENIAC (Electronic Numerical Integrator and Computer); the story of the ENIAC will be picked up again later in this chapter.

Digital electronics was a little-known field in those days. Flowers told me that at the outbreak of war with Germany, he was possibly the only person in Britain who realized that valves could be used for large-scale high-speed digital computing. So he turned out to be the right man in the right place when he was ordered to join Turing at Bletchley Park. Ironically, Flowers was brought into the code-breaking work because of his vast experience with relays, not because of his knowledge of electronics, and his first job was to design electromechanical equipment required for use against the Enigma cipher machine.

Flowers built Colossus to attack not Enigma but Tunny, a very different cipher machine. (Tunny was known to its German manufacturers, the Lorenz company, as the *Schlüsselzusatz* – "cipher attachment" – SZ40/42.) The story of Enigma is well known, but the story of Tunny has only recently been made public.[12] Tunny was considerably more sophisticated than Enigma. The Enigma machine was first marketed in 1923, and even though extensive modifications were introduced by the German military, Enigma was no longer state-of-the-art equipment by the time war broke out in 1939.

The Enigma machine was clumsy to use: three operators were required to encipher and send a message. The cipher clerk typed the ordinary German (the "plaintext") at the keyboard of the Enigma machine, while his assistant read the enciphered output (the "ciphertext") via an arrangement of twenty-six small lamp bulbs located near the keyboard. When a lamp bulb lit up, it illuminated a stenciled letter. As the letters of the ciphertext appeared one by one at the lamps, the clerk wrote them down on paper. Once encipherment was complete, the clerk handed the paper to the third player, the radio operator, who converted the enciphered message into Morse code and transmitted it.

Clearly there was room for improvement, and in 1940, the Lorenz company produced the thoroughly up-to-date Tunny machine. Only one operator was required at each end of the radio link. The sending operator typed the plaintext at the keyboard of a teleprinter (teletypewriter) that was attached to the Tunny machine, and the rest was automatic. The encrypted output of the Tunny machine went directly to air. Morse code was not used at all. Decryption was also automatic, and neither the sending nor the receiving operator ever even saw the "ciphertext." Unlike Enigma, Tunny carried only the highest grade of intelligence – messages from Hitler and the Army High Command to the generals in the field. The Tunny decrypts produced by Bletchley Park contained intelligence that changed the course of the war in Europe.

Messages from the unknown new machine were first intercepted by the British in June 1941. At first, nothing could be read, but then Tiltman scored a tremendous success, managing to break a message of about four thousand keyboard characters in length. Tiltman had very little to go on. Tunny was believed to encipher messages expressed in international binary teleprinter code (which was in widespread commercial use at that time). In teleprinter code, each keyboard character is expressed in the form

---

[12] In my 2006 book *Colossus* and in a two-volume report written at Bletchley Park in 1945 and declassified in 2000, *General Report on Tunny* (National Archives/Public Record Office, document reference HW 25/4 (vol. 1), HW 25/5 (vol. 2)). A digital facsimile of the complete *General Report on Tunny* is in The Turing Archive for the History of Computing: http://www.AlanTuring.net/tunny_report.

Figure 8.3. Tunny. When attached to a teletypewriter/teleprinter, the Tunny machine encrypted the text that the operator typed at the keyboard. The rims of the Tunny's encoding wheels are visible behind the see-through panel. Picture Library, Imperial War Museum, London. Photograph restored by Jack Copeland and Dustin Barrett. Reproduced with permission.

of five zeroes and ones (or dots and crosses, in Bletchley notation). Figure 8.4 shows a wartime memory aid listing the pattern of dots and crosses associated with each keyboard character. It also seemed likely that Tunny was an *additive* cipher machine. An additive machine would generate within itself a stream of letters and other keyboard characters, and this was automatically added to the plaintext to form the ciphertext. At Bletchley, this internally generated stream was called the "key." Symbolically, P + K = Z, where P is the plaintext, K is the key, and Z is the ciphertext.

An additive system should have the property that Z + K = P (i.e., (P + K) + K = P) so that the receiver's machine can decrypt the message by producing the same key and adding it to the ciphertext. For the receiving machine to generate the same key, the receiving operator had to know which settings of the machine had been used by the sender to generate the key. There were various ways of arranging this. For example, the sender might transmit these settings to the receiver (in a suitably masked form), or the sender and receiver might be issued identical instruction books that would tell them what settings should be used for their first message on a particular day, and then for their second message on that day, and so on. Initially, the former method was used by Tunny operators, but later the second, more secure method was introduced.

The property Z + K = P was achieved in Tunny by the fact that, at the bit level, letter addition was Boolean exclusive disjunction. That is to say, the teleprint equivalents of keyboard characters were added in accordance with the rules: dot plus dot is dot, cross plus cross is dot, dot plus cross is cross, and cross plus dot is cross. For example, A (**xx•••**) + B (**x••xx**) = G (**•x•xx**). (A + B) + B = A: if B is added to G, A is retrieved.

Sometimes the sending Tunny operator would foolishly use the same settings for two messages. At Bletchley, this was called a "depth." In the early days of the Tunny system, depths could easily be spotted by the British interceptors listening to the German radio transmissions, because the sending operator also broadcast the settings used to produce

CONVENTIONAL NAME	IMPULSE 1	2	3	4	5
/	.	.	.	.	.
9	.	.	x	.	.
H	.	.	x	.	x
T	.	.	.	.	x
O	.	.	.	x	x
M	.	.	x	x	x
N	.	.	x	x	.
3	.	.	.	x	.
R	.	x	.	x	.
C	.	x	x	x	.
V	.	x	x	x	x
G	.	x	.	x	x
L	.	x	.	.	x
P	.	x	x	.	x
I	.	x	x	.	x
4	.	x	.	.	.
A	x	x	.	.	.
U	x	x	x	.	.
Q	x	x	x	.	x
W	x	x	.	.	x
5 or +	x	x	.	x	x
8 or -	x	x	x	x	x
K	x	x	x	x	.
J	x	x	.	x	.
D	x	.	.	x	.
F	x	.	x	x	.
X	x	.	x	x	x
B	x	.	.	x	x
Z	x	.	.	.	x
Y	x	.	x	.	x
S	x	.	x	.	.
E	x	.	.	.	.

Figure 8.4. A wartime memory aide showing the teleprint code for each of the thirty-two keyboard characters used in the Tunny system. In modern notation, the dot would be written "0" and the cross "1." Recreated digitally from a typed card by Jack Copeland and Olwen Harrison.

the key (in masked form). Depths were often the result of something going wrong during the encryption and transmission of a message. The German operator would start again from the beginning of the message, using the same settings. If the message were repeated identically, the depth would be of no help to the code breakers. If, however, the sending operator introduced typing errors during the second attempt, or abbreviations or other variations, the depth would consist of two not-quite-identical plaintexts, each encrypted by means of exactly the same key – a code breaker's dream.

It was such a depth that Tiltman decrypted in late summer 1941, giving the code breakers their first entry into Tunny. On the hypothesis that Tunny was additive, he added the two ciphertexts. If the hypothesis were correct, this would have the effect of canceling out the key and would produce a sequence of keyboard characters consisting of the two plaintexts added together character by character (because $Z_1 + Z_2 = (P_1 + K) + (P_2 + K) = ((P_1 + P_2) + K) + K = P_1 + P_2$). Tiltman managed to prize the two individual plaintexts out of this sequence. It took him ten days. He had to guess at words of each message, and Tiltman was a very good guesser. Each time he guessed a word from one message, he added it to the characters at the relevant place in the $P_1 + P_2$ sequence, and if the guess was correct, an intelligible fragment of the second message would pop out. For example, adding the probable word *geheim* (secret) at a particular place in the sequence revealed the plausible fragment "eratta" (Bauer, 2006, 372). This short break could then be extended to the left and right. More letters of the second message are obtained by guessing that "eratta" is part of *militaerattache* (military attaché), and if these letters are added to their counterparts in the $P_1 + P_2$

sequence, further letters of the first message are revealed, and so on. Eventually, Tiltman achieved enough of these local breaks to realize that long stretches of each message were the same, and so he was able to decrypt the whole thing. By adding one of the resulting plaintexts to the ciphertext, he was then able to extract the four thousand or so characters of key that had been used to encrypt the message (since $P + Z = K$).

Breaking one message was a far cry from knowing how the Tunny machine worked. The British knew the internal workings of the military Enigma machine even before the war began (thanks to the Poles), and then a number of Enigmas were captured on land and sea after hostilities had commenced. Tunny, on the other hand, was a complete blank: no machine had been captured, nor would one be until the war was almost over. In January 1942, William Tutte, a member of the Research Section, was able to deduce the structure of the Tunny machine by studying the key that Tiltman had retrieved. It was one of the most astonishing pieces of cryptanalysis of the war.

Tutte deduced that the key was produced by the addition, within the machine, of two separate streams of characters. He inferred further that each of these two streams was produced by a different set of five wheels (five because there are five bits in the teleprint representation of each character). He called the first set the chi wheels and the second set the psi wheels (after the Greek letters χ and ψ). The chi wheels, he revealed, moved regularly, whereas the psi wheels moved in an irregular way, under the control of two further wheels ("motor" wheels). Tutte's secret deductions saved untold lives.

Breaking a cipher system requires two steps. First, it is necessary to discover the structure of the cipher machine, and this Tutte had now done. (This first step is today known as breaking the machine.) Second, a method must be found for breaking the messages as they are delivered by the interception operators, and the method must work swiftly enough to enable the message traffic to be deciphered before the intelligence is out of date. It was Turing who devised the first method used against Tunny, which became known simply as Turingery.[13] Turingery was the third of the three fundamental contributions that Turing made to the defeat of Nazi Germany, along with the Bombe and his unveiling of the special features of the system of Enigma used by the Atlantic U-boats.[14] Mathematician Jack Good, who worked on both Enigma and Tunny, observed, "I won't say that what Turing did made us win the war, but I daresay we might have lost it without him."[15]

Turingery depended on a technique, introduced by Turing, called "differencing" or "delta-ing" (after the Greek letter Δ), a process of sideways addition. To delta the four characters ABCD, one adds A to B, B to C, and C to D (producing three characters). It might easily be thought that the process of adding the letters of an encrypted message together in this way would scramble any information that was present, but Turing showed that delta-ing would in fact *reveal* information that was otherwise hidden. Delta-ing was the basis, not just of Turingery, but of every main algorithm used in Colossus – the entire computer-based attack on Tunny flowed from this basic insight of Turing's. Turingery, though, was itself a paper-and-pencil method, and it depended on insight – on what you "felt in your bones," Tutte said.

---

[13] Turingery is described in full in Copeland (2006, app. 6).
[14] See "Enigma" in Copeland (2004).
[15] Good in an interview with Pamela McCorduck (1979, 53).

Tutte described Turingery as "more artistic than mathematical" (Tutte, 2006, 360). He wanted a method for breaking Tunny messages that required no use of what you "felt in your bones" – the sort of method that a machine could carry out – and in November 1942, he found it, using an ingenious extension of delta-ing.[16] It was his second great contribution to the attack on Tunny. However, there was a snag. Tutte's method required a huge amount of calculation. If the method were carried out by hand, as Turingery was, it could take years to decrypt a single message. Tutte explained his new method to Newman, and Newman suggested using electronic counters to do the calculations. It was an Archimedean moment. Newman was aware that, before the war, electronic counters had been used in the Cavendish Laboratory at Cambridge to count radioactive emissions, and in a flash of inspiration, he saw that this technology could be applied to the very different problem of breaking Tunny messages. He sold his idea to the head of Bletchley Park (probably with Turing's assistance) and was put in charge of building a suitable machine.

The prototype of Newman's machine, installed in 1943, was soon dubbed "Heath Robinson," after a cartoonist who drew absurd devices. Heath Robinson was largely electromechanical – only the counters were electronic, and it contained no more than a few dozen valves. Newman's wonderful contraption proved the feasibility of carrying out Tutte's method by machine, but it was slow and prone to inaccuracy and break-down. Meanwhile, Flowers made a daring proposal: an all-electronic machine. He had been called in (at Turing's suggestion) to help with the design of Heath Robinson's electromechanical logic unit. Flowers was skeptical from the start about whether the Robinson's relay-based design could give the code breakers the accuracy and speed that they needed. He also knew that by using valves in large numbers, he could con-veniently store information that, in Heath Robinson, had to be supplied by means of wear-prone paper tape running through the machine. Flowers proposed an electronic monster containing about two thousand valves.

Newman took advice, and he was told that such a large installation of valves would never work reliably. Bletchley Park declined to support Flowers' proposal. In his laboratory at Dollis Hill, Flowers quietly got on with building the all-electronic machine that he could see the code breakers needed. He and his team of engineers worked day and night for ten months to build Colossus – worked until their "eyes dropped out," Flowers said. On January 18, 1944, Flowers' lads took Colossus to Bletchley Park in a truck. Flowers recalled:

> I don't think they really understood what I was saying in detail – I am sure they didn't – because when the first machine was constructed and working, they obviously were taken aback. They just couldn't believe it! . . . I don't think they understood very clearly what I was proposing until they actually had the machine.[17]

The name "Colossus" was apt – the computer weighed about a ton. At its trial runs, the code breakers were astonished by the speed of the machine and by the fact that, unlike Heath Robinson, Colossus would always produce the same result if given the

---

[16] Tutte's method is described in full in Copeland (2006, chap. 5 and app. 4).

[17] Flowers in an interview with Christopher Evans in 1977 (see Endnote 4); quoted in Copeland (2006, 75; transcription by Copeland).

same problem again. Colossus began real work on the German messages on February 5, 1944. (Flowers noted laconically in his diary for that day that "Colossus did its first job. Car broke down on way home.") The function of Colossus was to strip away from the ciphertext the component of the key that the five chi wheels had contributed (using Tutte's method). The resulting "de-chi" could almost always be broken rapidly by hand because of distinctive patterns in the component of the key that was contributed by the five psi wheels. These patterns were caused by the irregular movement of the psi wheels – presumably, the Tunny machine's designers had believed that this irregularity would strengthen the machine, but in fact, it introduced a crucial weakness.

By the end of the war in Europe, ten Colossi were working at Bletchley Park, in a section headed by Newman called simply the "Newmanry." The Newmanry was the world's first electronic computing facility. When the fighting ceased, two Colossi were retained for secret use by the code breakers, and the remainder were destroyed, on Churchill's orders.

That this reversal of scientific progress was unknown to the outside world hardly lessens the magnitude of the blow. The Colossi could have become part of public science. Flowers' engineers would quickly have adapted them for new applications, and they might have become the heart of a scientific research facility. This could have been in operation even before the ENIAC first ran in December 1945 (two years after the first Colossus). With eight massive electronic computers in the public arena from mid-1945, the story of modern scientific computing would have begun very differently. Who can say what changes this would have brought in its wake.

The secret of Colossus was a long time in coming out. Those associated with the story went to great lengths to obey their orders to disclose nothing of what they knew. Newmanry operator Catherine Caughey even feared going to the dentist in case she talked while under the anesthetic! Jerry Roberts, who broke the de-chis delivered by Colossus, regretted that when his parents died in the 1970s, they knew nothing of the work in which he was engaged during 1941–1945 at Bletchley – work of such importance that, in different circumstances, he might reasonably have expected a knighthood from the British crown. When the two-volume 1945 *General Report on Tunny* – which lays bare the whole incredible story of Colossus – was declassified in 2000, Caughey's great regret was that she could never tell her husband about her other life in the Newmanry; he had died in 1975, knowing nothing of his wife's extraordinary work as an operator of the first large-scale electronic computer. Helen Currie, whose job it was to produce the complete German plaintext after Roberts, or one of his colleagues, had broken part of a de-chi, described the burden of being unable to share her memories with her family: during the "years of silence," she said, her wartime experiences took on "a dream-like quality, almost as if I had imagined them."[18] Newman's son William (who himself worked in the computing industry) said that his father spoke to him only obliquely about his war work and died "having told little." Given the secrecy, it is not so surprising that books on the history of computing regularly state that the Eckert-Mauchly ENIAC was the first programmable electronic computer.

---

[18] Caughey, Roberts, Currie, and (William) Newman tell their stories in Copeland (2006, chaps. 13, 14, 18, 20).

Figure 8.5. The ENIAC. Commissioned by the U.S. Army, the computer's function was to calculate gunnery tables. The ENIAC was built at the Moore School of Electrical Engineering (part of the University of Pennsylvania) and was first worked in December 1945. Subsequently, this pioneering electronic computer was used for a wide variety of applications. Collections of the University of Pennsylvania Archives. Reproduced with permission.

In later life, Flowers was bitter about the secrecy. He said:

> When after the war ended I was told that the secret of Colossus was to be kept indefinitely I was naturally disappointed. I was in no doubt, once it was a proven success, that Colossus was an historic breakthrough, and that publication would have made my name in scientific and engineering circles – a conviction confirmed by the reception accorded to ENIAC, the U.S. equivalent made public just after the war ended. I had to endure all the acclaim given to that enterprise without being able to disclose that I had anticipated it. (Flowers, 2006, 82)

8.4 Next: The Stored Program

Colossus had no provision for storing a program of instructions internally (although Flowers had read "On Computable Numbers" at Newman's suggestion). It was programmed by means of plug boards and switches. This seems unbearably primitive from today's perspective, when Turing's glorious stored-program world is taken for granted. The larger ENIAC was programmed in much the same way: to set up the computer for a different job, operators would spend a day or more rerouting cables and setting switches.

The rest of the world did not know about Colossus, but its impact on Turing and Newman was colossal indeed: once they saw Flowers' racks of electronic equipment, it was, Flowers said, just a question of their waiting for an opportunity to present itself for putting the universal Turing machine into practice. Colossus was the connection between Turing's paper of 1936 and his and Newman's postwar projects to build a stored-program computer. However, historians have often assumed that the pioneers of computing uniformly drew their vision of electronic computation from the ENIAC.

The war over, Newman left Bletchley Park to take up the Fielden Chair of Mathematics at the University of Manchester. Shortly after his arrival in Manchester, he wrote to the Hungarian-American pioneer of computing John von Neumann. This letter made it clear that Newman had formed the intention to build a computer for peacetime applications by 1944 and that he had planned to embark on this project just as soon as he "got out" of Bletchley Park. Newman wrote:

I am . . . hoping to embark on a computing machine section here, having got very interested in electronic devices of this kind during the last two or three years. By about eighteen months ago I had decided to try my hand at starting up a machine unit when I got out. . . . I am of course in close touch with Turing.[19]

In the first weeks of 1946, Newman applied to the Royal Society of London for a large grant to construct an electronic computer "directed primarily to opening up new fields of research in pure mathematics."[20] The application was successful, and Newman founded his computing machine laboratory at Manchester in July 1946.

Von Neumann himself was at the center of the American effort to build an electronic stored-program computer. He had read Turing's "On Computable Numbers" before the war,[21] and when he became acquainted with the ENIAC project in 1944 (Goldstine, 1972, 182), it must have been obvious to him that the instructions making up a program should be stored internally in the form of numbers. He was well aware of the "great positive contribution of Turing" (as he described it in a 1946 letter to Norbert Wiener): the discovery that "one, definite mechanism can be 'universal.'"[22] Von Neumann put Turing's concept of an all-purpose stored-program computer into the hands of electronic engineers. (Newman would soon do the same at Manchester.) Von Neumann's colleague Stanley Frankel reported (in a letter to historian Brian Randell) that von Neumann "firmly emphasized to me, and to others I am sure, that the fundamental conception is owing to Turing."[23] Frankel added, "In my view von Neumann's essential role was in making the world aware of these fundamental concepts introduced by Turing." Von Neumann could hardly fail to have also been influenced by Gödel's demonstration (in his 1931 paper) that logical and arithmetical sentences can be expressed as numbers.

---

[19] Letter from Newman to von Neumann, February 8, 1946; in the von Neumann Archive at the Library of Congress (also quoted in Copeland, 2004, 209; a digital facsimile of the letter is in The Turing Archive for the History of Computing: http://www.AlanTuring.net/newman_vonneumann_8feb46).

[20] Minutes of the Council of the Royal Society of London, May 1946.

[21] Stanislaw Ulam in an interview with Christopher Evans in 1976 (see Endnote 4).

[22] Letter from von Neumann to Wiener, November 29, 1946; in the von Neumann Archive at the Library of Congress, Washington, D.C. (also quoted in Copeland, 2004, 209).

[23] Letter from Frankel to Brian Randell, 1972 (first published in Randell, 1972). I am grateful to Randell for giving me a copy of this letter.

Figure 8.6. John von Neumann beside the Princeton computer at the Institute for Advanced Study, Princeton, New Jersey. The computer's high-speed memory is located in the row of canisters level with von Neumann's left hand – each canister contains a Williams tube memory unit (named after its inventor, F. C. Williams). Photograph by Alan Richards. Archives of the Institute for Advanced Study. From The Shelby White and Leon Levy Archives Center, Institute for Advanced Study, Princeton, New Jersey. Reproduced with permission.

Von Neumann's (1945) "First Draft of a Report on the EDVAC" set out, in rather general terms, the first design for an electronic, all-purpose stored-program computer – a Turing machine in hardware. However, the ill-fated EDVAC (Electronic Discrete Variable Arithmetic Computer) was not operational until 1952 (Huskey, 1972, 702). The ENIAC-EDVAC group broke up when von Neumann and his collaborators Eckert and Mauchly fell out, following the appearance of "First Draft." The cause of the disagreement was that "First Draft" was distributed with only von Neumann's name on it (see, e.g., Stern, 1980). Eckert and Mauchly formed their own Electronic Control Company and began work on their EDVAC-like BINAC (Binary Automatic Computer), while von Neumann drew together a group of engineers at the Institute for Advanced Study in Princeton, New Jersey. He primed them by giving them Turing's "On Computable Numbers" to read.[24] Von Neumann's Princeton computer, which began working in 1951, was not the first of the early electronic machines, but it was the most influential.[25]

---

[24] Letter from Julian Bigelow to Copeland, April 12, 2002; see also Aspray (1990, 178).
[25] The Princeton computer is described in Bigelow (1980).

Meanwhile, in 1945, Turing joined London's National Physical Laboratory (NPL) to design an electronic, all-purpose stored-program computer. He was recruited by Womersley, head of NPL's newly formed Mathematics Division. Womersley had read "On Computable Numbers" shortly after it was published, and at the time, he had considered building a relay-based version of Turing's all-purpose machine. As early as 1944, he was advocating the potential of electronic computing.[26] Womersley named NPL's projected electronic computer the Automatic Computing Engine (ACE) – a deliberate echo of Babbage.

Turing studied "First Draft" but favored a very different type of design. To maximize the speed of the machine, he opted for a decentralized architecture, whereas von Neumann had described a centralized design that foreshadowed the modern central processing unit. Turing located different arithmetical and logical functions in different places in the hardware rather than following von Neumann's model of a single central unit in which "everything happens." His colleague James Wilkinson observed that Turing was "obsessed" with making the computations run as fast as possible,[27] and once a version of the ACE was operational, it could multiply at roughly twenty times the speed of its closest competitor.[28]

Turing described his design in a report titled "Proposed Electronic Calculator," completed by the end of 1945 (Turing, 1945; see also Copeland, 2005, chap. 20). The proposals in the report were much more concrete than those contained in von Neumann's rather abstract treatment in "First Draft." Von Neumann hardly mentioned electronics. The engineer whose job it was to draw up the first detailed hardware designs for the EDVAC, Harry Huskey, found "First Draft" to be of "no help."[29] Turing, on the other hand, gave detailed specifications of the various hardware units and even included sample programs in machine code.

Turing's plans caught the interest of the British press, with headlines such as "'ACE' May Be Fastest Brain," "Month's Work in a Minute," and "'ACE' Superior to U.S. Model." Behind the scenes, however, all was not well. Turing's plan was that Flowers and his assistants from the Colossus days should build the ACE at Dollis Hill, but this idea soon foundered. Flowers was under pressure to contribute to the restoration of Britain's war-ravaged telephone system, and despite his willingness to collaborate with Turing, he rapidly became "too busy to do other people's work."[30] The NPL's slow-moving bureaucracy proved unable to make alternative arrangements, and in April 1948, Womersley reported ruefully that hardware development was "probably as far advanced 18 months ago."[31]

While they waited for engineering developments to start, Turing and Wilkinson spent their time pioneering computer programming, preparing a large library of routines for the not-yet-existent machine. It was the availability of these ready-made programs that

---

[26] See "The Origins and Development of the ACE Project," in Copeland (2005, chap. 3).

[27] Wilkinson in an interview with Christopher Evans in 1976 (see Endnote 4).

[28] See the table by Martin Campbell-Kelly in Copeland (2005, 161).

[29] Letter from Huskey to Copeland, February 4, 2002.

[30] Flowers in an interview with Copeland, July 1998.

[31] Minutes of the Executive Committee of the National Physical Laboratory for April 20, 1948 (National Physical Laboratory Library; a digital facsimile is in The Turing Archive for the History of Computing: http://www.AlanTuring.net/npl_minutes_apr1948).

explained the success, once the hardware was working, of NPL's scientific computing service, which took on commissions from government, industry, and the universities – the first such service in the world.

In 1947, Huskey left the EDVAC project and joined Turing at the NPL. He suggested, very sensibly, that the NPL begin to construct a minimal version of the ACE on its own premises. Huskey's proposed machine soon came to be called the "Test Assembly." By about the middle of the year, circuit-block diagrams had been drawn up, a mainframe had been built, valve types had been determined, and the NPL workshops were making an experimental memory unit.[32] Huskey's first goal was to run a simple stored program using an absolute minimum of equipment; after that, he wanted to develop a small but substantial computer capable of solving practical problems.[33] Unfortunately, the director of the NPL, Sir Charles Darwin (grandson of Charles Darwin of evolutionary fame), summarily stopped the work on the Test Assembly in what was one of the worst administrative decisions in the history of computing. Had it not been for Darwin, the ACE Test Assembly might have been the first electronic stored-program computer to function.

In reality, however, the race was won in Manchester. Having secured his funding from the Royal Society and established his Computing Machine Laboratory – a large, empty room within which the proposed computer would be constructed – Newman explained the concept of the electronic stored-program computer to radar engineers F. C. "Freddie" Williams and Tom Kilburn.[34] Williams and his assistant Kilburn were just the men for the job. Not only had they spent the war designing innovative electronic circuits but Williams had an idea for solving what was the fundamental problem of computer engineering at that time: how to store large amounts of information economically and in such a way that the information would be available at electronic speeds. Williams and Kilburn pioneered a form of high-speed random-access memory in which digits were stored on the face of a cathode ray tube. Later known simply as the Williams tube, this form of memory was used in von Neumann's Princeton computer and thereafter in many early computers.

The first stored-program computer, the Manchester "Baby," came to life on June 21, 1948. This was a modest machine based around three Williams tubes (the Princeton computer had forty). "A small electronic digital computing machine has been operating successfully for some weeks in the Royal Society Computing Machine Laboratory," wrote Williams and Kilburn in the letter to *Nature* that announced their success to the world (1948, 487). Early stages of the construction of the Baby made use of a "bedstead" from one of the Colossi, the iron frame that held the message tape. "It reminds me of Adam's rib," said Jack Good, one of the Newmanry code breakers who moved with Newman to Manchester.[35]

Regrettably, the Manchester Baby is now remembered as the work of Williams and Kilburn alone (Copeland, 2001, 2004, 371–73). During the official celebrations

---

[32] Letter from Huskey to Copeland, June 3, 2003; Wilkinson in an interview with Christopher Evans in 1976 (see Endnote 4).

[33] Letter from Huskey to Copeland, January 18, 2004.

[34] See further "Colossus and the Dawning of the Computer Age" Copeland (2001) and the section on Newman in "A Lecture and Two Radio Broadcasts on Machine Intelligence by Alan Turing" Copeland (1999, 454–57).

[35] Letter from Jack Good to Copeland, March 5, 2004.

Figure 8.7. Freddie Williams (right) and Tom Kilburn in front of their tiny pioneering computer at the University of Manchester. The push-button switches were for entering the program digit by digit. The instructions were stored in the computer as patterns of dots on the surface of a cathode ray tube; this was similar in appearance to the small monitor tube in the center of the photo. School of Computer Science, University of Manchester. Reproduced with permission.

of the fiftieth anniversary of the Baby held at Manchester in June 1998, Newman's name was not even mentioned, and neither was Turing's. However, nearly thirty years earlier, Williams (who died in 1977) was unambiguous in attributing credit to Newman: "Now let's be clear before we go any further that neither Tom Kilburn nor I knew the first thing about computers when we arrived in Manchester University.... Newman explained the whole business of how a computer works to us."[36] Newman's experience with his wartime computing facility of ten Colossi, together with his appreciation of Turing's concept of a universal machine, placed him in a unique position to educate the engineers. Newman emphasized to Williams and others that it was in Turing's "On Computable Numbers" that the "idea of a truly universal computing machine was first clearly set out."[37]

Turing himself not only contributed the fundamental logical idea but also made a very practical contribution to the triumph at Manchester. He gave a series of lectures in London on computer design during the period December 1946 to February 1947, and

---

[36] Williams in an interview with Christopher Evans in 1976 (see Endnote 4). As far as I know, these words of Williams' first appeared in print in an article I published with Diane Proudfoot on the fiftieth anniversary of the Manchester Baby (Copeland and Proudfoot, 1998, 6).

[37] Report by Newman on the Royal Society Computing Machine Laboratory (1948), distributed to Williams and others (National Archive for the History of Computing, University of Manchester).

Figure 8.8. The EDSAC. Built by Maurice Wilkes in the Mathematical Laboratory at the University of Cambridge, this substantial computer was put to heavy use as soon as it was completed in 1949. Computer Laboratory, University of Cambridge. Reproduced with permission.

Kilburn attended these to learn how to design a computer.[38] Early plans for the Baby computer were ACE-like, following Turing's idea of a decentralized architecture (as I show in detail in Copeland, 2011a, 2011b). Subsequently, though, a centralized design was adopted, and this was implemented in the 1948 Baby (Williams and Kilburn, 1953).

As mentioned earlier, von Neumann's "First Draft" set out a centralized design. The nature of von Neumann's influence on the Manchester Baby has never been adequately recognized. The logical design of the Baby is, in fact, virtually identical to the design proposed at Princeton by von Neumann and his group (as I show in Copeland, 2011b). It was as electronic engineers, not computer architects, that Williams and Kilburn led the world in 1948. Newman (who had visited Princeton in 1946) and his mathematician colleagues transferred logical ideas from the von Neumann group to the Manchester engineers – although the engineers seem to have been largely unaware of the transatlantic origin of the information they acquired.

The next stored-program computer to run was the EDSAC (Electronic Delay Storage Automatic Calculator), at the University of Cambridge Mathematical Laboratory in 1949. The EDSAC's designer, Maurice Wilkes, followed the EDVAC proposal closely, as he indicated in his choice of name for the computer (Wilkes, 1985). His supreme skill as an electronic engineer, coupled with expert advice from Thomas Gold on how to design the computer's memory units (Gold is better known as the originator, with Fred Hoyle and Hermann Bondi, of the steady state theory in cosmology), gave Wilkes a lead on the various American projects.

---

[38] The notes of these lectures form Copeland (2005, chap. 22). Kilburn reported his presence at the lectures in Bowker and Giordano (1993, 17–32); see also Copeland (2004, 372–73).

In 1946, the NPL had attempted to recruit both Williams and Wilkes to assist with the engineering side of the bogged-down ACE. Williams (then at the Telecommunications Research Establishment in the west of England) turned down the NPL's offer once the Manchester project entered the picture. Wilkes, on the other hand, was willing to work with the NPL, but Turing gave the idea short shrift. In a memo to Womersley, Turing said:

> I have read Wilkes' proposals. . . . The "code" which he suggests is however very contrary to the line of development here, and much more in the American tradition of solving one's difficulties by means of much equipment rather than thought. I should imagine that to put his code (which is advertised as "reduced to the simplest possible form") into effect would require a very much more complex control circuit than is proposed in our full-size machine. Furthermore certain operations which we regard as more fundamental than addition and multiplication have been omitted.[39]

As the memo indicates, Turing's approach to computer design was very different from Wilkes's, and not only in the matter of centralization versus decentralization. Turing wanted to throw as much of the design burden as possible onto programming, in an approach reminiscent of the modern Reduced Instruction Set Computer (RISC).[40] He planned to minimize the amount of hardware by using complex programming. With this approach, special-purpose hardware – for multiplication, division, or floating-point arithmetic, for example – would ideally be dispensed with in favor of programs that performed the same functions.

The ACE's decentralized architecture led to an unusual style of programming. Instead of writing, say, "divide A by B and put the result in C," the ACE's programmer would write a series of low-level instructions achieving that effect. Programs were made up entirely of instructions such as "transfer the contents of unit 15 to unit 17" (usually written simply as "15–17"). The effect of the transfer was to subject the contents of the "source" to whatever operation was associated with the "destination" (in this case, unit 17). Because the nature of the operation was implicit in the destination address (or the source address, or both), no term specifying the operation was required in the instruction. The instruction "15–17" added the numbers stored in units 15 and 16.

The Pilot Model of Turing's ACE did not run until 1950. By then, Turing had lost patience with the NPL and had moved to Newman's Computing Machine Laboratory, where he used the Manchester computer to model biological growth.[41] The Pilot Model ACE was based on Turing's 1946 Version V of his ACE design[42] – although the Pilot Model had less than 5 percent of the memory capacity that Turing had originally specified (overambitiously, in the eyes of his colleagues[43]). With a clock speed of one megahertz, the Pilot Model left its competitors in the dust.

---

[39] Memorandum from Turing to Womersley, undated, ca. December 1946 (Woodger Papers, National Museum of Science and Industry, Kensington, London (catalog reference M15/77); a digital facsimile is in The Turing Archive for the History of Computing: http://www.AlanTuring.net/turing_womersley; also quoted in Copeland, 2005, 62).

[40] For a discussion of RISC, see Doran (2005).

[41] See "Artificial Life" and Turing's "The Chemical Basis of Morphogenesis" in Copeland (2005).

[42] Versions V–VII of Turing's ACE design are described in detail in Copeland (2005, chap. 22).

[43] Michael Woodger in an interview with Copeland, June 1998.

Figure 8.9. The Pilot Model of Turing's 1 MHz Automatic Computing Engine (ACE). The small control desk is to the left of the main frame. Standing next to the control desk, under the window, is the Hollerith punched-card input-output equipment. The computer's memory, a number of long tubes filled with mercury, is in the large wooden box (known to the engineers as the "coffin") atop the support at the right of the picture. National Physical Laboratory, Teddington. Crown copyright. Reproduced with permission.

The NPL called in the English Electric Company to produce a marketable version of the Pilot Model ACE. The result, the English Electric DEUCE (Digital Electronic Universal Computing Engine), went on sale in 1955. The DEUCE proved immensely successful and became a foundation stone of the developing computer industry in Britain. More than thirty DEUCEs were sold – confounding Darwin's suggestion, in 1946, that "it is very possible that . . . one machine would suffice to solve all the problems that are demanded of it from the whole country."[44] DEUCEs remained in service until about 1970. Turing's role in the development of hardware is rarely mentioned in orthodox histories of the computer, yet there is a very direct line leading from "On Computable Numbers," via Colossus, to one of the major workhorses of the first decades of the computer age.

Flowers' plan to build the ACE at Dollis Hill did eventually bear fruit. When the Post Office was contracted by the British government to supply a computer for top-secret military use, Flowers' right-hand men from the Colossus era, William Chandler and Allen Coombs, were put back on the job of building the ACE. Thanks to Colossus, they had unparalleled experience of large-scale digital electronics. Coombs and Chandler

---

[44] C. Darwin, "Automatic Computing Engine (ACE)," National Physical Laboratory, April 17, 1946 (National Archives/Public Record Office, document reference DSIR 10/385; a digital facsimile is in The Turing Archive for the History of Computing: http://www.AlanTuring.net/darwin_ace).

based their MOSAIC (Ministry of Supply Automatic Integrator and Computer) on Version VII of Turing's design for the ACE.[45] MOSAIC had approximately seven thousand valves, almost three times as many as the largest Colossus (or, for that matter, von Neumann's Princeton computer, which had about the same number of valves as the later Colossi). MOSAIC also contained two thousand germanium diodes, harbingers of semiconductor technology. Coombs and Chandler did not attempt the one-megahertz pulse rate that Turing had specified, settling for about half that (570 kilocycles per second). Apart from this reduction in speed, MOSAIC was much closer to Turing's original grand conception than were the relatively small Pilot Model and DEUCE. If Coombs and Chandler (who worked alone) had had access to anything like the resources thrown into Colossus during 1943, Turing's grand-scale ACE might have become a reality at about the time of the Pilot Model. As it was, MOSAIC first ran a trial program in 1952 or early 1953 (the exact date is not known) and was fully operational at the Radar Research and Development Establishment by early 1955.[46] Precise details of its use remain classified, but most likely, its function involved the guidance of antiaircraft weaponry, using data from a radar tracking system.

By no means have all the experimental stored-program computers intervening between the Manchester Baby and von Neumann's Princeton machine been mentioned so far. In 1949, Wilkes's EDSAC was followed in quick succession by three more computers: the Eckert-Mauchly BINAC in Philadelphia; Trevor Pearcey's CSIRAC (Commonwealth Scientific and Industrial Research Organisation Automatic Computer) at the University of Sydney in Australia; and Jay Forrester's Whirlwind I at the Massachusetts Institute of Technology.

In 1950, in Washington, D.C., the SEAC (National Bureau of Standards Eastern Automatic Computer) ran its first program just a few weeks before the Pilot Model ACE. The SEAC was built by Samuel Alexander and Ralph Slutz for the U.S. Bureau of Standards Eastern Division. Huskey, freshly returned to the United States from his year at the NPL, had initiated the Bureau of Standards computer project in 1948 and had proposed a design modeled on the ACE (Huskey, 2005). In the end, though, the Bureau of Standards opted for the home-grown approach, and SEAC was based on the EDVAC. Meanwhile, Huskey moved to the Bureau of Standards Western Division in Los Angeles, where he designed and built an experimental parallel computer using Williams tube memory. This, the SWAC (National Bureau of Standards Western Automatic Computer), also ran in 1950 (following the SEAC and the Pilot Model ACE).

Later, Huskey based his G15 computer on Turing's ACE design (although he used a rotating drum memory in place of Turing's acoustic, mercury-filled delay lines). The G15, which was of about the same dimensions as a jumbo-sized kitchen refrigerator, was the first personal computer. The prototype ran in 1954.[47] Huskey followed Turing's principle of substituting programming for hardware. The result was a cheap, compact

---

[45] A. W. M. Coombs (1954), MOSAIC, in *Automatic Digital Computation: Proceedings of a Symposium Held at the National Physical Laboratory*, London: Her Majesty's Stationery Office; Coombs in an interview with Christopher Evans in 1976 (see Endnote 4); and Copeland (2005, chap. 3).

[46] "Engineer-in-Chief's Report on the Work of the Engineering Department for the Year 1 April 1952 to 31 March 1953" and "Engineer-in-Chief's Report on the Work of the Engineering Department for the Year 1 April 1954 to 31 March 1955," Post Office Engineering Department.

[47] Letter from Huskey to Copeland, December 20, 2001. For more information on the Bendix G15, see Copeland (2005).

Figure 8.10. The MOSAIC. This highly secret computer was based on Version VII of Turing's design for the Automatic Computing Engine. Used in conjunction with radar equipment, the MOSAIC contributed to Britain's air defenses during the cold war period. Royal Mail Group, United Kingdom. Reproduced with permission.

machine requiring no air-conditioning and using a normal 220-volt electric supply. To increase the speed of the computer, Huskey adopted Turing's scheme known as optimum coding. Optimum coding, which was developed into a fine art at the NPL, involved storing instructions not sequentially but in a way that minimized the time the program would require to run. The programmer had to work out an optimal arrangement

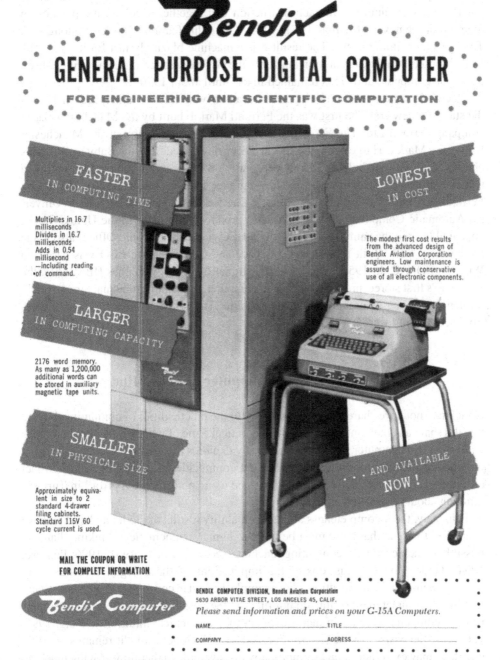

Figure 8.11. An advertisement for the Bendix G15 computer. The G15 was the first single-user desk-side computer – the first personal computer. Harry Huskey. Reproduced with permission.

for storing the instructions of each program. Turing's instruction format involved placing a pair of numbers in each instruction that together specified the position in memory of the next instruction. Huskey transferred this feature to the G15, along with Turing's other timing tricks. The result was a machine blazingly fast for its size. The Detroit-based Bendix Corporation marketed the G15 very successfully as a single-user, desk-side computer, and G15s remained in use until about 1970.

Electronic stored-program computers began arriving in the marketplace in 1951 – the start of a new era. The first was the Ferranti Mark I, built by the Manchester-based company Ferranti Ltd. The Ferranti was a mass-produced version of the Manchester University Mark I, a large computer that had been developed from the Baby by Williams and Kilburn, with the help of Turing, especially on the programming side,[48] and with some significant contributions from Newman (see further Copeland, 2011). The first Ferranti was installed in February 1951.[49] The Eckert-Mauchly UNIVAC I (Universal Automatic Computer), the earliest mass-produced computer in the United States, appeared later in the same year. The British LEO (Lyons Electronic Office) also made its debut in 1951. Built for caterers J. Lyons and Company, the LEO was based on Wilkes's EDSAC. In 1953 came the International Business Machines (IBM) 701, the company's first stored-program electronic computer, a Williams tube machine strongly influenced by von Neumann's Princeton design. The rough pioneering days of electronic computing were over.

8.5 Epilogue: The Computer and the Mind

Gödel may not have shared Turing's passion for the practicalities of computing, but the two did share an abundant interest in the theoretical issue of whether the human mind is equivalent to a computer. This epilogue briefly explores the views of Gödel and Turing on whether mental processes can go beyond computational processes. Gödel in fact criticized Turing sharply for what he regarded as a "philosophical error" in Turing's thinking about this matter, as I shall explain.

Famously, the incompleteness and undecidability results of Gödel and Turing have been used to argue that there must be more to human mathematical thinking than can possibly be achieved by a computing machine (see, e.g., Lucas, 1961, 1996; Penrose, 1994). Gödel's own discussions of the implications of these negative results were notably open-minded and cautious. For example, he said in conversation with Hao Wang that the "incompleteness results do not rule out the possibility that there is a theorem-proving computer which is in fact equivalent to mathematical intuition" (Wang, 1996, 186). On "the basis of what has been proved so far," Gödel said, "it remains possible that there may exist (and even be empirically discoverable) a theorem-proving machine which in fact *is* equivalent to mathematical intuition, but cannot be *proved* to be so,

---

[48] Turing's "Programmers' Handbook for Manchester Electronic Computer" (ca. 1950; available in The Turing Archive for the History of Computing: http://www.AlanTuring.net/programmers_handbook).

[49] Letter from Turing to Michael Woodger, undated, received February 12, 1951 (Woodger Papers, Science Museum, London; a digital facsimile is in The Turing Archive for the History of Computing: http://www.AlanTuring.net/turing_woodger_feb51).

nor even be proved to yield only *correct* theorems of finitary number theory" (Wang, 1996, 184–85). Gödel also said (but seemingly gave no clear indication whether, or to what extent, he himself agreed with what he called Hilbert's "rationalistic attitude"):

My incompleteness theorem makes it likely that mind is not mechanical, or else mind cannot understand its own mechanism. If my result is taken together with the rationalistic attitude which Hilbert had and which was not refuted by my results, then [we can infer] the sharp result that mind is not mechanical. This is so, because, if the mind were a machine, there would, contrary to this rationalistic attitude, exist number-theoretic questions undecidable for the human mind. (Gödel, as quoted in Wang, 1996, 186–87)

Hilbert's "rationalistic attitude" was summed up in his celebrated remark that "in mathematics there is no *ignorabimus*" (1902, 445) – no mathematical question that, in principle, the mind is incapable of settling.

In about 1970, Gödel wrote a brief note titled "A philosophical error in Turing's work."[50] This was, he said, to be regarded as a footnote to the postscript to his 1931 undecidability paper, which he had composed in 1964 (published as "Postscriptum" in Davis, 1965, 71). The main purpose of the 1964 postscript was to state generalized versions of incompleteness applicable to algorithms and formal systems. (It was in this postscript that Gödel officially adopted Turing's "analysis of the concept of 'mechanical procedure' . . . alias 'algorithm' or 'computation procedure'"; there he emphasized that it was "due to A. M. Turing's work [that] a precise and unquestionably adequate definition of the general concept of formal system can now be given . . . [a] formal system can simply be defined to be any mechanical procedure for producing formulas, called provable formulas.") In the 1964 postscript, Gödel also raised the intriguing "question of whether there exist finite *non-mechanical* procedures," and he observed that the generalized incompleteness results "do not establish any bounds for the powers of human reason, but rather for the potentialities of pure formalism in mathematics."

Gödel's (ca. 1970) retrospective footnote to his 1964 postscript attributed the view that "mental procedures cannot go beyond mechanical procedures" to Turing in "On Computable Numbers." Gödel criticized an argument for this view, which he claimed to find on page 250 of Turing's paper (Copeland, 2004, 75–76):

Turing [there] gives an argument which is supposed to show that mental procedures cannot go beyond mechanical procedures. However, this argument is inconclusive. What Turing disregards completely is the fact that *mind, in its use, is not static, but constantly developing.* (Gödel, ca. 1970)

Turing can readily be defended against Gödel's charge of philosophical error.[51] Gödel was too hasty in his claim that, in "On Computable Numbers," Turing put forward an argument supposed to show that mental procedures cannot go beyond mechanical procedures. No such argument is to be found in Turing's paper, nor is there even any

---

[50] Wang (1996, 195) reports that Gödel wrote the note "around 1970." The note is included, with an introduction by Judson Webb, in *Collected Works*, vol. 2 (1990), p. 306. In 1972, Gödel gave Wang a revised version of the note, which Wang (1974, 325–26) subsequently published.

[51] See further Copeland and Shagrir (forthcoming), which includes a discussion of Gödel's claim (made in the 1972 revision of the note; see Wang, 1974, 325) that Turing argues from "the supposition that a finite mind is capable of only a finite number of distinguishable states."

trace of a statement endorsing the conclusion of the supposed argument. Turing, on the page discussed by Gödel, was not talking about the general scope of mental procedures; he was addressing a different question: "What are the possible processes which can be carried out in computing a number?" (Turing, 1936, 249; see also Copeland, 2004, 74). Furthermore, a passage in "On Computable Numbers" seemingly runs counter to the view Gödel attributed to Turing. Having defined a certain infinite binary sequence δ, which he showed to be uncomputable, Turing said, "It is (so far as we know at present) possible that any assigned number of figures of δ can be calculated, but not by a uniform process. When sufficiently many figures of δ have been calculated, an essentially new method is necessary in order to obtain more figures" (Turing, 1936, 253; see also Copeland, 2004, 79). Gödel, on the other hand, considered Turing to have offered an "alleged proof that every mental procedure for producing an infinite series of integers is equivalent to a mechanical procedure" (Gödel, as quoted in Wang, 1996, 197).

In short, Turing's text does not support Gödel's interpretation. The situation becomes bleaker still for Gödel's interpretation when Turing's (1939) publication "Systems of Logic Based on Ordinals" is taken into account. There Turing emphasized the aspect of mathematical reasoning that he referred to as "intuition." He said:

> In pre-Gödel times it was thought by some that . . . all the intuitive judgments of mathematics could be replaced by a finite number of [formal] rules. . . . In consequence of the impossibility of finding a formal logic which wholly eliminates the necessity of using intuition, we naturally turn to "non-constructive" systems of logic with which not all the steps in a proof are mechanical, some being intuitive. (Turing, 1939, 215–16; see also Copeland, 2004, 192–93)

In Turing's view, the activity of what he called the "faculty" of intuition brings it about that mathematical "judgments" – again, his word – exceed what can be expressed by means of a single formal system (Turing, 1939, 214; see also Copeland, 2004, 192). "The activity of the intuition," he said, "consists in making spontaneous judgements which are not the result of conscious trains of reasoning" (Turing, 1939, 214–15; see also Copeland, 2004, 192). Turing's cheerful use of mentalistic vocabulary in this connection makes it very unlikely that Gödel was correct in finding an argument in the 1936 paper supposedly showing that mental procedures cannot go beyond mechanical procedures.

During the early part of the war, probably in 1940, Turing wrote a number of letters to Newman explaining his thinking about intuition. The following passage is illuminating:

> I think you take a much more radically Hilbertian attitude about mathematics than I do. You say "If all this whole formal outfit is not about finding proofs which can be checked on a machine it's difficult to know what it is about." When you say "on a machine" do you have in mind that there is (or should be or could be, but has not been actually described anywhere) some fixed machine on which proofs are to be checked, and that the formal outfit is, as it were about this machine. If you take this attitude (and it is this one that seems to me so extreme Hilbertian [sic]) there is little more to be said: we simply have to get used to the technique of this machine and resign ourselves to the fact that there are some problems to which we can never get the answer. On these lines my ordinal logics

would make no sense. However I don't think you really hold quite this attitude because you admit that in the case of the Gödel example one can decide that the formula is true, i.e. you admit that there is a fairly definite idea of a true formula which is quite different from the idea of a provable one. Throughout my paper on ordinal logics I have been assuming this too. . . .

If you think of various machines I don't see your difficulty. One imagines different machines allowing different sets of proofs, and by choosing a suitable machine one can approximate "truth" by "provability" better than with a less suitable machine, and can in a sense approximate it as well as you please. The choice of a . . . machine involves intuition . . .[52]

Turing's criticism of the "extreme Hilbertian" view is accompanied by what seems to be a cautious endorsement of the attitude that Gödel described (in the quotation given previously) as "rationalistic." The "sharp result" stated there by Gödel seems in effect to be that there is no *single* machine equivalent to the mind (at any rate, no more is justified by the reasoning that Gödel presented), and with this, Turing was evidently in agreement. Incompleteness, if taken together with an Hilbertian optimism, excludes the extreme Hilbertian position that the "whole formal outfit" corresponds to some one fixed machine. (Note, though, that Turing's discussion does not rule out the possibility raised by Gödel that there might in actuality be a machine that exhausts the mind, although it cannot be proved to do so.)

Turing's view, as he expressed it to Newman and in "Systems of Logic Based on Ordinals," appears to have been that mathematicians achieve progressive approximations to truth via a nonmechanical process involving intuition. This picture, in which minds devise and adopt successive, increasingly powerful mechanical formalisms in their quest for truth, is consonant with Gödel's view that "mind, in its use, is not static, but constantly developing." Gödel's own illustration of his claim that mind is constantly developing is certainly related to Turing's concerns. Gödel said, "This [that mind . . . is not static, but constantly developing] is seen, e.g., from the infinite series of ever stronger axioms of infinity in set theory, each of which expresses a new idea or insight" (from the 1972 revision of the note; see Wang, 1974, 325). Gödel's blunt criticism that Turing "disregards completely" this dynamic aspect of mind was simply misdirected. These two great founders of the study of computability were perhaps not quite as philosophically distant on the mind-machine issue as Gödel supposed.

References

Aiken, H. (1982 [1937]). Proposed automatic calculating machine. In *The Origins of Digital Computers: Selected Papers*, 3rd ed., ed. B. Randall, chap. 5.1. Berlin: Springer.

Aspray, W. (1990). *John von Neumann and the Origins of Modern Computing*. Cambridge, MA: MIT Press.

Atanasoff, J. V. (1982 [1940]). Computing machine for the solution of large systems of linear algebraic equations. In *The Origins of Digital Computers: Selected Papers*, 3rd ed., ed. B. Randall, chap. 7.2. Berlin: Springer.

---

[52] The complete letter is in Copeland (2004, 214–16).

Babbage, C. (1989 [1864]). Passages from the life of a philosopher. *The Works of Charles Babbage*, vol. 11, ed. M. Campbell-Kelly. London: William Pickering.

Bauer, F. L. (2006). The Tiltman break. In *Colossus: The Secrets of Bletchley Park's Codebreaking Computers*, B. J. Copeland, app. 5. Oxford: Oxford University Press.

Bigelow, J. (1980). Computer development at the Institute for Advanced Study. In *A History of Computing in the Twentieth Century*, ed. N. Metropolis, J. Howlett, and G. C. Rota. New York: Academic Press.

Bowker, G., and Giordano, R. (1993). Interview with Tom Kilburn. *IEEE Annals of the History of Computing*, **15**, 17–32.

Bromley, A. (1982). Charles Babbage's Analytical Engine, 1838. *IEEE Annals of the History of Computing*, **4**, 196–217.

Burks, A. R. (2003). *Who Invented the Computer? The Legal Battle That Changed Computing History*. Amherst, MA: Prometheus.

Coombs, A. W. M. (1954). MOSAIC. In *Automatic Digital Computation: Proceedings of a Symposium Held at the National Physical Laboratory*. London: Her Majesty's Stationery Office.

Copeland, B. J. (1999). A lecture and two radio broadcasts on machine intelligence by Alan Turing. In *Machine Intelligence 15*, ed. K. Furukawa, D. Michie, and S. Muggleton. Oxford: Oxford University Press.

———. (2001). Colossus and the dawning of the computer age. In *Action This Day*, ed. R. Erskine and M. Smith. London: Bantam.

———, ed. (2004). *The Essential Turing: Seminal Writings in Computing, Logic, Philosophy, Artificial Intelligence, and Artificial Life*. Oxford: Oxford University Press.

———, ed. (2005). *Alan Turing's Automatic Computing Engine: The Master Codebreaker's Struggle to Build the Modern Computer*. Oxford: Oxford University Press.

———. (2006). *Colossus: The Secrets of Bletchley Park's Codebreaking Computers*. Oxford: Oxford University Press. [Rev. ed. 2010.]

———. (2011a). The Manchester computer: A revised history. Part 1: The memory. *IEEE Annals of the History of Computing*, **3**, 4-21.

———. (2011b). The Manchester computer: A revised history. Part 2: The baby computer. *IEEE Annals of the History of Computing*, 22–37.

Copeland, B. J., and Proudfoot, D. (1998). Enigma variations. *Times Literary Supplement* [Information Technology], July 3.

Copeland, B. J., and Shagrir, O. (Forthcoming). Gödel on Turing on computability. In *Computability: Gödel, Turing, Church, and beyond*, ed. B. J. Copeland, C. Posy, and O. Shagrir. Cambridge, MA: MIT Press.

Davis, M. (1965). *The Undecidable: Basic Papers on Undecidable Propositions, Unsolvable Problems, and Computable Functions*. Hewlett, NY: Raven Press.

Doran, R. (2005). Computer architecture and the ACE computers. In *Alan Turing's Automatic Computing Engine: The Master Codebreaker's Struggle to Build the Modern Computer*, ed. B. J. Copeland, chap. 8. Oxford: Oxford University Press.

Flowers, T. H. (2006). D-Day at Bletchley Park. In *Colossus: The Secrets of Bletchley Park's Codebreaking Computers*, B. J. Copeland, chap. 6. Oxford: Oxford University Press.

Gandy, R. (1998). The confluence of ideas in 1936. In *The Universal Turing Machine: A Half-Century Survey*, ed. R. Herken. Oxford: Oxford University Press.

Gödel, K. (1931). Über formal unentscheidbare Sätze der *Principia Mathematica* und verwandter Systeme I. *Monatshefte für Mathematik und Physik*, **38**, 173–98. [Repr. with facing English trans. On formally undecidable propositions of *Principia Mathematica* and Related Systems. I. In *Collected Works*, vol. 1 (1986), pp. 145–95.]

———. (193?). Undecidable Diophantine propositions. In *Collected Works*, vol. 3 (1995), pp. 164–75.

————. (1951). Some basic theorems on the foundations of mathematics and their implications. In *Collected Works*, vol. 3 (1995), p. 304–23.

————. (1965). Postscriptum [to Gödel, 1931]. In *The Undecidable: Basic Papers on Undecidable Propositions, Unsolvable Problems, and Computable Functions*, ed. M. Davis. Hewlett, NY: Raven Press.

————. (ca. 1970). A philosophical error in Turing's work. In *Collected Works*, vol. 2 (1990), p. 306. [A revised version (1972) is in H. Wang (1974). *From Mathematics to Philosophy*. New York: Humanities Press, pp. 325–26.]

Goldstine, H. (1972). *The Computer from Pascal to von Neumann*. Princeton, NJ: Princeton University Press.

Hilbert, D. (1902). Mathematical problems. Lecture delivered before the International Congress of Mathematicians at Paris in 1900. *Bulletin of the American Mathematical Society*, **8**, 437–79.

Hodges, A. (1992). *Alan Turing: The Enigma*. London: Vintage.

Huskey, H. D. (1972). The development of automatic computing. In *Proceedings of the First USA-JAPAN Computer Conference*. Tokyo.

————. (2005). The ACE Test Assembly, the Pilot ACE, the Big ACE, and the Bendix G15. In *Alan Turing's Automatic Computing Engine: The Master Codebreaker's Struggle to Build the Modern Computer*, ed. B. J. Copeland, chap. 13. Oxford: Oxford University Press.

Kleene, S. C. (1981). Origins of recursive function theory. *IEEE Annals of the History of Computing*, **3**, 52–67.

————. (1986). Introductory note to 1930b, 1931 and 1932b. In *Collected Works*, vol. 1.

Lovelace, A. A., and Menabrea, L. F. (1953 [1843]). Sketch of the Analytical Engine invented by Charles Babbage, Esq. In *Faster Than Thought*, ed. B. V. Bowden. London: Sir Isaac Pitman.

Lucas, J. R. (1961). Minds, machines, and Gödel. *Philosophy*, **36**, 112–27.

————. (1996). Minds, machines and Gödel: A retrospect. In *Machines and Thought*, ed. P. Millican and A. Clark. Oxford: Clarendon Press.

McCorduck, P. (1979). *Machines Who Think*. New York: W. H. Freeman.

Murray, J. (1993). Hut 8 and Naval Enigma, part I. In *Codebreakers: The Inside Story of Bletchley Park*, ed. F. H. Hinsley and A. Stripp. Oxford: Oxford University Press.

Newman, M. H. A. (1954). Dr. A. M. Turing. *The Times*, June 16, p. 10.

Newman, M. H. A., and Turing, A. M. (1942). A formal theorem in Church's theory of types. *Journal of Symbolic Logic*, **7**, 28–33.

Newman, W. (2006). Max Newman: mathematician, codebreaker and computer pioneer. In *Colossus: The Secrets of Bletchley Park's Codebreaking Computers*, B. J. Copeland, chap. 14. Oxford: Oxford University Press.

Penrose, R. (1994). *Shadows of the Mind: A Search for the Missing Science of Consciousness*. Oxford: Oxford University Press.

Randell, B. (1972). On Alan Turing and the origins of digital computers. In *Machine Intelligence 7*, ed. B. Meltzer and D. Michie. Edinburgh: Edinburgh University Press.

————, ed. (1982). *The Origins of Digital Computers: Selected Papers*. 3rd ed. Berlin: Springer.

Stern, N. (1980). John von Neumann's influence on electronic digital computing, 1944–1946. *IEEE Annals of the History of Computing*, **2**, 349–62.

Swade, D. (2001). *The Difference Engine: Charles Babbage and the Quest to Build the First Computer*. New York: Viking.

————. (2010). Automatic computation: Charles Babbage and computational method. *Rutherford Journal for the History and Philosophy of Science and Technology*, **3**. http://www.rutherfordjournal.net.

Turing, A. M. (1936). On computable numbers, with an application to the *Entscheidungsproblem*. *Proceedings of the London Mathematical Society*, Series 2, **42**, 230–65. [Repr. B. J. Copeland (2004).

The Essential Turing: Seminal Writings in Computing, Logic, Philosophy, Artificial Intelligence, and Artificial Life. Oxford: Oxford University Press.]

———. (1939). Systems of logic based on ordinals. *Proceedings of the London Mathematical Society, Series 2,* **45,** 161–228. [Repr. B. J. Copeland (2004). *The Essential Turing: Seminal Writings in Computing, Logic, Philosophy, Artificial Intelligence, and Artificial Life.* Oxford: Oxford University Press.]

———. (1945). Proposed electronic calculator. Technical report. Teddington, UK: National Physical Laboratory. http://www.AlanTuring.net/proposed_electronic_calculator. [Repr. B. J. Copeland, ed. (2005). *Alan Turing's Automatic Computing Engine: The Master Codebreaker's Struggle to Build the Modern Computer.* Oxford: Oxford University Press, chap. 20.]

———. (1950). Computing machinery and intelligence. *Mind,* **59,** 433–60. [Repr. B. J. Copeland (2004). *The Essential Turing: Seminal Writings in Computing, Logic, Philosophy, Artificial Intelligence, and Artificial Life.* Oxford: Oxford University Press.]

———. (ca. 1950). Programmers' handbook for Manchester electronic computer. Computing Machine Laboratory, University of Manchester. http://www.AlanTuring.net/programmers_handbook.

———. (1952). The chemical basis of morphogenesis. *Philosophical Transactions of the Royal Society of London, Series B,* **237,** 37–72. [Repr. B. J. Copeland (2004). *The Essential Turing: Seminal Writings in Computing, Logic, Philosophy, Artificial Intelligence, and Artificial Life.* Oxford: Oxford University Press.]

Tutte, W. T. (2006). My work at Bletchley Park. In *Colossus: The Secrets of Bletchley Park's Codebreaking Computers,* B. J. Copeland, pp. 352–69. Oxford: Oxford University Press.

Von Neumann, J. (1945). First draft of a report on the EDVAC. Philadelphia: Moore School of Electrical Engineering, University of Pennsylvania. [Repr. 1993 *IEEE Annals of the History of Computing,* **15,** 27–75.]

Wang, H. (1974). *From Mathematics to Philosophy.* New York: Humanities Press.

———. (1987). *Reflections on Kurt Gödel.* Cambridge, MA: MIT Press.

———. (1996). *A Logical Journey: From Gödel to Philosophy.* Cambridge, MA: MIT Press.

Whitehead, A. N., and Russell, B. A. W. (1910–1913). *Principia Mathematica.* 3 vols. Cambridge: Cambridge University Press.

Wilkes, M. V. (1985). *Memoirs of a Computer Pioneer.* Cambridge, MA: MIT Press.

Williams, F. C., and Kilburn, T. (1948). Electronic digital computers. *Nature,* **162,** 487.

———. (1953). The University of Manchester computing machine. In *Faster Than Thought,* ed. B. V. Bowden. London: Sir Isaac Pitman.

Womersley, J. R. (2005 [1946]). A.C.E. project – origin and early history. In *Alan Turing's Automatic Computing Engine: The Master Codebreaker's Struggle to Build the Modern Computer,* ed. B. J. Copeland, pp. 38–39. Oxford: Oxford University Press.

Zuse, K. (1980). Some remarks on the history of computing in Germany. In *A History of Computing in the Twentieth Century,* ed. N. Metropolis, J. Howlett, and G. C. Rota. New York: Academic Press.

Gödelian Cosmology

Gödel, Einstein, Mach, Gamow, and Lanczos: Gödel's Remarkable Excursion into Cosmology[1]

Wolfgang Rindler

This chapter was written in the hope that it might serve a readership that may not necessarily be familiar with the details of Einstein's general relativity theory but that nevertheless may appreciate the gist of Gödel's contribution to relativistic cosmology. I have included all the technical details that general readers may wish to skip but that those familiar with relativity theory might wish to review. In particular, I give an elementary derivation of Gödel's metric and of its geodesics.

Gödel's brilliant burst into the world of physics in 1949 came as a surprise to those who knew him "only" as one of the greatest logicians of all time and thus as a very pure mathematician. However, to his colleagues at the Institute for Advanced Study (AIS) in Princeton, it was less surprising. At IAS, he had famously befriended Einstein, and much earlier, before switching over to mathematics, he had even entered the University of Vienna (in 1924) as a physics student and attended lectures by Hans Thirring, one of the earliest protagonists of Einstein's theories. Moreover, although this was not apparent from his published work, Gödel had maintained a lifelong interest in physics, attending the physics seminars at IAS and keeping abreast of ongoing developments. Then came the crucial trigger: the year 1949 brought Einstein's seventieth birthday, and Gödel was expected to contribute to the planned Festschrift for his friend. Not for the first time did pressure prove conducive to invention.

What Gödel (1949a) invented for the occasion was a model universe that was consistent with general relativity but that nevertheless exhibited two startlingly disturbing features: bulk rotation (but with respect to what, as there is no absolute space in general relativity?) and travel routes into the past (enabling one to witness or even prevent one's own birth?). Gödel did not claim that his model represented the actual universe in which we live – he knew well that general relativity permits much more appropriate models for that – but he nevertheless maintained that if general relativity permits such strange behavior, then that behavior should be studied in detail. In particular, he urged

---

[1] This chapter also appeared as a preprint article with the same title in the *American Journal of Physics*, 77, 2009; it is published in this volume (as it was in the journal) with the permission of Cambridge University Press.

astronomers to look for evidence of rotation and philosophers to rethink their ideas about time.

9.1 First Look at Gödel's Model

For the sake of concreteness, I shall begin with a brief preliminary description of Gödel's model universe based on general relativity. General relativity was Einstein's new (now a century old) theory of gravity, in which Newton's force of gravity is replaced by the curvature of four-dimensional space-time and where free matter moves along the natural rails of this curved space-time, namely, its geodesics. A geodesic is the closest analogue in any curved space to a straight line in flat Euclidean space. For example, if you march as straight as you can on the surface of a sphere, you will follow a great circle, and so great circles are the geodesics of a sphere. Unencumbered by extraneous concepts such as absolute space, general relativity is ideally applicable to whole universes. It determines, for example, how a universe moves under the action of its own gravity.

Our actual universe is, of course, lumpy, containing big blobs of matter separated by even bigger blobs of apparent emptiness. The exact dynamics of such lumpy systems cannot, in practice, be analyzed directly. Instead one studies the smoothed-out version of actual universes and makes the assumption that the dynamics are effectively the same. The smoothed-out counterpart of any universe is its substratum, and not only must the lumpiness be smoothed out, but so must the locally irregular motions. The actual galaxies then sit on this (generally expanding) substratum more or less uniformly distributed and with only relatively small irregular proper motions.

For the standard models of general relativity, as well as for Gödel's model, these substrata satisfy the so-called cosmological principle. This hypothesis, which is well supported by observation, asserts that the universe is regular and that our place in it, and in fact that of any other galaxy, is not special. Thus, for the sake of constructing the model, the substratum is assumed to be perfectly homogeneous at all times.

Additionally, our universe is also known to expand, and it is commonly believed to have originated in the big bang some fourteen billion years ago. A realistic substratum must therefore expand. Gödel's model, although homogeneous, ignores this expansion: it is stationary, the same at all times. There is yet another difference from the usual models of general relativity: Gödel's model is not isotropic. Its substratum is somewhat like a homogeneous crystal, having preferred directions at each point. We can picture it as a stack of identical layers, infinite in all directions (see Figure 9.1). Each layer, although it is drawn as a plane, is actually a Lobachevski plane, a two-dimensional space of constant negative curvature. (The circumferences of circles, centered anywhere, increase faster than in the Euclidean plane as we move away from the center. It is, of course, one of the features of general relativity that it permits, and often requires, curvature in both space and time.) On this layered spatial framework exists an overall time, indicated by identical standard clocks, one sitting on each galaxy and all ticking in unison forever. However, somewhat as in special relativity (the theory of flat vacuum space-time), the synchronization of these clocks is not unique and depends on which clock declares itself to be the "boss."

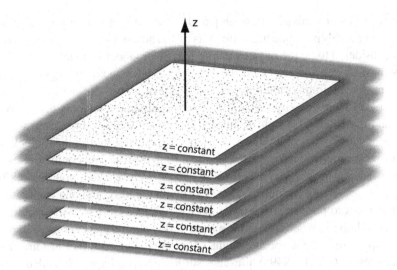

Figure 9.1. A spatial map of Gödel's universe; the map flattens the $z = const$ sections into planes.

So far, this all looks fairly harmless, but now come the surprises. Consider an inertial compass, also called a "gyrocompass." This instrument contains a number of gyroscopes and has the property of always pointing in the same direction in space. Install such a gyrocompass suitably in a stunt airplane and point it, for example, at the sun, and then fly any number of loops, twists, and turns. The gyrocompass ignores them all and keeps steadily pointing at the sun. Now fix such a gyrocompass to every galaxy in Gödel's universe, and behold: they all rotate in unison about the normals of the layers. This seems to indicate that the entire universe rotates rigidly in the opposite direction – but relative to what? As mentioned before, in general relativity, the only space is the space determined by the universe itself. Gödel (1949b) laconically commented:

> Evidently this state of affairs shows that the inertial field is to a large extent independent of the state of motion of the matter. This contradicts Mach's principle but it does not contradict relativity theory.

These two sentences seem to be the sum total of what Gödel has ever said about this paradoxical aspect of his model, and they occur in a lecture he never even published.[2]

Mach's principle, as formulated by Einstein in his early quest for general relativity, was supposed to explain the mysterious existence of the preferred set of inertial frames against which rotation and acceleration are measured in both Newton's theory and Einstein's special relativity theory. (Newton's absolute space, as an explanation, had already been repeatedly challenged, most recently by special relativity.) Mach's principle says that the local inertial frame, or inertial field, is actively determined by some average of the motions of all the masses in the universe. Einstein had hoped that general relativity would show in detail how this determination works, but for a number of reasons, he later (in the 1930s) discarded Mach's principle. Therefore the inertial

---

[2] It was published posthumously in the *Collected Works* (Gödel, 1949b).

properties of Gödel's model, although paradoxical, were not totally unacceptable to Einstein. Yet Mach's principle has a life of its own, and to its adherents, these properties are still considered to be the most troubling feature of Gödel's model.

Now for a second surprise. Consider a large circle in one of the layers of Gödel's substratum. (A minimum radius is required for this to work.) Now travel along this circle at a very large velocity. (Again, a certain minimum velocity is necessary, but it is less than the velocity of light.) Behold: you return to the galaxy from which you started at an earlier time than when you left, as indicated by the local clock, yet by your own reckoning, you have aged normally during the trip. You could now encounter your own father when he was a child, and, if you were wicked, you could kill him, thereby preventing your own birth. That is an awful paradox, and one would hope that nature has ways to prevent space-times such as Gödel's from actually materializing. (In special relativity, where a similar danger lurks, nature prevents it by imposing a universal speed limit – the speed of light.) That hope, indeed, was Einstein's reaction to Gödel's result. Gödel (1949a) himself – surprisingly, perhaps – defended his model on the grounds that it would cost impossible amounts of energy for a space traveler to accomplish such a journey. Later, he granted that one could simply send a light signal, guided by suitably placed mirrors, along a sufficiently large polygonal path to do the same damage, but the radius would have to be so immense as to render even this procedure impracticable (Gödel, 1949b).

9.2 How the Model Came into Being

In his Einstein Festschrift contribution, Gödel (1949a) stated that he was motivated to invent his model universe from sympathy for Kant's philosophy of time. It was to serve as the first counterexample on the cosmic scale to the objective view of time, which treats time as an infinity of layers "now" coming into existence successively. By 1905, Einstein had already shown this view to be problematic with his special theory of relativity. Indeed, one of the greatest shocks delivered by that theory was the discovery that simultaneity is relative, namely, the nows of different observers correspond to different sets of parallel slices through space-time, the slices of one observer being inclined to those of another. If the before-now already exists, and the after-now does not yet exist, existence itself would then be relative to the observer, which Gödel held to be nonsensical. He also pointed out that observers really play no essential role in this argument: the vacuum space-time of special relativity (Minkowski space) simply lacks distinguishing features between alternative parallel time slicings.

The situation becomes even worse with the irregular space-times of general relativity that correspond to real-life irregular matter distributions. Only in the idealized homogeneous-isotropic universes introduced by Friedman (1922, 1924), of which the 1917 static Einstein universe was a special case, do we find an absolutely (geometrically) determined worldwide time. These universes (except the Einstein universe) expand with a single expansion function, and their intrinsically determined time slices correspond to constant values of their steadily diminishing density. Thus the objective (or absolute) view of time got a reprieve from Friedmanian cosmology – which Gödel dismissed as accidental. His purported aim in the Festschrift essay (Gödel, 1949a) was to show that in more general cosmologies, no such objective time need exist.

Today we can learn a lot about the details of how Gödel arrived at his universe from the three different versions in which he presented it as well as from his correspondence with the Einstein Festschrift editor and with his mother. There is, first of all, his brief (six-page) Festschrift essay (Gödel, 1949a), completed by the end of March 1949, which contains not a single equation and concentrates mainly on the physics and philosophy of time. Almost simultaneously, Gödel's technical paper appeared in a special Einstein issue of *Reviews of Modern Physics* (Gödel, 1949c). This described, with similar brevity, the quantitative properties of the model as well as an outline of its construction. Most illuminating of all, though, was the lecture on rotating universes given by Gödel in May 1949 at IAS, which was published only posthumously in the *Collected Works* (Gödel, 1949b). The lecture began as follows:

> A few years ago, in a note in *Nature*, Gamow (1946) suggested that the whole universe might be in a state of uniform rotation and that this rotation might explain the observed rotation of the galactic systems.

Gamow's idea was that if the cells of primordial matter, which eventually collapsed under their own gravity to form galaxies, had no initial angular momentum, then all their matter would simply fall into the center and form a compact mass. (In fact, though, the initial rotation needed to form a galaxy is so small that it can be explained simply by natural random turbulence.) Importantly, however, Gamow (1946) ended his brief note (see Figure 9.2) with the conjecture that rotating universes can probably be constructed within general relativity.

In his lecture, Gödel immediately proceeded to exhibit a beautifully simple *Newtonian* version of a rotating universe, and only from there did he gradually build up his relativistic analogue. It is therefore tempting to contemplate the following route whereby Gödel might actually have arrived at his universe.

It is May 1946. Einstein's seventieth birthday is a little less than three years away. Paul Arthur Schilpp, a philosophy professor at Northwestern University and editor of the Library of Living Philosophers series, is already planning an Einstein Festschrift for the occasion. He visits Princeton and seeks out Gödel. Gödel promises a contribution. Soon thereafter, in correspondence with Schilpp, he offers to write about three pages under the title of "Some Remarks about the Relation between the Theory of Relativity and Kant" (Dawson, 1997; see also Gödel, 2003, 232–37). Although Schilpp presses him for a much longer paper, Gödel keeps insisting that three to five pages is all he needs for what he has to say. Evidently, he is not yet thinking of inventing a new universe.

Then, serendipitously, Gödel comes across Gamow's note in *Nature* with its challenge to find a rotating universe consistent with general relativity. Suddenly he has a problem worthy of his genius and the perfect gift for Einstein! However, that is not all. Gödel seems to have recognized quite early that in a rotating universe, there would be no absolute time so that his Kantian ambition of superseding Friedman would also come true. (Rotation implies twisting world lines of the galaxies and hence the nonexistence of preferred time slices.) Thus fortune placed into his hands not only a significant relativistic problem but one that even fell within his original Kantian program and that, when it was all done, turned out to be far more beautiful than he could possibly have foreseen.

No. 4016   October 19, 1946   N A T U R E   549

LETTERS TO THE EDITORS

The Editors do not hold themselves responsible for opinions expressed by their correspondents. No notice is taken of anonymous communications

Rotating Universe ?

ONE of the most mysterious results of the astronomical studies of the universe lies in the fact that all successive degrees of accumulation of matter, such as planets, stars and galaxies, are found in the state of more or less rapid axial rotation. In various cosmogonical theories the rotation of planets has been explained as resulting from the rotation of stars from which they were formed. The rotation of stars themselves (in particular that of *B*-stars) can be presumably reduced to their origin from the rotating gas-masses which form the spiral arms of various galaxies. But what is the origin of galactic rotation ?

If, according to the current theories, we consider the galaxies as the result of gravitational instability of the originally uniform distribution of matter in space, we will find it very difficult to understand why such condensations are in most cases found in the state of rather fast rotation. In fact, on the basis of statistical distribution of angular momentum, we would rather expect such condensations to show no more rotation than the water droplets in a fog formed from over-saturated vapour. Barring the possible explanation of the rotation of galaxies on the basis of the alleged irregular turbulent motion of the masses of the universe, we can ask ourselves whether it is not possible to assume that *all matter in the visible universe is in a state of general rotation around some centre located far beyond the reach of our telescopes ?*

The answer to such, at first sight fantastic, question need not wait until much larger telescopes shall have been built. It can be, in fact, settled by present means of observation. We know that the rotation of the stars of our system around the galactic centre can be proved by the study of the so-called Oort-effect in the radial velocities of comparatively near stars. In fact, due to the phenomenon of differential rotation, the mean radial velocities of stars located along the galactic plane show a double-sine periodicity with nodal axes directed parallel and perpendicular to the line connecting the sun with the centre of rotation. Thus if the realm of galaxies as seen through Mt. Wilson telescope represents only a small part of a much larger system (a 'super-galaxy' in the super-Shapley sense) rotating around a distant centre, careful observations of mean radial velocities of galaxies located in different regions of the sky should reveal similar periodicity.

The existence of this effect would prove general rotation of the universe and indicate the direction towards the rotation centre without, however, giving us its distance. Thus, it seems that the answer to the problem of universal rotation lies within the grasp of modern astronomical technique.

It must be added in conclusion that in the language of the general theory of relativity such a rotating universe can be probably represented by the group of anisotropic solutions of the fundamental equations of cosmology.

G. GAMOW

Department of Physics,
George Washington University,
Washington, D.C.
Sept. 13.

Conditions of Escape of Radio-frequency Energy from the Sun and the Stars

IN several communications in *Nature*[1],[2],[3] and elsewhere, various British, Australian and New Zealand workers have described experiments carried out during the War which prove conclusively that during times of solar disturbance there are large outbursts of radio-frequency energy from the sun. The wave-lengths measured vary from 1·5 metres to 30 metres (10 Mc. to 200 Mc.). On a rough estimate, the intensity of emission appears to be, as Appleton[1] has shown, 10^4 times the value calculated from the black-body formula taking $T = 6,000°$ K. If we assume that the radiation proceeds only from the active areas, as appears to be corroborated by the experiments now in progress at the Cavendish Laboratory, Cambridge[2], the emissivity of these regions for the range mentioned is increased nearly 10^7–10^8 times the black-body radiation.

There are certain difficulties in the escape of these radiations from the sun to which attention may be directed. It has been found that the quiescent sun has, like the earth, a magnetic field of the order of 50 gauss, but the spots show a field of much higher range, from 100 gauss in the case of tiny spots to 4,000 gauss for the largest ones[4]. If the radio waves are generated anywhere within the outer layers of the sun, then they must follow the physical laws of electro-magnetism. According to the magneto-ionic theory of Appleton, an electromagnetic wave of frequency f, generated anywhere on the earth's surface, can escape vertically from the earth only when the frequency of the waves exceeds certain limits, depending upon the maximum electron concentration above. The exact mathematical relations are

$$f_o{}^2 > \frac{4\pi Ne^2}{m} > 8 \cdot 0 \times 10^7 \cdot N$$

$$f_e(f_e + f_h) > \frac{4\pi Ne^2}{m} > 8 \cdot 0 \times 10^7 \cdot N.$$

Here N is maximum number of electrons per c.c. in the ionosphere, f_o is frequency of the o-wave, f_e is frequency of the two extraordinary waves, f_h the characteristic gyro-frequency of the electrons under the total field H, $f_h = eH/4\pi$ cm. $= 1·32\,H$ Mc. These conditions set a lower limit to the frequency of the radiations which can escape from the earth, and their validity has been verified by innumerable experiments.

If we apply these conditions to the sun, and also to the stars, we find at once that severe physical conditions have to be imposed on the emission of radio-waves from these bodies. Taking first the o-wave, we should have

$$N < 1 \cdot 25 \times 10^{-8} \cdot f^2$$
$$< 1 \cdot 25 \times 10^6 \text{ for } f = 10 \text{ Mc.}$$
$$< 5 \times 10^8 \text{ for } f = 200 \text{ Mc.}$$

The concentration of electrons in the different layers of the sun has been found by well-tried astrophysical methods[5] to have the mean values of 10^{11} per c.c. for the reversing layer, 4×10^{11} per c.c. for the mean chromosphere, and 4×10^8 per c.c. for the base of the inner corona. It is, therefore, obvious that o-radiations of radio-frequency range which we obtain from the sun cannot have their origin either in the reversing layer or the chromosphere, but only in the corona, and that also progressively in the outer layers as the wave-length is increased. But the corona has been shown to be a purely 'electron atmosphere' without any heavier atomic particles, excepting very small concentrations of heavily ionized *Fe*, *Ni* and *Ca* which produce the coronal lines. The mechanism of origin contemplated by Greenstein, Henyey and Keenan[7] which ascribes the radio-waves to recombination between protons and electrons therefore appears to fall to the ground in the case of the sun.

The e-waves. For the e-waves, the value of f_h is decisive, and this varies from 66 Mc. for the quiescent sun to roughly 4,000 Mc. for the spot, taking $H = 3,000$. These are frequencies of an order which are not contemplated in Appleton's theory, but a little work shows that whatever has been said regarding the o-wave also applies to *that* e-wave which corresponds to the condition $f_e(f_e - f_h) > 8 \cdot 10^7 \times N$ with greater emphasis. In fact, this wave cannot escape unless f_h has very high values, > 66 Mc. The e-wave corresponds to the condition $f_e(f_e + f_h) > 8 \times 10^7\,N$.

The possibility of reception of this wave on the earth has generally been ignored by European and American workers, but it has been obtained distinctly on several occasions by Toshniwal[8] at Allahabad, and his findings have been confirmed by Leiv Harang[9]. Recently, Saha and B. K. Banerjee[10] have shown that any radio-wave generated on the earth would be decomposed into three waves as in inverse Zeemann effect, the p-component corresponding to the o-wave, and the S-components to the e-waves. If this deduction be accepted, we at once see that for the spots, the e-wave of this type has a far greater probability of escape ; for now we should have

$$N < 1 \cdot 25 \times 10^8\, f_e(f_e + f_h)$$
$$< 1 \cdot 25 \times 10^8\, f_e f_h, \text{ taking } f_h \gg f_e$$
$$< 5 \times 10^8 \text{ for } 10 \text{ Mc. waves, and } < 10^{10} \text{ for } 200 \text{ Mc. waves ;}$$

taking $f_h = 4,000$ Mc., corresponding to the field-strength of 3,000 gauss. For a quiescent sun, the figures are $N < 8 \times 10^6$ and $1 \cdot 4 \times 10^8$ respectively. Hence the probability of escape of these waves from the quiescent sun continues to be very small, if the wave originates in the deeper layers. For larger spots, the field generally increases and has been known to reach values as high as 4,000 gauss.

From these arguments, it is fair to draw the conclusion that the large spots are just the regions whence the e-waves of the frequency range 10–200 Mc. can escape. The value of the fields given above corresponds to the level where the atomic lines originate, but Chapman[11] thinks that fields might increase to even 10,000 gauss in the deeper layers. If this be true, the e-waves can originate even from much deeper layers. Further, it is well known that the spot is a region of far lower temperature, and the electron concentration in the spot is much lower than on the general surface of the sun ; this circumstance also helps the escape of the e-waves.

If these considerations be on the right line, the radio-waves received on the earth when a big spot is in the centre of the sun's disk should be circularly polarized, and its sense of polarization will be determined by the sign of the field.

These considerations apply equally well to the stars comprising the Milky Way region, from which waves in the metre range have been observed[*]. They cannot be emitted from the surface of the hotter stars, but from cooler stars of *G*-, *K*- and *M*-type, and probably the escape of the radiation is facilitated by the development of spots in these stars, analogous to the case of the sun. The difficulties of the dilution factor pointed out by Greenstein *et al.*[7] are therefore eased to a large extent, as, according to Dunham[13], the disk area covered by *K*- and *M*-stars is nearly 10^4 times that of *B*-stars.

M. N. SAHA

University College of Science,
Calcutta.
Aug. 30.

*I am indebted to Dr. J. A. Ratcliffe for showing me these experiments during my recent visit to Cambridge.

[1] Appleton, *Nature*, **156**, 534 (1945).
[2] Hey, Phillips, Parsons, *Nature*, **157**, 297 (1946).
[3] Hey, *Nature*, **157**, 47 (1946).
[4] Pawsey, Payne-Scott, and McCready, *Nature*, **157**, 158 (1946).
[5] Nicholson, *Pub. Astro. Soc. Pacific*, **45**, 51 (1933).
[6] See for reference, Unsold, "Sternatmosphare", 82, 436, 440.
[7] Greenstein, Henyey, Keenan, *Nature*, **157**, 806 (1946).
[8] Toshniwal, *Nature*, **135**, 471 (1935).
[9] Harang, *Terr. Mag.*, **41**, 143 (1936).
[10] Saha and Banerjee, *Ind. J. Phys.*, **19**, 159 (1945).
[11] Chapman, *Nature*, **124**, 19 (1929).
[12] Dunham, *Proc. Amer. Phil. Soc.*, **81**, 277 (1939).

Figure 9.2. The "trigger": Gamow's 1946 letter to the editor of *Nature*. (Reprinted by permission from Macmillan Publishers Ltd: "Rotating Universe?" *Nature*, **158**, 549; 1946 [Oct. 19].)

No wonder Gödel soon immerses himself happily in this work. On July 15, 1947, he writes to Schilpp regretting the delay but saying that there is still an important point to settle that depends "on the solution of a mathematical problem, at which I am working now" (Gödel, 2003, 232–37). By September 1947, Gödel seems to have the outline of his model, but more problems keep cropping up. Only on May 10, 1948, can he correspond with his mother in Vienna, saying that he had intended to write long before but that for several weeks, he had been beset by a problem that had driven everything else out of his mind and that at last he had "settled the matter enough to be able to sleep well again" (Dawson, 1997). What he had just found was that in his universe, one could travel into the past – yet Schilpp had to wait another ten months for the final manuscript.

9.3 Gödel's Newtonian Rotating Universe

In his May 1949 lecture, Gödel, after quoting Gamow, recalled that "Newtonian physics gives a surprisingly good approximation for the expanding (nonrotating Friedman) universes." He then proceeded ingeniously to construct a Newtonian rotating (but non-expanding) universe. I shall outline his arguments here, partly because this Newtonian model is intrinsically interesting and surprising and partly because it still seems to be the best approach to the full Gödel model.[3]

In Newton's theory, we start with absolute space. We pick one fixed axis about which the universe is to rotate uniformly and rigidly. Its density ρ must then be constant in time, and for the sake of the cosmological principle, we take it to be constant in space as well. As in the Friedman case, we use Newton's law only in its differential form:

$$\nabla^2 \Phi = 4\pi G \rho, \tag{1}$$

where G is Newton's constant of gravity. The solution we want is

$$\Phi = \pi G \rho (x^2 + y^2) = \pi G \rho r^2, \tag{2}$$

if the field is radially away from the axis, as symmetry demands.

The gravitational force toward the axis,

$$\vec{f}_{grav} = -\nabla \Phi = -2\pi G \rho \vec{r}, \tag{3}$$

is then precisely balanced by the centrifugal force

$$\vec{f}_{cent} = \Omega^2 \vec{r}, \tag{4}$$

provided that the angular velocity Ω satisfies

$$\Omega^2 = 2\pi G \rho. \tag{5}$$

At first sight, this may seem to be an unlikely model universe because, in defiance of the cosmological principle, it has a center, or at least a central axis. However, on closer inspection, it turns out that there is complete *empirical* symmetry among the

---

[3] Incidentally, this Newtonian model was foreshadowed in van Stockum (1937), where Equation (5) had already appeared.

galaxies. Each moves freely; that is, each sits still without constraint on the rigidly moving substratum. No observable gravitational force relative to the substratum exists anywhere. At each point of the substratum, however, there is the same Coriolis force,

$$\vec{f}_{Cor} = 2\vec{v} \times \vec{\Omega}, \tag{6}$$

acting on any particle that moves relative to the substratum with velocity $\vec{v}$. Empirically, therefore, each galaxy can consider itself to be at rest on the axis of a universe that rotates rigidly at angular velocity $\vec{\Omega}$ around it.

Because of the close analogy with the later relativistic model, it is worth noting that all free orbits in the Newtonian model are circles, if started in a horizontal plane $z = const$, and circular helices otherwise, with respect to the substratum. This follows from the uniformity of the Coriolis force (6). Suppose, first, that a particle is projected with some velocity $\vec{v}$ relative to the substratum in a plane $z = const$. Because there is no force on it in the direction of $\vec{\Omega}$, it stays in its original plane, and because it experiences only a sideways force, its speed remains constant. Consequently, the magnitude of the sideways Coriolis force also remains constant, and the particle traces out a circle of radius

$$r = v/2\Omega. \tag{7}$$

It is easy to see that this circle is described in the direction opposite to that of $\vec{\Omega}$. If, on the other hand, the particle's initial velocity also has a component in the direction of $\vec{\Omega}$, that component stays constant, while only the horizontal velocity component determines the radius of the resulting helix, according to (7).

We may note from (7) that the magnitude ω of the angular velocity of the orbiting particle is given by

$$\omega = 2\Omega, \tag{8}$$

and this is to be expected: if the mass of some central vertical cylinder of the substratum (itself rotating at angular velocity Ω) allows a free substratum particle on its surface to orbit its axis at angular velocity $\vec{\Omega}$, it must also allow a particle to orbit its axis at angular velocity $-\vec{\Omega}$, and that is $-2\vec{\Omega}$ relative to the substratum.

Gödel's Newtonian universe, as discussed here and in his lecture, is not quite the correct Newtonian analogue of his relativistic universe. The reason is that his relativistic universe is predicated on a negative cosmological constant Λ, for which allowance can be made in Newtonian theory – but such allowance was not made by Gödel. The Newtonian field equation to which Einstein's field equation with Λ reduces in first approximation is[4]

$$\nabla^2 \Phi + \Lambda c^2 = 4\pi G \rho \tag{9}$$

instead of (1), c being the speed of light and Φ being the joint potential for gravity and the Λ force. Gödel's relativistic universe, it turns out, crucially needs

$$\Lambda = -4\pi G \rho / c^2. \tag{10}$$

---

[4] See Rindler (2006, eq. (14.20)) and Ozsvath and Schuecking (2001, sect. 3(c)).

If we use this value in (9), then instead of (2), which satisfies (1), we need

$$\Phi = 2\pi G\rho(x^2 + y^2),\tag{11}$$

which satisfies (9) and (10). Consequently, instead of (5), we now find

$$\Omega^2 = 4\pi G\rho,\tag{12}$$

which is actually the precise relation that holds in Gödel's (1949b.) relativistic model. Gödel considered (5) as already a "surprisingly good approximation."

It should be noted, however, that whereas in the case of Einstein's static universe of 1917, the Newtonian analogue "explains" the need for a positive Λ (a repulsive force is needed to counteract gravity), the Newtonian rotating universe seems to throw no light on the need for a negative Λ in Gödel's universe. Somehow, via Einstein's field equation, it must contribute to the puzzling equilibrium in the z-dimension (Ozsvath and Schuecking, 2001, sect. 3(c)).

9.4 Stationary Metrics in General and the Form of Gödel's Metric in Particular

Our next step here, as it was in Gödel's lecture, is to construct the relativistic model in close analogy to the Newtonian model. However, Gödel's original method is geometrically quite demanding, depending, as it does, on clever tricks with Clifford parallels in hyperspheres, quaternions, and so on, all of which modern relativists are not so familiar with. Here, therefore, I shall give an elementary derivation that uses only standard results from the general theory of stationary space-times (Rindler, 2006, chap. 9).

Loosely speaking, a gravitational field is stationary if it does not change with time, and it is said to be static if, additionally, "there is no rotation." One can think of the fixed points in the stationary field as forming a rigid (and generally curved) three-dimensional lattice. Relative to this lattice is generally a permanent gravitational field, and if the field is only stationary, but not static, there is also a unique permanent rotation of the "compass of inertia" at each lattice point.

Four-dimensionally speaking, each point of the lattice has its world line in space-time, and static space-times are characterized by the property that the set of these "fundamental" world lines is irrotational. This is equivalent to the existence of a unique set of identical (isometric) hypersurfaces (three-dimensional subspaces) cutting orthogonally across all the fundamental world lines. It is these hypersurfaces that constitute the unique simultaneities in static space-times.

In merely stationary space-times, the fundamental world lines twist and are not hypersurface-orthogonal. Infinitely many sets of parallel and isometric (and, of course, generally curved) hypersurfaces can then be drawn across the fundamental world lines, forming different sets of equally admissible simultaneities. Gödel's universe is of this kind.

It can be shown that for the most general stationary space-time, one can always find an "adapted" global time t (nonuniquely) of which the metric coefficients are independent. It is then usual to write the metric in the following canonical form[5]:

$$ds^2 = e^{2\Phi/c^2}(cdt - c^{-2}w_i dx^i)^2 - h_{ij}dx^i dx^j, \qquad (13)$$

where Φ is called the scalar potential, w_i is the vector potential ($i = 1, 2, 3$), and x^i are the coordinates for the lattice, whose metric tensor is h_{ij}; Φ, w_i, and h_{ij} are functions of the x^i only. One of the advantages of this canonical metric is that it allows us to "read off" the effective gravitational field and rotation. The effective gravitational field $\vec{g}$ is given by

$$\vec{g} = -grad\Phi, \qquad (14)$$

and the proper (i.e., with respect to proper time at a given lattice point) rotation rate of the lattice relative to the local gyrocompass is given by[6]

$$\vec{\Omega} = \frac{1}{2c}e^{\Phi/c^2}curl\ \vec{w}; \qquad (15)$$

both *grad* and *curl* refer to the metric h_{ij}.

In the case of Gödel's universe, where the lattice is free-floating, we need $\vec{g} = 0$ and can therefore set $\Phi = 0$. This still leaves the following "gauge" freedom, namely, the freedom to transform the metric (13) into an equivalent one:

$$x^i \rightarrow x^{i'} = x^{i'}(x^i) \qquad (16a)$$

$$t \rightarrow t' = t + f(x^i), \qquad (16b)$$

with $x^{i'}$ and f being any well-behaved functions of the x^i. The effect of (16b) on Φ and w_i is as follows:

$$\Phi' = \Phi, \quad w_i \rightarrow w_i' = w_i + c^3 f_{,i}, \qquad (17)$$

where ", i" denotes partial differentiation with respect to x^i. (I shall return to this gauge freedom presently.) Note that if we know the metric of the lattice, plus the gravitational field and the rotation rate at each point, then the metric (13) is uniquely determined up to gauge transformations. (This argument can eventually be used to verify that Gödel's universe is indeed homogeneous.)

To construct the lattice for Gödel's universe, we need, first of all, a set of parallel lines (in the sense of constant orthogonal distance between neighboring members of the set) to serve as local rotation axes for the universe. The 2-spaces orthogonal to these lines must be homogeneous if we wish to make the whole lattice homogeneous, but in accordance with general relativity, they need not be flat. They can, in fact, be 2-spaces of constant curvature. There are three types of such 2-spaces, having positive, negative, or zero curvature: k/a^2 ($k = 1, -1,$ or 0). Their respective metrics are[7]

$$dl^2 = a^2(dr^2 + \Sigma^2 d\varphi^2), \qquad (18)$$

with

$$\Sigma = \sin r, \ \Sigma = \sinh r, \ \Sigma = r \ \text{for} \ \ k = 1, -1, 0, \ \text{respectively.} \qquad (19)$$

---

[5] See Rindler (2006, eq. (9.13)). Einstein's summation convention applies here and in all subsequent formulas.
[6] See Rindler (2006, 197).
[7] See Rindler (2006, eq. (16.19)).

Here φ is the angle around the origin, and ar measures radial ruler distance from the origin. (There are other forms for these metrics, but for our purposes, these are the most convenient.) If z is distance along the rotation lines, the 3-metric of the lattice becomes

$$dl^2 = a^2(dr^2 + \Sigma^2 d\varphi^2) + dz^2. \tag{20}$$

By the rotational symmetry about the central z-line, all the coefficients of the full 4-metric must be independent of φ, and by homogeneity in the z- and t-dimensions, they must also be independent of z and t. We have already justified setting the scalar potential equal to zero: $\Phi = 0$. The vector potential w_i, now reduced to dependence on r only, cannot have a z-component because its *curl* is to point in the z-direction, and a possible r-component can be transformed away by a gauge transformation (17). Consequently, the full metric may be written in the form

$$ds^2 = [dt - w(r)d\varphi]^2 - a^2(dr^2 + \Sigma^2 d\varphi^2) - dz^2, \tag{21}$$

where here, and from now on, unless otherwise stated, units are chosen so as to make $c = 1$.

Following Gödel, we now redefine t, w, and z,

$$t \to at, \quad w \to aw, \quad z \to az, \tag{22}$$

so that (21) takes on the conformal form

$$ds^2 = a^2 \left\{ [dt - w(r)d\varphi]^2 - (dr^2 + \Sigma^2 d\varphi^2 + dz^2) \right\}. \tag{23}$$

Let us first deal with the rotation rate of the lattice, as calculated from (15). For the canonical metric (13) (with $c = 1$), the magnitude Ω of the rotation vector is given explicitly by[8]

$$\Omega = \frac{1}{2\sqrt{2}} e^{\Phi} \left[h^{ik} h^{jl} (w_{i,j} - w_{j,i})(w_{k,l} - w_{l,k}) \right]^{1/2}. \tag{24}$$

The metric (23) falls under the category (13), with

$$a^2 = e^{2\varphi} = h_{rr} = h_{zz} = h_{\varphi\varphi}/\Sigma^2. \tag{25}$$

If we now set

$$(r, \varphi, z) = (x^1, x^2, x^3), \tag{26}$$

we see that $w' := dw/dr = w_{2,1}$ is the only nonvanishing derivative $w_{i,j}$, and with that, formula (24) yields

$$\Omega = \frac{w'}{2a\Sigma}. \tag{27}$$

Because the coordinates (26) are right-handed if φ is measured in the counterclockwise direction, it can be seen from (15) and (27) that $\vec{\Omega}$ acts in the direction of increasing or decreasing φ, accordingly to w' being positive or negative.

---

[8] See Rindler (2006, eq. (9.23)).

For a homogeneous universe, we clearly need $\Omega = const$, and thus, by (27), $w' = \alpha\Sigma$ for some constant α. We could now integrate this and substitute the explicit function $w(r)$ into (23) before applying the field equations to that metric. In practice, however, the field equations look a little simpler if w is left generic. Because of the symmetry of all the other ingredients, the field equations themselves will imply $w' = \alpha\Sigma$.

At this point, we still have enough gauge freedom left to add a constant to $w(r)$, and simultaneously, a suitable multiple of φ to t. We can use this freedom to associate a geometrically preferred – or canonical – global time with any galaxy X. The obvious choice is the time that makes the hypersurface $t = const$ cut X's world line orthogonally so that t then coincides locally with X's inertial time. If X is the origin galaxy of the metric (23), this simply requires that

$$w(0) = 0, \tag{28}$$

a condition we shall now impose. Of course, the canonical time slices determined by the galaxy X differ from those of another galaxy Y, just as they do in special relativity from one inertial observer to another. However, that is precisely what made this universe so attractive to Gödel. It stands in contrast to the situation in the Friedman universes, where the canonical time slices cut orthogonally across all the fundamental world lines and are thus shared by all the galaxies. In rotating universes, the fundamental world lines twist, making such unicity impossible.

9.5 Do the Field Equations Permit a Gödel Universe?

So far, we have used a wish list of properties – mainly suggested by the Newtonian analogy – to arrive at the metric (23), and we even know $w(r)$ up to a constant (see after (27)). We also know that to be on the same footing as the Friedman models, the source matter should be pressureless dust, so almost no wiggle room is left; only the sign of the curvature index k (cf. (19)) is still free. It is time to take our metric before the judge – will it pass?

The judge is Einstein, and his field equations are the law. According to general relativity, matter (in the form of the energy tensor) and geometry (in the form of the metric) must jointly satisfy this law. In units that make $c = 1$, Einstein's field equations with cosmological constant Λ read as follows:

$$G_{\mu\nu} = -8\pi G T_{\mu\nu} - \Lambda g_{\mu\nu}. \tag{29}$$

Here $G_{\mu\nu}$ stands for the Einstein tensor $R_{\mu\nu} - (1/2)g_{\mu\nu}R$, which is built out of the metric and represents the geometry. We adopt the same sign convention as Gödel did (which also coincides with that in Rindler, 2006). Greek indices run from 1 to 4, and we augment (26) by

$$t = x^4. \tag{30}$$

On the right-hand side of (29), $T_{\mu\nu}$ represents the sources. It is usual to assume that the mechanical properties of the substratum – the smoothed-out universe – are equivalent to those of dust, the technical term for a pressureless perfect fluid. Its energy tensor is then given by the alternative formulae[9]

$$T^{\mu\nu} = \rho U^{\mu}U^{\nu} \quad \text{or} \quad T_{\mu\nu} = \rho g_{\mu\alpha}g_{\nu\beta}U^{\alpha}U^{\beta}, \tag{31}$$

where ρ is the proper density (a constant in Gödel's universe) and $U^{\mu} = dx^{\mu}/d\tau$ is the 4-velocity of the substratum ($\tau =$ proper time). The substratum satisfies $x^i = const$, whence, from (23), $d\tau = adt$ so that

$$U^{\mu} = (0, 0, 0, a^{-1}). \tag{32}$$

Then, because for (23) we have

$$g_{\mu\nu} = a^2 \begin{pmatrix} -1 & 0 & 0 & 0 \\ 0 & w^2 - \Sigma^2 & 0 & -w \\ 0 & 0 & -1 & 0 \\ 0 & -w & 0 & 1 \end{pmatrix}, \tag{33}$$

we find from (31) the following to be the only nonzero components of $T_{\mu\nu}$:

$$T_{22} = \rho a^2 w^2, \quad T_{24} = -\rho a^2 w, \quad T_{44} = \rho a^2. \tag{34}$$

To calculate the Einstein tensor $G_{\mu\nu}$ for the metric (33) – a formidable task if done by hand – it is nowadays best to use a computer program.[10] The following values are found:

$$G_{12} = G_{13} = G_{14} = G_{23} = G_{34} = 0, \tag{35}$$

$$G_{11} = -\frac{1}{4}\frac{w'^2}{S^2}, \tag{36}$$

$$G_{33} = \frac{1}{4}\frac{w'^2}{S^2} + k, \tag{37}$$

$$G_{44} = -\frac{3}{4}\frac{w'^2}{S^2} - k, \tag{38}$$

$$G_{24} = \frac{3}{4}\frac{ww'^2}{S^2} + \frac{1}{2}w'' - \frac{1}{2}\frac{C}{S}w' + kw, \tag{39}$$

$$G_{22} = -\frac{3}{4}\frac{w^2w'^2}{S^2} - \frac{1}{4}w'^2 - ww'' + \frac{C}{S}ww' - kw^2. \tag{40}$$

In these formulas, the meanings of S and C depend on the curvature index, as follows:

$$\begin{aligned} &\text{if } k = 1: S = \sin r, \quad C = \cos r; \\ &\text{if } k = -1: S = \sinh r, \quad C = \cosh r; \\ &\text{if } k = 0: S = 1, \quad C = 0. \end{aligned} \tag{41}$$

We note that all the components $G_{\mu\nu}$ are independent of the constant scale factor a.

---

[9] See Rindler (2006, eq. (7.86)).

[10] Donato Bini and Andrea Geralico of Consiglio Nazionale delle Ricerche (National Research Council), Rome, provided valuable help with this.

We are now ready to look at the Einstein field equations (29) explicitly, using the metric components from (33), the Einstein tensor components from (35)–(40), and the energy tensor components from (34). For $\mu\nu = 12, 13, 14, 23, 34$, these equations are trivially satisfied, each term vanishing separately. For $\mu\nu = 11$, the field equation reads

$$-\frac{1}{4}\frac{w'^2}{S^2} = a^2\Lambda, \tag{42}$$

which shows that we need the cosmological constant Λ and that it must be negative. Then, for $\mu\nu = 33$, we find

$$\frac{1}{4}\frac{w'^2}{S^2} + k = a^2\Lambda, \tag{43}$$

which, when added to (42), yields $k = 2a^2\Lambda$. Thus, because of the negativity of Λ, we must have

$$k = -1 \tag{44}$$

and then

$$\Lambda = -\frac{1}{2a^2}. \tag{45}$$

According to (44), we must choose the second line in (41). Then, with (45), (42) yields the expected relation (see after (27))

$$w' = \pm\sqrt{2}\sinh r, \tag{46}$$

and consequently, with (28),

$$w = \pm\sqrt{2}(\cosh r - 1). \tag{47}$$

Using (44), (45), and (46), the field equation for $\mu\nu = 44$ now yields

$$\rho = \frac{1}{8\pi G a^2}. \tag{48}$$

At this stage, no more freedom is left, and the last two field equations (for $\mu\nu = 22$ and 24) either are or are not satisfied by the values for k, Λ, ρ, and w as already found. Happily (and miraculously?), it turns out that they *are* satisfied. As Gödel pointed out, it is sufficient to check the field equations at a single point (most conveniently the origin) because the space-time is homogeneous. At the origin, from (46) and (47), we have

$$w = w' = 0, \quad w'/S = w'' = \pm\sqrt{2}, \tag{49}$$

which makes the checking trivial.

One more point needs clarification here. In equations (46) and (47), we may pick the positive or the negative sign. The result will be a universe that rotates counterclockwise, that is, in the same direction as φ ($w > 0$) or clockwise ($w < 0$): both, obviously, are equally possible. We prefer the choice $w > 0$, although Gödel chose $w < 0$, nevertheless asserting, apparently erroneously, that the rotation was in the direction of increasing φ.

Thus we have established the metric for Gödel's universe in the following form (cf. (23)):

$$ds^2 = a^2 \left\{ \left[dt - \sqrt{2} \, (\cosh r - 1) d\varphi \right]^2 - (dr^2 + \sinh^2 r \, d\varphi^2 + dz^2) \right\}. \qquad (50)$$

The proper rotation rate of the lattice relative to the compass of inertia (cf. (27)) is now given in full units by

$$\Omega = \frac{c}{\sqrt{2}a}. \qquad (51)$$

Summarizing our findings, we thus have, returning to full units,

$$-\Lambda = -\frac{1}{2} K = 4\pi G \rho / c^2 = \Omega^2 / c^2 = 1/2a^2, \qquad (52)$$

where K is the curvature of the layers $z = const$ of the lattice.

Note the required relation between Λ and ρ (which is actually the same, except for the sign, as in Einstein's static universe of 1917). If Λ is indeed a constant of nature, Gödel's universe, just like Einstein's, would need a highly tuned creation – unless one regards Λ as determined by the ρ of the universe.

9.6 Light Cones, Time Travel, and Geodesics in Gödel's Universe

The basic absolute structure in the (Minkowskian) space-time of special relativity is the set of parallel-oriented hourglass-like light cones, one at each event. They are closely related to the invariance of the speed of light and to the well-known special-relativistic speed limit $v \leq c$ for all particles and signals needed to ensure nonparadoxical causality.

Let us consider this speed limit pictorially (see Figure 9.3) under the usual suppression of one spatial dimension, say that of z. In an x,y,t space-time diagram (see Figure 9.3a), where t is always drawn vertically, consider a portion (dx, dy, dt) of a particle's world line, starting at some event P (see Figure 9.3b). Let the distance traveled be $dr = (dx^2 + dy^2)^{1/2}$. The inclination dr/dt of this world line away from the t-axis measures the particle's speed. In ordinary units, the speed of light is very large so that the world line of a photon would make almost a right angle with the t-axis. For visual and dimensional convenience, it is therefore preferable to choose relativistic units such that the speed of light is unity. Then photon world lines are inclined at 45 degrees to the vertical. All the 45 degree lines through an event P constitute the light cone at P (see Figure 9.3c), and ordinary particles passing through P must have their world lines within the cone ($v < c$). An extended particle world line must be inclined everywhere to the vertical at less than 45 degrees and so must lie within the light cones all along its extent (see Figure 9.3d). All events in the top half of the light cone at P can be influenced by signals from P, so this region is called the "absolute future" of P (see Figure 9.3e). Similarly, all events in the lower half of the cone can influence P, so this region is called the "absolute past" of P. All events, such as Q, outside the cone occur before P in some inertial frames, after P

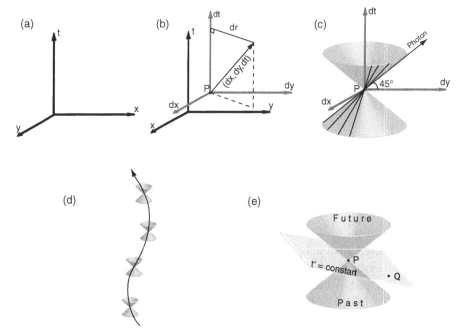

Figure 9.3 The space-time aspect of special relativity in a nutshell.

in other inertial frames, and simultaneously with P in yet others because any plane through P outside the cone represents a simultaneity $t' = const$ in *some* inertial frame (x',y',z',t').

In full Minkowski space-time (x,y,z,t), the speed restriction $dr/dt < c$ on a portion $(dx,\ dy,\ dz,\ dt)$ of a particle's world line reads, in full units,

$$dx^2 + dy^2 + dz^2 < c^2 dt^2 \tag{53a}$$

or

$$ds^2 := c^2 dt^2 - dx^2 - dy^2 - dz^2 > 0, \tag{53b}$$

whereas the local equation of a light cone is

$$dx^2 + dy^2 + dz^2 = c^2 dt^2 \text{ or } ds^2 = 0. \tag{54}$$

In Minkowski's language, ds^2 stands for the squared displacement between neighboring points (events) in the flat space-time of special relativity, and it is the primary law of special relativity that every inertial observer obtains the same value for it. Evidently, it is an analogue of $dx^2 + dy^2 + dz^2$ – which represents squared distance between neighboring points in Euclidean space – except that Minkowski's ds^2 can be negative as well as positive or zero for perfectly ordinary event pairs. As we have seen, it is zero for neighboring events on a photon world line and negative for events such as P and Q in Figure 9.3e.

According to general relativity, the space-time in the presence of gravitating sources is curved, although locally Minkowskian. This is analogous to the situation in differential geometry, where, for example, the surface of a sphere is curved but sufficiently

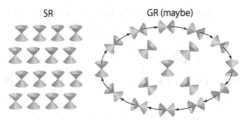

Figure 9.4. The invariant orientation of the light cones in special relativity (SR) and a permissible orientation of the light cones in general relativity (GR).

small portions of it can be treated as flat and Euclidean to a high degree of accuracy. In fact, in every suitably small region of space-time, we can release an "Einstein cabin" – a freely falling nonrotating box, like a severed elevator cabin or an astronaut's capsule – inside of which gravity has disappeared and special relativity holds. The ds^2 between neighboring events in the curved space-time is the same as that measured in the local Einstein cabin. Hence, $ds^2 = 0$ is still the equation of the local light cone, and $ds^2 \geq 0$ is still the local speed restriction on particle and signal world lines.

Now, whereas in the flat space-time of special relativity, the light cones are all aligned in parallel like soldiers on parade (see Figure 9.4a), and no world line can loop back on itself, general-relativistic space-times are curved, and the light cones can be all over the place (as in Figure 9.4b). A year before he had even completed his general theory of relativity, Einstein (1914) already worried about the possible existence of curved space-times that allowed closed particle world lines, as in Figure 9.3b, even though the light cones permit a consistent past-future labeling throughout. Einstein wrote, "This conflicts strongly with my physical intuition. But I am unable to prove that the theory excludes the occurrence of such orbits." He was right. Much earlier than 1949, space-times had indeed been discovered where such orbits exist.[11] But Gödel's universe was the cleanest example, certainly the one that caught the widest attention and possibly the first in which time loops were explicitly recognized.

On the scale of the universe, an observer is but a particle. Observers' world lines must therefore also lie everywhere within the light cone. For an observer to travel back to an event already experienced, his or her world line must be a closed loop, lying everywhere within the light cone. In other words, the observer's world line must satisfy $ds^2 > 0$ all along its length. Observe also, from (53b), that if you momentarily fly along with an orbiting particle in a free Einstein cabin, its world line relative to the cabin will satisfy $dx = dy = dz = 0$ and thus $ds^2 = c^2 dt^2$, so time elapsed at the particle is given by $\int ds/c$.

In Gödel's universe, the existence of closed particle world lines is almost trivially easy to read off directly from the metric (50). Consider a circular path, $r = const$, in one of the layers $z = const$, and let it be traced out so that $t = const$. (This last condition is not nonsensical: time cannot stand still on an orbiting particle, nor on a given galaxy along its path, but the cosmic time coordinate t *can* be the same on successive galaxies

[11] Anti-de Sitter space, Lanczos (1924) space, van Stockum (1937) space, etc. For a discussion of anti–de Sitter space, see Synge (1960) or Rindler (2006, sect. 14.6). I was unable to determine when its time loops were first noticed.

passed by the particle.) If we accordingly set $dr = dz = dt = 0$ in (50), we are left with

$$ds^2 = a^2 \left[2(\cosh r - 1)^2 - \sinh^2 r \right] d\varphi^2, \tag{55}$$

which is obviously positive for sufficiently large r. Specifically, (55) can be transformed into

$$ds^2 = a^2 \left[(\cosh r - 2)^2 - 1 \right] d\varphi^2 \tag{56}$$

so that, provided that

$$\cosh r > 3, \tag{57}$$

such circles are possible particle world lines, and they clearly return to precisely the initial event after one complete revolution. With equality instead of inequality in (57), we would have $ds^2 = 0$, and the world line would be that of a photon. Note, however, that none of these orbits are geodesics (free-fall paths), as we shall see subsequently; the photon would have to be guided by mirrors, and the particle would have to be propelled by rocket motors.

As far as the requirement $ds^2 > 0$ is concerned, a circle satisfying the condition (57) could be described by a particle in the positive or negative direction of φ. However, in any properly *causal* space-time – namely, one throughout which the two halves of the local light cones can be consistently labeled "past" and "future" (and Gödel's universe is of this kind, as we shall presently see) – all world lines along their entire length must point into the future cones. This will require the closed world lines of the positively rotating universe to be described in the negative direction.

For the purpose of establishing this result, let us examine the light-cone structure of the metric (50) all over a surface $t = const$ in space-time. Such a surface is not a plane in the geometrical sense, but we are at liberty to draw a map in which it is a plane (see Figure 9.5). Nor do the fundamental world lines $t = var$ (except that of the central galaxy) cut such a surface orthogonally, but we are at liberty to map them as vertical lines cutting $t = const$ orthogonally. We shall not restrict the generality of the argument by looking only at events satisfying $z = const$. In the plane $t = const$, let us now go out from the center $r = 0$ along a radius $\varphi = const$ and there look at directions satisfying $r = const$. For them to lie on the light cone, they must satisfy $ds^2 = 0$ (with $dr = dz = 0$), which, together with (50), implies

$$dt - \sqrt{2}(\cosh r - 1)d\varphi = \pm(\sinh r)d\varphi$$

or

$$dt/d\varphi = \sqrt{2}(\cosh r - 1) \pm \sinh r. \tag{58}$$

Let us assume that we are sufficiently far from the origin to make the first term in the square brackets in (55) greater than the second. Then both values for $dt/d\varphi$ in (58) are positive. We know that the fundamental (galaxy) world line $t = var$ at the event in question must point into the future cone. Hence the light cone there looks like the third one out in Figure 9.5.

At the origin $r = 0$, the metric (50), to first order in r, reduces to a Minkowskian form. Hence the central light cone has the usual special-relativistic appearance with

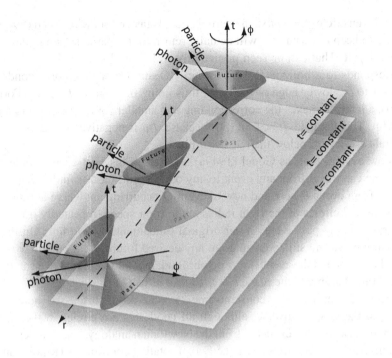

Figure 9.5. A space-time map of Gödel's universe; the map flattens the $t = const$ sections into planes and progressively widens the intrinsically 90 degree angle of the light cones as we go away from the chosen center.

a 45 degree semivertical angle. As we move away from the center, the cone tilts and widens until at $\cosh r = 3$, one of its edges touches the plane $t = const$ because one of the values of $dt/d\varphi$ in (58) becomes zero. (The reason that its semiangle is no longer 45 degrees is that our map distorts angles.) Farther out still, the direction $\varphi = var$ lies in its interior, and as Figure 9.5 shows, its future direction is that of decreasing φ. This is what we set out to establish.

Let us next inquire into the particle speeds needed to describe the orbits under discussion. Consider the ds^2 of a portion of such an orbit, satisfying $r,z,t = const$. It is given by (55). In terms of the local inertial coordinates associated with whatever galaxy the particle just passes, it is also given by $ds^2 = d\tilde{t}^2 - dl^2$, where $\tilde{t}$ is the local inertial time and, by (50), $dl = a(\sinh r)d\varphi$ is the ruler distance traveled. (Note that whereas at coincides with the proper time along any galaxy world line, the hyperplane elements $t = const$ are inclined to the hyperplane elements $\tilde{t} = const$ everywhere, except at the center.) The local speed of the particle is then given by

$$v = \frac{dl}{d\tilde{t}} = \frac{dl}{\sqrt{ds^2 + dl^2}} = \frac{\sinh r}{\sqrt{2}(\cosh r - 1)}. \tag{59}$$

For the critical circle, we have $\cosh r = 3$ (and thus $\sinh r = 2\sqrt{2}$), and this gives us $v = 1$, the speed of light. For larger circles, the required v gradually decreases to $1/\sqrt{2}$ the speed of light, which is therefore the minimum speed for such journeys into the past.

If, on a given circle, $r = const$, the particle travels faster than with speed (59), which is required to keep t constant; it will actually return to its spatial starting point earlier than when it left. That can be seen particularly well from Figure 9.5, where the v of (59) corresponds to a horizontal orbit to the left, and a greater v corresponds to an orbit closer to the lower edge of the cone, thus going into earlier t-values. Formally, from (59), a larger v makes the corresponding ds^2 smaller for a given dl and thus for a given $d\varphi$. This leads to $dt < 0$ in (50) because $d\varphi$ is negative. The corresponding orbit spirals downward below the $t = const$ plane.

It is now evident that, as Gödel (1949a) asserted in his Festschrift article, in his universe one can travel from any event to any other event, past or future. Suppose we wish to go from event P here and now to event Q occurring anywhere at time $t = t_o$. By going around circles of the right radius at the right speed, repeatedly if necessary, we can travel to any past event on our original fundamental world line. On the other hand, simply by sitting still, we can go to any desired future event on this world line. Once we have reached the event $t = t_o$ on our fundamental world line, we can find any number of circular arcs in the plane $t = t_o$ to take us to Q.

As mentioned earlier, the constant-time circular orbits discussed here are not geodesics in the Gödel universe, although all geodesics are circles in the lattice or circular helices, as is to be expected from the Newtonian analogy. The simplest method of determining these geodesics is the method of rotating coordinates (Rindler and Perlick, 1990). According to this method, we take a uniformly rotating "vertical coordinate plane" $\varphi = \omega t, (\omega = const)$ as the base plane $\tilde{\varphi} = 0$ for a new angular coordinate $\tilde{\varphi}$:

$$\tilde{\varphi} = \varphi - \omega t \tag{60a}$$

$$d\varphi = d\tilde{\varphi} + \omega dt. \tag{60b}$$

The canonical stationary metric (50) is thereby transformed into another canonical stationary metric, one whose lattice rotates relative to the old lattice at coordinate angular velocity $d\varphi/dt = \omega$. For our present purposes, we need not calculate that entire metric; all we need is the coefficient of dt^2. Replacing $d\varphi$ by $d\tilde{\varphi}$ according to (60) transforms (50) into

$$ds^2 = a^2 \left\{ \left[1 - \sqrt{2}(\cosh r - 1)\omega \right]^2 - (\sinh^2 r)\omega^2 \right\} dt^2 + \cdots . \tag{61}$$

A fixed point P on this new lattice will move geodesically if the gravitational field, relative to this lattice, vanishes at P. According to (14), we thus need only set the r-derivative of the brace in (61) equal to zero. A simple calculation then yields a quadratic equation for ω of the form $A\omega + B\omega^2 = 0$, in which one of the roots is zero (not surprisingly – particles sitting still in the original lattice also trace out geodesics). The other root is

$$\omega = \frac{-\sqrt{2}}{2 - \cosh r}. \tag{62}$$

Hence, within a coordinate plane $z = const$, a circle traced out at radius r with this coordinate angular velocity is a geodesic; the negative sign means that it must be traced out in the clockwise (negative) direction if φ increases in the counterclockwise direction. Note that as r increases from zero to $\cosh^{-1} 2$, ω numerically increases from $1/\sqrt{2}$ to infinity. But a particle (or photon) orbit must be timelike (or null) and hence must satisfy $ds^2 \geq 0$. Because the ellipses in (61) stand for terms involving the new spatial differentials, we now set them equal to zero for the rotating lattice and substitute for ω from (62). After some manipulation, this leads to

$$ds^2 = \frac{2 - \cosh^2 r}{(2 - \cosh r)^2} a^2 dt^2 \tag{63}$$

and hence to the following condition for $ds^2 \geq 0$:

$$\cosh r \leq \sqrt{2}. \tag{64}$$

Equality here corresponds to a photon geodesic, which is thus the largest circular geodesic. We may note that (64) is equivalent to $\cosh 2r \leq 3$; comparison with (57) then shows that the largest circular geodesic is half as big as the smallest time loop.

To determine the local velocity of a particle following a circular geodesic, we repeat the argument leading to (59), except that in the last step, we now take ds^2 from (63) and in dl replace $d\varphi$ by ωdt. This leads to the pleasant result

$$v = \sqrt{2} \tanh r, \tag{65}$$

in terms of which the condition (64) reads $v \leq 1$, just as one would expect.

The preceding analysis, in fact, determines all plane geodesics, namely, those in coordinate planes $z = const$. For, given any initial particle velocity at any point of such a plane, there is a unique geodesic with that initial vector velocity, and it will be a circle: we need only to proceed orthogonally to the given velocity, along a geodesic of the spatial lattice, through a ruler distance ar, where r is given by (65), to find the center of the circular geodesic that fits the given initial data. It is well to bear in mind this relation of the coordinate r to radial ruler distance – implicit in the metric (50) – in all the preceding formulas.

We must also remember that not t but at in (50) corresponds to proper time at the galaxies. If we wish, therefore, to compare the ω of geodesics with the Ω of the entire Gödel universe, we need a factor a in the denominator of (62). Then, for small values of r, we find, from (62) and (51), that $\omega = 2\Omega$ numerically, just as in the Newtonian analogue (cf. (8)).

To complete our discussion of the geodesics in Gödel's universe, we would expect from the Newtonian analogy that in the most general case, they are helices in the lattice, and this is indeed so. For proof, we simply augment the rotation (60) by an additional uniform translation (up or down) in the z-direction,

$$\tilde{z} = z - ut, \quad dz = d\tilde{z} + udt, \quad (u = const), \tag{66}$$

so that each lattice point of the $(r, \tilde{\varphi}, \tilde{z})$ system now traces out a helix around the central axis $r = 0$. Algebraically, this merely brings an additional term $-u^2$ into the brace in (61), which leaves (62) unchanged. Consequently, any point fixed in the new lattice, now spiraling in the z-direction, still describes a geodesic when (62) is satisfied. However, the condition for $ds^2 \geq 0$ is now obtained from the obvious modification of

(63), which yields

$$\frac{2 - \cosh^2 r}{(2 - \cosh r)^2} \geq u^2, \tag{67}$$

with equality again corresponding to photons.

The local horizontal velocity component of the geodesic is still given by (65). The local vertical velocity component is not u but must again be found by a suitable variation of (59) (putting $dl = u\,dz$). If we denote it by $\tilde{u}$, we find that

$$\tilde{u} = \frac{u}{1 + [2(\cosh r - 1)/(2 - \cosh r)]}. \tag{68}$$

With that, (67) is equivalent to $\tilde{u}^2 + v^2 \leq 1$, as one would expect.

The helicoidal geodesics discussed here are, in fact, the most general geodesics in the Gödel universe in the sense that every possible geodesic is such a helix around some suitable axis. Suppose that at any point in the lattice, we are given an initial local velocity having a horizontal component v and a vertical component $\tilde{u}$. Then v allows us, as before, to find r and ω and hence the axis of the helix; and finally, $\tilde{u}$ together with r determines the u of the helix via (68).

In any universe, the propagation of light is of special importance for interpreting observations. In Gödel's universe, as we have just seen, photons propagate along circles or, more generally, along helices with axes in the z-direction; each satisfies (65) and equality in (67). The larger the radius, the larger the horizontal velocity component (65), and therefore the smaller the vertical velocity component $\tilde{u}$ (because $v^2 + \tilde{u}^2 = 1$) and with it also the pitch $\tilde{u}/v$ of the helix. The largest possible radius, by (64), is $r = \cosh^{-1}\sqrt{2}$ when $v = 1$, and the helix reduces to a circle. Also, the larger the radius, the larger the angular velocity ω (by (62)) and hence the shorter the period. On the other hand, the pitch increases without bound as $r \to 0$, and the path ultimately becomes a straight line in the z-direction.

From the preceding remarks, it is clear that galaxies on the same z-line can see one another directly along that straight line, traversable in either direction. All other light paths are traversable in one direction only, namely, clockwise. The farthest one can see from any galaxy horizontally is to a ruler distance $2a \cosh^{-1}\sqrt{2}$. A galaxy at that distance sends us light along a semicircle of half that radius and, in turn, sees us along the complementary semicircle. That same full circle, of course, shows us the backs of our heads (and that galaxy, incidentally, lies on the smallest time loop around ourselves). Other galaxies see us along portions of suitable helices, and we see them along the complementary helices, the two being mirror images in the coordinate plane $\varphi = const$, joining us to the galaxy in question. All galaxies that lie strictly within a cylinder of radius $2a \cosh^{-1}\sqrt{2}$ around us can receive light from us, and we from them. However, galaxies with large z-values relative to us may have to wait a while for the signal because many turns of the required helix may be necessary. Also, there will generally be multiple connecting paths, and thus multiple images, having left the source at different ages. This is most obvious for galaxy pairs on the same vertical, which can be connected not only directly but also via an infinite number of helicoidal light paths: the z-distance between successive loops of the helices need only be whole fractions of the z-difference between the galaxies.

(Think of the cylinders on which the helices lie as touching the z-line joining the two galaxies.)

9.7 Stability

An important topic on which we have not yet touched is the stability, or lack of it, of Gödel's universe. Einstein's static model of 1917 is patently unstable (in hindsight). The slightest contraction leads to total collapse because the contraction decreases the Λ repulsion (which is proportional to distance) and increases the gravity (which is inversely proportional to the square of the distance). Conversely, the slightest expansion becomes unstoppable.

An analogous Newtonian argument suggests that Gödel's universe, on the contrary, is stable with regard to contraction or expansion. Consider a slight deformation $r_o \mapsto r_o + dr$ of a central cylinder $r \le r_o$ in Gödel's model. The density ρ will decrease like $1/r^2$ (by mass conservation), as will the rotation rate Ω (by angular momentum conservation). Consequently, we have

$$\rho = \rho_o r_o^2 / r^2, \quad \Omega = \Omega_o r_o^2 / r^2$$

so that, with $k = 2\pi G \rho_o$, (3), (4), and (12) yield

$$f_{grav} = -kr_o^2/r, \quad f_{cent} = 2kr_o^4/r^3.$$

Λc^2, of course, stays constant at its equilibrium value $-4\pi G \rho_o$ so that, by (2), (3), and (11), we have, for the Λ force,

$$f_\Lambda = -kr.$$

At r_o, these three forces balance. Thereafter we have

$$df_{grav} + df_\Lambda + df_{cent} = k(r_o^2/r^2)dr - kdr - 6k(r_o^4/r^4)dr = -6kdr,$$

when $r = r_o$. The changes in f_{grav} and f_Λ just cancel each other out, and a restoring force is provided by a change in centrifugal force. This shows the radial equilibrium to be stable. (A very meticulous discussion of the stability problem can be found in Barrow and Tsagas (2004).)

9.8 Comments

This is as far as I shall here go in discussing the technicalities of Gödel's stationary model universe. In summary, this model provides a prime example of an anti-Machian distribution of matter, that is, one in which the rotation of the compass of inertia bears no relation to the mass distribution. This goes against our intuition but need not be fatal to the model. Gödel's universe also neatly exhibits the two main difficulties with time in general relativity. The first is the existence of closed time loops, which play havoc with causality. A reasonable response to this is essentially Einstein's: one hopes that nature has some as yet undiscovered mechanism to prevent the formation of such universes, analogously perhaps to the speed limit in special relativity, which also serves to preserve causality. The second difficulty is the nonunicity of simultaneity in relativity, even in some relativistic cosmologies, although not in others. However, though this dependence of the objectivity of time on the cosmic mass distribution may well be a problem for philosophers, it presents neither physical nor logical problems for physicists, who have long learned to live and work with this state of affairs.

We might ask one last question: how original were the remarkable features of Gödel's model? There is little doubt that both Gödel and Einstein regarded Gödel's discovery of time loops as unprecedented, yet solutions with time loops undoubtedly existed before 1949 such as anti–de Sitter space, Lanczos space (Lanczos, 1924), and the van Stockum spaces (van Stockum, 1937). In anti–de Sitter space, the time loops can be removed by simple topological extension (see, for example, Rindler 2006, 314), but in the Lanczos and van Stockum spaces, they are irremovable, just as in Gödel space. On the other hand, an admittedly incomplete literature search failed to uncover any evidence that the time loops in those earlier solutions had actually been recognized as such until Gödel's work made people look again. Rotating solutions, on the other hand, were definitely known and recognized as such much earlier (van Stockum, 1937), although certainly not rotating homogeneous universes.

In the epilogue below, I briefly report on the model by Lanczos, simply because of the historical irony that Lanczos came so uncannily close to discovering Gödel's universe exactly twenty-five years earlier.

Gödel himself continued his interest in rotating universes well beyond 1949. However, only one more publication resulted from this work: the text of a lecture delivered in September 1950 titled "Rotating Universes in General Relativity Theory" (Gödel, 1952). In this lecture, Gödel said that he was "setting forth the main results (for the most part without proofs) to which my investigations on rotating universes have led me so far." The models he examines are spatially finite, homogeneous, and expanding.

Here I shall content myself with reporting only one very interesting result from that paper. Because in an expanding universe, the density ρ must steadily decrease, we can consider the set of hypersurfaces $\rho = const$ cutting across the fundamental world lines as determining a kind of global time. However, they cannot cut across those world lines orthogonally, for then we would have a Friedman universe and no rotation. Imagine therefore a locally parallel set of such constant-density surfaces, going from greater to lesser density, and the world line of a fundamental observer (a substratum particle) cutting across them obliquely. The observer's proper surface of simultaneity is orthogonal to his or her world line and therefore cuts across the constant-density surfaces. At any given instant, therefore, the observer sees more galaxies in one half of the sky (where his or her simultaneity dips into the greater density) than in the other half. In fact, Gödel long persisted in searching for this effect, both through personal calculations from published data and by urging astronomers onward (see Dawson, 1997, 182). However, no such indication of rotation has ever been found.

E. Schuecking, who later talked with Gödel, has conjectured that the absence of proofs from this paper may have been because of Gödel's dissatisfaction with the inelegance of his private calculations (Ellis, 1996). However, as described in detail by Ellis, this paper contains a number of truly seminal ideas and results that later became part and parcel of the various extensions of theoretical cosmology beyond the Friedman models. In fact, it is probably fair to say that Gödel's 1952 paper was one of the main impulses for many of these extensions.

9.9 Epilogue: Lanczos' Model Universe of 1924

Twenty-five years before Gödel, Kornel (later Cornelius) Lanczos, one of the distinguished theoreticians of the twentieth century, invented a model universe (Lanczos, 1924) possessing many of the striking features of Gödel's later model. Excusably for that period, however, Lanczos failed to recognize the two most striking features: rotation and time loops. His universe was rediscovered by van Stockum (1937), who at least recognized its rotation, if not its time loops. The great defect of Lanczos' model as a cosmology is that it is not homogeneous.

Lanczos, like Einstein, still took it for granted that the universe is unchanging in time. Friedman's groundbreaking work, in which he introduced expanding universes (in 1922, in the same journal in which Lanczos would later publish), did not percolate into the scientific consciousness of the day, nor would it for many years to come. (It took Hubble's forceful personality finally to convince Einstein in 1931 that the universe was really expanding.) Lanczos had only Einstein's 1917 universe at which to look, and he found it unsatisfactory. It required a specific relation, $\Lambda c^2 = 4\pi G \rho$ (numerically the same as Gödel's), to hold between the density and the cosmological constant. Such a relation, Lanczos wrote, "would be a mere coincidence and is therefore not a satisfactory assumption." He therefore set out to construct a stationary universe without Λ. The field equations with dust sources do not permit such a universe to be spherically symmetric, so Lanczos opted for the next best thing: rotational symmetry about an axis.

Guided partly by symmetry, partly by simplicity, he then arrived at the following *Ansatz*:

$$ds^2 = dt^2 - G(dr^2 + dz^2) - 2Q d\varphi\, dt - P d\,\varphi^2, \tag{69}$$

where G, Q, and P are functions of r only. (For comparison purposes, I reversed the sign of Lanczos' metric and wrote r, φ, z for his x, ψ, y.) It wasn't too late to get Gödel's metric (50) out of this, but Lanczos proceeded down another road. Setting $\Lambda = 0$ in the field equations, he now necessarily found

$$ds^2 = dt^2 - e^{-r^2}(dr^2 + dz^2) - 2r^2 d\varphi\, dt - (r^2 - r^4)d\,\varphi^2, \tag{70}$$

apart from an overall dimension-giving constant factor, which he set equal to unity. The density function corresponding to this metric is given by

$$\rho = \frac{e^{r^2}}{2\pi G}, \tag{71}$$

where I have written ρ for Lanczos' μ and used Gödel's units, in which $c = 1$ but $8\pi G \neq 1$. Note that $\rho \to \infty$ as $r \to \infty$. However, Lanczos' r is only a conventional coordinate, related to ruler distance l by

$$l = \int_0^r e^{-r^2/2} dr \tag{72}$$

so that $l \to \sqrt{\pi/2}$ as $r \to \infty$. Lanczos' universe is thus recognized as a cylinder, infinite in the z-dimension but with a singular edge and infinite density at ruler radius $\sqrt{\pi/2}$.

For ease of comparison, we recast Gödel's metric (50) into an alternative form, which results when we set

$$t = 2\tilde{t}, \quad r = 2\tilde{r}, \quad z = 2\tilde{z} \text{ (and } \tilde{S} = \sinh \tilde{r}), \tag{73}$$

namely,

$$ds^2 = 4a^2 \left\{ d\tilde{t}^2 - (d\tilde{r}^2 + d\tilde{z}^2) - 2\sqrt{2}\tilde{S}^2 d\varphi dt - (\tilde{S}^2 - \tilde{S}^4) d\varphi^2 \right\}. \tag{74}$$

The analogy with Lanczos' metric is now quite striking.

As is easily verified today (cf. (24)), Lanczos' universe rotates locally (but not rigidly) relative to the compass of inertia with proper angular velocity

$$\Omega = e^{r^2/2}, \tag{75}$$

although he did not know it. Even so, he determined the quasi-helicoidal shape of the light paths in his model (which could have suggested rotation). Together, (75) and (71) imply $\Omega^2 = 2\pi G\rho$, as compared with Gödel's $\Omega^2 = 4\pi G\rho$.

Lanczos' universe also has closed time loops. He actually went so far as to remark that the simultaneity surfaces $t = const$ eventually cut into the light cone (just as in Figure 9.5), in fact, when $r > 1$. However, he also remarked that this was nothing to worry about. Evidently he was only a hair away from realizing that all loops t, z, $r = const$ with $r > 1$ in (70) are possible closed-particle world lines (as follows immediately from inspection of the metric: $ds^2 > 0$). Alas.

Later in his paper, Lanczos did include Λ but, unfortunately and inexplicably, took it as an axiom that $\Lambda \geq 0$, so nothing significantly new emerged. Alas once more.

Acknowledgments

I wish to thank Chris Allison for his fine execution of the diagrams, Donato Bini and Andrea Geralico for help with computing the field equations, and Mustapha Ishak for extricating me from a bad computer crash. My special thanks go to Pamela Contractor, the book's developmental editor, for selflessly smoothing the process all along the way.

References

Barrow, J. D., and Tsagas, C. G. (2004). Dynamics and stability of the Gödel universe. *Classical and Quantum Gravity*, **21**, 1773–89. [This paper, with its numerous references, will be useful to anyone desiring a deeper immersion in the literature.]

Dawson, J. W., Jr. (1997). *Logical Dilemmas: The Life and Work of Kurt Gödel*. Wellesley, MA: A K Peters.

Einstein, A. (1914). Die formale Grundlage der allgemeinen Relativitaetstheorie. *Sitzungsberichte d.kon.preuss.Ak. Wissenschaften*, XLI, 1030–85. [See p. 1079. I am indebted to J. Ehlers for drawing my attention to this reference.]

Ellis, G. F. R. (1996). Contributions of K. Gödel to relativity and cosmology. University of Cape Town preprint, Institute of Theoretical Physics and Astrophysics.

Friedman, A. (1922). Über die Kruemmung des Raumes. *Zeitschrift für Physik*, **10**, 377–86.

———. (1924).

Gamow, G. (1946). Rotating universe? *Nature*, 158, 549.

Gödel, K. (1949a). A remark about the relation between relativity theory and idealistic philosophy. In *Albert Einstein: Philosopher-Scientist*, ed. P. A. Schilpp, pp. XX–XX. Library of Living Philosophers 7. Evanston, IL: MJF Books. [Also in *Collected Works*, vol. 2 (1990), pp. 202–7.]

———. (1949b). Lecture on rotating universes given at the Institute for Advanced Study, Princeton, May 7. [In *Collected Works*, vol. 3 (1995), pp. 269–87.]

———. (1949c). An example of a new type of cosmological solution of Einstein's field equations of gravitation. *Reviews of Modern Physics*, **21**, 447–50. [Also in *Collected Works*, vol. **2** (1990), pp. 190–98.]

———. (1952). Rotating universes in general relativity theory. In *Proceedings of the International Congress of Mathematicians*, vol. 1, ed. L. M. Graves et al., pp. 175–81. Cambridge, MA: American Mathematical Society. [Also in *Collected Works*, vol. 2 (1990), pp. 208–16.]

———. (2003). *Collected Works*. Vol. 5. *Correspondence H–Z*. Oxford: Clarendon Press.

Lanczos, K. (1924). Über eine stationaere Kosmologie im Sinne der Einsteinschen Gravitationstheorie. *Zeitschrift für Physik*, **21**, 73–110.

Ozsvath, I., and Schuecking, E. L. (2001). Approaches to Gödel's rotating universe. *Classical and Quantum Gravity*, **18**, 2243–52.

Rindler, W. (2006). *Relativity: Special, General, and Cosmological*. 2nd ed. Oxford: Oxford University Press.

Rindler, W., and Perlick, V. (1990). Rotating coordinates as tools for calculating circular geodesics and gyroscopic precession. *General Relativity and Gravitation*, **22**, 1067–81.

Synge, J. L. (1960). *Relativity: The General Theory*. New York: North-Holland Interscience.

Van Stockum, W. J. (1937). The gravitational field of a distribution of particles rotating about an axis of symmetry. *Proceedings of the Royal Society of Edinburgh*, **A57**, 135.

CHAPTER 10

Physical Unknowables

Karl Svozil

As we know, there are known knowns;
there are things we know we know.
We also know there are known unknowns;
that is to say we know there are some things we do not know.
But there are also unknown unknowns –
the ones we don't know we don't know.
> – United States Secretary of Defense Donald H. Rumsfeld
> at a Department of Defense news briefing on
> February 12, 2002

Ei mihi, qui nescio saltem quid nesciam!
(Alas for me, that I do not at least know the extent of my own ignorance!)
> – Aurelius Augustinus, 354–430, "Confessiones"
> (Book XI, chapter 25)

10.1 Rise and Fall of Determinism

In what follows, a variety of physical unknowables will be discussed. Provable lack of physical omniscience, omnipredictability and omnipotence is derived by reduction to problems that are known to be recursively unsolvable. "Chaotic" symbolic dynamical systems are unstable with respect to variations of initial states. Quantum unknowables include the random occurrence of single events, complementarity, and value indefiniteness.

From antiquity onward, various waves of (in)determinism have influenced human thought. Regardless of whether they were shaped by some *Zeitgeist,* or whether, as Goethe's *Faust* puts it, "what you the Spirit of the Ages call, is nothing but the spirit of you all, wherein the Ages are reflected," their proponents have sometimes vigorously defended their stance in irrational, unscientific, and ideologic ways. Indeed, from an emotional point of view, may it not appear frightening to be "imprisoned" by remorseless, relentless predetermination, even in a dualistic setup (Descartes, 1641); and, equally frightening, to accept that one's fate depends on total arbitrariness and chance? Does determinism expose freedom, self-determination and human dignity as an idealistic illusion? On the other extreme, what kind of morale, merits and efforts appear worthy in a universe governed by pure chance? Is there some reasonable in-between straddling those extreme positions that may also be consistent with science?

We shall, for the sake of separating the scientific debate from emotional overtones and possible bias, adopt a contemplative strategy of *evenly-suspended attention* outlined by Freud (1999), who admonishes analysts to be aware of the dangers caused by "temptations to project, what [the analyst] in dull self-perception recognizes as the peculiarities of his own personality, as generally valid theory into science." Nature is thereby treated as a client-patient, and whatever findings come up are accepted as is without any immediate emphasis or judgment.

10.1.1 Toward Explanation and Feasibility

Throughout history, the human desire to foresee and manipulate the physical world for survival and prosperity, and in accord with personal wishes and fantasies, has been confronted with the inability to predict and manipulate large portions of the habitat. As time passed, people have figured out various ways to tune ever increasing fragments of the world according to their needs. From a purely behavioral perspective, this is brought about in the way of pragmatic quasi-causal conditional rules of the following kind, "if one does this, one obtains that." A typical example of such a rule is "if I rub my hands, they get warmer."

How does one arrive at those kinds of rules? Guided by suspicions, thoughts, formalisms and by pure chance, inquiries start by roaming around, inspecting portions of the world and examining their behavior. Repeating phenomena or patterns of behavior are observed and pinned down by reproducing and evoking them. A physical behavior is anything that can be observed and thus operationally obtained and measured; for example, the rise and fall of the sun, the ignition of fire, the formation and melting of ice (in principle even time series of financial entities traded at stock exchanges or over-the-counter).

As physical behaviors are observed, people attempt to understand them by trying to figure out some cause (Frank, 1932; Schlick, 1932) or reason for their occurrences. Researchers invent virtual parallel worlds of thoughts and intellectual concepts such as "electric field" or "mechanical force" to explain and manipulate the physical behaviors, calling these creations of their minds "physical theories." Contemporary physical theories are heavily formalized and spelled out in the language of mathematics. A good theory provides people with the feeling of a key unlocking new ways of world comprehension and manipulation. Ideally, an explanation should be as compact as possible and should apply to as many behavioral patterns as possible.

Ultimately, theories of everything (Barrow, 1991; Kragh, 1999; Schlick, 1935) should be able to predict and manipulate all phenomena. In the extreme form, science becomes omniscient and omnipotent, and we envision ourselves almost as becoming empowered with magic: we presume that our ability to manipulate and tune the world is limited by our fantasies alone, and any constraints whatsoever can be bypassed or overcome one way or another. Indeed, some of what in the past has been called "supernatural," "mystery," and "the beyond" has been realized in everyday life. Many wonders of witchcraft have been transferred into the realm of the physical sciences. Take, for example, our abilities to fly, to transmute mercury into gold (Sherr et al., 1941), to listen and speak to far away friends, or to cure bacterial diseases with a few pills of antibiotics.

Until about 1900, the fast-growing natural sciences, guided by rational (Descartes, 1637) and empirical (Hume, 1748; Locke, 1690) thinking, and seconded by the European Enlightenment, prospered under the assumption of physical determinism. Under the *aegis* of physical determinism, all incapacities to predict and manipulate physical behavior were interpreted to be merely *epistemic* in nature, purporting that, with growing precision of measurements and improvements of theory, all physical unknowables will eventually be overcome and turned into knowables; that is, everything should in principle be knowable. Even statistical quantities would describe underlying deterministic behaviors. Consequently, there could not exist any physical behavior or entity without a cause stimulating or pushing it into existence.

The uprise of determinism culminated in the following statement by Laplace (1998, chap. 2):

> Present events are connected with preceding ones by a tie based upon the evident principle that a thing cannot occur without a cause which produces it. This axiom, known by the name of the principle of sufficient reason, extends even to actions which are considered indifferent . . .

> We ought then to regard the present state of the universe as the effect of its anterior state and as the cause of the one which is to follow. Given for one instant an intelligence which could comprehend all the forces by which nature is animated and the respective situation of the beings who compose it an intelligence sufficiently vast to submit these data to analysis it would embrace in the same formula the movements of the greatest bodies of the universe and those of the lightest atom; for it, nothing would be uncertain and the future, as the past, would be present to its eyes.

The invention of (analytic) functions reflects this paradigm quite nicely: some dispersionless point coordinate $x(t)$ of infinite precision serves as the representation (Hertz, 1894) of a physical state as a (unique) function of physical time t.

Indeed, the possibility to formulate theories *per se*, and in particular, the applicability of formal, mathematical models, comes as a mind-boggling surprise and cannot be taken for granted; there appears to be what Wigner (1960) called an "unreasonable effectiveness of mathematics in the natural sciences." Even today, there is a Pythagorean consensus that there is no limit to dealing with physical entities in terms of mathematical formalism. And, as mathematics increasingly served as a proper representation of reality, and computational deduction systems were increasingly introduced to delineate formalizable truth, algorithmics started to become a metaphor for physics. In algorithmic terms, nature computes, and can be (re)programmed to perform certain tasks.

The natural sciences continued to be uninhibited by any sense of limits until about *fin-de-siècle*, around 1900. In parallel, the formalization of mathematics progressed in an equally uninhibited way. Hilbert (1926, 170) argued that nobody should ever expel mathematicians from the paradise created by Cantor's set theory and posed a challenge (Hilbert, 1902) to search for a consistent, finite system of formal axioms which would be able to render all mathematical and physical truths; just like quasi-finitistic ways to cope with infinitesimal calculus had been found.

This type of belief system that claims omniscience could be called "deterministic conjecture" because no proof for its validity can be given, nor is there any way of falsification (Popper, 1959). Alas, from a pragmatic point of view, omniscience can be effectively disproved on a daily basis by tuning in to local weather forecasts.

Furthermore, it seems to be an enduring desire of human nature to be able not merely to trust the rules and theories syntactically and operationally (Bridgman, 1934) but also to be able to semantically interpret them as implying and carrying some ontological significance or truth – as if reality would communicate with us, mediated through our senses, thereby revealing the laws governing nature. Stated pointedly, we not only wish to accept physical theories as pure abstractions and constructions of our own mind (Berkeley, 1710) but we associate meaning and truth to them so much so that only very reluctantly do we admit their preliminary, transient, and changing character (Lakatos, 1978).

10.1.2 Rise of Indeterminism

Almost unnoticed, the tide of indeterminism started to build toward the end of the nineteenth century (Kragh, 1999; Purrington, 1997). At that time, mechanistic theories faced an increasing number of anomalies: Poincaré's discovery of instabilities of trajectories of celestial bodies (which made them extremely sensible to initial conditions), radioactivity (Kragh, 1997, 2009), X-rays, specific heats of gases and solids, emission and absorption of light (in particular, blackbody radiation), the (ir)reversibility dichotomy between classical reversible mechanics and Boltzmann's statistical-mechanical theory of entropy *versus* the second law of thermodynamics, and the experimental refutation of classical constructions of the ether as a medium for the propagation of light waves.

After the year 1900 followed a short period of revolutionary new physics, in particular, quantum theory and relativity theory, without any strong inclination toward (in)determinism. Then indeterminism erupted with Born's claim that quantum mechanics has it both ways: the quantum state evolves strictly deterministically, whereas the individual event or measurement outcome occurs indeterministically. Born also stated that he *believed* that there is no cause for an individual quantum event; that is, such an outcome occurs irreducibly at random.

There followed a fierce controversy, with many researchers such as Born, Bohr, Heisenberg, and Pauli taking the indeterministic stance, whereas others, like Planck (Born, 1955), Einstein (Einstein, 1938; Einstein et al., 1935), Schrödinger, and De Brogli, leaning toward determinism. This latter position was pointedly put forward by Einstein's *dictum* in a letter to Born, dated December 12, 1926 (Born, 1969, 113): "In any case I am convinced that he [the Old One] does not throw dice." At present, indeterminism is clearly favored, the canonical position being expressed by Zeilinger (2005): "The discovery that individual events are irreducibly random is probably one of the most significant findings of the twentieth century. . . . For the individual event in quantum physics, not only do we not know the cause, there is no cause."

The last quarter of the twentieth century saw the rise of yet another form of physical indeterminism, originating in Poincaré's aforementioned discovery of instabilities of the motion of classical bodies against variations of initial conditions (Campbell, and Garnett, 1882; Diacu, and Holmes, 1996; Poincaré, 1914). This scenario of *deterministic chaos* resulted in a plethora of claims regarding indeterminism that resonated with a general public susceptible to fables and fairy tales (Bricmont, 1996).

In parallel, Gödel's incompleteness theorems (Davis, 1958, 1965; Gödel, 1931; Smullyan, 1992a; Tarski, 1932), as well as related findings in the computer sciences (Calude, 2002; Chaitin, 1987a; Grünwald, and Vitányi, 1987; Turing, 1937), put an end to Hilbert's program of finding a finite axiom system for all mathematics. Gödel's incompleteness theorems also established formal bounds on provability, predictability, and induction. (The incompleteness theorems also put an end to philosophical contentions expressed by Schlick (1935, 101) that, beyond epistemic unknowables and the "essential incompetence of human knowledge," there is "not a single real question for which it would be *logically* impossible to find a solution.")

Alas, just like determinism, physical indeterminism cannot be proved, nor can there be given any reasonable criterion for its falsification. After all, how can one check against all laws and find none applicable? Unless one is willing to denote any system whose laws are currently unknown or whose behavior is hard to predict with present techniques as indeterministic, there is no scientific substance to such absolute claims, especially if one takes into account the bounds imposed by the theory of recursive functions discussed later. So, just as in the deterministic case, this position should be considered conjectural.

In discussing the present status of physical (in)determinism, we shall first consider provable unknowables through reduction to incompleteness theorems of recursion theory, then discuss classical deterministic chaos, and finally deal with the three types of quantum indeterminism: the occurrence of certain single events, complementarity, and value indefiniteness. The latter quantum unknowables are not commonly accepted by the entire community of physicists; a minority is still hoping for a more complete quantum theory than the present statistical theory.

10.2 Provable Physical Unknowables

In the past century, unknowability has been formally defined and *derived* in terms of a precise, formal notion of unprovability (Davis, 1958; Gödel, 1931; Odifreddi, 1989; Rogers, Jr., 1967; Smullyan, 1992a; Tarski, 1932, 1956; Turing, 1937). This is a remarkable departure from informal suspicions and observations regarding the limitations of our worldview. No longer is one reduced to informal, heuristic contemplations and comparisons about what one knows and can do *versus* one's ignorance and incapability. Formal unknowability is about formal proofs of unpredictability and impossibility.

There are several pathways to formal undecidability. For contemporaries accustomed to computer programs (and their respective codes), a straight route may be algorithmic. What is an algorithm? In Turing's (1968, 34) own words,

> a man provided with paper, pencil and rubber, and subject to strict discipline [carrying out a set of rules of procedure written down] is in effect a universal computer.

From a purely syntactic point of view, formal systems in mathematics can be identified with computations and *vice versa*. Indeed, as stated by Gödel (1986, 369–370) in a *postscript,* dated from June 3, 1964:

> due to A. M. Turing's work, a precise and unquestionably adequate definition of the general concept of formal system can now be given, the existence of undecidable arithmetical propositions and the non-demonstrability of the consistency of a system in the same system can now be proved rigorously for *every* consistent formal system containing a certain amount of finitary number theory.

> Turing's work gives an analysis of the concept of "mechanical procedure" (alias "algorithm" or "computation procedure" or "finite combinatorial procedure"). This concept is shown to be equivalent with that of a "Turing machine." A formal system can simply be defined to be any mechanical procedure for producing formulas, called provable formulas.

Almost since its discovery, attempts (Popper, 1950a,b) have been made to translate formal incompleteness into physics, mostly by reduction to some provable undecidable problem of recursion theory such as the halting problem (Barrow, 1998; Casti, and Karlquist, 1996; Casti, and Traub, 1994; Costa, and Doria, 1991; da Costa, and Doria, 1991; Hole, 1994; Kanter, 1990; Moore, 1990; Suppes, 1993; Svozil, 1993; Wolfram, 1984, 1985). Here the term *reduction* indicates that physical undecidability is linked or reduced to logical undecidability. A typical example is the embedding of a Turing machine or any type of computer capable of universal computation into a physical system. As a consequence, the physical system inherits any type of unsolvability derivable for universal computers such as the unsolvability of the halting problem: because the computer is part of the physical system, so are its behavioral patterns [and *vice versa* (Bridgman, 1934; Landauer, 1986, 1991)].

Note that these logical and recursion-theoretical types of physical unknowables are only derivable within deterministic systems that are strong enough to express *self-reference, substitution* (Smullyan, 1992a, chap. 1), and *universal computation.* Indeterministic systems are not deterministic by definition, and too-weak forms of expressibility are trivially incomplete (Brukner, 2003), as they are incapable of expressing universal computation or self-reference and substitution.

Gödel himself did not believe that his incompleteness theorems had any relevance for physics, especially not for quantum mechanics. The author was told by professor Wheeler that Gödel's resentments [also mentioned in Bernstein (1991, 140–141)] may have been due to Einstein's negative opinion about quantum theory, because Einstein may have brainwashed Gödel into believing that all efforts in this direction were in vain.

10.2.1 Intrinsic Self-Referential Observers

Embedded (Toffoli, 1978), intrinsic observers (Svozil, 1994) cannot leave their Cartesian prison (Descartes, 1641, Meditation 1.12) and step outside the universe examining it from some Archimedean point (Boskovich, 1966, sect. 11, 405–409). Thus every physical observation is reflexive (Nagel, 1986; Sosa, 2009) and circular (Kauffman,

1987). The self-referential and substitution capability of observers results in very diverse, unpredictable forms of behavior and in provable unknowables.

For the sake of the further analysis, suppose that there exist observers measuring objects and that observers and objects are distinct from one another, separated by a cut. Through that cut, information is exchanged. Symbolically, we may regard the object as an agent contained in a black box, whose only relevant emanations are representable by finite strings of zeroes and ones appearing on the cut, which can be modeled by any kind of screen or display. According to this purely syntactic point of view, a physical theory should be able to render identical symbols like the ones appearing through the cut; that is, a physical theory should be able to mimic or emulate the black box to which it purports to apply. This view is often adapted in quantum mechanics (Fuchs, and Peres, 2000), where the question regarding any meaning of the quantum formalism is notorious (Feynman, 1965, 129).

A sharp distinction between a physical object and an extrinsic outside observer is a rarely affordable abstraction. Mostly the observer is part of the system to be observed. In such cases, the measurement process is modeled symmetrically, and information is exchanged between observer and object bidirectionally. This symmetrical configuration makes a distinction between observer and object purely conventional (Svozil, 2002a). The cut is constituted by the information exchanged. We tend to associate with the measurement apparatus one of the two subsystems that, in comparison, is larger, more classical, and up-linked with some conscious observer (Wigner, 1961). The rest of the system can then be called the measured object.

Intrinsic observers face all kinds of paradoxical self-referential situations. These have been expressed informally as puzzling amusement and artistic perplexity, and as a formalized, scientifically valuable resource. The *liar paradox*, for instance, is already mentioned in the Bible's Epistle to Titus 1:12, stating that "one of Crete's own prophets has said it: 'Cretans are always liars, evil brutes, lazy gluttons.' He has surely told the truth." In what follows, paradoxical self-referentiality will be applied to argue against the solvability of the general induction problem as well as for a pandemonium of undecidabilities related to physical systems and their behaviors. All are based on intrinsic observers embedded in the systems they observe.

It is not totally unreasonable to speculate that the limits of intrinsic self-expression seems to be what Gödel himself considered the gist of his incompleteness theorems. In a reply to a letter by Burks [reprinted in von Neumann (1966, 55); see also Feferman (1984, 554)], Gödel states:

> that a complete epistemological description of a language A cannot be given in the same language A, because the concept of truth of sentences of A cannot be defined in A. It is this theorem which is the true reason for the existence of undecidable propositions in the formal systems containing arithmetic.

One of the first researchers to become interested in the application of paradoxical self-reference to physics was the philosopher Popper (1950a,b) who published two almost forgotten papers discussing, among other issues, Russell's paradox of Tristram Shandy (Sterne, 1767): In volume 1, chapter 14, Shandy finds that he could publish two volumes of his life every year, covering a time span far shorter than the time it took him to write these volumes. This de-synchronization, Shandy concedes, will rather increase than diminish as he advances; one may thus have serious doubts about whether he will ever complete his autobiography. This relates to a question of whether there can be a physical computer that can be assured of correctly *processing information faster than the universe does.* Wolpert (2001, 016128-1) states that [see also Calude et al. (1995, sect. 5)] "In a certain sense, the universe is more powerful than any information-processing system constructed within it could be. This result can alternatively be viewed as a restriction on the computational power of the universe – the universe cannot support the existence within it of a computer that can process information as fast as it can."

10.2.2 Unpredictability

For any deterministic system strong enough to support universal computation, the general forecast or prediction problem is provable unsolvable. This proposition will be argued by reduction to the halting problem, which is provable unsolvable. A straightforward embedding of a universal computer into a physical system results in the fact that, owing to the reduction to the recursive undecidability of the halting problem, certain future events cannot be predicted and are thus provable indeterministic. Here reduction again means that physical undecidability is linked or reduced to logical undecidability.

A clear distinction should be made between *determinism* (such as *computable evolution laws*) and *predictability* (Suppes, 1993). Determinism does not exclude unpredictability in the long run. The local (temporal), step-by-step evolution of the system can be perfectly deterministic and computable, whereas recursion-theoretic unknowables correspond to global observables at unbounded time scales. Indeed, (nontrivial) provable unpredictability requires determinism, because formalized proofs require formal systems or algorithmic behavior.

Unpredictability in indeterministic systems is tautological and trivial. At the other extreme, one should also keep in mind that there exist rather straightforward pre-Gödelian impossibilities (Brukner, 2003) to express certain mathematical truths in weak systems that are incapable of representing universal computation or Peano arithmetic.

For the sake of exploring (algorithmically) what paradoxical self-reference is like, one can consider the sketch of a proof by contradiction of the unsolvability of the halting problem. The halting problem is about whether or not a computer will eventually halt on a given input, that is, will evolve into a state indicating the completion of a computation task or will stop altogether. Stated differently, a solution of the halting problem will be an algorithm that decides whether another arbitrary algorithm on arbitrary input will finish running or will run forever.

The scheme of the proof by contradiction is as follows: the existence of a hypothetical halting algorithm capable of solving the halting problem will be *assumed.* This could, for instance, be a subprogram of some suspicious supermacro library that takes the code of an arbitrary program as input and outputs 1 or 0, depending on whether or not the program halts. One may also think of it as a sort of oracle or black box analyzing an arbitrary program in terms of its symbolic code and outputting one of two symbolic states, say, 1 or 0, referring to termination or nontermination of the input program, respectively.

On the basis of this *hypothetical halting algorithm* one constructs another *diagonalization program* as follows: on receiving some arbitrary *input program* code as input, the diagonalization program consults the *hypothetical halting algorithm* to find out whether or not this input program halts; on receiving the answer, it does the *opposite:* If the hypothetical halting algorithm decides that the input program *halts,* the diagonalization program does *not halt* (it may do so easily by entering an infinite loop). Alternatively, if the hypothetical halting algorithm decides that the input program does *not halt,* the diagonalization program will *halt* immediately.

The diagonalization program can be forced to execute a paradoxical task by receiving *its own program code* as input. This is so because, by considering the diagonalization program, the hypothetical halting algorithm steers the diagonalization program into *halting* if it discovers that it *does not halt;* conversely, the hypothetical halting algorithm steers the diagonalization program into *not halting* if it discovers that it *halts.*

The contradiction obtained in applying the *diagonalization program* to its own code proves that this program and, in particular, the hypothetical halting algorithm cannot exist. A slightly revised form of the proof (using quantum diagonalizaton operators that are equivalent to a classical *derangement* or *subfactorial*) holds for quantum diagonalization (Svozil, 2009b), as quantum information could be in a fifty-fifty fixed-point halting state. Procedurally, in the absence of any fixed-point halting state, the aforemetioned task might turn into a nonterminating alteration of oscillations between halting and nonhalting states (Kauffman, 1987).

A universal computer can in principle be embedded into, or realized by, certain physical systems designed to universally compute. An example of such a physical system is the computer on which I am currently typing this chapter. Assuming unbounded space [i.e., memory (Calude, and Staiger, 2010)] and time, it follows by reduction (Barrow, 1998; Calude et al., 1995; Casti, and Karlquist, 1996; Casti, and Traub, 1994; Costa, and Doria, 1991; da Costa, and Doria, 1991; Hole, 1994; Kanter, 1990; Moore, 1990; Suppes, 1993; Svozil, 1993; Wolfram, 1984, 1985) that there exist physical observables, in particular, forecasts about whether or not an embedded computer will ever halt in the sense sketched earlier, that are provably undecidable.

10.2.3 The Busy Beaver Function as the Maximal Recurrence Time

The busy beaver function (Brady, 1988; Chaitin, 1974; Dewdney, 1984; Rado, 1962) addresses the following question: suppose one considers all programs (on a particular computer) up to length (in terms of the number of symbols) n. What is the *largest number* producible by such a program before halting? (Note that non-halting programs, possibly producing an infinite number, e.g., by a non-terminating loop, do not apply.) This number may be called the *busy beaver function* of n. The first values of a certain universal computer's busy beaver function with two states and n symbols are, for $n = 2, 3, 4, 5, 7$ and 8, known to be, or estimated by (Brady, 1988; Dewdney, 1984), 4, 6, 13, greater than 10^3, greater than 10^4, and greater than 10^{44}.

Consider a related question: what is the upper bound of running time – or, alternatively, recurrence time – of a program of length n bits before terminating or, alternatively, recurring? An answer to this question will explain just how long we have to wait for the most time-consuming program of length n bits to halt. That, of course, is a worst-case scenario. Many programs of length n bits will have halted long before the maximal halting time. We mention without proof (Chaitin, 1974, 1987b) that this bound can be represented by the busy beaver function.

Knowledge of the maximal halting time would solve the halting problem quantitatively because if the maximal halting time were known and bounded by any computable function of the program size of n bits, one would have to wait just a little longer than the maximal halting time to make sure that every program of length n – also this particular program, if it is destined for termination – has terminated. Otherwise, the program would run forever. Hence, because of the recursive unsolvability of the halting problem the maximal halting time cannot be a computable function. Indeed, for large values of n, the maximal halting time explodes and grows faster than any computable function of n.

By reduction, upper bounds for the recurrence of any kind of physical behavior can be obtained; for deterministic systems representable by n bits, the maximal recurrence time grows faster than any computable number of n. This bound from below for possible behaviors may be interpreted quite generally as a measure of the impossibility to predict and forecast such behaviors by algorithmic means.

10.2.4 Undecidability of the Induction Problem

Induction, in physics, is the inference of general rules dominating and generating physical behaviors from these behaviors alone. For any deterministic system strong enough to support universal computation, the general induction problem is provable unsolvable. Induction is thereby reduced to the unsolvability of the rule inference problem (Adleman, and Blum, 1991; Angluin, and Smith, 1983; Blum, and Blum, 1975; Gold, 1967; Li, and Vitányi, 1992) of identifying a rule or law reproducing the behavior of a deterministic system by observing its input-output performance by purely algorithmic means (not by intuition).

Informally, the algorithmic idea of the proof is to take any sufficiently powerful rule or method of induction and, by using it, to define some functional behavior that is not identified by it. This amounts to constructing an algorithm which (passively) fakes the guesser by simulating some particular function until the guesser pretends to be able to guess the function correctly. In a second, diagonalization step, the faking algorithm then switches to a different function to invalidate the guesser's guess.

One can also interpret this result in terms of the recursive unsolvability of the halting problem, which in turn is related to the busy beaver function; there is no recursive bound on the time the guesser has to wait to make sure that the guess is correct.

10.2.5 Impossibility

Physical tasks which would result in paradoxical behavior (Hilbert, 1926) are impossible to perform. One such task is the solution of the general halting problem, as discussed earlier. Thus omnipotence appears infeasible, at least as long as one sticks to the usual formal rules opposing inconsistencies (Hilbert, 1926, 163).

Another such paradoxical task (requiring substitution and self-reference) can be forced upon *La Bocca della Veritá* (Mouth of Truth), located in the *portico* of the church of *Santa Maria in Cosmedin* in Rome. It is believed that if one tells a lie with one's hand in the mouth of the sculpture, the hand will be bitten off; another less violent legend has it that anyone sticking a hand in the mouth while uttering a false statement will never be able to pull the hand back out. Rucker (1982, 178) once allegedly put in his hand in the sculpture's mouth uttering, "I will not be able to pull my hand back out." The author leaves it to the reader to imagine *La Bocca della Verità*'s confusion when confronted with such as statement!

There is a *pandemonium* of conceivable physical tasks (Barrow, 1998), some quite entertaining (Smullyan, 1992b), which would result in paradoxical behavior and are thus impossible to perform. Some of these tasks are pre-Gödelian and merely require *substitution*.

For the sake of demonstrating paradoxical substitution and the resulting impossibility, consider the following *printing task* discussed by Smullyan (1992a, 2–4). Let the expressions (not), (printable), (self-substitute), have a standard interpretation in terms of negation, printing, and self-reference by substitution [i.e., if X is some expression formed by the earlier three expressions and brackets, then (self-substitute)$(X) = X(X)$], respectively, and define (not)(printable)(X) for arbitray expressions X to be true if and only if X cannot be printed. Likewise, (not)(printable)(self-substitute)(X) is defined to be true if and only if (self-substitute)X cannot be printed. Whatever the rules deriving expressions (subject to the notion of truth defined earlier) may be, as long as the system is consistent and produces only true propositions (and no false ones), within this small system, the following proposition is *true but unprintable:* (not)(printable)(self-substitute)[(not)(printable)(self-substitute)]. By definition, this proposition is true if and only if (self-substitute)[(not)(printable)(self-substitute)] cannot be printed. As per definition, (self-substitute)[(not)(printable)(self-substitute)] is just (not)(printable)(self-substitute)[(not)(printable)(self-substitute)], the proposition is true if and only if it is not printable. Thus the proposition is either true and cannot be printed, or it is printable and thus false. The latter alternative is excluded by the assumption of consistency. Thus one is left with the only consistent alternative that the proposition (not)(printable)(self-substitute)[(not)(printable)(self-substitute)] is true but unprintable. Note also that, since its negation (printable)(self-substitute)[(not)(printable)(self-substitute)] is false, it is also not printable (by the consistency assumption), and hence (printable)(self-substitute)[(not)(printable)(self-substitute)] is an example of a proposition which is undecidable within the system – neither it nor its negation will ever be printed in a consistent formalized system with the notion of truth defined earlier.

10.2.6 Results in Classical Recursion Theory with Implications for Theoretical Physics

The following theorems of recursive analysis (Aberth, 1980; Weihrauch, 2000) have some implications for theoretical physics (Kreisel, 1974): (1) There exist recursive monotone bounded sequences of rational numbers whose limit is no computable number (Specker, 1949). A concrete example of such a number is Chaitin's Omega number (Calude, 2002; Calude, and Dinneen, 2007; Chaitin, 1987a), the halting probability for a computer (using prefix-free code), which can be defined by a sequence of rational numbers with no computable rate of convergence. (2) There exist a recursive real function which has its maximum in the unit interval at no recursive real number (Specker, 1959). This has implications for the principle of least action. (3) There exists a real number r such that $G(r) = 0$ is recursively undecidable for $G(x)$ in a class of functions which involves polynomials and the sine function (Wang, 1974). This, again, has some bearing on the principle of least action. (4) There exist incomputable solutions of the wave equations for computable initial values (Bridges, 1999; Pour-El, and Richards, 1989). (5) On the basis of theorems of recursive analysis (Richardson, 1968; Scarpellini, 1963), many questions in dynamical systems theory are provable undecidable (Calude et al., 2010; da Costa et al., 1993; Hirsch, 1985; Stewart, 1991).

10.3 Deterministic Chaos

The wording *deterministic chaos* appears to be a *contradictio in adjecto*, indicating a hybrid form of chaotic behavior in deterministic systems (Anishchenko et al., 2007; Lichtenberg, and Lieberman, 1983). Operationally, it is characterized by the practical impossibility of forecasting the future because the system is unstable (Lyapunov, 1992) and very sensitive to tiny variations of the initial state. Because the initial state can only be determined with finite accuracy, its evolution will soon become totally unpredictable.

10.3.1 Instabilities in Classical Motion

In 1885 King Oscar II of Sweden and Norway, stimulated by Weierstrass, Hermite, and Mittag-Leffler, offered a prize to anybody contributing toward the solution of the so-called *n-body problem* (Weierstrass et al., 1885, 2):

> Given a system of arbitrarily many mass points that attract each according to Newton's law, try to find, under the assumption that no two points ever collide, a representation of the coordinates of each point as a series in a variable that is some known function of time and for all of whose values the series converges uniformly.

The prize-winning work was expected to render systematic techniques toward a solution to *stable* motion such that systems whose states start out close together will stay close together forever (Diacu, and Holmes, 1996, 69). To everyone's surprise, the exciting course of events (Diacu, 1996; Diacu, and Holmes, 1996; Peterson, 1993) resulted in Poincaré's prize-winning centennial revised contribution (Poincaré, 1890), which predicted unexpected and irreducible *instabilities* in the mechanical motion of bodies. Poincaré was led to the conclusion that sometimes small variations in the initial state could lead to huge variations in the evolution of a physical system at later times. In Poincaré's own words (Poincaré, 1914, chapt. 4, sect. 2, 56–57):

> If we would know the laws of nature and the state of the Universe precisely for a certain time, we would be able to predict with certainty the state of the Universe for any later time. But ... it can be the case that small differences in the initial values produce great differences in the later phenomena; a small error in the former may result in a large error in the latter. The prediction becomes impossible and we have a "random phenomenon."

Note that Poincaré adheres to a Laplacian-type determinism but recognizes the possibility that systems whose states start out close together will stay close together *for a while* (Diacu, and Holmes, 1996, 69) and then diverge into totally different behaviors. Today such behaviors are subsumed under the name *deterministic chaos*. In chaotic systems, it is practically impossible to specify the initial value precise enough to allow long-term predictions.

Already in 1873, Maxwell mentioned (Campbell, and Garnett, 1882, 211–212)

> When an infinitely small variation in the present state may bring about a finite difference in the state of the system in a finite time, the condition of the system is said to be unstable. It is manifest that the existence of unstable conditions renders impossible the prediction of future events, if our knowledge of the present state is only approximate, and not accurate.

Maxwell also discussed unstable states of high potential energy whose spontaneous (Frank, 1932) decay or change (Campbell, and Garnett, 1882, 212) "requires an expenditure of work, which in certain cases may be infinitesimally small, and in general bears no definite proportion to the energy developed in consequence thereof."

Today, after more than a century of research into unstable chaotic motion, *symbolic dynamics* identified the *Poincaré map near a homocyclic orbit*, the *horseshoe map* (Smale, 1967), and the *shift map* as equivalent origins of classical deterministic chaotic motion, which is characterized by a *computable evolution law* and the *sensitivity* and instability with respect to variations of the *initial value* (Anishchenko et al., 2007; Lichtenberg, and Lieberman, 1983; Shaw, 1981).

This scenario can be demonstrated by considering the shift map σ as it pushes up dormant information residing in the successive bits of the initial state represented by the sequence $s = 0.(\text{bit } 1)(\text{bit } 2)(\text{bit } 3) \ldots$, thereby truncating the bits before the comma; that is, $\sigma(s) = 0.(\text{bit } 2)(\text{bit } 3)(\text{bit } 4) \ldots$, $\sigma(\sigma(s)) = 0.(\text{bit } 3)(\text{bit } 4)(\text{bit } 5) \ldots$, and so on. Suppose a measurement device operates with a precision of, say, two bits after the comma, indicated by a two bit window of measurability; thus intially all information beyond the second bit after the comma is hidden to the experimenter. Consider two initial states $s = [0.(\text{bit } 1)(\text{bit } 2)](\text{bit } 3) \ldots$ and $s' = [0.(\text{bit } 1)(\text{bit } 2)](\text{bit } 3)' \ldots$, where the square brackets indicate the boundaries of the window of measurability (two bits in this case). Initially, as the representations of both states start with the same two bits after the comma $[0.(\text{bit } 1)(\text{bit } 2)]$, these states appear operationally identical and cannot be discriminated experimentally. Suppose further that, after the second bit, when compared, the successive bits (bit i) and (bit i)$'$ in both state representations at identical positions $i = 3, 4, \ldots$ are totally independent and uncorrelated. After just two iterations of the shift map σ, s and s' may result in totally different, diverging observables $\sigma(\sigma(s)) = [0.(\text{bit } 3)(\text{bit } 4)](\text{bit } 5) \ldots$ and $\sigma(\sigma(s')) = [0.(\text{bit } 3)'(\text{bit } 4)'](\text{bit } 5)'. \ldots$

If the initial values are *defined* to be elements of a continuum, then almost all (of measure one) of them are not representable by any algorithmically compressible number; in short, they are random (Calude, 2002; Martin-Löf, 1966). Classical deterministic chaos results from the assumption of such a random initial value – drawn somehow [one needs the *axiom of choice* (Svozil, 1995b; Wagon, 1986) for doing this] from the continuum urn – and the unfolding of the information contained therein by a recursively enumerable (computable), deterministic (temporal evolution) function. Of course, if one restricts the initial values to finite sets, or, say, to the rationals, then the behavior will be periodic. The randomness of classical, deterministic chaos resides in the *assumption of the continuum*; an assumption which might be considered a convenience (for the sake of applying the infinitesimal calculus), as it is difficult to conceive of any convincing physical operational evidence supporting the full structure of continua. If the continuum assumption is dropped, then what remains is Maxwell's and Poincaré's observation of the unpredictability of the behavior of a deterministic system due to instabilities and diverging evolutions from almost identical initial states (Lyapunov, 1992).

10.3.2 Rate of Convergence

The connections between symbolic dynamical systems and universal computation result in provable unknowables (da Costa et al., 1993; Stewart, 1991). These symbolic dynamic unknowables are different in type from the dynamical instabilities, and should be interpreted recursion theoretically, as outlined in Section 10.2.2.

Let us come back to the original n-body problem. About one hundred years after its formulation, as quoted earlier, the n-body problem has been solved (Babadzanjanz, 1969, 1979, 1993; Babadzanjanz, and Sarkissian, 2006; Diacu, 1996; Wang, 1991, 2001). The three-body problem was already solved by Sundman (1912). The solutions are given in terms of convergent power series.

Yet, to be practically applicable, the rate of convergence of the series must be computable and even reasonably good. One might already expect from symbolic dynamics, in particular, from chaotic motion, that these series solutions could converge very slowly. Even the short-term prediction of future behaviors may require the summation of a huge number of terms, making these series unusable for all practical purposes (Diacu, 1996; Rousseau, 2004).

Alas, the complications regarding convergence may be more serious. Consider a universal computer based on the n-body problem. This can, for instance, be achieved by ballistic computation, such as the "Billiard Ball" model of computation (Fredkin, and Toffoli, 1982; Margolus, 2002) that effectively embeds a universal computer into an n-body system (Svozil, 2007). It follows by reduction that certain predictions, say, for instance, the general halting problem, are impossible.

What are the consequences of this reduction for the convergence of the series solutions? It can be expected that not only do the series converge very slowly, like in deterministic chaos, but that, in general, there does not exist any computable rate of convergence for the series solutions of particular observables. This is very similar to the busy beaver function or to Chaitin's Omega number (Calude, 2002; Chaitin, 1987a), representing the halting probability of a universal computer. The Omega number can be enumerated by series solutions from quasi-algorithms computing its very first digits (Calude, and Dinneen, 2007). Yet, because of the incomputable growth of the time required to determine whether certain summation terms corresponding to halting programs possibly contribute, the series lack any computable rate of convergence.

Though it may be possible to evaluate the state of the n bodies by Wang's power series solution for any finite time with a computable rate of convergence, global observables, referring to (recursively) unbounded times, may be incomputable. Examples of global observables correspond to solutions of certain decision problems such as the stability of some solar system (we do not claim that this is provable incomputable), or the halting problem.

This, of course, stems from the metaphor and robustness of universal computation and the capacity of the n-bodies to implement universality. It is no particularity or peculiarity of Wang's power series solution. Indeed, the troubles reside in the capacity to implement substitution, self-reference, universal computation, and Peano arithmetic by n-body problems. Because of this capacity, there cannot exist other formalizable methods, analytic solutions, or approximations capable of deciding and computing certain decision problems or observables for the n-body problem.

10.4 Quantum Unknowables

In addition to provable physical unknowables by reduction to recursion-theoretic ones, and chaotic symbolic dynamic systems, a third group of physical unknowables resides in the quantum domain. Although it has turned out to be a highly successful theory, quantum mechanics, in particular, its interpretation and meaning, has been controversially received within the physics community. Some of its founding fathers, like Schrödinger and, in particular, Einstein, considered quantum mechanics to be an unsatisfactory theory: Einstein, Podolsky and Rosen (1938; 1935) argued that there exist counterfactual (Svozil, 2009d; Vaidman, 2007) ways to infer observables from experiment that, according to quantum mechanics, cannot coexist simultaneously; hence quantum mechanics cannot predict what experiment can (counterfactually) measure. Thus quantum mechanics is *incomplete* and should eventually be substituted by a more complete theory. Others, among them Born, Bohr, and Heisenberg, claimed that unknowability in quantum mechanics is irreducible, is ontic, and will remain so forever. Over the years, the latter view seems to have prevailed (Bub, 1999; Fuchs, and Peres, 2000), although not totally unchallenged (Jammer, 1966, 1974, 1992). Already Sommerfeld warned his students not to get into the meaning behind quantum mechanics, and as mentioned by Clauser (2002), not long ago, scientists working in that field had to be very careful not to become discredited as *quacks*. Richard Feynman (Feynman, 1965, 129) once mentioned the

> perpetual torment that results from [the question], "But how can it be like that?" which is a reflection of uncontrolled but utterly vain desire to see [quantum mechanics] in terms of an analogy with something familiar.... Do not keep saying to yourself, if you can possibly avoid it, "But how can it be like that?" because you will get "down the drain," into a blind alley from which nobody has yet escaped.

This antirationalistic postulate of irreducible indeterminism and meaninglessness came after a period of fierce debate on the quantum foundations, followed by decades of vain attempts to complete quantum mechanics in any operationally testable way, and after the discovery of proofs of the incompatibility of local, realistic, context-independent ways to complete quantum mechanics (Clauser, and Shimony, 1978; Mermin, 1993).

In what follows, we shall discuss three realms of quantum unknowables: (1) randomness of single events, (2) complementarity, and (3) value indefiniteness.

10.4.1 Random Individual Events

In 1926, Born (1926b, 866) [see an English translation in Wheeler, and Zurek (1983, 54)] postulated that

> "from the standpoint of our quantum mechanics, there is no quantity which in any individual case causally fixes the consequence of the collision; but also experimentally we have so far no reason to believe that there are some inner properties of the atom which condition a definite outcome for the collision. Ought we to hope later to discover such properties ... and determine them in individual cases? Or ought we to believe that the agreement of theory and experiment – as to the impossibility of prescribing conditions? I myself am inclined to give up determinism in the world of atoms."

Furthermore, Born suggested that, though *individual particles behave irreducibly indeterministic*, the *quantum state evolves deterministically* in a strictly Laplacian causal way. Indeed, between (supposedly irreversible) measurements the (unitary) quantum state evolution is even reversible, that is, one-to-one, and amounts to a generalized (distance preserving) rotation in complex Hilbert space. In Born's (1926a, 804) [see an English translation in Jammer (1989, 302)] own words,

> the motion of particles conforms to the laws of probability, but the probability itself is propagated in accordance with the law of causality. [This means that knowledge of a state in all points in a given time determines the distribution of the state at all later times.]

This distinction between a reversible, deterministic evolution of the quantum state, on one hand, and the irreversible measurement, on the other hand, has left some physicists with an uneasy feeling; in particular, because of the possibility to erase (Chapman et al., 1995; Greenberger, and YaSin, 1989; Herzog et al., 1995; Kwiat et al., 1992; Peres, 1980; Pfau et al., 1994; Scully, and Drühl, 1982; Scully et al., 1991; Zajonc et al., 1991) measurements by reconstructing the quantum state, accompanied by a complete loss of the information obtained from the quantum state before the (undone) measurement – unlike in classical reversible computation (Bennett, 1973, 1982; Leff, and Rex, 1990), which still allows copying, that is, one-to-many operations, the quantum state evolution is strictly one-to-one. If the possibility to undo measurements on quantum states is merely limited by the experimenter's technological capacities (and not bound by any fundamental principle), then, one could speculate, Born's statement seems to suggest that the deterministic state evolution uniformly prevails. Pointedly stated, if, at least in principle, there is no such thing as an irreversible measurement, and the quantum state evolves uniformly deterministically, why should there exist indeterministic individual events? In this view, the insistence in irreversible measurements as well as in an irreducible indeterminism associated with individual quantum events appears to be an idealistic, subjective illusion – in fact, this kind of indeterminism depends on measurement irreversibility and decays into thin air if the latter is denied.

Similar arguments have been brought forth by Everett (1957) and Schrödinger (1995). Note that it is not entirely clear [and indeed remains conventional (Svozil, 2002a)] where exactly the measurement cut (Rössler, 1998; Wigner, 1961) between the observer and the object is located. By assuming the universal applicability of quantum mechanics, the object and the measurement apparatus could be uniformly *combined* into a larger system whose quantum mechanical evolution should be deterministic;

otherwise quantum mechanics would not be universally valid. Such frameworks hardly offer objective opportunities for indeterminism besides subjective ones – in the *many worlds* resolution (Everett, 1957), every one of many simultaneous observers branching off to different universes subjectively experiences the arbitrariness of the occurrence of events as indeterminism. (This resembles the perception of a particular sequence of bits as compared to all possible ones.)

Alas, the deterministic evolution of the quantum state could result in the *super-position* of classically contradictory states. One of the mind-boggling, perplexing and counterintuitive consequences associated with this coexistence of classical contradictions is Schrödinger's (1935a, 812) cat paradox implying the simultaneous coexistence of death and life of a macroscopic object such as a mammal. Another one is Everett's (1957) aforementioned many-worlds interpretation suggesting that our universe perpetually branches off into zillions of consistent alternatives.

Thus one is faced with a *dilemma:* either to accept a somehow spurious nonuniformity in the evolution of the quantum state during (irreversible) measurement processes – an *ad hoc* assumption challenged by quantum erasure experiments – or being confronted with the counterintuitive decay of quantum states into superpositions of classically mutually exclusive states – a sort of jelly – not backed by our everday experience as conscious beings (although often ambivalent we usually dont reside in mental ambiguity for too long). Schrödinger (1995, 19–20) sharply addressed the difficulties of a quantum theorist coping with this aspect of the quantum formalism:

> The idea that [the alternate measurement outcomes] be not alternatives but *all* really happening simultaneously seems lunatic to [the quantum theorist], just *impossible*. He thinks that if the laws of nature took *this* form for, let me say, a quarter of an hour, we should find our surroundings rapidly turning into a quagmire, a sort of a featureless jelly or plasma, all contours becoming blurred, we ourselves probably becoming jelly fish. It is strange that he should believe this. For I understand he grants that unobserved nature does behave this way – namely according to the wave equation. ... according to the quantum theorist, nature is prevented from rapid jellification only by our perceiving or observing it.

If, however, an additional irreducible irreversible evolution or some other, possibly environmental (Peres, 1980; Zurek, 2003), effect associated with measurements (and the collapse of the quantum wave function) is postulated or somehow emerges, individual events may occur indeterministically. The considerations might appear to be sophistries, but they have direct consequences for the supposedly most advanced random number generators of our time. These devices operate with beam splitters (Calude et al., 2010; Jennewein et al., 2000; Rarity et al., 1994; Stefanov et al., 2000; Svozil, 1990; Wang et al., 2006), which are strictly reversible (Greenberger et al., 1993; Ou et al., 1987; Svozil, 2005c; Zeilinger, 1981) – one could demonstrate reversibility on beam splitters by forming a Mach-Zehnder interforemeter with two serially connected ones – or parametric down-conversions and entanglement (Fiorentino et al., 2007; Hai-Qiang et al., 2004; Pironio et al., 2010).

Born did not address these questions, nor did he specify the formal notion of indeterminism to which he was relating. So far, no mathematical characterization of quantum randomness has been proved (Calude, and Svozil, 2008). In the absence of any indication to the contrary, it is mostly implicitly assumed that quantum randomness is of the strongest possible kind, which amounts to postulating that the symbolic sequences associated with measurement outcomes are uncomputable or even algorithmically incompressible.

Indeed, the quantum formalism does not predict the outcome of single events when there is a mismatch between the context in which a state was prepared, and the context in which it is measured. Here, the term *context* (Svozil, 2009a,d) denotes a maximal collection of comeasurable observables, or, more technically, the maximal operator from which all commuting operators can be functionally derived (Halmos, 1974, sect. 84). Ideally, a quantized system can be prepared to yield exactly one answer in exactly one context (Donath, and Svozil, 2002; Svozil, 2002b; Zeilinger, 1999). Other outcomes associated with other contexts occur indeterministically (Calude, and Svozil, 2008).

Furthermore, the quantum formalism is incapable of predicting deterministically the radioactive decay of individual particles. Attempts to find causal laws lost steam (Kragh, 1997, 2009) at the time of Born's suggestion of the indeterministic interpretation of individual measurement outcomes, and nobody has come up with a operationally satisfactory deterministic prediction since then.

In the absence of other explanations, it is not too unreasonable to pragmatically presume that these single events occur without any causation and thus at random. Presently, this appears to be the prevalent opinion among physicists. Such random quantum coin tosses (Fiorentino et al., 2007; Hai-Qiang et al., 2004; Jennewein et al., 2000; Pironio et al., 2010; Rarity et al., 1994; Stefanov et al., 2000; Svozil, 1990, 2009e; Wang et al., 2006) have been used for various purposes, such as delayed choice experiments (Jennewein et al., 2000; Weihs et al., 1998a).

Note that randomness of this type (Calude, 2005; Calude, and Dinneen, 2005) is postulated rather than proved and thus, unless disproved, remains conjectural. This is necessarily so, for any claim of randomness can only be corroborated *relative* to, and *with respect* to, a more or less large class of laws or behaviors; it is impossible to inspect the hypothesis against an infinity of – and even less so all – conceivable laws. To rephrase a statement about computability (Davis, 1958, 11), how can we ever exclude the possibility of our presented, some day (perhaps by some extraterrestrial visitors), with a (perhaps extremely complex) device that computes and predicts a certain type of hitherto random physical phenomenon?

10.4.2 Complementarity

Complementarity is the impossibility of measuring two or more complementary observables with arbitrary precision simultaneously. In 1933, Pauli (1958, 7) gave the first explicit definition of *complementarity* stating that [see the partial English translation in (Jammer, 1989, 369)]

> in the case of an indeterminacy of a property of a system at a certain configuration (at a certain state of a system), any attempt to measure the respective property (at least partially) annihilates the influence of the previous knowledge of the system on the (possibly statistical) propositions about possible later measurement results. ... The impact on the

system by the measurement apparatus for momentum (position) is such that within the limits of the uncertainty relations the value of the knowledge of the previous position (momentum) for the prediction of later measurements of position and momentum is lost.

Einstein, Podolsky, and Rosen (1935) challenged quantum complementarity (and doubted the completeness of quantum theory) by utilizing a configuration of two entangled (Schrödinger, 1935a,b, 1936) particles. They claimed to be able to empirically infer two different complementary contexts counterfactually simultaneously, thus circumventing quantum complementarity. Thereby, one context is measured on one side of the setup, whereas the other context is measured on the other side of it. By the uniqueness property (Svozil, 2006a) of certain two-particle states, knowledge of a property of one particle entails the certainty that, if this property were measured on the other particle as well, the outcome of the measurement would be a unique function of the outcome of the measurement performed.

This makes possible the measurement of one context *as well as* the *simultaneous counterfactual inference* of a different complementary context. Because, one could argue, although one has actually measured on one side a different, incompatible context compared to the context measured on the other side, if, on both sides, the same context *would be measured*, the outcomes on both sides *would be uniquely correlated.* (This can indeed be verified in another experiment.) Hence, the Einstein, Podolsky, and Rosen argument continues, measurement of one context per side is sufficient, for the outcome could be counterfactually inferred on the other side. Thus, effectively two complementary contexts are knowable. Based on this argument, Einstein, Podolsky, and Rosen suggested that quantum mechanics must be considered incomplete, because it cannot predict what can be measured; thus a more complete theory is needed.

Complementarity was first encountered in quantum mechanics, but it is a phenomenon also observable in the classical world. To get better intuition of complementarity, we shall consider generalized urn models (Wright, 1978, 1990) or, equivalently (Svozil, 2005b), finite deterministic automata (Calude et al., 1997; Dvurečenskij et al., 1995; Moore, 1956; Schaller, and Svozil, 1996; Svozil, 1993) in an unknown initial state. Both quasi-classic examples mimic complementarity to the extent that even quasi-quantum cryptography can be performed with them (Svozil, 2006c) as long as value indefiniteness is not a feature of the protocol (Bechmann-Pasquinucci, and Peres, 2000; Svozil, 2010a), that is, for instance, the Bennett and Brassard (1984) protocol (Bennett et al., 1992) can be implemented with generalized urn models, whereas the Ekert protocol (Ekert, 1991) cannot.

A generalized urn model is characterized by an ensemble of balls with black background color. Printed on these balls are some color symbols. Every ball contains just one symbol per color. Further assume some filters or eyeglasses that are perfect because they totally absorb light of all other colors but a particular one. In that way, every color can be associated with a particular pair of eyeglasses and *vice versa.*

When a spectator looks at a ball through such a particular pair of eyeglasses, the only operationally recognizable symbol will be the one in the particular color that is transmitted through the eyeglasses. All other colors are absorbed, and the symbols printed on them will appear black and therefore will not be differentiable from the black background. Hence the ball will appear to carry a different message or symbol, depending on the color with which it is viewed.

For the sake of demonstration, let us consider a generalized urn model with four ball types, two colors, say red and green, and two symbols, say "0" and "1," per color, that is, ball type 1: (red 0 green 0), ball type 2: (red 0 green 1), ball type 3: (red 1 green 0), and ball type 4: (red 1 green 1). The green pair of eyeglasses associated with the green observable allows the observer to differentiate between ball types 1 or 3 (associated with the green symbol "0"), and ball types 2 or 4 (associated with the green symbol "1"). The red pair of eyeglasses associated with the red observable allows the observer to differentiate between ball types 1 or 2 (associated with the green symbol "0"), and ball types 3 or 4 (associated with the green symbol "1"). [Without going into details in general this yields sets of partitions of the set of ball types resulting in partition logics (Svozil, 1993, chapt. 10).]

The difference between the balls and the quanta is the possibility of viewing all the different symbols on the balls in all different colors by taking off the eyeglasses; also, one can consecutively look at one and the same ball with differently colored pair of eyeglasses, thereby identifying the ball completely. Quantum mechanics does not provide us with a possibility to look across the quantum veil, as it allows neither a global, simultaneous measurement of all complementary observables nor a measurement of one observable without disturbing the measurement of another complimentary observable (with the exception of Einstein, Podolsky, and Rosen counterfactual measurements discussed earlier). On the contrary, there are strong formal arguments suggesting that the assumption of a simultaneous physical coexistence of such complementary observables yields a complete contradiction. These issues will be discussed next.

10.4.3 Value Indefiniteness *versus* Omniscience

Still another quantum unknowable results from the fact that no global (in the sense of all or at least certain finite sets of complementary observables) classical truth assignment exists which is consistent with even a finite number of local (in the sense of comeasurable) ones, that is, no consistent classical truth table can be given by pasting together the possible outcomes of measurements of certain complementary observables. This phenomenon is also known as *value indefiniteness* or, by an option to interpret this result, *contextuality* (see later). Here the term *local* refers to a particular context (Svozil, 2009a) that, operationally, should be thought of as the collection of all comeasurable or copreparable (Zeilinger, 1999) observables. The structure of quantum propositions (Birkhoff, and von Neumann, 1936; Kalmbach, 1983, 1986; Kochen, and Specker, 1965; Navara, and Rogalewicz, 1991; Pták, and Pulmannová, 1991; Svozil, 1998) can be obtained by pasting contexts together.

As by definition, only *one* such context is directly measurable, arguments based on more than one context must necessarily involve counterfactuals (Svozil, 2009d; Vaidman, 2007). A *counterfactual* is a would-be-observable or *contrary-to-fact conditional* (Chisholm, 1946) which has not been measured but potentially could have been measured if an observer would have decided to do so; alas the observer decided to measure a different, presumably complementary, observable.

Already scholastic philosophy, for instance, Thomas Aquinas, considered similar questions such as whether God has knowledge of non-existing things (Aquinas, 1981, part one, question 14, article 9) or things that are not yet (Aquinas, 1981, part one, question 14, article 13); see also Specker's (1960, 243) reference to *infuturabilities*. Classical omniscience, at least its naive expression that, if a proposition is true, then an omniscient agent (such as God) knows that it is true, is plagued by controversies and paradoxes. Even without evoking quantum mechanics, there exist bounds on omniscience because of the self-referential perception of intrinsic observers endowed with free will; if such an observer is omniscient and has absolute predictive power, then free will could counteract omniscience and, in particular, the observer's own predictions. Within a consistent formal framework, the only alternative is to either abandon free will, stating that it is an idealistic illusion, or accept that omniscience and absolute predictive power is bound by paradoxical self-reference.

The empirical sciences implement classical omniscience by assuming that in principle, all observables of classical physics are comeasurable without any restrictions, regardless of whether they are actually measured. No ontological distinction is made between an observable obtained by an actual and a potential or counterfactual measurement. [In contrast, compare Schrödinger's (1935a, sect. 7) own epistemological interpretation of the wave function as a *catalog of expectations*.] Classically, precision and comeasurability are limited only by the technical capacities of the experimenter. The principle of empirical classical omniscience has given rise to the realistic believe that all observables exist regardless of their observation, that is, regardless and independent of any particular measurement.

Physical (co-)existence is thereby related to the realistic assumption [sometimes referred to as the "ontic" (Atmanspacher, and Primas, 2005) viewpoint] that (Stace, 1934) "some entities sometimes exist without being experienced by any finite mind." With regards to such unexperienced counterfactual entities, Stace (1934, 364, 365, 368) questions their existence (compare also Schrödinger's remark quoted earlier):

> In front of me is a piece of paper. I assume that the realist believes that this paper will continue to exist when it is put away in my desk for the night, and when no finite mind is experiencing it. . . . I will state clearly at the outset that I cannot prove that no entities exist without being experienced by minds. For all I know completely unexperienced entities may exist, but what I shall assert is that . . . there is absolutely no reason for asserting that these non-mental, or physical, entities ever exist except when they are being experienced, and the proposition that they do so exist is utterly groundless and gratuitous, and one which ought not to be believed. . . . As regards [a] unicorn on Mars, the correct position, as far as logic is concerned, is obviously that if anyone asserts that there is a unicorn there, the onus is on him to prove it; and that until they do prove it, we ought not to believe that they exist.

One might criticize Stace's idealistic position by responding that suppose an experimenter can choose which observable among a collection of different, complementary, observables is actually measured. Regardless of this choice, a measurement of any observable that *could* be measured *would* produce some result. This contrary-to-fact conditional could be interpreted as an existing *element of physical reality*. Furthermore, according to the argument of Einstein, Podolsky and Rosen (1935, 777), even certain sets of *complementary* counterfactual elements of physical reality coexist "if, without in any way disturbing a system, we can predict with certainty (i.e., with probability equal to unity) the value of [these] physical quantit[ies]." The idealist might repond that these arguments are unconvincing because they are merely based on conterfactual inference and are thus empirically "utterly groundless and gratuitous."

The formal expression of classical omniscience is the Boolean algebra of observable propositions (Boole, 1958), in particular the abundance of two-valued states interpretable as omniscience about the system. Thereby, any such dispersionless quasi-classical two-valued state – associated with a truth assignment – can be defined for all observables, regardless of whether they have been actually observed.

After the discovery of complementarity, a further indication against quantum omniscience came from Boole's (1862) *conditions of possible (classical) experience* which are bounds for the occurrence of (classical) events that are derivable within classical probability theory (Pitowsky, 1989a,b, 1994; Pitowsky, and Svozil, 2001) for quantum probabilities and quantum expectation functions. Bell (1966) pointed out that experiments based on counterfactually inferred observables discussed by Einstein, Podolsky and Rosen (1935) discussed earlier violate these conditions of possible (classical) experience and thus seem to indicate the impossibility of a faithful embedding (i.e., preserving the logical structure) of quantum observables into classical Boolean algebras. Stated pointedly, under some (presumably mild) side assumptions, *unperformed experiments have no results* (Peres, 1978); that is, there cannot exist a table enumerating all actual and hypothetical context independent (see later) experimental outcomes consistent with the observed quantum frequencies (Svozil, 2010b; Weihs et al., 1998b). As any such table could be interpreted as omniscience with respect to the observables in the Boole-Bell-Einstein-Podolsky-Rosen-type experiments, the impossibility to consistently enumerate such tables (under the noncontextual assumption) appears to be a very serious indication against omniscience in the quantum domain.

The quantum nonlocal (i.e., the particles are spatially separated) correlations among observables in the Boole-Bell-Einstein-Podolsky-Rosen-type experiments are stronger than classical in the sense that *ex post facto,* when the two outcomes are communicated and compared, in the case of dichotomic observables, say "0" and "1," for some measurement parameter regions, there appear to be *more equal* occurrences "00" or "11" and thus *fewer unequal* occurrences "01" or "10" than could be classically accounted for; likewise, for other measurement parameter regions, there appear to be *fewer equal* occurrences "00" or "11" and thus *more unequal* occurrences "01" or "10" than could be classically accounted for. These conclusions can only be drawn *in retrospect*, that is, after bringing together and comparing the outcomes. Individual outcomes occur indeterministically and, in particular, independently of the measurement parameter regions [but not of outcomes (Shimony, 1984)] of other distant, measurements. No faster-than-light signaling can occur. Indeed, even stronger-than-quantum correlations would, in this scenario, not violate relativistic causality (Krenn, and Svozil, 1998; Popescu, and Rohrlich, 1994, 1997; Svozil, 2005a).

The reason that it is impossible to describe all quantum observables simultaneously by classical tables of experimental outcomes can be understood in terms of a stronger conclusion that, for quantum systems whose Hilbert space is of dimension greater than two, there does not exist any dispersionless quasi-classical, two-valued state interpretable as truth assignment. This conclusion, which is known as the Kochen-Specker theorem (Alda, 1980, 1981; Cabello et al., 1996; Kamber, 1964, 1965; Kochen, and Specker, 1967; Mermin, 1993; Specker, 1960; Svozil, 1998, 2009a; Svozil, and

Tkadlec, 1996); Zierler, and Schlessinger, 1965), has a finitistic proof by contradiction. Proofs of the Kochen-Specker theorem amount to brain teasers in graph coloring resulting in the fact that, for the geometric configurations considered, there does not exist any possibility to consistently and context independently enumerate and tabulate the values of all the observables occurring in a Kochen-Specker-type argument (Cabello et al., 1996).

The violations of conditions of possible classical experience in Boole-Bell-type experiments or the Kochen-Specker theorem do not exclude realism restricted to a single context but (noncontextual) realistic omniscience beyond it. It may thus not be totally unreasonable to suspect that the assumption of (pre-)determined observables outside a single context may be unjustified (Svozil, 2004).

If one nevertheless *insists* in the simultaneous physical coexistence of counterfactual observables, any *forced* tabulation (Peres, 1978; Svozil, 2010b) of truth values for Boole-Bell-type or Kochen-Specker-type configurations would either result in a complete contradiction or in *context dependence*, also termed *contextuality*, that is, the outcome of a measurement of an observable would depend on what other comeasurable observables are measured alongside it (Bell, 1966; Bohr, 1949; Heywood, and Redhead, 1983; Redhead, 1990; Svozil, 2009a).

Indeed, the current mainstream interpretation of the Boole-Bell-type or Kochen-Specker-type theorems is in terms of contextuality, that is, by assuming a dependence of the outcome of a single observable *on what other observables* are actually measured or at least what could have been consistently known alongside it. This insistence in the coexistence of complementary observables could be interpreted as an attempt to rescue classical omniscience accompanied by ontological realism at the price of accepting contextuality. The realist Bell (1966, 451) suggested that "the result of an observation may reasonably depend . . . on the complete disposition of the apparatus." (Already Bohr (1949) mentioned "the impossibility of any sharp separation between the behaviour of atomic objects and the interaction with the measuring instruments which serve to define the conditions under which the phenomena appear.")

For the sake of demonstrating contextuality (Svozil, 2010b) consider a dichotomic observable (with outcomes "0" or "1"). Contextuality predicts that, when measured together with some particular set of observables, this observable yields a certain outcome, say "0," whereas when measured together with another, complementary, set of other observables, the observable may yield a different outcome, say "1."

However, statistically the quantum probability and expectation value of this observable is noncontextual and thus *independent* of the set of co-observables. Thus contextuality is a hypothetical (counterfactual) phenomenon regarding complementary measurements on an individual particle, making it inaccessible for direct tests. Alas, as far as Einstein-Podolsky-Rosen-type measurements might reproduce such contextual behavior for individual particles, quantum mechanics predicts noncontextuality (Svozil, 2009c) and thus contradicts the assumption of quantum contextuality. (Often claims of experimental evidence of quantum contextuality do not deal with its individual particle character but deal with statistical violations of Boole-Bell-type or Kochen-Specker-type configurations. The terms which contribute to (in)equalities are not measured on one and the same particle; operationally they even originate in very different measurement setups.) One may argue that contextuality occurs only when absolutely necessary, that is, when the set of observables allows only an insufficient number of two-valued states for a homeomorphic embedding into (classical) Boolean algebras; but in view of the fact that quantum noncontextuality for single events occurs for configurations which can be pasted together to construct a Kochen-Specker-type scheme, any such argument might appear *ad hoc*.

On the basis of the aforementioned lack of quantum omniscience, it is possible to postulate the existence of absolute sources of indeterminism; if there are no (preexisting) observables, and no causal laws yielding individual outcomes, the occurrence of any such outcome can only be unpredictable and incomputable (Calude, and Svozil, 2008). This quantum dice approach has first been proposed (Rarity et al., 1994; Svozil, 1990; Zeilinger, 1999) and realized (Hai-Qiang et al., 2004; Jennewein et al., 2000; Stefanov et al., 2000; Wang et al., 2006) in setups which utilize *complementarity*, yet still allow omniscience. More recently, it was suggested (Pironio et al., 2010; Svozil, 2009e) to utilize quantum systems with more than two exclusive outcomes that are are subject to *value indefiniteness* (two-dimensional systems cannot be proven to be value indefinite). The additional advantage over devices utilizing merely complementarity is that these new type of quantum oracles (Fiorentino et al., 2007; Paterek et al., 2010; Pironio et al., 2010) are "quantum mechanically certified" by Boole-Bell-type, Kochen-Specker-type, and Greenberger-Horne-Zeilinger-type (Greenberger et al., 1990) theorems not to allow omniscience. Of course, all these devices operate under the assumption that there are no hidden variables that could complete the quantum mechanical description of nature, especially no contextual ones, as well as no quasi-indeterminism caused by environmental influences [such as in the context translation principle (Svozil, 2004)]. Thus, ultimately, these sources of quantum randomness are grounded in our belief that quantum mechanics is the most complete representation of physical phenomenology.

10.5 Miracles Due to Gaps in Causal Description

A different issue, discussed by Frank (1932), is the possible occurrence of miracles in the presence of *gaps* of physical determinism. Already Maxwell has considered *singular points* (Campbell, and Garnett, 1882, 212–213), "where prediction, except from absolutely perfect data, and guided by the omniscience of contingency, becomes impossible." One might perceive individual events occurring outside the validity of classical and quantum physics without any apparent cause as miracles. For if there is no cause to an event, why should such an event occur altogether rather than not occur?

Although such thoughts remain highly speculative, miracles could be the basis for an operator-directed evolution in otherwise deterministic physical systems. Similar models have been applied to dualistic models of the mind (Eccles, 1986, 1990; Popper, and Eccles, 1977). The objection that this scenario is unnecessarily complicating an otherwise monistic model should be carefully reevaluated in view of computer-generated *virtual realities* (Descartes, 1641; Putnam, 1981; Svozil, 1995a). In such algorithmic universes, there are computable evolution laws as well as inputs from interfaces. From the intrinsic perspective (Svozil, 1994), the inputs cannot be causally accounted for, and hence they remain irreducibly transcendental with respect to the otherwise algorithmic universe.

10.6 Concluding Thoughts

10.6.1 Metaphysical Status of (In)determinism

Hilbert's (1902) sixth problem is about the axiomatization of physics. Regardless of whether this goal is achievable, omniscience cannot be gained via the formalized, syntactic route, which will remain blocked forever by the paradoxical self-reference to which intrinsic observers and operational methods are bound. Even if the universe were a computer (Fredkin, 1990; Svozil, 2006b; Wolfram, 2002; Zuse, 1970), we would intrinsically experience unpredictability and complementarity.

With regard to conjectures about the (in)deterministic evolution of physical events, the situation is unsettled and can be expected to remain unsettled forever. The reason for this is the provable impossibility to formally prove (in)determinism: it is not possible to ensure that physical behaviors are causal and will remain so forever, nor is it possible to exclude all causal behaviors.

The postulate of indeterministic behavior in physics or elsewhere is impossible to *prove* by considering a finite operationally obtained encoded phenotype such as a finite sequence of (supposedly random) bits from physical experiments alone. Furthermore, recursion theory and algorithmic information theory (Calude, 2002; Chaitin, 1987a; Grünwald, and Vitányi, 1987) imply that an unbounded system of axioms is required to prove the unbounded algorithmic information content of an unbounded symbolic sequence. There also exist irreducible complexities in pure mathematics (Chaitin, 2004, 2007).

The opportunistic approach that (as historically, many ingenious scientists have failed to come up with a causal description) indeterminism will prevail appears to be anecdotal, at best, and misleading, at worst. Likewise, the advice of authoritative researchers to avoid asking questions related to completing a theory, or to avoid thinking about the meaning of quantum mechanics or any kind of rational interpretation, and to avoid searching for causal laws for phenomena which are, at the same time, postulated to occur indeterministically by the same authorities – even wisely and benevolently posted – hardly qualify as proof.

Any kind of lawlessness can thus be claimed only *with reference to,* and *relative to,* certain criteria, laws, or quantitative statistical or algorithmic tests. For instance, randomness could be established merely *with respect to* certain tests, such as some batteries of tests of randomness, for instance, *diehard* (Marsaglia, 1995), *NIST* (Rukhin et al., 2001), *TestU01* (L'Ecuyer, and Simard, 2007), or algorithmic (Calude, and Dinneen, 2005; Calude et al., 2010) tests. Note, however, that even the decimal expansion of π, the ratio between the circumference and the diameter of an ideal circle (Bailey et al., 1997; Bailey, and Borwein, 2005), behaves reasonably random (Calude et al., 2010); π might even be a good source of randomness for many Monte Carlo calculations.

Thus, both from a formal as well as from an operational point of view, any rational investigation into, or claim of, absolute (in)determinism is metaphysical and can only be proved *relative to* a limited number of statistical or algorithmic tests which some specialists happen to choose; with very limited validity for the formal and the natural sciences.

10.6.2 Harnessing Unknowables and Indeterminism

Physical indeterminism need not necessarily be perceived negatively as the absence of causal laws but rather as a *valuable resource*. Indeed, ingenious quasi-programs to compute the *halting probability* (Calude, and Chaitin, 2007; Calude, and Dinneen, 2007; Chaitin, 1987a) through summation of series without any computable rate of convergence could, at least in principle, and in the limit of unbounded computational resources, be interpreted as generating provable random sequences. However, as has already been expressed by von Neumann (1951, 768), "anyone who considers arithmetical methods of producing random digits is, of course, in a state of sin."

Besides recursion-theoretic undecidability, there appear to be at least two principal sources of indeterminism and randomness in physics: (1) one scenario is associated with instabilities of classical physical systems and with a strong dependence of future behaviors on the initial value, and (2) quantum indeterminism, which can be subdivided into three subcategories, including random outcomes of individual events, complementarity, and value indefiniteness.

The production of random numbers by physical generators has a long history (The RAND Corporation, 1955). The similarities and differences between classical and quantum randomness can be conceptualized in terms of two black boxes: the first of them, called the *"Poincaré box,"* containing a classical, deterministic, chaotic source of randomness and the second, called the *"Born box,"* containing a quantum source of randomness.

A Poincaré box could be realized by operating a classical dynamical system in the shift map region. Major principles for Born boxes utilizing beam splitters or parametric down conversion include the following: (1) there should be at least three mutually exclusive outcomes to ensure value indefiniteness (Bechmann-Pasquinucci, and Peres, 2000; Calude, and Svozil, 2008; Paterek et al., 2010; Pironio et al., 2010; Svozil, 2009e); (2) the states prepared and measured should be pure and in mutually [possibly interlinked (Svozil, 2009c)] unbiased bases or contexts; and (3) events should be independent to be able to apply proper normalization procedures (Samuelson, 1968; von Neumann, 1951).

Suppose an agent is being presented with both boxes without any label on, or hint about, them; that is, the origin of indeterminism is unknown to the agent. In a modified Turing test, an agent's task would be to find out which is the Born and which is the Poincaré box solely by observing their output. In the absence of any criteria, there should not exist any operational method or procedure capable of discriminating among these boxes. Moreover, both types of indeterminism appear to be based on speculative assumptions: in the classical case, it is the existence of continua and the possibility to randomly choose elements thereof, representing the initial values; in the quantum case, it is the irreducible indeterminism of single events.

10.6.3 Personal Remarks

It is perpetually amazing, perplexing and mind-boggling how many laws and mathematical formæ can be found to express and program or induce physical behavior with high precision. There definitely is substance to the Pythagorean belief that, at least in a restricted manner, nature is numbers and God computes; maybe also throwing dice sometimes.

The apparent impossibility to explain certain phenomena by any causal law should be perceived carefully and cautiously in a historic, transient perspective. The author has the impression that in their attempts to canonize beliefs in the irreducible randomness of (quantum) mechanics, many physicists, philosophers, and communicators may have prematurely thrown out a thorough rationalistic worldview with the provably unfounded claims of total omniscience and omnipotence.

Let me sketch some very speculative attempts to undo the *Goridan Knot* that haunts the perception of randomness in the classical and quantum domains in recent times. (1) Gödel-Turing-Tarski-type undecidability will remain with us forever, at least as long one allows substitution, self-reference, and universal computation. (2) Most classical as well quantum unknowables might be epistemic and not ontic. (3) The classical continua might be convenient abstractions that will have to be abandoned in favor of granular, course-graining structures eventually. As a consequence, classical randomness originating from deterministic chaos might turn out to be formally computable but for all practical purposes impossible to predict. (4) Space and time might turn out to be intrinsic constructions to represent dichotomic events in a world dominated by one-to-one state evolution. (5) There might only exist pure quantum states that can be associated with a unique (measurement and preparation) context. Mixed quantum states might turn out to be purely epistemic, that is, based on our ignorance of the pure state we are dealing with. (6) Kochen-Specker and Boole-Bell-type arguments should be interpreted to indicate value indefiniteness beyond a single context. The idea that there is physical existence beyond a single context at a time (and, associated with it, contextuality) might be misleading. (7) Quantum randomness originate in the process of context translation between different, mismatching preparation and measurement contexts. It might thus be induced by the environment of the measurement apparatus and our technologic inability to maintain universal coherence. (8) Dualistic operator controlled scenarios might present an option that are consistent or at least in *peaceful coexistence* with a certain type of determinism (leaving room for miracles or gaps of causality). The information flow from and through the interface might either be experienced as miracle, or, within the statistical bounds, as incomputable event or input. Whether these specutations and feelings are justified only generations to come will know.

Acknowledgements

The author gratefully acknowledges discussions with Matthias Baaz, Norbert Brunner, Cristian S. Calude, Elena Calude, John Casti, Gregory Chaitin, Robert K. Clifton, Michael J. Dinneen, Paul Adrien Maurice Dirac Anatolij Dvurečenskij, Klaus Ehrenberger, Kurt Rudolf Fischer, Daniel Greenberger, Hans Havlicek, Gudrun Kalmbach, Günther Krenn, Eckehart Köhler, Alexander Leitsch, David Mermin, Mirko Navara, Pavel Pták, Sylvia Pulmannová, Werner DePauli Schimanovich, Ernst Specker, Friedrich Stadler Johann Summhammer, Josef Tkadlec, John Archibald Wheeler, Ron Wright, and Anton Zeilinger, None of these persons should be blamed for my ignorance.

Bibliography

Aberth, O. (1980). *Computable Analysis*. New York: McGraw-Hill.

Adamatzky, A. (2002). *Collision-based Computing*. London: Springer.

Adleman, L. M., and Blum, M. (1991). Inductive inference and unsolvability. *Journal of Symbolic Logic*, 56, 891–900.

Alda, V. (1980). On 0-1 measures for projectors I. *Aplikace matematiky (Applications of Mathematics)*, 25, 373–374.

Alda, V. (1981). On 0-1 measures for projectors II. *Aplikace matematiky (Applications of Mathematics)*, 26, 57–58.

Angluin, D., and Smith, C. H. (1983). A survey of inductive inference: Theory and methods. *Computing Surveys*, 15, 237–269.

Anishchenko, V. S., Astakhov, V., Neiman, A., Vadivasova, T., and Schimansky-Geier, L. (2007). *Nonlinear Dynamics of Chaotic and Stochastic Systems Tutorial and Modern Developments. Tutorial and Modern Developments* (second ed.). Berlin, Heidelberg: Springer.

Aquinas, T. (1981). *Summa Theologica. Translated by Fathers of the English Dominican Province.* Grand Rapids, MI: Christian Classics Ethereal Library.

Atmanspacher, H., and Primas, H. (2005). Epistemic and ontic quantum realities. In Khrennikov, A. (Ed.), *Foundations of Probability and Physics – 3, AIP Conference Proceedings Volume 750*, (pp. 49–62)., New York, Berlin. Springer.

Babadzanjanz, L. K. (1969). Analytical methods of computing perturbations of coordinates of the planets. *Leningradskii Universitet Vestnik Matematika Mekhanika Astronomiia*, 7, 121–132.

Babadzanjanz, L. K. (1979). Existence of the continuations in the n-body problem. *Celestial Mechanics and Dynamical Astronomy*, 20(1), 43–57.

Babadzanjanz, L. K. (1993). On the global solution of the n-body problem. *Celestial Mechanics and Dynamical Astronomy*, 56, 427–449.

Babadzanjanz, L. K., and Sarkissian, D. R. (2006). Taylor series method for dynamical systems with control: Convergence and error estimates. *Journal of Mathematical Sciences*, 139, 7025–7046.

Bailey, D., Borwein, P., and Plouffe, S. (1997). On the rapid computation of various polylogarithmic constants. *Mathematics of Computation*, 66, 903–913.

Bailey, D. H., and Borwein, J. M. (2005). Experimental mathematics: Examples, methods and implications. *Notices of the American Mathematical Society*, 52, 502–514.

Barrow, J. D. (1991). *Theories of Everything*. Oxford: Oxford University Press.

Barrow, J. D. (1998). *Impossibility. The Limits of Science and the Science of Limits*. Oxford: Oxford University Press.

Bechmann-Pasquinucci, H., and Peres, A. (2000). Quantum cryptography with 3-state systems. *Physical Review Letters*, 85(15), 3313–3316.

Bell, J. S. (1966). On the problem of hidden variables in quantum mechanics. *Reviews of Modern Physics*, 38, 447–452. Reprinted in Ref. (Bell, 1987, 1–13).

Bell, J. S. (1987). *Speakable and Unspeakable in Quantum Mechanics*. Cambridge: Cambridge University Press.

Bennett, C. H. (1973). Logical reversibility of computation. *IBM Journal of Research and Development*, 17, 525–532. Reprinted in Ref. (Leff, and Rex, 1990, 197–204).

Bennett, C. H. (1982). The thermodynamics of computation – a review. In Leff, and Rex (1990), (pp. 905–940). Reprinted in Ref. (Leff, and Rex, 1990, 213–248).

Bennett, C. H., Bessette, F., Brassard, G., Salvail, L., and Smolin, J. (1992). Experimental quantum cryptography. *Journal of Cryptology*, 5, 3–28.

Bennett, C. H., and Brassard, G. (1984). Quantum cryptography: Public key distribution and coin tossing. In *Proceedings of the IEEE International Conference on Computers, Systems, and Signal Processing, Bangalore, India*, (pp. 175–179). IEEE Computer Society Press.

Berkeley, G. (1710). *A Treatise Concerning the Principles of Human Knowledge.*

Bernstein, J. (1991). *Quantum Profiles.* Princeton, NJ: Princeton University Press.

Birkhoff, G., and von Neumann, J. (1936). The logic of quantum mechanics. *Annals of Mathematics*, 37(4), 823–843.

Blum, L., and Blum, M. (1975). Toward a mathematical theory of inductive inference. *Information and Control*, 28(2), 125–155.

Bohr, N. (1949). Discussion with Einstein on epistemological problems in atomic physics. In P. A. Schilpp (Ed.), *Albert Einstein: Philosopher-Scientist* (pp. 200–241). Evanston, Ill.: The Library of Living Philosophers.

Boole, G. (1862). On the theory of probabilities. *Philosophical Transactions of the Royal Society of London*, 152, 225–252.

Boole, G. (1958). *An Investigation of the Laws of Thought.* New York: Dover.

Born, M. (1926a). Quantenmechanik der Stoßvorgänge. *Zeitschrift für Physik*, 38, 803–827.

Born, M. (1926b). Zur Quantenmechanik der Stoßvorgänge. *Zeitschrift für Physik*, 37, 863–867.

Born, M. (1955). Ist die klassische Mechanik tatsächlich deterministisch? *Physikalische Blätter*, 11, 49–54. English translation "Is classical mechanics in fact deterministic?" Reprinted in Ref. (Born, 1969, 78–83).

Born, M. (1969). *Physics in my generation* (second ed.). New York: Springer.

Boskovich, R. J. (1966). De spacio et tempore, ut a nobis cognoscuntur. In J. M. Child (Ed.), *A Theory of Natural Philosophy* (pp. 203–205). Cambridge, MA: Open Court (1922) and MIT Press.

Brady, A. H. (1988). The busy beaver game and the meaning of life. In R. Herken (Ed.), *The Universal Turing Machine. A Half-Century Survey* (pp. 259). Hamburg: Kammerer und Unverzagt.

Bricmont, J. (1996). Science of chaos or chaos in science? *Annals of the New York Academy of Sciences*, 775, 131–175.

Bridges, D. S. (1999). Can constructive mathematics be applied in physics? *Journal of Philosophical Logic*, 28(5), 439–453.

Bridgman, P. W. (1934). A physicist's second reaction to Mengenlehre. *Scripta Mathematica*, 2, 101–117, 224–234. Discussed in Landauer (1994).

Brukner, Č. (2003). Quantum experiments can test mathematical undecidability. In C. S. Calude, J. F. G. Costa, R. Freund, M. Oswald, and G. Rozenberg (Eds.), *Unconventional Computing. Unconventional Computation. Proceedings of the 7th International Conference, UC 2008, Vienna, Austria, August 25–28, 2008. Lecture Notes in Computer Science. Volume 5204/2008* (pp. 1–5). Berlin: Springer.

Bub, J. (1999). *Interpreting the Quantum World.* Cambridge: Cambridge University Press.

Cabello, A., Estebaranz, J. M., and García-Alcaine, G. (1996). Bell-Kochen-Specker theorem: A proof with 18 vectors. *Physics Letters A*, 212(4), 183–187.

Calude, C. (2002). *Information and Randomness – An Algorithmic Perspective* (second ed.). Berlin: Springer.

Calude, C., Calude, E., Svozil, K., and Yu, S. (1997). Physical versus computational complementarity I. *International Journal of Theoretical Physics*, 36(7), 1495–1523.

Calude, C., Campbell, D. I., Svozil, K., and Ştefănescu, D. (1995). Strong determinism vs. computability. In Schimanovich, W. D., Köhler, E., and Stadler, F. (Eds.), *The Foundational Debate. Complexity and Constructivity in Mathematics and Physics. Vienna Circle Institute Yearbook, Vol. 3*, (pp. 115–131). Dordrecht, Boston, London. Kluwer.

Calude, C. S. (2005). Algorithmic randomness, quantum physics, and incompleteness. In M. Margenstern (Ed.), *Proceedings of the Conference "Machines, Computations and Universality" (MCU'2004)* (pp. 1–17). Berlin: Lectures Notes in Comput. Sci. 3354, Springer.

Calude, C. S., Calude, E., and Svozil, K. (2010). The complexity of proving chaoticity and the Church–Turing thesis. *Chaos: An Interdisciplinary Journal of Nonlinear Science*, 20(3), 037103.

Calude, C. S., and Chaitin, G. J. (2007). What is . . . a halting probability? *Notices of the AMS*, 57(2), 236–237.

Calude, C. S., and Dinneen, M. J. (2005). Is quantum randomness algorithmic random? a preliminary attack. In Bozapalidis, S., Kalampakas, A., and Rahonis, G. (Eds.), *Proceedings of the 1st International Conference on Algebraic Informatics*, (pp. 195–196). Thessaloniki, Greece. Aristotle University of Thessaloniki.

Calude, C. S., and Dinneen, M. J. (2007). Exact approximations of omega numbers. *International Journal of Bifurcation and Chaos*, 17, 1937–1954. CDMTCS report series 293.

Calude, C. S., Dinneen, M. J., Dumitrescu, M., and Svozil, K. (2010). Experimental evidence of quantum randomness incomputability. *Phys. Rev. A*, 82(2), 022102.

Calude, C. S., and Staiger, L. (2010). A note on accelerated turing machines. *Mathematical Structures in Computer Science*, 20(Special Issue 06), 1011–1017.

Calude, C. S., and Svozil, K. (2008). Quantum randomness and value indefiniteness. *Advanced Science Letters*, 1(2), 165–168.

Campbell, L., and Garnett, W. (1882). *The life of James Clerk Maxwell. With a selection from his correspondence and occasional writings and a sketch of his contributions to science*. London: MacMillan.

Casti, J. L., and Karlquist, A. (1996). *Boundaries and Barriers. On the Limits to Scientific Knowledge*. Reading, MA: Addison-Wesley.

Casti, J. L., and Traub, J. F. (1994). *On Limits*. Santa Fe, NM: Santa Fe Institute. Report 94-10-056.

Chaitin, G. J. (1974). Information-theoretic limitations of formal systems. *Journal of the Association of Computing Machinery*, 21, 403–424. Reprinted in Ref. Chaitin (1990).

Chaitin, G. J. (1987a). *Algorithmic Information Theory*. Cambridge: Cambridge University Press.

Chaitin, G. J. (1987b). Computing the busy beaver function. In T. M. Cover, and B. Gopinath (Eds.), *Open Problems in Communication and Computation* (pp. 108). New York: Springer. Reprinted in Ref. Chaitin (1990).

Chaitin, G. J. (1990). *Information, Randomness and Incompleteness* (second ed.). Singapore: World Scientific. This is a collection of G. Chaitin's early publications.

Chaitin, G. J. (2004). Irreducible complexity in pure mathematics. eprint math/0411091.

Chaitin, G. J. (2007). The halting probability Omega: Irreducible complexity in pure mathematics. *Milan Journal of Mathematics*, 75(1), 291–304.

Chapman, M. S., Hammond, T. D., Lenef, A., Schmiedmayer, J., Rubenstein, R. A., Smith, E., and Pritchard, D. E. (1995). Photon scattering from atoms in an atom interferometer: Coherence lost and regained. *Physical Review Letters*, 75(21), 3783–3787.

Chisholm, R. M. (1946). The contrary-to-fact conditional. *Mind*, 55(220), 289–307. reprinted in (Chisholm, 1949, 482–497).

Chisholm, R. M. (1949). The contrary-to-fact conditional. In H. Feigl, and W. Sellars (Eds.), *Readings in Philosophical Analysis* (pp. 482–497). New York: Appleton-Century-Crofts.

Clauser, J. (2002). Early history of Bell's theorem. In R. Bertlmann, and A. Zeilinger (Eds.), *Quantum (Un)speakables. From Bell to Quantum Information* (pp. 61–96). Berlin: Springer.

Clauser, J. F., and Shimony, A. (1978). Bell's theorem: experimental tests and implications. *Reports on Progress in Physics*, 41, 1881–1926.

Costa, N. C. A., and Doria, F. A. (1991). Classical physics and Penrose's thesis. *Foundations of Physics Letters*, 4, 363–373.

da Costa, N. C. A., and Doria, F. A. (1991). Undecidability and incompleteness in classical mechanics. *International Journal of Theoretical Physics*, 30, 1041–1073.

da Costa, N. C. A., Doria, F. A., and do Amaral, A. F. F. (1993). Dynamical system where proving chaos is equivalent to proving Fermat's conjecture. *International Journal of Theoretical Physics*, 32(11), 2187–2206.

Davis, M. (1958). *Computability and Unsolvability*. New York: McGraw-Hill.

Davis, M. (1965). *The Undecidable. Basic Papers on Undecidable, Unsolvable Problems and Computable Functions*. Hewlett, N.Y.: Raven Press.

Descartes, R. (1637). *Discours de la méthode pour bien conduire sa raison et chercher la verité dans les sciences (Discourse on the Method of Rightly Conducting One's Reason and of Seeking Truth)*.

Descartes, R. (1641). *Meditation on First Philosophy*.

Dewdney, A. K. (1984). Computer recreations: A computer trap for the busy beaver, the hardest-working Turing machine. *Scientific American*, 251(2), 19–23.

Diacu, F. (1996). The solution of the n-body problem. *The Mathematical Intelligencer*, 18(3), 66–70.

Diacu, F., and Holmes, P. (1996). *Celestial Encounters – the Origins of Chaos and Stability*. Princeton, NJ: Princeton University Press.

Donath, N., and Svozil, K. (2002). Finding a state among a complete set of orthogonal ones. *Physical Review A*, 65, 044302.

Dvurečenskij, A., Pulmannová, S., and Svozil, K. (1995). Partition logics, orthoalgebras and automata. *Helvetica Physica Acta*, 68, 407–428.

Eccles, J. C. (1986). Do mental events cause neural events analogously to the probability fields of quantum mechanics? *Proceedings of the Royal Society of London. Series B. Biological Sciences*, 227(1249), 411–428.

Eccles, J. C. (1990). The mind-brain problem revisited: The microsite hypothesis. In J. C. Eccles, and O. Creutzfeldt (Eds.), *The Principles of Design and Operation of the Brain* (pp. 549–572). Berlin: Springer.

Einstein, A. (1938). Reply to criticism: Remarks concerning the essays brought together in this co-operative volume. In P. A. Schilpp (Ed.), *Albert Einstein: Philosopher-Scientist (New York: Library of Living Philosophers, 1949), on p. 668.* (pp. 665–688). New York, NY: Harper and Brothers Publishers.

Einstein, A., Podolsky, B., and Rosen, N. (1935). Can quantum-mechanical description of physical reality be considered complete? *Physical Review*, 47(10), 777–780.

Ekert, A. K. (1991). Quantum cryptography based on Bell's theorem. *Physical Review Letters*, 67, 661–663.

Everett, H. (1957). 'relative state' formulation of quantum mechanics. *Reviews of Modern Physics*, 29, 454–462. Reprinted in Ref. (Wheeler, and Zurek, 1983, 315–323).

Feferman, S. (1984). Kurt Gödel: conviction and caution. *Philosophia Naturalis*, 21, 546–562.

Feynman, R. P. (1965). *The Character of Physical Law*. Cambridge, MA: MIT Press.

Fiorentino, M., Santori, C., Spillane, S. M., Beausoleil, R. G., and Munro, W. J. (2007). Secure self-calibrating quantum random-bit generator. *Physical Review A*, 75(3), 032334.

Frank, P. (1932). *Das Kausalgesetz und seine Grenzen*. Vienna: Springer. English translation in Ref. Frank, and R. S. Cohen (Editor) (1997).

Frank, P., and R. S. Cohen (Editor) (1997). *The Law of Causality and its Limits (Vienna Circle Collection)*. Vienna: Springer.

Fredkin, E. (1990). Digital mechanics. an informational process based on reversible universal cellular automata. *Physica*, D45, 254–270.

Fredkin, E., and Toffoli, T. (1982). Conservative logic. *International Journal of Theoretical Physics*, 21(3-4), 219–253. Reprinted in Ref. (Adamatzky, 2002, Part I, Chapter 3).

Freud, S. (1999). Ratschläge für den Arzt bei der psychoanalytischen Behandlung. In Freud, A., Bibring, E., Hoffer, W., Kris, E., and Isakower, O. (Eds.), *Gesammelte Werke. Chronologisch geordnet. Achter Band. Werke aus den Jahren 1909–1913*, (pp. 376–387). Frankfurt am Main. Fischer.

Fuchs, C. A., and Peres, A. (2000). Quantum theory needs no 'interpretation'. *Physics Today*, 53(4), 70–71. Further discussions of and reactions to the article can be found in the September issue of Physics Today, *53*, 11–14 (2000).

Gödel, K. (1931). Über formal unentscheidbare Sätze der Principia Mathematica und verwandter Systeme. *Monatshefte für Mathematik und Physik*, 38(1), 173–198. Reprint and English translation in Ref. (Gödel, 1986, 144–195), and in Ref. (Davis, 1965, 5–40).

Gödel, K. (1986). In S. Feferman, J. W. Dawson, S. C. Kleene, G. H. Moore, R. M. Solovay, and J. van Heijenoort (Eds.), *Collected Works. Publications 1929–1936. Volume I.* Oxford: Oxford University Press.

Gold, M. E. (1967). Language identification in the limit. *Information and Control*, 10, 447–474.

Greenberger, D. M., Horne, M. A., Shimony, A., and Zeilinger, A. (1990). Bell's theorem without inequalities. *American Journal of Physics*, 58, 1131–1143.

Greenberger, D. M., Horne, M. A., and Zeilinger, A. (1993). Multiparticle interferometry and the superposition principle. *Physics Today*, 46, 22–29.

Greenberger, D. M., and YaSin, A. (1989). "Haunted" measurements in quantum theory. *Foundation of Physics*, 19(6), 679–704.

Grünwald, P. D., and Vitányi, P. M. B. (1987). Algorithmic information theory. In P. Adriaans, and J. F. van Benthem (Eds.), *Handbook of the Philosophy of Information* (pp. 281–320). New York, NY: Freeman. a volume in the Handbook of the Philosophy of Science, ed. by Dov Gabbay, Paul Thagard, and John Wood.

Hai-Qiang, M., Su-Mei, W., Da, Z., Jun-Tao, C., Ling-Ling, J., Yan-Xue, H., and Ling-An, W. (2004). A random number generator based on quantum entangled photon pairs. *Chinese Physics Letters*, 21(10), 1961–1964.

Halmos, P. R. (1974). *Finite-Dimensional Vector Spaces*. New York, Heidelberg, Berlin: Springer.

Hertz, H. (1894). *Prinzipien der Mechanik*. Leipzig: Barth.

Herzog, T. J., Kwiat, P. G., Weinfurter, H., and Zeilinger, A. (1995). Complementarity and the quantum eraser. *Physical Review Letters*, 75(17), 3034–3037.

Heywood, P., and Redhead, M. L. G. (1983). Nonlocality and the Kochen-Specker paradox. *Foundations of Physics*, 13(5), 481–499.

Hilbert, D. (1902). Mathematical problems. *Bulletin of the American Mathematical Society*, 8(10), 437–479.

Hilbert, D. (1926). Über das Unendliche. *Mathematische Annalen*, 95(1), 161–190.

Hirsch, M. W. (1985). The chaos of dynamical systems. In *Chaos, fractals, and dynamics (Guelph, Ont., 1981/1983)*, volume 98 of *The chaos of dynamical systems. Lecture notes in pure and applied mathematics* (pp. 189–196). New York: Dekker.

Hole, A. (1994). Predictability in deterministic theories. *International Journal of Theoretical Physics*, 33, 1085–1111.

Hooker, C. A. (1975). *The Logico-Algebraic Approach to Quantum Mechanics. Volume I: Historical Evolution*. Dordrecht: Reidel.

Hume, D. (1748). *An Enquiry Concerning Human Understanding*.

Jammer, M. (1966). *The Conceptual Development of Quantum Mechanics*. New York: McGraw-Hill Book Company.

Jammer, M. (1974). *The Philosophy of Quantum Mechanics*. New York: John Wiley, and Sons.

Jammer, M. (1989). *The Conceptual Development of Quantum Mechanics. The History of Modern Physics, 1800–1950; v. 12* (second ed.). New York: American Institute of Physics.

Jammer, M. (1992). John Steward Bell and the debate on the significance of his contributions to the foundations of quantum mechanics. In A. van der Merwe, F. Selleri, and G. Tarozzi (Eds.), *Bell's Theorem and the Foundations of Modern Physics* (pp. 1–23). Singapore: World Scientific.

Jennewein, T., Achleitner, U., Weihs, G., Weinfurter, H., and Zeilinger, A. (2000). A fast and compact quantum random number generator. *Review of Scientific Instruments*, 71, 1675–1680.

Kalmbach, G. (1983). *Orthomodular Lattices*. New York: Academic Press.

Kalmbach, G. (1986). *Measures and Hilbert Lattices*. Singapore: World Scientific.

Kamber, F. (1964). Die Struktur des Aussagenkalküls in einer physikalischen Theorie. *Nachrichten der Akademie der Wissenschaften in Göttingen, Mathematisch-Physikalische Klasse*, 10, 103–124.

Kamber, F. (1965). Zweiwertige Wahrscheinlichkeitsfunktionen auf orthokomplementären Verbänden. *Mathematische Annalen*, 158(3), 158–196.

Kanter, I. (1990). Undecidability principle and the uncertainty principle even for classical systems. *Physical Review Letters*, 64, 332–335.

Kauffman, L. H. (1987). Self-reference and recursive forms. *Journal of Social and Biological Structures*, 10(1), 53–72.

Kochen, S., and Specker, E. P. (1965). The calculus of partial propositional functions. In *Proceedings of the 1964 International Congress for Logic, Methodology and Philosophy of Science, Jerusalem*, (pp. 45–57). Amsterdam. North Holland. Reprinted in Ref. (Specker, 1990, 222–234).

Kochen, S., and Specker, E. P. (1967). The problem of hidden variables in quantum mechanics. *Journal of Mathematics and Mechanics (now Indiana University Mathematics Journal)*, 17(1), 59–87. Reprinted in Ref. (Specker, 1990, 235–263).

Kragh, H. (1997). The origin of radioactivity: from solvable problem to unsolved non-problem. *Archive for History of Exact Sciences*, 50, 331–358.

Kragh, H. (1999). *Quantum Generations: A History of Physics in the Twentieth Century*. Princeton, NJ: Princeton University Press.

Kragh, H. (2009). Subatomic determinism and causal models of radioactive decay, 1903–1923. RePoSS: Research Publications on Science Studies 5. Department of Science Studies, University of Aarhus.

Kreisel, G. (1974). A notion of mechanistic theory. *Synthese*, 29, 11–26.

Krenn, G., and Svozil, K. (1998). Stronger-than-quantum correlations. *Foundations of Physics*, 28(6), 971–984.

Kwiat, P. G., Steinberg, A. M., and Chiao, R. Y. (1992). Observation of a "quantum eraser:" a revival of coherence in a two-photon interference experiment. *Physical Review A*, 45(11), 7729–7739.

Lakatos, I. (1978). *Philosophical Papers. 1. The Methodology of Scientific Research Programmes*. Cambridge: Cambridge University Press.

Landauer, R. (1986). Computation and physics: Wheeler's meaning circuit? *Foundations of Physics*, 16, 551–564.

Landauer, R. (1991). Information is physical. *Physics Today*, 44(5), 23–29.

Landauer, R. (1994). Advertisement for a paper I like. In J. L. Casti, and J. F. Traub (Eds.), *On Limits* (pp. 39). Santa Fe, NM: Santa Fe Institute Report 94-10-056.

Laplace, P.-S. (1995,1998). *Philosophical Essay on Probabilities. Translated from the fifth French edition of 1825*. Berlin, New York: Springer.

L'Ecuyer, P., and Simard, R. (2007). TestU01: A C library for empirical testing of random number generators. *ACM Transactions on Mathematical Software (TOMS)*, 33(4), Article 22, 1–40.

Leff, H. S., and Rex, A. F. (1990). *Maxwell's Demon*. Princeton, NJ: Princeton University Press.

Li, M., and Vitányi, P. M. B. (1992). Inductive reasoning and Kolmogorov complexity. *Journal of Computer and System Science*, 44, 343–384.

Lichtenberg, A. J., and Lieberman, M. A. (1983). *Regular and Stochastic Motion*. New York: Springer.

Locke, J. (1690). *An Essay Concerning Human Understanding*.

Lyapunov, A. M. (1992). The general problem of the stability of motion. *International Journal of Control*, 55(3), 531–534.

Margolus, N. (2002). Universal cellular automata based on the collisions of soft spheres. In A. Adamatzky (Ed.), *Collision-based Computing* (pp. 107–134). London: Springer.

Marsaglia, G. (1995). The Marsaglia random number CDROM including the diehard battery of tests of randomness. available from URL www.stat.fsu.edu/pub/diehard/.

Martin-Löf, P. (1966). The definition of random sequences. *Information and Control*, 9(6), 602–619.

Mermin, N. D. (1993). Hidden variables and the two theorems of John Bell. *Reviews of Modern Physics*, 65, 803–815.

Moore, C. D. (1990). Unpredictability and undecidability in dynamical systems. *Physical Review Letters*, 64, 2354–2357. Cf. Ch. Bennett, Nature, **346**, 606 (1990).

Moore, E. F. (1956). Gedanken-experiments on sequential machines. In C. E. Shannon, and J. McCarthy (Eds.), *Automata Studies* (pp. 129–153). Princeton, NJ: Princeton University Press.

Nagel, T. (1986). *The View from Nowhere*. New York and Oxford: Oxford University Press.

Navara, M., and Rogalewicz, V. (1991). The pasting constructions for orthomodular posets. *Mathematische Nachrichten*, 154, 157–168.

Odifreddi, P. (1989). *Classical Recursion Theory, Vol. 1*. Amsterdam: North-Holland.

Ou, Z., Hong, C., and Mandel, L. (1987). Relation between input and output states for a beam splitter. *Optics Communications*, 63(2), 118–122.

Paterek, T., Kofler, J., Prevedel, R., Klimek, P., Aspelmeyer, M., Zeilinger, A., and Brukner, Č. (2010). Logical independence and quantum randomness. *New Journal of Physics*, 12(1), 013019.

Pauli, W. (1958). Die allgemeinen Prinzipien der Wellenmechanik. In S. Flügge (Ed.), *Handbuch der Physik. Band V, Teil 1. Prinzipien der Quantentheorie I* (pp. 1–168). Berlin, Göttingen and Heidelberg: Springer.

Peres, A. (1978). Unperformed experiments have no results. *American Journal of Physics*, 46, 745–747.

Peres, A. (1980). Can we undo quantum measurements? *Physical Review D*, 22(4), 879–883.

Peterson, I. (1993). *Chaos in the Solar System*. New York: W. H. Freeman and Company.

Pfau, T., Spälter, S., Kurtsiefer, C., Ekstrom, C. R., and Mlynek, J. (1994). Loss of spatial coherence by a single spontaneous emission. *Physical Review Letters*, 73(9), 1223–1226.

Pironio, S., Acín, A., Massar, S., Boyer de la Giroday, A., Matsukevich, D. N., Maunz, P., Olmschenk, S., Hayes, D., Luo, L., Manning, T. A., and Monroe, C. (2010). Random numbers certified by Bell's theorem. *Nature*, 464, 1021–1024.

Pitowsky, I. (1989a). From George Boole to John Bell: The origin of Bell's inequality. In M. Kafatos (Ed.), *Bell's Theorem, Quantum Theory and the Conceptions of the Universe* (pp. 37–49). Dordrecht: Kluwer.

Pitowsky, I. (1989b). *Quantum Probability – Quantum Logic*. Berlin: Springer.

Pitowsky, I. (1994). George Boole's 'conditions of possible experience' and the quantum puzzle. *The British Journal for the Philosophy of Science*, 45, 95–125.

Pitowsky, I., and Svozil, K. (2001). New optimal tests of quantum nonlocality. *Physical Review A*, 64, 014102.

Poincaré, H. (1890). Sur le probléme des trois corps et les équations de la dynamique. *Acta Mathematica*, 13(1), A3–A270.

Poincaré, H. (1914). *Wissenschaft und Hypothese*. Leipzig: Teubner.

Popescu, S., and Rohrlich, D. (1994). Quantum nonlocality as an axiom. *Foundations of Physics*, 24(3), 379–358.

Popescu, S., and Rohrlich, D. (1997). Action and passion at a distance: an essay in honor of professor Abner Shimony. *Boston Studies in the Philosophy of Science*, 194, 197–206.

Popper, K. R. (1950a). Indeterminism in quantum physics and in classical physics I. *The British Journal for the Philosophy of Science*, 1, 117–133.

Popper, K. R. (1950b). Indeterminism in quantum physics and in classical physics II. *The British Journal for the Philosophy of Science*, 1, 173–195.

Popper, K. R. (1959). *The Logic of Scientific Discovery*. New York: Basic Books.

Popper, K. R., and Eccles, J. C. (1977). *The Self and Its Brain*. Berlin, Heidelberg, London, New York: Springer.

Pour-El, M. B., and Richards, J. I. (1989). *Computability in Analysis and Physics*. Berlin: Springer.

Pták, P., and Pulmannová, S. (1991). *Orthomodular Structures as Quantum Logics*. Dordrecht: Kluwer Academic Publishers.

Purrington, R. D. (1997). *Physics in the Nineteenth Century*. New Brunswick, NJ: Rutgers University Press.

Putnam, H. (1981). *Reason, Truth and History*. Cambridge: Cambridge University Press.

Rado, T. (1962). On non-computable functions. *The Bell System Technical Journal*, *XLI(41)*(3), 877–884.

Rarity, J. G., Owens, M. P. C., and Tapster, P. R. (1994). Quantum random-number generation and key sharing. *Journal of Modern Optics*, 41, 2435–2444.

Redhead, M. (1990). *Incompleteness, Nonlocality, and Realism: A Prolegomenon to the Philosophy of Quantum Mechanics*. Oxford: Clarendon Press.

Richardson, D. (1968). Some undecidable problems involving elementary functions of a real variable. *Journal of Symbolic Logic*, 33(4), 514–520.

Rogers, Jr., H. (1967). *Theory of Recursive Functions and Effective Computability*. New York: MacGraw-Hill.

Rössler, O. E. (1998). *Endophysics. The World as an Interface*. Singapore: World Scientific. With a foreword by Peter Weibel.

Rousseau, C. (2004). Divergent series: Past, present, future . . . preprint.

Rucker, R. (1982). *Infinity and the Mind*. Boston: Birkhäuser.

Rukhin, A., Soto, J., Nechvatal, J., Smid, M., Barker, E., Leigh, S., Levenson, M., Vangel, M., Banks, D., Hekert, A., Dray, J., and Vo, S. (2001). *A Statistical Test Suite for Random and Pseudorandom Number Generators for Cryptographic Applications. NIST Special Publication 800-22*. Gaithersburg, MD: National Institute of Standards and Technology (NIST).

Samuelson, P. A. (1968). Constructing an unbiased random sequence. *Journal of the American Statistical Association*, *63*(324), 1526–1527.

Scarpellini, B. (1963). Zwei unentscheidbare Probleme der Analysis. *Zeitschrift für Mathematische Logik und Grundlagen der Mathematik*, 9, 265–289.

Schaller, M., and Svozil, K. (1996). Automaton logic. *International Journal of Theoretical Physics*, 35(5), 911–940.

Schlick, M. (1932). Causality in everday life and in recent science. *University of California Publications in Philosophy*, 15, 99–125. reprinted in (Schlick, 1949, 515–533) and (Schlick, 2008a, 415–445).

Schlick, M. (1935). Unanswerable questions? *The Philosopher*, 13, 98–104. reprinted in (Schlick, 2008b, 621–634).

Schlick, M. (1949). Causality in everday life and in recent science. In H. Feigl, and W. Sellars (Eds.), *Readings in Philosophical Analysis* (pp. 515–533). New York: Appleton-Century-Crofts. previously published in Schlick (1932).

Schlick, M. (2008a). Causality in everday life and in recent science. In F. Stadler, and H. J. Wendel (Eds.), *Gesamtausgabe. Die Wiener Zeit. Aufsätze, Beiträge, Rezensionen 1926–1936*, volume 6 (pp. 415–445). Wien and New York: Springer.

Schlick, M. (2008b). Unanswerable questions? In F. Stadler, and H. J. Wendel (Eds.), *Gesamtausgabe. Die Wiener Zeit. Aufsätze, Beiträge, Rezensionen 1926–1936*, volume 6 (pp. 621–634). Wien and New York: Springer.

Schrödinger, E. (1935a). Die gegenwärtige Situation in der Quantenmechanik. *Naturwissenschaften*, 23, 807–812, 823–828, 844–849. English translation in Ref. Trimmer (1980) and in Ref. (Wheeler, and Zurek, 1983, 152–167).

Schrödinger, E. (1935b). Discussion of probability relations between separated systems. *Mathematical Proceedings of the Cambridge Philosophical Society*, 31(04), 555–563.

Schrödinger, E. (1936). Probability relations between separated systems. *Mathematical Proceedings of the Cambridge Philosophical Society*, 32(03), 446–452.

Schrödinger, E. (1995). *The Interpretation of Quantum Mechanics. Dublin Seminars (1949–1955) and Other Unpublished Essays*. Woodbridge, Connecticut: Ox Bow Press.

Scully, M. O., and Drühl, K. (1982). Quantum eraser: A proposed photon correlation experiment concerning observation and "delayed choice" in quantum mechanics. *Physical Review A*, 25(4), 2208–2213.

Scully, M. O., Englert, B.-G., and Walther, H. (1991). Quantum optical tests of complementarity. *Nature*, 351, 111–116.

Shaw, R. S. (1981). Strange attractors, chaotic behavior, and information flow. *Zeitschrift für Naturforschung A*, 36, 80–112.

Sherr, R., Bainbridge, K. T., and Anderson, H. H. (1941). Transmutation of mercury by fast neutrons. *Physical Review*, 60(7), 473–479.

Shimony, A. (1984). Controllable and uncontrollable non-locality. In Kamefuchi, S., and Gakkai, N. B. (Eds.), *Proceedings of the International Symposium… Proceedings of the International Symposium Foundations of Quantum Mechanics in the Light of New Technology*, (pp. 225–230). Tokyo. Physical Society of Japan. See also J. Jarrett, Bell's Theorem, Quantum Mechanics and Local Realism, Ph. D. thesis, Univ. of Chicago, 1983; Nous, **18**, 569 (1984).

Smale, S. (1967). Differentiable dynamical systems. *Bulletin of the American Mathematical Society*, 73, 747–817.

Smullyan, R. M. (1992a). *Gödel's Incompleteness Theorems*. New York, New York: Oxford University Press.

Smullyan, R. M. (1992b). *What Is the Name of This Book?* Englewood Cliffs, NJ: Prentice-Hall, Inc.

Sosa, E. (2009). *Reflective Knowledge. Apt Belief and Reflective Knowledge, Volume II*. Oxford: Clarendon Press.

Specker, E. (1949). Nicht konstruktiv beweisbare Sätze der Analysis. *The Journal of Smbolic Logic*, 14, 145–158. Reprinted in Ref. (Specker, 1990, 35–48); English translation: *Theorems of Analysis which cannot be proven constructively*.

Specker, E. (1959). Der Satz vom Maximum in der rekursiven Analysis. In Heyting, A. (Ed.), *Constructivity in mathematics: proceedings of the colloquium held at Amsterdam, 1957*, (pp. 254–265). Amsterdam. North-Holland Publishing Company. Reprinted in Ref. (Specker, 1990, 148–159); English translation: *Theorems of Analysis which cannot be proven constructively*.

Specker, E. (1960). Die Logik nicht gleichzeitig entscheidbarer Aussagen. *Dialectica*, 14(2–3), 239–246. Reprinted in Ref. (Specker, 1990, 175–182); English translation: *The logic of propositions which are not simultaneously decidable*, Reprinted in Ref. (Hooker, 1975, 135–140).

Specker, E. (1990). *Selecta*. Basel: Birkhäuser Verlag.

Stace, W. T. (1934). The refutation of realism. *Mind*, 43(170), 145–155. reprinted in (Stace, 1949, 364–372).

Stace, W. T. (1949). The refutation of realism. In H. Feigl, and W. Sellars (Eds.), *Readings in Philosophical Analysis* (pp. 364–372). New York: Appleton-Century-Crofts. previously published in *Mind* **53**, 349–353 (1934).

Stefanov, A., Gisin, N., Guinnard, O., Guinnard, L., and Zbinden, H. (2000). Optical quantum random number generator. *Journal of Modern Optics*, 47, 595–598.

Sterne, L. (1760–1767). *The Life and Opinions of Tristram Shandy, Gentleman*. London.

Stewart, I. (1991). Deciding the undecidable. *Nature*, 352, 664–665.

Sundman, K. F. (1912). Memoire sur le problème de trois corps. *Acta Mathematica*, 36, 105–179.

Suppes, P. (1993). The transcendental character of determinism. *Midwest Studies In Philosophy*, 18(1), 242–257.

Svozil, K. (1990). The quantum coin toss – testing microphysical undecidability. *Physics Letters A*, 143, 433–437.

Svozil, K. (1993). *Randomness, and Undecidability in Physics*. Singapore: World Scientific.

Svozil, K. (1994). Extrinsic-intrinsic concept and complementarity. In H. Atmanspacher, and G. J. Dalenoort (Eds.), *Inside versus Outside* (pp. 273–288). Heidelberg: Springer.

Svozil, K. (1995a). A constructivist manifesto for the physical sciences – Constructive re-interpretation of physical undecidability. In Schimanovich, W. D., Köhler, E., and Stadler, F. (Eds.), *The Foundational Debate, Complexity and Constructivity in Mathematics and Physics*, (pp. 65–88)., Dordrecht, Boston, London. Kluwer.

Svozil, K. (1995b). Set theory and physics. *Foundations of Physics*, 25, 1541–1560.

Svozil, K. (1998). *Quantum Logic*. Singapore: Springer.

Svozil, K. (2002a). Conventions in relativity theory and quantum mechanics. *Foundations of Physics*, 32, 479–502.

Svozil, K. (2002b). Quantum information in base n defined by state partitions. *Physical Review A*, 66, 044306.

Svozil, K. (2004). Quantum information via state partitions and the context translation principle. *Journal of Modern Optics*, 51, 811–819.

Svozil, K. (2005a). Communication cost of breaking the Bell barrier. *Physical Review A*, 72(9), 050302(R).

Svozil, K. (2005b). Logical equivalence between generalized urn models and finite automata. *International Journal of Theoretical Physics*, 44, 745–754.

Svozil, K. (2005c). Noncontextuality in multipartite entanglement. *J. Phys. A: Math. Gen.*, 38, 5781–5798.

Svozil, K. (2006a). Are simultaneous Bell measurements possible? *New Journal of Physics*, 8, 39, 1–8.

Svozil, K. (2006b). Computational universes. *Chaos, Solitons, and Fractals*, 25(4), 845–859.

Svozil, K. (2006c). Staging quantum cryptography with chocolate balls. *American Journal of Physics*, 74(9), 800–803.

Svozil, K. (2007). Omega and the time evolution of the n-body problem. In Calude, C. S. (Ed.), *Randomness and Complexity, from Leibniz to Chaitin*, (pp. 231–236). Singapore. World Scientific. eprint arXiv:physics/0703031.

Svozil, K. (2009a). Contexts in quantum, classical and partition logic. In K. Engesser, D. M. Gabbay, and D. Lehmann (Eds.), *Handbook of Quantum Logic and Quantum Structures* (pp. 551–586). Amsterdam: Elsevier.

Svozil, K. (2009b). On the brightness of the Thomson lamp: A prolegomenon to quantum recursion theory. In Calude, C. S., Costa, J. F., Dershowitz, N., Freire, E., and Rozenberg, G. (Eds.), *UC '09: Proceedings of the 8th International Conference on Unconventional Computation*, (pp. 236–246). Berlin, Heidelberg. Springer.

Svozil, K. (2009c). Proposed direct test of a certain type of noncontextuality in quantum mechanics. *Physical Review A*, 80(4), 040102.

Svozil, K. (2009d). Quantum scholasticism: On quantum contexts, counterfactuals, and the absurdities of quantum omniscience. *Information Sciences*, 179, 535–541.

Svozil, K. (2009e). Three criteria for quantum random-number generators based on beam splitters. *Physical Review A*, 79(5), 054306.

Svozil, K. (2010a). Chocolate cryptography. eprint arXiv:0903.0231.

Svozil, K. (2010b). Quantum value indefiniteness. *Natural Computing*, in print. eprint arXiv:1001.1436.

Svozil, K., and Tkadlec, J. (1996). Greechie diagrams, nonexistence of measures in quantum logics and Kochen–Specker type constructions. *Journal of Mathematical Physics*, 37(11), 5380–5401.

Tarski, A. (1932). Der Wahrheitsbegriff in den Sprachen der deduktiven Disziplinen. *Akademie der Wissenschaften in Wien. Mathematisch-naturwissenschaftliche Klasse, Akademischer Anzeiger*, 69, 9–12.

Tarski, A. (1956). *Logic, Semantics and Metamathematics*. Oxford: Oxford University Press.

The RAND Corporation (1955). *A Million Random Digits with 100,000 Normal Deviates Free Press Publishers*. Glencoe, Illinois: Knolls Atomic Power Lab. Report KAPL-3147.

Toffoli, T. (1978). The role of the observer in uniform systems. In G. J. Klir (Ed.), *Applied General Systems Research, Recent Developments and Trends* (pp. 395–400). New York, London: Plenum Press.

Trimmer, J. D. (1980). The present situation in quantum mechanics: a translation of Schrödinger's "cat paradox". *Proceedings of the American Philosophical Society*, 124, 323–338. Reprinted in Ref. (Wheeler, and Zurek, 1983, 152–167).

Turing, A. M. (1936–7 and 1937). On computable numbers, with an application to the Entscheidungsproblem. *Proceedings of the London Mathematical Society, Series 2, 42, 43*, 230–265, 544–546. Reprinted in Ref. Davis (1965).

Turing, A. M. (1968). Intelligent machinery. In C. R. Evans, and A. D. J. Robertson (Eds.), *Cybernetics: Key Papers* (pp. 27–52). London: Butterworths.

Vaidman, L. (2007). Counterfactuals in quantum mechanics. In D. Greenberger, K. Hentschel, and F. Weinert (Eds.), *Compendium of Quantum Physics* (pp. 132–136). Berlin, Heidelberg: Springer.

von Neumann, J. (1951). Various techniques used in connection with random digits. *National Bureau of Standards Applied Math Series*, 12, 36–38. Reprinted in John von Neumann, Collected Works, (Vol. V), A. H. Traub, editor, MacMillan, New York, 1963, pp. 768–770.

von Neumann, J. (1966). *Theory of Self-Reproducing Automata*. Urbana: University of Illinois Press. A. W. Burks, editor.

Wagon, S. (1986). *The Banach-Tarski Paradox*. Cambridge: Cambridge University Press.

Wang, P. S. (1974). The undecidability of the existence of zeros of real elementary functions. *Journal of the Association for Computing Machinery (JACM)*, 21, 586–589.

Wang, P. X., Long, G. L., and Li, Y. S. (2006). Scheme for a quantum random number generator. *Journal of Applied Physics*, 100(5), 056107.

Wang, Q. D. (1991). The global solution of the n-body problem. *Celestial Mechanics*, 50, 73–88.

Wang, Q. D. (2001). Power series solutions and integral manifold of the n-body problem. *Regular, and Chaotic Dynamics*, 6(4), 433–442.

Weierstrass, C., Hermite, C., and Mittag-Leffler, G. (1885). Mittheilung, einen von König Oscar II gestifteten mathematischen Preis betreffend. Communication sur un prix de mathématiques fondé par le roi Oscar II. *Acta Mathematica*, 7(1), I–VI.

Weihrauch, K. (2000). *Computable Analysis. An Introduction*. Berlin, Heidelberg: Springer.

Weihs, G., Jennewein, T., Simon, C., Weinfurter, H., and Zeilinger, A. (1998a). Violation of Bell's inequality under strict Einstein locality conditions. *Physical Review Letters*, 81, 5039–5043.

Weihs, G., Jennewein, T., Simon, C., Weinfurter, H., and Zeilinger, A. (1998b). Violation of Bell's inequality under strict Einstein locality conditions. *Physical Review Letters*, 81, 5039–5043.

Wheeler, J. A., and Zurek, W. H. (1983). *Quantum Theory and Measurement*. Princeton, NJ: Princeton University Press.

Wigner, E. P. (1960). The unreasonable effectiveness of mathematics in the natural sciences. Richard Courant Lecture delivered at New York University, May 11, 1959. *Communications on Pure and Applied Mathematics*, 13, 1–14.

Wigner, E. P. (1961). Remarks on the mind-body question. In I. J. Good (Ed.), *The Scientist Speculates* (pp. 284–302). London and New York: Heinemann and Basic Books. Reprinted in Ref. (Wheeler, and Zurek, 1983, 168–181).

Wolfram, S. (1984). Computation theory of cellular automata. *Communications in Mathematical Physics*, 96(1), 15–57.

Wolfram, S. (1985). Undecidability and intractability in theoretical physics. *Physical Review Letters*, 54(8), 735–738.

Wolfram, S. (2002). *A New Kind of Science*. Champaign, IL: Wolfram Media, Inc.

Wolpert, D. H. (2001). Computational capabilities of physical systems. *Physical Review E*, 65(1), 016128.

Wright, R. (1978). The state of the pentagon. A nonclassical example. In A. R. Marlow (Ed.), *Mathematical Foundations of Quantum Theory* (pp. 255–274). New York: Academic Press.

Wright, R. (1990). Generalized urn models. *Foundations of Physics*, 20(7), 881–903.

Zajonc, A. G., Wang, L. J., Zou, X. Y., and Mandel, L. (1991). Quantum eraser. *Nature*, 353, 507–508.

Zeilinger, A. (1981). General properties of lossless beam splitters in interferometry. *American Journal of Physics*, 49(9), 882–883.

Zeilinger, A. (1999). A foundational principle for quantum mechanics. *Foundations of Physics*, 29(4), 631–643.

Zeilinger, A. (2005). The message of the quantum. *Nature*, 438, 743.

Zierler, N., and Schlessinger, M. (1965). Boolean embeddings of orthomodular sets and quantum logic. *Duke Mathematical Journal*, 32, 251–262.

Zurek, W. H. (2003). Decoherence, einselection, and the quantum origins of the classical. *Reviews of Modern Physics*, 75(3), 715–775.

Zuse, K. (1970). *Calculating Space. MIT Technical Translation AZT-70-164-GEMIT*. Cambridge, MA: MIT (Proj. MAC).

A Wider Vision: The Interdisciplinary, Philosophical, and Theological Implications of Gödel's Work

On the Unknowables

CHAPTER 11

Gödel and Physics

John D. Barrow

In this chapter, I introduce some early considerations of physical and mathematical impossibility as preludes to Gödel's incompleteness theorems. I consider some informal aspects of these theorems and their underlying assumptions and discuss some of the responses to these theorems by those seeking to draw conclusions from them about the completability of theories of physics. Also, I argue that there is no reason for us to expect Gödel incompleteness to handicap the search for a description of the laws of nature, but we do expect it to limit what we can predict about the outcomes of those laws, and I have provided some examples. I then discuss the "Gödel universe" – a solution to Einstein's equations describing a rotating universe where time travel is possible, as demonstrated by Gödel in 1949 – and the role it played in exposing the full spectrum of possibilities that a global understanding of space-time would reveal. Finally, I show how recent studies of so-called supertasks – doing an infinite number of things in a finite amount of time – have shown how global space-time structure determines the ultimate capability of computational devices within them.

11.1 Some Historical Background

11.1.1 Physical Impossibilities

Scientific and philosophical consideration of physical impossibilities has a long history (Barrow, 1998). The Aristotelian worldview outlawed the possibility that physical infinities or local physical vacua could be created or observed (Barrow, 2000). During the Middle Ages, physicists devised ingenious thought experiments to try to imagine how nature could be tricked into allowing an instantaneous vacuum to form and then argued about how this possibility was stopped from occurring by natural processes – or, if that failed, by the invocation of a Cosmic Censor to prevent its appearance (Grant, 1981).[1]

---

[1] Earlier examples of some of these challenging thought experiments can be found in Lucretius, *De Rerum Natura*, Book 1.

Chemistry had its own ongoing alchemical debate about the possibility or impossibility of making gold from base metals, and engineering retained an enduring attachment to the quest for a perpetual motion machine that only fully abated when the consequences of the laws of thermodynamics were systematically understood during the nineteenth century. Subtle examples, like Maxwell's sorting demon, still remained, until they were eventually fully exorcised by the application of the modern thermodynamic theory of computation in 1961 (Leff and Rex, 1990).[2]

11.1.2 Mathematical Impossibilities

Mathematicians also occasionally considered the question of impossibility in the context of several fundamental problems of arithmetic, geometry, and algebra. Supposedly, in about 550 B.C., the Pythagoreans first encountered the so-called irrationality of numbers such as $\sqrt{2}$, which cannot be expressed as the ratio of two integers (*irrational* originally meaning simply "not a ratio" rather than "beyond reason," as it does today) (Guthrie, 1987). Legend has it that this discovery was such a scandal that the discoverer, Hippasos, was drowned by the members of the Pythagorean brotherhood for his trouble. This gives us the first glimpse of operations and questions that have no answers given a particular set of rules. In the first quarter of the nineteenth century, the problem of finding an explicit form for the solution of a general quintic algebraic equation in terms of its coefficients was proved to have no solution involving ordinary arithmetic operations and radicals by the young Norwegian mathematician Henrik Abel (Pesic, 2003). Unlike the case of quadratic, cubic, or quartic equations, the general quintic cannot be solved by any exact formula. Just a few years later, in 1837, rigorous proofs were given that an angle of 60 degrees could not be trisected by just the use of a straight edge and pair of compasses. These examples revealed for the first time, to those who looked at them in the right way, some hints as to the limitations of particular axiomatic systems.

In light of the ongoing impact of Gödel's work on speculations about the limitations of the human mind, it is interesting to reflect briefly on the sociological and psychological effects of some of these early results. The existence of irrational numbers was of the deepest concern to the Pythagoreans; however, as far as we can judge, no deep philosophical questions about the limitations of mathematical reasoning were raised by the demonstration that the quintic could not be solved. Yet something changed. Previously, many things had been thought impossible that could not be so proved, despite many efforts to do so, but now there were proofs that something could not be done.

11.1.3 Axiomatics

The development of understanding of what constructions and proofs could be carried out by limited means, such as ruler and compass construction or the use of only

---

[2] This work contains an overview with reprints of all the crucial papers.

arithmetic operations and radicals, showed that axioms mattered. The power and scope of a system of axioms determined what its allowed rules of reasoning could encompass.

Until the nineteenth century, the archetypal axiomatic system was that of Euclidean geometry, but it is important to appreciate that this system was not then viewed, as it is today, as just one among many axiomatic possibilities. Euclidean geometry was how the world really *was*. It was part of the absolute truth about the universe. This gave it a special status, and its constructions and elucidation, largely unchanged for more than two thousand years, provided a style that was mimicked by many works of philosophy and theology. The widespread belief in its absolute truth provided an important cornerstone for the beliefs of theologians and philosophers that human reason could grasp something of the ultimate nature of things. If challenged that this was beyond the power of our minds to penetrate, the ancients could always point to Euclidean geometry as a concrete example of how and where this type of insight into the ultimate nature of things had already been possible. As a result, the discoveries by Bolyai, Lobachevskii, Gauss, and Riemann that other geometries existed, but in which Euclid's parallel postulate was not included, had a major impact outside mathematics (Richards, 1979). The existence of other logically consistent geometries meant that Euclid's geometry was not *the* truth: it was simply a model for some parts of the truth. As a result, new forms of relativism sprang up, nourished by the demonstration that even Euclid's ancient foundational system was merely one of many possible geometries – and indeed, one of these alternatives was a far more appropriate model for describing the geometry of the earth's surface than Euclid's. Curious books appeared about non-Euclidean models of government and economics. *Non-Euclidean* became a byword for new and relative truth, the very latest intellectual fashion (Barrow, 1992, 8–15). Later, new logics would be created, as well, by changing the axioms of the classical logical system that Aristotle had defined.

Out of these studies emerged a deeper appreciation of the need for axioms to be consistently defined and clearly stated. The traditional realist view of mathematics as a description of how the world was had to be superseded by a more sophisticated view that recognized mathematics to be an unlimited system of patterns that arise from the infinite number of possible axiomatic systems that can be defined. Some of these patterns appear to be made use of in nature, but most are not. Mathematical systems like Euclidean geometry had been assumed to be part of the absolute truth about the world and uniquely related to reality, but the development of non-Euclidean geometries and nonstandard logics meant that mathematical existence now meant nothing more than logical self-consistency (i.e., it must not be possible to prove that $0 = 1$). It no longer had any necessary requirement of physical existence.

11.1.4 Hilbert's Program

The careful study of axiomatic systems revealed that even Euclid's beautiful development of plane geometry made use of unstated axioms. In 1882, Moritz Pasch gave a very simple example of an intuitively "obvious" property of points and lines that could not be proved from Euclid's classical axioms. If the points A, B, C, and D lie on a straight line such that B lies between A and C and C lies between B and D, then it is not possible to prove that B lies between A and D. The picture of the setup made this appear inevitable, but this is not a substitute for a proof:

$$A \qquad\quad B \qquad\quad C \qquad\quad D$$

Pasch wanted to distinguish between the logical consequences of the axioms of geometry and those properties that we just assumed were intuitively true. For him, mathematical argumentation should not depend on any physical interpretation or visualization of the quantities involved. He was concerned that axiomatic systems should be complete and has been described as the "father of rigor in geometry" by Freudenthal (1962).

David Hilbert, the greatest mathematician of the day, felt the influence of Pasch's writings both directly and through their effects on Peano's work (Kennedy, 1972) from 1882 to 1899, and he began his systematic program in 1899 to place mathematics on a formal axiomatic footing (Gray, 1986; Toepell, 1986).[3] This was a new emphasis, conveyed by Hilbert's remark that in mathematics, "one must be able to say . . . – instead of points, straight lines and planes – tables, chairs, and beer mugs" (Reid, 1970, 57). He believed that it would be possible to determine the axioms underlying each part of mathematics (and hence of the whole), demonstrate that these axioms are self-consistent, and then show that the resulting system of statements and deductions formed from these axioms is both complete and decidable. More precisely, a system is *consistent* if we cannot prove that a statement, S, and its negation, $\sim$S, are both true theorems. It is *complete* if, for every statement S we can form in its language, either S or its negation $\sim$S is a true theorem. It is *decidable* if, for every statement S that can be formed in its language, we can prove whether S is true or false. Thus if a system is decidable, it must be complete.

Hilbert's formalistic vision of mathematics was of a tight web of deductions spreading out with impeccable logical connections from the defining axioms. Indeed, mathematics was *defined* to be the collection of all those deductions. Hilbert set out to complete this formalization of mathematics with the help of others and believed that it would ultimately be possible to extend the scope of mathematics to include sciences, such as physics (Corry, 1997), that were built on applied mathematics. He began with Euclidean geometry and succeeded in placing it on a rigorous axiomatic basis. His program then imagined strengthening the system by adding additional axioms, showing at each step that consistency and decidability remained, until eventually, the system had become large enough to encompass the whole of arithmetic.

Hilbert's program began confidently, and he believed that it would just be a matter of time before all of mathematics was corralled within its formalistic web. Alas, the world was soon turned on its head by the young Kurt Gödel. Gödel had completed one of the early steps in Hilbert's program as part of his doctoral thesis by proving the consistency and completeness of first-order logic (later, Alonzo Church and Alan Turing would show that it was not decidable). However, the next steps that he took have ensured his fame as the greatest logician of modern times. Far from extending Hilbert's program to achieve its key objective – a proof of the completeness of arithmetic – Gödel proved that any system rich enough to contain arithmetic must be incomplete and undecidable. This

---

[3] Gray (1986) contains the text of Hilbert's address to the International Congress of Mathematicians in 1900 on pp. 240–82. See also Toepell (1986).

Table 11.1. *Summary of the Results Established about the Consistency, Completeness, and Decidability of Simple Logical Systems*

Theory	Is It Consistent?	Is It Complete?	Is It Decidable?
Propositional calculus	Yes	Yes	Yes
Euclidean geometry	Yes	Yes	Yes
First-order logic	Yes	Yes	No
Arithmetic (+, −) only	Yes	Yes	Yes
Arithmetic in full (+, −, ×, ÷)	??	No	No

took almost everyone by surprise, including John von Neumann, who was present at the conference in Königsberg (Hilbert's home town) on September 7, 1930, when Gödel briefly communicated his results – von Neumann quickly appreciated them, and even extended them (only to find that Gödel had already made the extension in a separate paper), realizing that Gödel had effectively killed Hilbert's program with one stroke. Also surprised was Paul Finsler, who had tried, unsuccessfully, to convince Gödel that he had discovered these results before Gödel did. Table 11.1 contains a summary of the results established about the consistency, completeness, and decidability of simple logical systems.

11.2 Some Mathematical Jujitsu

11.2.1 The Optimists and the Pessimists

Gödel's monumental demonstration – that systems of mathematics have limits – gradually infiltrated the way in which philosophers and scientists viewed the world and our quest to understand it. Some commentators claimed that it showed that all human investigations of the universe must be limited. Science is based on mathematics; mathematics cannot discover all truths; therefore science cannot discover all truths. One of Gödel's contemporaries, Hermann Weyl (1946), described Gödel's discovery as exercising "a constant drain on the enthusiasm and determination with which I pursued my research work." He believed that this underlying pessimism, so different from the rallying cry that Hilbert had issued to mathematicians in 1900, was shared "by other mathematicians who are not indifferent to what their scientific endeavors mean in the context of man's whole caring and knowing, suffering and creative existence in the world." Turning to more recent times, one writer on theology and science with a traditional Roman Catholic stance, Stanley Jaki (1980, 49), believes that Gödel's theorem prevents us from gaining an understanding of the cosmos as a necessary truth:

> Clearly then no scientific cosmology, which of necessity must be highly mathematical, can have its proof of consistency within itself as far as mathematics goes. In the absence of such consistency, all mathematical models, all theories of elementary particles, including the theory of quarks and gluons . . . fall inherently short of being that theory which shows in virtue of its a priori truth that the world can only be what it is and nothing else. This is true even if the theory happened to account with perfect accuracy for all phenomena of the physical world known at a particular time.

It constitutes a fundamental barrier to understanding of the universe, for it seems that on the strength of Gödel's theorem that the ultimate foundations of the bold symbolic constructions of mathematical physics will remain embedded forever in that deeper level of thinking characterized both by the wisdom and by the haziness of analogies and intuitions. For the speculative physicist this implies that there are limits to the precision of certainty, that even in the pure thinking of theoretical physics there is a boundary. . . . An integral part of this boundary is the scientist himself, as a thinker. (Jaki, 1966, 129)

Intriguingly, and just to show the important role human psychology plays in assessing the significance of limits, other scientists, such as Freeman Dyson, acknowledge that Gödel places limits on our ability to discover the truths of mathematics and science but interprets this as ensuring that science will go on forever. Dyson (2004) sees the incompleteness theorem as an insurance policy against the scientific enterprise, which he admires so much, coming to a self-satisfied end, for

> Gödel proved that the world of pure mathematics is inexhaustible; no finite set of axioms and rules of inference can ever encompass the whole of mathematics; given any set of axioms, we can find meaningful mathematical questions which the axioms leave unanswered. I hope that an analogous situation exists in the physical world. If my view of the future is correct, it means that the world of physics and astronomy is also inexhaustible; no matter how far we go into the future, there will always be new things happening, new information coming in, new worlds to explore, a constantly expanding domain of life, consciousness, and memory.

Thus we see epitomized the optimistic and the pessimistic responses to Gödel. The optimists, such as Dyson, see Gödel's result as a guarantor of the never-ending character of human investigation. They see scientific research as part of an essential part of the human spirit that, if it were completed, would have a disastrous demotivating effect on us – just as it did on Weyl. The pessimists, such as Jaki (1966), Lucas (1970), and Penrose (1989),[4] by contrast, interpret Gödel as establishing that the human mind cannot know all (maybe not even most) of the secrets of nature.

Gödel's (1964 [1947]) own view was as unexpected as ever. He thought that intuition, by which we can see truths of mathematics and science, was a tool that would one day be valued just as formally and reverently as logic itself:

> I don't see any reason why we should have less confidence in this kind of perception, i.e., in mathematical intuition, than in sense perception, which induces us to build up physical theories and to expect that future sense perceptions will agree with them and, moreover, to believe that a question not decidable now has meaning and may be decided in the future.

However, it is easy to use Gödel's theorem in ways that play fast and loose with the underlying assumptions of his theorem. Many speculative applications can be found spanning the fields of philosophy, theology, and computing, and they have been examined in a lucid, critical fashion by the late Torkel Franzén (2005).

Gödel was not so minded as to draw any strong conclusions for physics from his incompleteness theorems. He made no connections with the uncertainty principle of quantum mechanics, which was advertised as another great deduction that limited

---

[4] Also see Chapter 16 of this volume.

our ability to know and which was discovered by Heisenberg just a few years before Gödel made his discovery. In fact, Gödel was rather hostile to any consideration of quantum mechanics at all. Those who, like Gödel, worked at the Institute for Advanced Study (no one really worked *with* Gödel) believed that this was a result of his frequent discussions with Einstein, who, in the words of John Wheeler (who knew them both), "brainwashed Gödel" into disbelieving quantum mechanics and the uncertainty principle. Greg Chaitin[5] records this account of Wheeler's attempt to draw Gödel out on the question of whether there is a connection between Gödel incompleteness and Heisenberg's uncertainty principle:

> Well, one day I was at the Institute for Advanced Study, and I went to Gödel's office, and there was Gödel. It was winter and Gödel had an electric heater and had his legs wrapped in a blanket. I said "Professor Gödel, what connection do you see between your incompleteness theorem and Heisenberg's uncertainty principle?" And Gödel got angry and threw me out of his office!

The claim that mathematics contains unprovable statements and that therefore physics – which is based on mathematics – will not be able to discover everything that is true has been around for a long time. More sophisticated versions of it have been constructed that exploit the possibility that uncomputable mathematical operations would be required to make predictions about observable quantities. From this vantage point, Stephen Wolfram (1994) has conjectured:

> One may speculate that undecidability is common in all but the most trivial physical theories. Even simply formulated problems in theoretical physics may be found to be provably insoluble.

Indeed, it is known that undecidability is the rule rather than the exception among the truths of arithmetic (see Calude, 1994; Calude et al., 1994; Svozil, 1996, 112).

11.2.2 Drawing the Line between Completeness and Incompleteness

With these worries in mind, let us look a little more closely at what Gödel's result might have to say about physics. The situation is not so clear-cut as some commentators would often have us believe. It is useful to lay out the precise assumptions that underlie Gödel's deduction of incompleteness.

Gödel's theorem says that if a formal system is

1. finitely specified
2. large enough to include arithmetic
3. consistent

then it is *incomplete*. Condition 1 means that there is not an uncomputable infinity of axioms. We could not, for instance, choose our system to consist of all the true statements about arithmetic because this collection cannot be finitely listed in the required sense. Condition 2 means that the formal system includes all the symbols and axioms used in arithmetic. The symbols are 0 (zero), S (successor of), $+$, $\times$, and $=$.

---

[5] Quoted in Bernstein (1991, 140–41); see also Svozil (1993) and Barrow (1998, chap. 8).

Hence the number 2 is the successor of the successor of 0, written as the *term* SS0, and
2 and $+ 2 = 4$ is expressed as $SS0 + SS0 = SSSS0$.

The structure of arithmetic plays a central role in the proof of Gödel's theorem.
Special properties of numbers, like their factorizations and the fact that any number
can be factored in only one way as the product of prime divisors (e.g., $130 = 2 \times 5
\times 13$), were used by Gödel to establish a crucial correspondence between statements
of mathematics and statements about mathematics. Thereby, linguistic paradoxes such
as that of the liar could be embedded, like Trojan horses, within the structure of
mathematics itself. Only logical systems that are rich enough to include arithmetic
allow this incestuous encoding of statements about themselves to be made within their
own language.

Again, it is instructive to see how these requirements might fail to be met. If we
picked a theory that consisted of references to (and relations between) only the first
ten numbers (0, 1, 2, 3, 4, 5, 6, 7, 8, 9) with arithmetic modulo 10, then condition
2 fails, and such a miniarithmetic is complete. Arithmetic makes statements about
individual numbers or terms (like SS0). If a system does not have individual terms
like this but, like Euclidean geometry, only makes statements about a continuum of
points, circles, and lines, in general, then, it cannot satisfy condition 2. Accordingly, as
Alfred Tarski first showed, Euclidean geometry is complete. There is nothing magical
about the flat, Euclidean nature of the geometry, either: the non-Euclidean geometries
on curved surfaces are also complete. Completeness can be long-winded, though. A
statement of geometry involving n symbols can take up to exp [exp [n]] computational
steps to have its truth or falsity determined (Harel, 2000). For just $n = 10$, this number
amounts to a staggering 9.44×10^{9565}; for comparison, there have only been about
10^{27} nanoseconds since the apparent beginning of the universe's expansion history.
Similarly, if we had a logical theory dealing with numbers that used only the concept
of "greater than" without referring to any specific numbers, then it would be complete:
we can determine the truth or falsity of any statement about real numbers involving
just the "greater than" relationship.

Another example of a system that is smaller than arithmetic is arithmetic without the
multiplication, $\times$, operation. This is called "Presburger arithmetic" (Presburger, 1929;
Cooper, 1972; the full arithmetic is called "Peano arithmetic," after the mathematician
Giuseppe Peano, who first expressed it axiomatically in 1889). At first, this sounds
strange – in our everyday encounters with multiplication, it is nothing more than a
shorthand way of doing addition (e.g., $2 + 2 + 2 + 2 + 2 + 2 = 2 \times 6$) – but in the
full logical system of arithmetic, in the presence of logical quantifiers such as "there
exists" or "for any," multiplication permits constructions that are not merely equivalent
to a succession of additions.

Presburger arithmetic is complete: all statements about the addition of natural numbers can be proved or disproved; all truths can be reached from the axioms (Fischer
and Rabin, 1974).[6] Similarly, if we create another truncated version of arithmetic that

---

[6] The decision procedure is in general double-exponentially long, though; that is, the computational time required
to carry out N operations grows as $(2^N)^N$. Presburger arithmetic allows us to talk about positive integers and
variables whose values are positive integers. If we enlarge the system by permitting the concept of sets of
integers to be used, then the situation becomes almost unimaginably intractable. It has been shown that this

does not have addition but retains multiplication, this is also complete. It is only when addition and multiplication are simultaneously present that incompleteness emerges. Extending the system further by adding extra operations, such as exponentiation, to the repertoire of basic operations makes no difference. Incompleteness remains, but no intrinsically new form of it is found. Arithmetic is the watershed in complexity.

The use of Gödel to place limits on what a mathematical theory of physics (or anything else) can ultimately tell us seems a fairly straightforward consequence. However, as one looks more carefully into the question, things are not quite so simple. Suppose, for the moment, that all the conditions required for Gödel's theorem to hold are in place. What would incompleteness look like in practice? We are familiar with the situation of having a physical theory that makes accurate predictions about a wide range of observed phenomena: we might call it the "standard model." One day, we may be surprised by an observation about which it has nothing to say. It cannot be accommodated within its framework. Examples are provided by some so-called grand unified theories in particle physics. Some early editions of these theories had the property that all neutrinos must have zero mass. Now, if a neutrino is observed to have a nonzero mass (as experiments have now confirmed), then we know that the new situation cannot be accommodated within our original theory. What do we do? We have encountered a certain sort of incompleteness, but we respond to it by extending or modifying the theory to include the new possibilities. Thus, in practice, incompleteness looks very much like inadequacy in a theory. It would become more like Gödel incompleteness if we could find no extension of the theory that could predict the new observed fact.

An interesting example of an analogous dilemma is provided by the history of mathematics. During the sixteenth century, mathematicians started to explore what happened when they added together infinite lists of numbers. If the quantities in the list get larger, then the sum will diverge; that is, as the number of terms approaches infinity, so does the sum. An example is the sum

$$1 + 2 + 3 + 4 + 5 + \cdots = \text{infinity.}$$

However, if the individual terms get smaller and smaller sufficiently rapidly,[7] then the sum of an infinite number of terms can get closer and closer to a finite limiting value, which we shall call the sum of the series, for example,

$$1 + 1/9 + 1/25 + 1/36 + 1/49 + \cdots = \pi^2/8 = 1.2337005\ldots.$$

This left mathematicians to worry about a most peculiar type of unending sum:

$$1 - 1 + 1 - 1 + 1 - 1 + 1 - \cdots = ?????.$$

If you divide up the series into pairs of terms, it looks like $(1 - 1) + (1 - 1) + \cdots$, and so on. This is just $0 + 0 + 0 + \cdots = 0$, and the sum is zero. However, think of the series as $1 - \{1 - 1 + 1 - 1 + 1 - \cdots,\}$, and it looks like $1 - \{0\} = 1$. We seem to have proved that $0 = 1$.

---

system does not admit even a K-fold exponential algorithm for any finite K. The decision problem is said to be nonelementary in such situations. The intractability is unlimited; see Fischer and Rabin (1974).

[7] That the terms in the sum get progressively smaller is a necessary, but not a sufficient, condition for an infinite sum to be finite. For example, the sum $1 + 1/2 + 1/3 + 1/4 + 1/5 + \cdots$ is infinite.

Mathematicians had a variety of choices when they were faced with ambiguous sums like this. They could reject infinities in mathematics and deal only with finite sums of numbers, or, as Cauchy showed in the early nineteenth century, the sum of a series like the last one could be defined by specifying more closely what is meant by its sum. The limiting value of the sum must be specified together with the procedure used to calculate it. The contradiction $0 = 1$ arises only when one fails to specify the procedure used to work out the sum. In both cases, it is different, and so the two answers are not the same because they arise in different axiomatic systems. Thus here we see a simple example of how a limit is sidestepped by enlarging the concept that seems to create limitations. Divergent series can be dealt with consistently so long as the concept of a sum for a series is suitably extended (Rosen, 1996).

Another possible way of evading Gödel's theorem is if the physical world only makes use of the decidable part of mathematics. We know that mathematics is an infinite sea of possible structures. Only some of those structures and patterns appear to find existence and application in the physical world. It may be that they are all from the subset of decidable truths. Things may be even better protected than that: perhaps only computable patterns are instantiated in physical reality?

It is also possible that the conditions required to prove Gödel incompleteness do not apply to physical theories. Condition 1 requires the axioms of the theory to be listable. It might be that the laws of physics are not listable in this predictable sense. This would be a radical departure from the situation that we think exists, where the number of fundamental laws is believed to be not just listable but finite (and very small). However, it is always possible that we are just scratching the surface of a bottomless tower of laws, only the top of which has significant effect on our experience. If there were an unlistable infinity of physical laws, though, we would face a more formidable problem than incompleteness.

An equally interesting issue is finiteness. It may be that the universe of physical possibilities is finite, although astronomically large (Barrow, 2005).[8] However, no matter how large the number of primitive quantities to which the laws refer, so long as they are finite, the resulting system of interrelationships will be complete. I should stress that although we habitually assume that there is a continuum of points of space and time, this is just an assumption that is very convenient for the use of simple mathematics. We have no deep reason to believe that space and time are continuous, rather than discrete, at their most fundamental microscopic levels; in fact, some theories of quantum gravity assume that they are not. Quantum theory has introduced discreteness and finiteness in a number of places where once we believed in a continuum of possibilities. Curiously, if we give up this continuity, there is not necessarily another point between any two sufficiently close points you care to choose. Space-time structure becomes infinitely more complicated because continuous functions can be defined by their values on the rationals. Many more things can happen. This question of finiteness might also be bound up with the question of whether the universe is finite in volume and whether the number of elementary particles (or whatever the most elementary entities might be) of nature are finite or infinite in number. Thus only a finite number of terms may exist

---

[8] This could be because it has positive spatial curvature or a compact spatial topology like that of the 3-torus, even though it has nonpositive curvature; see Barrow (2005, chap. 7).

to which the ultimate logical theory of the physical world applies. Hence it would be complete.

A further possibility with regard to the application of Gödel to the laws of physics is that condition 2 of the incompleteness theorem might not be met. How could this be? Although we seem to make wide use of arithmetic, and much larger mathematical structures, when we carry out scientific investigations of the laws of nature, this does not mean that the inner logic of the physical universe needs to employ such a large structure. It is undoubtedly convenient for us to use large mathematical structures together with concepts such as infinity, but this may be an anthropomorphism. The deep structure of the universe may be rooted in a much simpler logic than that of full arithmetic and hence be complete. All that this would require would be for the underlying structure to contain either addition or multiplication but not both. Recall that all the sums that you have ever done have used multiplication simply as a shorthand for addition. They would be possible in Presburger arithmetic as well. Alternatively, a basic structure of reality that made use of simple relationships of a geometrical variety, or that derived from "greater than" or "less than" relationships, or subtle combinations of them all, could also remain complete (Misner et al., 1973).[9] That Einstein's theory of general relativity replaces many physical notions, such as force and weight, with *geometrical* distortions in the fabric of space-time may well hold some clue about what is possible here.

There is a surprisingly rich range of possibilities for a basic representation of mathematical physics in terms of systems that might be decidable or undecidable. Tarski showed that unlike Peano's arithmetic of addition and multiplication of natural numbers, the first-order theory of real numbers under addition and multiplication is decidable. This is rather surprising and may give some hope that theories of physics based on the real or complex numbers will evade undecidability in general. Tarski also went on to show that many mathematical systems used in physics, such as lattice theory, projective geometry, and abelian group theory, are decidable, whereas others, notably nonabelian group theory, are not (Tarski et al., 1953). Little consideration seems to have been given to the consequences of these results on the development of ultimate theories of physics.

We should keep another important aspect of the situation in view: even if a logical system is complete, it always contains unprovable "truths." These are the axioms that are chosen to define the system, and after they are chosen, all the logical system can do is deduce conclusions from them. In simple logical systems, such as Peano arithmetic, the axioms seem reasonably obvious because we are thinking backward – formalizing something that we have been doing intuitively for thousands of years. When we look at a subject such as physics, we find parallels and differences. The axioms, or laws, of physics are the prime targets of physics research. They are by no means intuitively obvious because they govern regimes that can lie far outside our experience. The outcomes of those laws are unpredictable in certain circumstances because they involve

---

[9] John A. Wheeler has speculated about the ultimate structure of space-time being a form of "pregeometry" obeying a calculus of propositions restricted by Gödel incompleteness. We are proposing that this pregeometry might be simple enough to be complete; see Misner et al. (1973, 1211–12).

symmetry breakings. Trying to deduce the laws from the outcomes is not something that we can ever do uniquely and completely by means of a computer program.

Thus we detect a completely different emphasis in the study of formal systems and in physical science. In mathematics and logic, we start by defining a system of axioms and laws of deduction, then we might try to show that the system is complete or incomplete and deduce as many theorems as we can from the axioms. In science, we are not at liberty to pick any logical system of laws that we choose. We are trying to find the system of laws and axioms (assuming there is one – or more than one, perhaps) that will give rise to the outcomes we see. As I stressed earlier, it is always possible to find a system of laws that will give rise to any set of observed outcomes, but it is the very set of unprovable statements that the logicians and mathematicians ignore – the axioms and laws of deduction – that the scientist is most interested in discovering rather than simply assuming. The only hope of proceeding as the logicians do would be if, for some reason, there were only one possible set of axioms or laws of physics. So far, this does not seem likely[10]; even if it were, we would not be able to prove it.

11.3 Laws versus Outcomes

11.3.1 Symmetry Breaking

The structure of modern physics presents us with an important dichotomy. It is important to appreciate this division to understand the significance of Gödel incompleteness for physics. The fundamental laws of nature governing the weak, strong, electromagnetic, and gravitational forces are all local gauge theories derived from the maintenance of particular mathematical symmetries. As these forces become unified, the number of symmetries involved will be reduced until ultimately (perhaps) there is only one overarching symmetry dictating the form of the laws of nature – a "theory of everything," of which M theory is the current candidate. Thus the laws of nature are in a real sense simple and highly symmetrical. The ultimate symmetry that unites them must possess a number of properties to accommodate all the low-energy manifestations of the separate forces – the states that look like elementary particles with all their properties – and it must be big enough for them all to fit in.

There is no reason why Gödel incompleteness should hamper the search for this all-encompassing symmetry governing the *laws* of nature. This search is, at root, a search for a pattern, perhaps a group symmetry or some other mathematical prescription. It need not be complicated, and it probably has a particular mathematical property that makes it especially (or even uniquely) fitted for this purpose.

In reality, we never "see" laws of nature: they inhabit a Platonic realm; rather we witness their *outcomes*. This is an important distinction because the outcomes are quite different from the laws that govern them. They are asymmetrical and complicated and need possess none of the symmetries displayed by the laws. This is fortunate

---

[10] The situation in superstring theory is still very fluid. Many different, logically self-consistent superstring theories appear to exist, but there are strong indications that they may be different representations of a much smaller number (maybe even just one) theory.

because, were it not so, we could not exist. If the outcomes of the laws of nature possessed all the symmetries of the laws, then nothing could happen that did not respect them. No structures could be located at particular times and places, no directional asymmetries could exist, and nothing could be happening at any one moment. All would be unchanging and empty.

This dichotomy between laws and outcomes is what I would call the "secret of the universe." It is what enables a universe to be governed by a very small number (perhaps just one) of simple and symmetrical laws yet give rise to an unlimited number of highly complex, asymmetrical states – of which we are one variety (Barrow, 2007).

Thus, whereas we have no reason to worry about Gödel incompleteness frustrating the search for the mathematical descriptions of the laws of nature, we might well expect Gödel incompleteness to arise in our attempts to describe some of the complicated sequences of events that arise as outcomes of the laws of nature.

11.3.2 Undecidable Outcomes

Specific examples have been given of physical problems in which the outcomes of their underlying laws are undecidable. As one might expect from what we have just said, they do not involve an inability to determine something fundamental about the nature of the laws of physics or even the most elementary particles of matter; rather they involve an inability to perform some specific mathematical calculation, which inhibits our ability to determine the course of events in a well-defined physical problem. However, although the problem may be mathematically well defined, this does not mean that it is possible to create the precise conditions required for the undecidability to exist.

An interesting series of examples of this sort have been created by the Brazilian mathematicians Da Costa and Doria (1991a, 1991b). Responding to a challenge problem posed by the Russian mathematician Vladimir Arnold, they investigated whether it was possible to have a general mathematical criterion that would decide whether any equilibrium was stable. A stable equilibrium is a situation like a ball sitting in the bottom of a basin – displace it slightly, and it returns to the bottom; an unstable equilibrium is like a needle balanced vertically – displace it slightly, and it moves away from the vertical.[11] When the equilibrium is of a simple nature, this problem is very elementary; first-year science students learn about it. However, when the equilibrium exists in the face of more complicated couplings between the different competing influences, the problem soon becomes more complicated than the situation studied by science students.[12] So long as there are only a few competing influences, the stability of the equilibrium can still be decided by inspecting the equations that govern the situation. Arnold's challenge was to discover an algorithm that tells us whether this can always be done, no matter how many competing influences there are and no matter how complex their interrelationships. By *discover*, he meant to find a formula into which you can feed the equations that govern the equilibrium along

---

[11] Actually, other, more complicated possibilities are clustered around the dividing line between these two simple possibilities, and it is these that provide the indeterminacy of the problem in general.

[12] It may not be determined at linear order if eigenvalues are zero or are purely imaginary.

with your definition of stability and out of which will pop the answer "stable" or "unstable."

Da Costa and Doria discovered that no such algorithm can exist. Equilibria exist that are characterized by special solutions of mathematical equations whose stability is undecidable. For this undecidability to have an impact on problems of real interest in mathematical physics, the equilibria have to involve the interplay of very large numbers of different forces. Though such equilibria cannot be ruled out, they have not arisen yet in real physical problems. Da Costa and Doria went on to identify similar problems for which the answer to a simple question such as "will the orbit of a particle become chaotic?" is Gödel undecidable. They can be viewed as physically grounded examples of the theorems of Rice (1953) and Richardson (1968), which show, in a precise sense, that only trivial properties of computer programs are algorithmically decidable. Others have also tried to identify formally undecidable problems. Geroch and Hartle have discussed problems in quantum gravity that predict the values of potentially observable quantities as a sum of terms whose listing is known to be a Turing-uncomputable operation (Geroch and Hartle, 1986).[13] Pour-El and Richards (1979, 1981, 1982) showed that very simple differential equations widely used in physics, such as the wave equation, can have uncomputable outcomes when the initial data are not very smooth. This lack of smoothness gives rise to what mathematicians call an "ill-posed" problem. It is this feature that gives rise to the uncomputability. However, Traub and Wozniakowski (1991) have shown that every ill-posed problem is well posed on the average under rather general conditions. Wolfram (1982, 1985a, 1985b) gives examples of intractability and undecidability arising in condensed-matter physics and even believes that undecidability is typical in physical theories.

The study of Einstein's general theory of relativity also produces an undecidable problem if the mathematical quantities involved are unrestricted (Buchberger and Loos, 1983).[14] When one finds an exact solution of Einstein's equations, it is always necessary to discover whether it is just another, known solution that is written in a different form. Usually, one can investigate this by hand, but for complicated solutions, computers can help. For this purpose, we require computers programmed to carry out algebraic manipulations. A computer can check various quantities to discover whether a given solution is equivalent to one already sitting in its memory bank of known solutions. In practical cases encountered so far, this checking procedure comes up with a definite result after a small number of steps. However, in general, the comparison is an undecidable process equivalent to another famous undecidable problem of pure mathematics, the word problem of group theory first posed by Max Dehn (1911) and later shown to be undecidable (Boone, 1959).

The tentative conclusion we should draw from this discussion is that, just because physics makes use of mathematics, it is by no means required that Gödel places any limit

---

[13] The problem is that the calculation of a wave function for a cosmological quantity involves the sum of quantities evaluated on every four-dimensional compact manifold in turn. The listing of this collections of manifolds is uncomputable.

[14] If the metric functions are polynomials, then the problem is decidable but is computationally double-exponential. If the metric functions are allowed to be sufficiently smooth, then the problem becomes un-decidable; see Buchberger and Loos (1983). I am grateful to Malcolm MacCallum for supplying these details.

on the overall scope of physics to understand the laws of nature.[15] The mathematics of which nature makes use may be smaller and simpler than is needed for undecidability to rear its head.

11.4 Gödel and Space-Time Structure

11.4.1 Space-Time in a Spin

Although Kurt Gödel is famous among logicians for his incompleteness theorems, he is also famous among cosmologists, but for a quite different reason. In 1949, inspired by his many conversations with Einstein about the nature of time and Mach's principle, Gödel found a new and completely unsuspected type of solution to Einstein's equations of general relativity (Gödel, 1949; Hawking and Ellis, 1973; Barrow and Tsagas, 2004a).[16] Gödel's solution was a universe that rotates and permits time travel to occur into the past.

This was the first time that the possibility of time travel (into the past) had emerged in the context of a theory of physics. The idea of time travel first appeared in H. G. Wells' famous 1895 story *The Time Machine*, but it was widely suspected that backward time travel would in some way be in conflict with the laws of nature.[17] Gödel's universe showed that was not necessarily so: it could arise as a consequence of a theory obeying all the conservation laws of physics. Time travel into the future is a relatively uncontroversial matter and is just another way of describing the observed effects of time dilation in special relativity.

Gödel's universe is not the one in which we live. For one thing, Gödel's universe is not expanding; for another, there is no evidence that our universe is rotating – and if it is, then its rate of spin must be at least 10^5 times slower than its expansion rate because of the isotropy of the microwave background radiation (Barrow et al., 1985). Nonetheless, Gödel's universe was a key discovery in the study of space-time and gravitation. If time travel were possible, perhaps it could arise in other universes that are viable descriptions of our own?

However, the influence of Gödel's solution on the development of the subject was more indirect. It revealed for the first time the subtlety of the global structure of space-time, particularly when rotation is present. Previously, the cosmological models that were studied tended to be spatially homogeneous, with simple topologies and high degrees of global symmetry that ruled out or disguised global structure. Later, in 1965, Roger Penrose would apply powerful new methods of differential topology to this problem and prove the first singularity theorems in cosmology. The possibility of closed, timelike curves in space-time that Gödel had revealed meant that specific

---

[15] For a fuller discussion of the whole range of limits that may exist on our ability to understand the universe, see Barrow (1998), on which the discussion here is based.

[16] For details of the global structure of the Gödel universe, see Hawking and Ellis (1973); for a study of its stability and a discussion of its Newtonian counterpart, see Barrow and Tsagas (2004a).

[17] Forward time travel is routinely observed and is a simple consequence of special relativity. For example, in the case of the twin paradox, the twin who has undergone accelerations returns to find that he is younger than his twin: he has in effect time traveled into his twin's future.

vetoes had to be included in some of these theorems to exclude their presence; otherwise past incompleteness of geodesics could be avoided by periodically reappearing in the future. It was Gödel's universe that first showed how unusual space-times could be while remaining physically and factually consistent. Before its discovery, physicists and philosophers of science regarded time travel into the past as the necessary harbinger of factual contradictions. But Gödel's solution shows that there exist self-consistent histories that are periodic in space and time (Barrow, 2005). It continues to be studied as a key example of an intrinsically general-relativistic effect, and its full stability properties have been elucidated only recently (Barrow and Tsagas, 2004b). Some of its unusual properties are explained in Chapter 9 of this volume.

In recent years, Gödel's study of space-time structure and his work on the incompleteness of logical systems have been pulled together in a fascinating way. It has been shown that there is a link between the global structure of space-time and the sorts of computations that can be completed within them. This unexpected link arises from a strange old problem with a new name: is it possible to do an infinite number of things in a finite amount of time? The new name for such a remarkable old activity is a "supertask."

11.4.2 Supertasks

The ancients, beginning with Zeno, were challenged by the paradoxes of infinities on many fronts (Grünbaum, 1973a, 1973b, chap. 18), but what about philosophers today? What sorts of problems do they worry about? Live issues at the interface between science and philosophy are concerned with whether it is possible to build an "infinity machine" that can perform an infinite number of tasks in a finite amount of time. Of course, this simple question needs some clarification: what exactly is meant by *possible, tasks, number, infinite, finite*, and, by no means least, *time*? Classical physics appears to impose few physical limits on the functioning of infinity machines because the speed at which signals can travel or switches can move has no limit. Newton's laws allow an infinity machine. This can be seen by exploiting a discovery about Newtonian dynamics made in 1971 by the U.S. mathematician Jeff Xia (1992; Saari and Xia, 1989).[18] First, take four particles of equal mass and arrange them in two binary pairs orbiting with equal but oppositely directed spins in two separate parallel planes so that the overall angular momentum is zero. Now, introduce a fifth, much lighter particle that oscillates back and forth along a perpendicular line joining the mass centers of the two binary pairs. Xia showed that such a system of five particles will expand to infinite size in a *finite* time!

How does this happen? The little oscillating particle runs back and forth between the binary pairs, each time creating an unstable meeting of three bodies. The lighter particle then gets kicked back, and the binary pair recoils outward to conserve momentum. The lighter particle then travels across to the other binary, and the same ménage à trois is repeated there. This continues without end, accelerating the binary pairs apart so

---

[18] The connection between Xia's remarkable construction is pointed out in Barrow (2005, chap. 10). Note, however, that the initial conditions needed for Xia's specific example are special (a Cantor set) and will not arise in practice.

strongly that they become infinitely separated, while the lighter particle undergoes an infinite number of oscillations in the finite time before the system achieves infinite size.

Unfortunately (or perhaps fortunately), this behavior is not possible when relativity is taken into account. No information can be transmitted faster than the speed of light, and gravitational forces cannot become arbitrarily strong in Einstein's theory of motion and gravitation, nor can masses get arbitrarily close to each other and recoil – there is a limit to how close separation can get, after which an "event horizon" surface encloses the particles to form a black hole.[19] Their fate is then sealed – no such infinity machine could send information to the outside world.

However, this does not mean that all relativistic infinity machines are forbidden.[20] Indeed, the Einsteinian relativity of time that is a requirement of all observers, no matter what their motion, opens up some interesting new possibilities for completing infinite tasks in finite time. Could it be that one moving observer could see an infinite number of computations occurring, even though only a finite number had occurred according to someone else? Misner (1969) and Barrow and Tipler (1986, chap. 10) have shown that there are examples of entire universes in which an infinite number of oscillations occur on approach to singularities in space-time, but it is necessary for the entire universe to hit the singularity; in effect, the whole universe is the infinity machine. It still remains to ask whether a local infinity machine could exist and send us signals as a result of completing an infinite number of operations in a finite amount of time.

The famous motivating example of this sort of temporal relativity is the so-called twin paradox. Two identical twins are given different future careers. Tweedlehome stays at home, while Tweedleaway goes away on a space flight at a speed approaching that of light. When they are eventually reunited, relativity predicts that Tweedleaway will find Tweedlehome to be much older. The twins have experienced different careers in space and time because of the acceleration and deceleration that Tweedleaway underwent on his round trip.

So can we ever send a computer on a journey so extreme that it could accomplish an infinite number of operations by the time it returns to its stay-at-home owner? Itamar Pitowsky (1990) first argued that if Tweedleaway could accelerate his spaceship sufficiently strongly, then he could record a finite amount of the universe's history on his own clock, while his twin records an infinite amount of time on his clock. Does this, he wondered, permit the existence of a Platonist computer – one that could carry out an infinite number of operations along some trajectory through space and time and print out answers that we could see back home? Alas, there is a problem – for the receiver to stay in contact with the computer, he also has to accelerate dramatically to maintain the flow of information. Eventually, the gravitational forces become stupendous, and he is torn apart.

These problems notwithstanding, a checklist of properties has been compiled for universes that can allow an infinite number of tasks to be completed in finite time – supertasks. These are called Malament-Hogarth (MH) universes, after David Malament, a University of Chicago philosopher, and Mark Hogarth (1992), a former Cambridge University research student who, in 1992, investigated the conditions under which

---

[19] Once two masses, each of mass M, fall inside a radius $4GM/c^2$, an event horizon forms around them.

[20] We could be living inside a very large black hole along with the infinity machine.

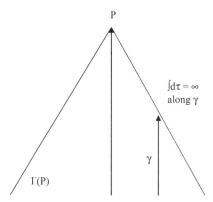

Figure 11.1. A Malament-Hogarth space-time with time mapped vertically and space (compressed to one dimension) mapped horizontally. We are located at P and our causal past, $I^-(P)$ consists of all the events that can influence us. There is a path in our past, γ, such that there is an infinite amount of its own time passing on approach to the space-time point where it intersects our past light cone, the edge of $I^-(P)$.

they were theoretically possible. Supertasks (Earman and Norton, 1993) offer the fascinating prospect of finding or creating conditions under which an infinite number of things can be seen to be accomplished in a finite amount of time.[21] This has all sorts of consequences for computer science and mathematics because it would remove the distinction between computable and uncomputable operations. It is something of a surprise that MH universes (see Figure 11.1) are self-consistent mathematical possibilities, but unfortunately, they have properties that suggest they are not realistic physical possibilities unless we embrace some disturbing notions such as the prospect of things happening without causes and travel backward through time. Figure 11.1 shows an MH universe.

The most serious by-product of being allowed to build an infinity machine is rather more alarming, though. Observers who stray into bad parts of these universes will find that being able to perform an infinite number of computations in a finite time also means that any amount of radiation, no matter how small, gets compressed to zero wavelength and amplified to infinite frequency and energy along the infinite computational trail. Thus any attempt to transmit the output from an infinite number of computations will zap the receiver and destroy him or her. So far, these dire problems seem to rule out the practicality of engineering a relativistic infinity machine in such a way that we could safely receive and store the information. However, the universes in which infinite tasks are possible in finite time include a type of space that plays a key role in the structure of the very superstring theories that looked so appealingly finite.

If you could see the output from an infinity machine that completes supertasks, then you have the possibility of deciding undecidable problems by a direct search through the infinite catalog of possibilities: Turing's uncomputable operations seem to become completable in a finite amount of our wristwatch time. Is this really possible? Remarkably, Hogarth (2004) showed that in some space-times, it was possible to decide

---

[21] See also Barrow (2005, chap. 10).

Gödel undecidable questions through a direct search by sending a computer along a certain space-time path, γ, so that it could print out and send you the answer to the question. Now, create a hierarchy of n space-time structures of ascending complexity such that the nth in the sequence allows a supertask to be completed that can check the truth of any arithmetical assertion made in the nth, but not in the (n + 1)st, quantifier arithmetic in Kleene's logical hierarchy, by which logicians calibrate the complexity of possible logical expressions. There is a neat one-to-one correspondence between the list of space-times and the complexity of the logical statements that they can decide. Subsequently, Etesi and Németi (2002) showed that some relations on natural numbers that are neither universal nor couniversal can be decided in Kerr space-times. Welch (2008) recently generalized these results to show that the computational capability of space-times could be raised beyond that of arithmetic to hyperarithmetics and showed that there is an upper bound on the computational ability in any space-time that is a universal constant defined by the space-time.

Thus, in conclusion, we find that Gödel's ideas are still provoking new research programs and unsuspected properties of the worlds of logical and physical reality. His incompleteness theorems should not be a drain on our enthusiasm to seek out and codify the laws of nature: there is no reason for them to limit the search for the fundamental symmetries of nature in any significant way. By contrast, though, in situations of sufficient complexity, we do expect to find that Gödel incompleteness places limits on our ability to use those laws to predict the future, carry out specific computations, or build algorithms: incompleteness besets the outcomes of very simple laws of nature. Finally, if we study universes, then Gödel's impact will always be felt as we try to reconcile the simple local geometry of space and time with the extraordinary possibilities that its exotic global structure allows. Space-time structure defines what can be proved in a universe.

Acknowledgments

I would like to thank Philip Welch for discussions and Christos Papadimitriou for many helpful suggestions.

References

Barrow, J. D. (1992). *Pi in the Sky.* Oxford: Oxford University Press.

———. (1998). *Impossibility.* Oxford: Oxford University Press.

———. (2000). *The Book of Nothing.* London: Cape.

———. (2005). *The Infinite Book.* London: Cape.

———. (2007). *New Theories of Everything.* Oxford: Oxford University Press.

Barrow, J. D., and Tipler, F. J. (1986). *The Anthropic Cosmological Principle.* Oxford: Oxford University Press.

Barrow, J. D., and Tsagas, C. G. (2004a). Dynamics and stability of the Gödel universe. *Classical and Quantum Gravity*, **21**, 1773–89.

———. (2004b). Gödel brane. *Physical Review D*, 69, 064007.

Barrow, J. D., Juszkiewicz, R., and Sonoda, D. H. (1985). Universal rotation: How large can it be? *Monthly Notices of the Royal Astronomical Society*, **213**, 917.

Bernstein, J. (1991). *Quantum Profiles*. New York: Basic Books.

Boone, W. (1959). The word problem. *Annals of Mathematics*, **70**, 207–65.

Buchberger, B., and Loos, R. (1983). Algebraic simplification. In *Computer Algebra: Symbolic and Algebraic Computation*, 2nd ed., ed. B. Buchberger, R. Loos, G. E. Collins et al., New York: Springer.

Calude, C. (1994). *Information and Randomness: An Algorithmic Perspective*. Berlin: Springer.

Calude, C., Jürgensen, H., and Zimand, M. (1994). Is independence an exception? *Applied Mathematics and Computation*, 66, 63.

Cooper, D. C. (1972). Theorem proving in arithmetic without multiplication. In *Machine Intelligence*, ed. B. Meltzer and D. Michie, pp. 91–100. Edinburgh: Edinburgh University Press.

Corry, L. (1997). David Hilbert and the axiomatization of physics (1894–1905). *Archive for History of Exact Sciences*, **51**, 83.

Da Costa, N. C., and Doria, F. (1991a). Undecidability and incompleteness in classical mechanics. *International Journal of Theoretical Physics*, **30**, 1041.

———. (1991b). Classical physics and Penrose's thesis. *Foundations of Physics Letters*, 4, 363.

Dehn, M. (1911). Fiber unendliche diskontinuierliche Gruppen. *Mathematische Annalen*, 71, 73.

Dyson, F. (2004). The world on a string. *New York Review of Books*, May 13.

Earman, J., and Norton, J. D. (1993). Forever is a day: Supertasks in Pitowsky and Malament-Hogarth spacetimes. *Philosophy of Science*, **60**, 22.

Etesi, G., and Németi, I. (2002). Non-Turing computability via Hogarth-Malament spacetimes. *International Journal of Theoretical Physics*, **41**, 341.

Fischer, M. J., and Rabin, M. O. (1974). Super-exponential complexity of Presburger arithmetic. *Proceedings of the Symposium in Applied Mathematics of the American Mathematical Society and the Society for Industrial and Applied Mathematics*, **7**, 27–41.

Franzén, T. (2005). *Gödel's Theorem: An Incomplete Guide to Its Use and Abuse*. Wellesley, MA: A K Peters.

Freudenthal, H. (1962). *The Main Trends in the Foundation of Geometry in the 19th Century*. Stanford, CA: Stanford University Press. [Also quoted in H. C. Kennedy (1972). *American Mathematical Monthly*, 79, 133.]

Geroch, R., and Hartle, J. (1986). Computability and physical theories. *Foundations of Physics*, **16**, 533.

Gödel, K. (1964 [1947]). What is Cantor's continuum problem? *American Mathematical Monthly*, **54**, 515–25. [Rev. version P. Benacerraf and H. Putnam, eds. (1984 [1964]). *Philosophy of Mathematics*. Englewood Cliffs, NJ: Prentice Hall, p. 483. Also in *Collected Works*, vol. 2 (1990), pp. 176–87 (1947 version); pp. 254–70 (1964 version).]

———. (1949). An example of a new type of cosmological solution of Einstein's field equations of gravitation. *Reviews of Modern Physics*, **21**, 447–50. [Also in *Collected Works*, vol. 2 (1990), pp. 190–98.]

Grant, E. (1981). *Much Ado about Nothing: Theories of Space and Vacuum from the Middle Ages to the Scientific Revolution*. Cambridge: Cambridge University Press.

Gray, J. (1986). *The Hilbert Challenge*. Oxford: Oxford University Press.

Grünbaum, A. (1973a). *Modern Science and Zeno's Paradoxes*. London: Allen and Unwin.

———. (1973b). *Philosophical Problems of Space and Time*. 2nd ed. Dordrecht, Netherlands: D. Reidel.

Guthrie, K., ed. (1987). *The Pythagorean Sourcebook and Library*. Grand Rapids, MI: Phares Press.

Harel, D. (2000). *Computers Ltd*. Oxford: Oxford University Press.

Hawking, S. W., and Ellis, G. F. R. (1973). *The Large Scale Structure of Space-Time*. Cambridge: Cambridge University Press.

Hogarth, M. L. (1992). Does general relativity allow an observer to view an eternity in a finite time? *Foundations of Physics Letters*, **5**, 173.

———. (2004). Deciding arithmetic using SAD computers. *British Journal of the Philosophy of Science*, **55**, 681.

Jaki, S. (1966). *The Relevance of Physics*. Chicago: University of Chicago Press.

———. (1980). *Cosmos and Creator*. Edinburgh: Scottish Academic Press.

Kennedy, H. C. (1972). *American Mathematical Monthly*, **79**, 133–36.

Leff, H., and Rex, A., eds. (1990). *Maxwell's Demon*. Princeton, NJ: Princeton University Press.

Lucas, J. R. (1961). Minds, machines, and Gödel. *Philosophy*, **36**, 112–27. [Repr. A. R. Anderson, ed. (1964). *Minds and Machines*. Englewood Cliffs, NJ: Prentice Hall.]

———. (1970). *The Freedom of the Will*. New York: Oxford University Press.

Misner, C. W. (1969). Absolute zero of time. *Physical Review*, **186**, 1328.

Misner, C., Thorne, K., and Wheeler, J. A. (1973). *Gravitation*. San Francisco: W. H. Freeman.

Pasch, M. (1882). *Vorlesungen über neuere Geometrie*. Leipzig, Germany: Teubner.

Penrose, R. (1989). *The Emperor's New Mind: Concerning Computers, Minds, and the Laws of Physics*. Oxford: Oxford University Press. [Repr. 1999 with new preface.]

Pesic, P. (2003). *Abel's Proof: An Essay on the Sources and Meaning of Mathematical Unsolvability*. Cambridge, MA: MIT Press. [This book also contains a new translation of Abel's 1824 discovery paper.]

Pitowsky, I. (1990). The physical Church thesis and physical computational complexity. *Iyyun*, **39**, 81.

Pour-El, M. B., and Richards, I. (1979). A computable ordinary differential equation which possesses no computable solution. *Annals of Mathematical Logic*, **17**, 61.

———. (1981). The wave equation with computable initial data such that its unique solution is not computable. *Advances in Mathematics*, **39**, 215.

———. (1982). Non-computability in models of physical phenomena. *International Journal of Theoretical Physics*, **21**, 553–55.

Presburger, M. (1929). *Comptes Rendus du Congrès de Mathématiciens des Pays Slaves*. Warsaw.

Reid, C. (1970). *Hilbert*. New York: Springer.

Rice, H. D. (1953). Classes of recursively enumerable sets and their decision problems. *Transactions of the American Mathematical Society*, **74**, 358.

Richards, J. (1979). The reception of a mathematical theory: Non-Euclidean geometry in England 1868–1883. In *Natural Order: Historical Studies of Scientific Culture*, ed. B. Barnes and S. Shapin, pp. XX–XX. Beverly Hills, CA: Sage.

Richardson, D. (1968). Some unsolvable problems involving elementary functions. *Symbolic Logic*, **33**, 514.

Rosen, R. (1996). On the limitations of scientific knowledge. In *Boundaries and Barriers: On the Limits to Scientific Knowledge*, ed. J. L. Casti and A. Karlqvist, pp. XX–XX. Reading, MA: Addison-Wesley.

Saari, D. G., and Xia, Z. (1989). Oscillatory and superhyperbolic solutions in Newtonian systems. *Differential Equations*, **82**, 342.

Svozil, K. (1993). *Randomness and Undecidability in Physics*. Singapore: World Scientific.

———. (1996). Undecidability everywhere. In *Boundaries and Barriers: On the Limits to Scientific Knowledge*, ed. J. L. Casti and A. Karlqvist, pp. XX–XX. Reading, MA: Addison-Wesley.

Tarski, A., Mostowski, A., and Robinson, R. M. (1953). *Undecidable Theories*. Amsterdam, Netherlands: North-Holland.

Toepell, M. (1986). On the origins of David Hilbert's *Grundlagen der Geometrie. Archive for History of Exact Sciences*, **35**, 329.

Traub, J. F., and Wozniakowski, H. (1991). Information-based complexity: New questions for mathematicians. *The Mathematical Intelligencer*, **13**, 34.

Welch, P. D. (2008). The extent of computation in Malament-Hogarth spacetimes. *British Journal of the Philosophy of Science,* **59**, 659.

Weyl, H. (1946). Mathematics and logic. *American Mathematical Monthly,* **53**, 13.

Wolfram, S. (1982). Physics and computation. *International Journal of Theoretical Physics*, **21**, 165.

———. (1985a). Undecidability and intractability in theoretical physics. *Physical Review Letters*, **54**, 735.

———. (1985b). Origins of randomness in physical systems. *Physical Review Letters*, **55**, 449.

———. (1994). *Cellular Automata and Complexity: Collected Papers*. Reading, MA: Addison-Wesley.

Xia, Z. (1992). The existence of non-collision singularities in Newtonian systems. *Annals of Mathematics*, **135**, 411.

CHAPTER 12

Gödel, Thomas Aquinas, and the Unknowability of God

Denys A. Turner

12.1 Incompleteness, Arithmetical and Theological

The invitation to discuss issues of "indeterminacy" and "unknowability" across so diverse a range of interests as that represented at the 2006 "Horizons of Truth" conference at the University of Vienna, and then to contribute a chapter to this volume, was, for me, a jobbing student of medieval theology, far too good an opportunity to turn down. I should perhaps explain why, briefly and in personal terms. Although a philosopher by background and training, my principal academic interests have for some years now been in the study of forms of medieval Christian theology, wherein can be found a cluster of notions of what might be called "systematic indeterminacy," or perhaps better, "demonstrable unknowability," and perhaps most relevantly to the purposes of this chapter, systematic and "demonstrable incompleteness." These are notions that bear teasingly like-sounding analogies with parallel notions today in philosophy, mathematics, natural science, and literary theory, and although by long-acquired academic instinct, I am skeptical of grand theory purporting to link intellectual traditions across epochs, even within historically continuous disciplines, let alone between very diverse disciplines to boot, those like-sounding analogies do have a capacity to tease intellectually. For as with flirts, such analogies as often evade the advances they elicit; evasions notwithstanding, though, they still exert their power to attract.

My interest in these theological topics was particularly renewed on the occasion of the one-hundredth anniversary of the birth of Kurt Gödel, but not in the first instance by the most obvious point of connection for a theologian, namely, his one explicitly theological intervention by way of an attempted reformulation of the so-called ontological argument (Gödel, 1941).

For I doubt if his contribution to the long history of attempts to offer a deductive proof of God's existence gets us much further with the difficulties identified immediately on its first presentation by Anselm, as I will explain later. More intriguing for a theologian, if in rather more general terms, is the question of what the implications are for theology as a discipline, if any, of Gödel's two famous incompleteness theorems, at any rate, in the extended and generalized form demonstrated by Rosser.[1] The first of these shows that every formal system of arithmetic is in this sense incomplete in that, on the condition that it is consistent, it is possible to construct some statement in number theory that can neither be proved nor disproved in that system; that is, that it will contain some undecidable propositions. The second, Gödel says, shows "a surprising consequence" of the first, namely, that on the condition of the consistency of a number-theoretic formal system, that consistency cannot be proved within the formal system. In short, what Gödel-Rosser (i.e., Gödel as amplified by Rosser) shows in these two theorems is the significance for a consistent number theory of two broad, and apparently paradoxical, notions: that of the demonstrably undecidable and that of the indemonstrability of its consistency.

How far may the theologian indulge the temptation to further generalize from Gödel-Rosser's highly specific theorems about the incompleteness of number theory, thereon to exploit analogies for theological incompleteness? Later, we will examine more closely the plausibility of such analogies, but for the time being, let us indulge the temptation so far as would appear seemly and note that as embodying a general set of metamathematical implications about undecidability and consistency, Gödel-Rosser seem tantalizingly, if also elusively, to promise points of connection with a variety of metatheological propositions about theological incompleteness with which I am personally familiar from my own studies of medieval thought, in particular of Thomas Aquinas. For although there are many Christian (and also many Jewish and Muslim) theologians in the premodern traditions for whom the incompleteness of theology, given consistency, is a core theological topos, it is a distinctive feature of the theology of Thomas Aquinas that that theological incompleteness is connected specifically, as in Gödel-Rosser, with the nature of *proof*.

Of course, I should insist that it is at best only on the strength of some (highly risky) generalizations from Gödel-Rosser's results that analogies of any such kind have the capacity to attract. All the same, the temptation thus to explore those generalized analogies is especially acute for those, like myself, who are sympathetic to Thomas Aquinas's view, almost universally unpopular among theologians today, that, in principle, the existence of God is rationally demonstrable but that – in a paradox prima facie parallel to Gödel-Rosser's – a rational proof of God's existence also proves the unknowability of the God thus proved. Now this view indeed differs sharply in considerations of rationality and proof from what, since Kant in one way, and since Karl Barth in another, is nowadays pretty much the theological consensus. For it seems generally agreed among theologians today that any purported rational demonstration of God's existence would place God all too knowably, and therefore idolatrously, "within the bounds of reason," where reason is conceived of as a complete, closed, and bounded discourse and as such would *displace* faith. As Pascal so famously put it (I paraphrase),

---

[1] Gödel's original statement and proof of the incompleteness theorem requires the assumption that the theory is not just consistent but ω-consistent. Rosser (1936) strengthened the incompleteness theorem by finding a variation of the proof that requires the theory to be only consistent rather than ω-consistent. I am much indebted to Professor Hilary Putnam for some comments on an earlier draft of this chapter, which saved me from some egregious errors of interpretation of Gödel's two incompleteness theorems. Also see Chapter 14 on Gödel's ontological proof and its variants.

there is no room for the "God of the philosophers" in the religion of "Abraham, Isaac, and Jacob"; as we might better say, inverting the proposition, a God of the philosophers leaves no room at all for the God of faith. Hence, if, speaking from within my own theological standpoint, Gödel tantalizes the theologian, it will be if there is anything to be gained by way of theological implications or analogies from his proof that the very nature of proof in arithmetic shows its incompleteness, and if there is anything to be gained by such analogical generalizations, it will be insofar as by them, a challenge might be put to modern theologians concerning their conception of reason as a closed system, a challenge that would, interestingly, set Gödel-Rosser's propositions about incompleteness in arithmetic in closer relation with some premodern theologies than it would with the theological consensus of our times. The analogy would look like this: just as in Gödel-Rosser, we are forced at the limits of number theory into the meta-mathematical proposition that those limits are not closed, so in Thomas Aquinas, we are forced at the limits of reason into the metaphysical proposition that reason is open to that which altogether transcends it, that is, to God. It is this intriguing possibility of a parallelism between Gödel and Thomas that will principally occupy me in this chapter.

Just here, a secondary objective turns the tables on the expectation that the interest the theologian might be expected to have in Gödel's work will be found in his revision of the ontological argument. For it is my view that on the contrary, his proof fails for much the same reason that all previous versions of the ontological argument fail, namely, on account of the formal requirements of its deductive character. Among those formal requirements is the dependence, as I will explain, of a deductive argument for God's existence on the provision of some definition of *God*, the provision of which is exactly that which is ruled out by the argument strategy of Thomas Aquinas's proof – on the grounds, that is to say, that the nature of God is unknowable and beyond any definition usable as a premise within a deductive inference. If that is so, then exactly that which offers us theological prospects of connection between Gödel's incompleteness theorems and Thomas Aquinas – namely, that proof of God's existence rules out the knowability of God's nature – is that which also rules out in principle the argument strategy of Gödel's ontological argument. In short, if the Gödel-Rosser of the incompleteness theorems is of interest to the theologian, it will be on grounds that reduce, if not eliminate, the interest of Gödel in the ontological argument. Such, then, are the two speculations that I offer in a provisional and general way in this chapter.

12.2 The Variables of Unknowability

I begin with some general remarks about what one might call the "systematic indeterminacy" of theology, or, as one might also call it, the "demonstrable unknowability" of its proper object, namely, God. I was first caused to wonder about analogies between theological notions of unknowability and those in much contemporary philosophy, literary theory, and science by Wittgenstein's (1961) famous remarks about the "mystical" at the end of the *Tractatus*. I first read them many years ago as a philosophy undergraduate, then making no particular connections with anything but the picture theory of language that formed the substance of that work. It was only many years later that I found myself supervising the research of a PhD student at Cambridge who had undertaken to explore in detail the claim, recycled as often as it is left unin-vestigated, that in this matter of the mystical, and of others, Wittgenstein was much influenced by Schopenhauer. When that student, Andy King (2005), digging deeper into Schopenhauer's sources, was able to show that he, in turn, was deeply indebted to the fourteenth-century Dominican friar known as Meister Eckhart, on whom I was working at the time, it was then that there seemed to be better reasons than mere whim to wonder at a potential analogy between Wittgenstein and Eckhart. For now it seemed that there were grounds for exploring links in a traceable line of intellectual continu-ity. Of course, Schopenhauer was an atheist, and when Wittgenstein tells us that the logical conditions that underlie the possibility of language cannot themselves be stated *within* language – they can be "shown," he says, somewhat opaquely, but not "said" (Wittgenstein, 1961, 4.022) – he is not in the least inclined to attach the name "God" in any way to those unsayable conditions, as Eckhart did. However, when, for his part, Eckhart demonstrates that this "God" has not, and cannot, have a "name" – for "God," he insists, is not, properly speaking, a name at all but is, as it were, a "place-holder" for the unnameable "ground" of that which we can name (Colledge and McGinn, 1981, 180) – then the analogy continued to tease, for all that Wittgenstein himself refused any satisfyingly theological consummation. For was not Eckhart thereby concluding in his own way that even if God, as the extralinguistic condition of the possibility of language, can be "shown," most certainly he cannot be "said"?

Then Schopenhauer's near-contemporary, Nietzsche, came into it along another route, but again through the work of a Cambridge research student, working this time on the relationship between theology and music. Dr. Stone-Davis and I used to argue over whether music – at any rate, classical music of an "absolute" sort – could properly be described as a "language." I was inclined to think that it could, she only with qualifications, but she did persuade me to agree at least to this much: that Kant had entirely missed the point in ranking music as lowest among the arts on the grounds that of all the arts, music was least like formal verbal discourse (Kant, 2000, 202). For if there is something to be said in favor of music's being construed on the model of language – it is, after all, a form of communication by means of structured arrangements of sound – musical utterance is not, as the philosophers say, *constative*; there is nothing that music is *about* (Stone-Davis, 2005). If, she argued, it is right to concede that music in any way bears an analogy with language, then what is wrong is to do so on that model of the constative utterance, of reference-bearing speech, for on that account, music is bound to turn out to rank lowest on the scale of approximations to that linguistic ideal. For if in any way music is a language, it is a language whose utterances *tell* you nothing.

That was how I was brought back to Nietzsche, who simply reversed Kant's priorities. Rather than seeing music as a secondary and degenerate form of communication, falling as it were away from a paradigm of formal language, Nietzsche thought of music as a primary form of communication precisely in its character of *not* being constative, of not being bound up in the binary oppositions of language as representational and referential. As if supposing there were some real possibility of excising it, Nietzsche, of course, saw reason to regret the binding of formal language into binary oppositions, into those structures that he calls, generically, "grammar" – whether that grammar divides language internally into the truth-falsity binary of classical logic or whether, from the standpoint of its constative character, it divides into the epistemological binary of utterer-subject in its relation to object-uttered. Music seemed to Nietzsche to be a more elemental form of human communication precisely because, in lacking all character of the constative, and so in not being bound into the truth-falsity binary, it cuts below language-as-referential. For music is sound structured internally in relations of pitch, harmony, and rhythm in a free play of difference, difference, that is to say, that is free of the zero-sum constraints of formal language. Relations of pitch are not true or false. Harmonic progressions do not submit to the principle of contradiction – they do not have to because truth and falsity are irrelevant in music. Because, in music, you therefore cannot tell anyone anything, in music, both the "you" as subject telling and the "anything" by way of object told are set aside. As Nietzsche says, "the whole opposition between the subjective and the objective . . . is absolutely inappropriate in aesthetics" (Nietzsche, 1999, 32), generally. As for music, like the Cheshire cat's smile, it is all smile and no cat, expression unembodied in anything except its character as form. Expression *is*, as it were, its "body."

Because Nietzsche, as I say, thought of music as a more elemental form of communication than formal language, it was his purpose to deconstruct – the word is ours, not his – formal speech into a convergence on the paradigm of music rather than, as in Kant, to miscall music on a logocentric paradigm. Therein lies the theological significance of his philosophy, for God stands in the way of his deconstructive purpose. The stability, the givenness, of the grammar of the constative – the formal laws of logic (Nietzsche, 2002, I.3, 6–7), the claims to objectivity of the empirical laws of scientific causality (Nietzsche, 2002, I.14, 15–16) – depend, he thinks, on a condition's obtaining that could not itself fall within those structures, the name of which, were there any such being to instantiate that condition, would have to be "God." It is God alone who could hold "grammar" in place. God, of course, is dead, but dead as God is, like the shadow on the wall of the Buddha's cave (Nietzsche, 2001, 109), the shadow of God long outlives the death of its original – and that shadow is long, for it consists in our continuing faith in grammar. Hence, he says, "we are not getting rid of God because we still believe in grammar" (Nietzsche, 1998, 19).

Now this conclusion intrigued me. I was still reading Meister Eckhart, and Thomas Aquinas, and Nicholas of Cusa, and what I saw in them was the mirror image of Nietzsche; that is to say, I saw in them all the elements of relationship in Nietzsche between logic, language, and God, but, as it were, in horizontally reversed order. You get to God, they seemed to say, not by holding grammar in place but only at the point of grammar's breakdown, by letting your grip on grammar slip. They seemed to say not, as Nietzsche did, that deconstruction pressed subversively all the way down through grammar is the gain you make from the denial of God but rather that deconstruction is the price to be paid at the limits of language for the assertion of God's existence. That is to say, they do not think that their theological deconstruction visits generalized ungrammatical mayhem across "language" *as such*, for they do seem to think that language gets a grip on the world, and its grammar is in order therein: they do not deny, as Nietzsche did, the possibility of theological proofs. In fact, they maintain that you could, through argument from the nature of the world, prove that God exists and that reason in its characteristically grammatical activity of proof holds thus far. However, what you show to exist by such measures of proof must lie beyond the reach of the reason that got you there. Hence, just because all three maintain that it is only on the world that language secures its grip, when it comes to God, that grip is finally lost, for reason, functioning in proper order on the "inside" of grammar, proves that it rests on an "outside" that that grammar cannot contain. So to say, the apotheosis of language about the world is, they maintain, its *failure* as theology; in other words, it would not be God we had a grip on if, as of God, we still believed in grammar's grip.

More of that in due course. At this point, my reading of Nietzsche against the background of the high medieval theologians had caused me to be much puzzled by how exactly one was to get clear about the distinction between the atheistic deconstructions of Nietzsche and the theistic deconstructions of the medieval theologians. These perplexities were only intensified on discovering from yet another graduate student in Cambridge, Mary-Jane Rubenstein, that that postmodernist guru Jacques Derrida shared them (Rubenstein, 2003), for he, like Nietzsche, thought that all theology was – in the end – but an inevitably failed attempt to eat your cake and have it, to say both that God is at the center, holding grammar firmly in place, and at the same time is evading the clutches of the grammar it "centers," situated impossibly, therefore, both "inside" and "outside" language. In the meantime, fainter echoes of those like-sounding analogies came to my ears, if not to my understanding, from the world of science, echoes that I was personally incompetent to interpret. For scientific colleagues would tell me of how Heisenberg's uncertainty principle, and more generally quantum mechanics, had set a cat among the epistemological pigeons, at any rate in respect of just how stable *the* relation between observer and observed could be counted on to be. It began to seem that perhaps, after all, Nietzsche was not so far off the mark; at any rate, he began to seem not so paradoxical today as he had seemed to be to his contemporaries in letting cognitive instability run amok, even among those human discourses most unproblematically credited with objectivity and scientificity, those of natural science – if, at any rate, the scientists themselves were running scared of such grammatical notions. Of course, as I say, I was unable to judge as between the competing accounts forthcoming from the realists and conceptualists concerning how to explain what the disruption of that stable relationship between knower and known entailed or was occasioned by: whether to put it down to the real properties of the elementary particles or whether to say that the problem lay on the side of the observer's position. Whichever it entailed, though, these faint echoes confirmed an instinct that possibly in our times some sort of line of intellectual history was turning full circle on an older medieval theological story so as to revisit it, and this possibility had a prima facie attraction to it, even if, to be sure, for the most part, the circle was turning in three dimensions and was a spiral gyrating away in atheist directions from the theological point that those medieval theologians were occupying on it.

Then came my invitation to speak on the unknowability of God, occasioned by the centenary of Kurt Gödel's birth. Satisfied that I needed to know no more than the general principle of Gödel's theorem, and not its formidable mathematical proof, I agreed, for it does seem to me worth testing out whether there is anything theologically significant in Gödel's conception of "systematic" or provable "incompleteness," but now violating the boundaries of that, as one might have supposed it to be, most securely defended citadel of consistency and completeness – and so of "rationality" – arithmetic. It seemed worth asking whether there is anything more to it than a "like-sounding analogy" with what in high and late medieval theology was thought of as the demonstrable existence of its own object, namely, God, simultaneously with the demonstrable unknowability of the nature of that object. I knew that all I would be able to do is explain what those medieval theologians said about the divine unknowability, and what they said about the logic of their theological discipline, whose task they conceived of as therefore consisting primarily in demonstrating the paradox of its own incompleteness. Beyond that, I should be happy to listen to the mathematicians and logicians speaking on their own ground, as I do on mine here and now, and let it transpire one way or the other whether our deliberations might cast any light on those questions about "analogies" between scientific and theological disciplines.

12.3 Systematic Indeterminacy

I said just now that those medieval theologians Thomas Aquinas, Meister Eckhart, and Nicholas of Cusa conceived of their discipline as being "systematically incomplete" and open to a region beyond its power to determine, and I suppose that this is some sort of paradox. For a discipline's being *systematic* – speaking just of colloquial uses of the word – would seem generally to imply its closedness and determination, a fixity of boundary externally, and, as regards the internal relations of the elements of a system, a consistency of method and a definite logic governing those relations. On the other hand, *indeterminacy* would appear to imply the opposite of all these.

As I have said, there is an old theological story here concerning the relationship, whether of contradiction or consistency, between completeness and indeterminacy – it goes back at least as far as the Cappadocian Gregory of Nyssa in the fourth century of the Christian era, and that is just as far as I happen to know. But it is also a complex story, and the issues arise in very different historical and intellectual contexts. For which reason I shall have to be selective and oversimplify, confining my remarks to just three very different formulations: those of Gregory of Nyssa, Immanuel Kant, and Thomas Aquinas.

12.3.1 Gregory of Nyssa

To begin with Gregory, then, his problem arose from within very specific historical constraints and was perhaps not immediately a problem with the relation between the determinate and the indeterminate within the discipline of theology as such. It is in any case questionable whether he thought of there being any sort of occupation to be called "theology," not at any rate in that clearly articulated sense in which it first came to be conceived in the high Middle Ages as a distinct academic discipline within the division of academic labors in the university schools. Gregory thought that what the theologian did was to engage in *theoria*, or as it came to be translated in the West, *contemplatio*; and whether in Greek or Latin, the term was far more generous in its range of meanings than our *theology*. For in either case, *theoria* or *contemplatio* embraced all sorts of cognitive relations with God, from liturgical worship to formal doctrinal statement, and much personal piety in between. Moreover, in the form in which Gregory encountered the problem, it was as inherited from the dominant, and consciously Platonist, philosophical culture with which he was attempting to wed his

Christian theological beliefs. In short, Gregory's problem concerned how consistently to speak of that *theoria*'s object, namely, God, both as "infinite" and as "perfect" (Gregory of Nyssa, 1978).

Now this was a problem for Gregory because, on the accounts of the "infinite" and of the "perfect" that he inherited from that Platonist tradition, the "infinite" implied boundlessness and so lack of form, hence incompleteness and imperfection, whereas "perfection" implied form, and so determination, hence not infinity. Gregory therefore had to provide an account of *theoria* within which the infinity and perfection of God, properties of the deity inherited from his Hebraic biblical sources, could be, if not exactly reconciled, then at least represented as not standing in formal contradiction with one another. For if Gregory's conception of *theoria* owes little to later, Western conceptions of theology as a "system" of consistent propositions, nonetheless Gregory did think that *theoria* had debts to pay to logic to this extent at least: that its utterances must be, if not *demonstrably free of internal inconsistency,* then at least *free of demonstrable internal inconsistency.* That minimal debt to logic entailed, therefore, a theological responsibility: if he nowhere claims that formal consistency could be demonstrated across *theoria* as a whole, then at least claims to formal contradictoriness had better in principle be rebuttable, else *theoria* would collapse into nothing more exalted or worthy of assent than the merest gobbledygook – hence the need to resolve the Platonist aporia.

Gregory's theological strategy, therefore, is the dialectical product of two converging energies, each embodying complex relations between determinacy and incompleteness. If, in the first place, *theoria* – conceived of minimalistically as the sum total of propositions about God – is incapable of demonstrating its own formal consistency, in the second, it must be in a position to counter claims to its formal inconsistency. Those conditions being met, but only if they are, the way is open for Gregory to inscribe paradoxicality into the very notion of theology as "systematic." Theology, he thinks, *generates* paradox: that is its method. God's being infinite is something of which we can more or less, if only negatively, give an account. God's being perfect is something of which we can more or less give an account, if only by denial of what we understand of the limitedness of creatures. Hence God is both infinite and perfect, and so "infinitely perfect." But of that whole expression we can give no positive account, except as the conjunction of statements for each of which we can, disjunctively, provide the justification. Theology cannot resolve the paradoxes it generates, for, on the contrary, it *shows* that we occupy no epistemological ground on which they can be resolved. The tension between the perfection and the infinity of God remains unresolved and unresolvable by a finite mind. In that sense, theology is "systematically indeterminate" in that it demonstrates its own incompleteness: at best it can show what it cannot say. Indeed, it is properly speaking "theology" just insofar as it does that; moreover, it is just insofar as it does so that theology ends in mystery, that is to say, it is just in *that* sense – but in that sense *alone* – that theology is mystical.[2]

Of course, it might well be said that Gregory has this problem and is compelled thus to resort to its dialectical, "apophatic" solution – not because there is something

---

[2] Although nowadays there are plenty of other senses of that word that have few enough connections with Gregory of Nyssa.

in the nature of the theological project as such that requires it but only because he is trapped into it by the particular exigencies of the Platonist starting point he inherits. If you preferred to say that, I do not think you would be entirely wrong, so far as the argument goes in the source to which I am here referring, his *Life of Moses* (Gregory of Nyssa, 1978). However, there is more to it than that, and if it is not stretching historical credibility too far for tolerance, we can see why if, taking a regrettably vast historical leap to Immanuel Kant, we turn to his construction of what he calls the "antinomies" of speculative reason.

12.3.2 Immanuel Kant

Kant's critical strategy in this connection was principally negative – to show that "speculative" reason could not by any means of its own generate knowledge of a transcendental object, that is to say, an object transcending the forms of sensibility, space, and time and the categories of understanding that organize the sensory, spatiotemporal manifold into scientifically determinate objects of knowledge. His method of thus limiting the "claims of speculative reason" was to show that from within different speculative standpoints, you could demonstrate with equal soundness a proposition, p, and its contradictory, $\sim p$, thus showing the necessity of a standpoint that "transcended" both.[3] Thus, in Kant's statement of the "third antinomy" in *The Critique of Pure Reason,* arguments are presented on the "dogmatic rationalist" side that there is an absolutely "spontaneous" freedom, whether as cause *of* the universe or as cause *in* the universe – and on the empiricist side, that such freedom is impossible. Now, on one hand, just because both p and $\sim p$ are demonstrable with equal soundness from within their respective "standpoints," the paradoxical tension cannot be reduced by, as it were, any concessions on either half of it: the truth of p and the truth of $\sim p$ are apodictically demonstrated *each within its standpoint.* On the other hand, the resulting paradox forces the mind to admit to the possibility of other ground – that of "transcendental idealism" – whereon the paradox is resolved, but only on the condition that the constraint of contradictoriness is not abandoned. As Kant says, there is an "absolute necessity of a solution" of the antinomies (Kant, 1964, B504, 430); that is to say, you cannot, for Kant, take the Nietzschean way out, which is to throw logic and the principle of contradiction to the winds and simply say, as Walt Whitman is reported to have declared, "I contradict myself? So I do. I am large, I contain multitudes."[4] It is only if both conditions hold, first that p and $\sim p$ are demonstrable and second that their contradictoriness is as binding as it is intolerable, that the mind is forced by virtue of the antinomy to conclude that there is – there *must* be – the transcendental standpoint beyond the range of speculative reason whereon the contradiction is resolved, whereon,

---

[3] His purpose, of course, was to clear space free of "speculative" intrusion into the transcendental concepts of God, freedom, and immortality so that they might reemerge as presuppositions of practical reason, curtailing thereby the claims of (speculative) reason, as he puts it, "in order to leave room for faith" (Kant, 1964, 29).

[4] With what truth, or in what context, he is said to have so remarked I do not happen to know, although "I am large, I contain multitudes" is the famous line, of course, from his poem "Song of Myself" in *Leaves of Grass* (1855).

as Hegel was later to put it, the "negation" – that is, between p and $\sim p$ – "is negated," or, as Kant himself puts it at the very outset of *Critique of Pure Reason*:

> Human reason . . . has this peculiar fate that in one species of its knowledge it is burdened
> by questions which, as prescribed by the very nature of reason itself, it is not able to
> ignore, but which, as transcending all of its powers, it is also not able to answer. (Kant,
> 1929, preface to 1st ed., A vii, 7)

Now I think that Gregory of Nyssa's theological strategy is, in point of its obedience to the exigencies of logic, not thus far different from Kant's: no more than any other discipline of thought may theology simply throw logical caution to the winds, and in that minimal sense, at least, theology is "systematic." Nothing in Gregory of Nyssa lends support for that irrationalist Christian equivalent of Walt Whitman, Tertullian, who was in his way as proud as Whitman was to proclaim that he believed the Christian faith *because* it was absurd. If, then, Gregory finds himself knowing within different standpoints that there are two propositions true of God that contradict each other, then, like Kant, he has grounds for deriving from the contradiction that there must be another standpoint within which the conflict between them is resolved: behind such contradictions lies not the *stasis* of the irrational and nonsensical but the *movement* of reason, compelling the mind to go beyond them. Moreover, like Kant, although Gregory knows that there has to be such a standpoint, he knows equally that there can be no occupation of it available within the constraints of human knowledge: we cannot stand where God is, there where the paradox is dissolved and is there known in its dissolution.

It is here, then, that we get to the heart of the matter, to the proposition that theology, whether in Gregorian or Kantian form, is necessarily "incomplete." For in either case, the demonstration that the conflicting propositions are both true *demands* that there is that in which the contradictory conjunction is resolved; but just that same demonstration excludes the possibility of knowing the nature of that in which their conjunction is true. For if, on the grounds on which we know the conflicting propositions to be true, we could know the nature of that *of* which they are true, then that of which they are true could not be God. For just insofar as we could know that nature, that nature would be something that fell within the domain of our language, failing thus to transcend it; and, thus enmeshed in its antinomies, it would fall from strictly the sublime into strictly the ridiculous. So it would seem that the logic of the positions of both Gregory and Kant is this: that theological "completeness" would entail its being internally contradictory, whereas any strategy that would save its consistency would require as a condition its incompleteness.

12.4 Theology as Essentially Uncompletable

So far, then, we have one form of "systematic incompleteness" characteristic of a certain theological tradition: "systematic" in that the paradoxes thrown up by the multiplicity of properties predicable of God demand – because the constraints of logic demand it – resolution; "incompleteness" because the resolution of the paradox is demonstrably beyond our powers. It is worth emphasizing, furthermore, that Gregory of Nyssa is far from seeing in this systematic failure of *theoria* some sort of deficiency as knowledge. On the contrary, *theoria* encounters this double constraint of the equal necessity and impossibility of resolving its paradoxes precisely insofar as it meets with its own highest object, God, in its own subjective fulfillment as knowledge. This encounter is, for Gregory, knowledge, *theoria*, reaching out to its own limits, there to know that it cannot know what must lie beyond them. Moreover, this conception of theology's predicament is rooted, he thinks, in a biblical paradigm in the encounter of Moses with Yahweh on Mount Sinai, where, meeting the divine presence in a dark cloud, he asks to see the *face* of God and is denied it: "You may see me from behind, but my face you may not see," Yahweh tells him, "for no one may see my face and live" (Exodus 33:20). You can, at the limits of human knowledge, know that there is that the nature of which you cannot know, and this is God. It is, then, precisely in its theological failure that knowledge reaches its fulfillment as *theoria*; it is knowledge "realized," to employ a Hegelianism, by its abolition. The theological is knowledge's *Aufhebung*; it is reason's self-transcendence.

However, if it might be said that Gregory's problem arises only on account of his Platonist presuppositions about "infinity" and "perfection," it might equally be said that Kant's antinomies are generated not out of any absolute character of the nature of knowledge as such but only out of an artificially contrived construction of the dogmatic rationalist and empiricist standpoints as being not only mutually exclusive but also as collectively exhaustive. For there are in principle restatements of the epistemological options that do not trap the philosopher between the Scylla of dogmatic rationalism and the Charybdis of empiricism, just as there are conceptions of infinity and of perfection that need not entail Gregory's Platonist aporia. And if such restatements are possible, the dialectical necessity with which the philosopher is required to postulate a higher, and to human understanding unknowable, standpoint is to that extent reduced, if not altogether removed.

Perhaps. However, even if neither Gregory nor Kant succeeds in demonstrating absolutely this apophatic conclusion – that language has an "outside" that sustains it, although equally we cannot inhabit it – this should not distract us from the fact that that is what they attempt to prove. In acknowledging at least their intent, we might feel that there are, at least in the general terms of an epistemically structural analogy, some resonances with Gödel's second incompleteness theorem in both in that due regard is being made toward basic consistency within a system – in Gregory's case, *theoria*, in Kant's, the constraints of reason – while acknowledging that consistency is in the nature of the case indemonstrable. That said, whatever may be the truth concerning the general applicability of their conclusions across *theoria* or reason as such, there are epistemologically tougher statements of this "apophaticism" in the later Western medieval traditions, indebted as the latter are to Gregory, at least indirectly. Tougher than Gregory's is that of Thomas Aquinas, for if it is after all true that Gregory's conclusion of the inherent paradoxicality of theology derived from the tensions embodied in his classical Greek sources, for Thomas, the paradoxicality of the theological bites deeper into language as such, and not on some assumptions derived from a particular theological culture. For it bites into the very notion of proof of God's existence and of existence itself as predicated of God, and that paradox may be most summarily presented in this form: Thomas thinks that it is possible to prove that God exists, but, he goes on to say, just those arguments that prove there to be a God also show that the predicate form "x exists" has, as of God, itself passed beyond our power of understanding. In short, if Thomas Aquinas's proofs prove *that* God is, they also prove that *what* God is is unknowable to us. For Thomas, if unknowability is the outcome, it is paradoxes of proof that drive the mind to it.

12.5 Thomas Aquinas, Unknowability, Proof, and Proof of Unknowability

Deadpan as is Thomas Aquinas's default theological style, there is something at once artless, and just for that reason, intensely dramatic, about the transition he effects from question 2 to question 3 of his *Summa Theologiae*. In question 2, he has presented, in highly abbreviated form, what one might call five "argument-strategies" of proof for the existence of God. Leaving aside the issue of whether any or all of the arguments, in their compressed form as there presented, are in fact sound, or even whether, considered just as formal argument-schemas, they are in principle valid, it is at least certain – although I grant that some dispute it (Milbank, 1999) – that Thomas thought of them in some strict sense as "proofs": formally valid inferences from true premises whose truth we could know about independently of any theological presuppositions. These are, he says, five ways in which *Deum esse probari potest*, and in terms of strict apodicity, there is no stronger word in his logical vocabulary than *probare*. But if we can, for today's purposes, leave this controverted question on one side, it is not possible to sidestep the issue of what it is that Thomas thinks the proofs prove. That God exists, you'll say. Just so, says Thomas, adding that the proposition "God exists" is analyzable out – as we would nowadays say – by the existential quantifier "something or other is God," or, as he puts it, whatever it is that these proofs prove is *quod omnes dicunt Deum* (*Summa Theologiae*, 1a q2 a3 *corp*).

Now, it is not entirely clear how best to translate *quod omnes dicunt Deum*. It goes without saying that Thomas does not suppose that whatever it is that his proofs prove is what "all people" *mean* when they use the name "God." In my view, the phrase is best translated, as I have argued elsewhere (Turner, 2004, 17–20), as "that is what people *refer to* when they use the name 'God.'" Manifestly "the first cause" or "necessary being" or whatever is not the God "all people" have in mind when they pray to Him – or for that matter when they curse Him. It would be all the same with blessings and curses, of course, because, as Thomas quotes Aristotle as saying, *eadem est scientia oppositorum*[5] (one and the same is the knowledge of contraries), and you fail to pull off an adequately blasphemous imprecation unless it is one and the same God to whom people pray that you curse. What Thomas thinks the proofs yield is a minimum grip on the noun *God*, showing the word *God* to have reference, and that, as I have put it, there is something or other answering to that word and that you *could* show that whatever thus answers to it is extensionally equivalent to the God "all people" praise or curse when they do – or, for that matter, deny. You would, he well knows, have to *show* this

---

[5] Thomas frequently appeals to this Aristotelian principle, e.g., in *Summa Theologiae* 1a q14 a8 *corp* and 1a q58 a4 ad2; see also Aristotle, *De Interpretatione* 6, 17a 33–35, and *Metaphysics*, IX, 1048 a11.

extensional equivalence, and in fact, he takes his time over it, occupying some 149 articles of his *Summa Theologiae* engaged in the demonstration that the God of the proofs is all the same with the Trinitarian God in whose name pious Christians bless themselves.

However, the one thing Thomas rules out as a means to doing so in the demonstration of this extensional equivalence is some definition of God. Of course, he says, providing a definition would seem to be his obvious next move, or even that he should have begun his account of proofs of God's existence, as Anselm does, with a definition, namely, "that than which nothing greater can be thought."

For how else than by means of a definition of God would we know what the proofs are meant to prove? Thomas acknowledges the force of the objection: as he remarks in the prologue to question 3, in the normal case, "once you know whether or not something exists, it remains to consider its manner of existence, so that we may know of it what it is." Alas for the vanity of theological hopes, he stops us short just there: the urge for a definition must be resisted, he says, for in this case, no such thing is possible. He goes on: "but since we cannot know of God what he is, but [only] what he is not, we cannot enquire into the manner of God['s] existence, but only into how he is not. So, next we must consider this 'how God is not'" (*Summa Theologiae*, 1a q.3 prol). And so he does: everything he contrives to show we can say about God places God further beyond our comprehension, for the nature of God is to us, he says, *omnino ignota* (*Summa Theologiae*, 1a q12 a13 ad1).

Therefore Thomas cannot begin his proofs in what might seem to be the obvious way, with a definition of God, for, simply, there is no definition of God, although here it is important not to become confused by ignorance of Thomas Aquinas's Latin. *Deus*, Thomas says, is a *nomen*, which is not to be translated as "proper name." Grammatically, he says, *Deus* is not the proper name of an individual, like "Peter Smith" functioning as a subject term but never as predicate, but a common noun, and so a predicative term, what he calls a *nomen naturae*, like "human." Thomas maintains that a person who mistakenly thinks there are many Gods is not making a grammatical mistake, as one does who wonders if there might be many individual Peter Smiths (as distinct from the undoubted fact that in the telephone directory, there are many individuals *called* "Peter Smith"). The polytheist is committing a rational error, not a grammatical one. For although it is demonstrable that there is and could be only one God and false that there are many, polytheism is not simply incoherent in the way it is to wonder if there might be many one and the same Peter Smiths. Being a common noun, *God* therefore functions grammatically as a predicate, which is the reason why Thomas agrees with the analysis of the proposition that forms the conclusion of an argument for the existence of God in terms of the existential quantifier: "God exists" is analyzable as "something or other is God" (*Summa Theologiae*, 1a q3 a4 ad2).

Now it might be asked how Thomas could consistently maintain both that *God* is a grammatically descriptive term, a common noun, and that we cannot know the nature it describes. In any case, one can repeat the question, how, without some initial definition of what God is, would it be possible to know that a purported proof of God's existence has got to its destination? But the question confuses grammar with semantics. Knowing the grammar of talk about God does not depend on having a definition of its topic in any formal sense. To adapt an example of Herbert McCabe (2002, 37–88), I am afraid to say that I do not know what a computer is; I am nowhere near being in possession of a definition. However, I know well enough how to use one, and, being old enough to have painful experience of mechanical typewriters, carbon copies, and correcting fluids, I am well aware of the difference it makes to be able to cut and paste text, to be able to send papers by e-mail, and to repaginate at the click of a button. I do not need to know what a computer is to be able to use one, and I can know how to use the word *computer* in appropriately truth-bearing sentences, knowing only that whatever a computer is, it is what accounts for my being able to cut and paste, send e-mails, change pagination at the click of a button, and so forth: a *computer* needs only to mean whatever it is that accounts for my being able to do these things. I can know the grammar of the word *computer* from its effects on my writing, without knowing what the word *computer* means.

Just so, Thomas says, with God: in place of a definition of God, arguments for God's existence rely on knowledge of the divine effects. If you interrogate the world with sufficient persistence, you will, he thinks, come to know that it is created, and that is all the grip you need on the word *God* to know its grammar. So we can know that *God* is a common noun grammatically, that is to say, that it *functions* as a description without knowing otherwise what, as Thomas puts it, "God is in himself." That knowledge, he insists, is altogether beyond us. Nothing of which you could give a definition could be God. Moreover, that is the reason why Thomas thinks Anselm's argument is invalid: not just, that is, for Kant's reason, namely, that *existence* is not a predicate and so could not be included in the concept of God. For Thomas indicates that he could be happy neither with Anselm's proof nor with Kant's critique of it: existence neither follows nor does not follow from the concept of God because there is no concept of God for Her existence to follow or not follow from (*Summa Theologiae* 1a q2 a2 *corp*).

12.6 The Failure of Gödels' Ontological Argument

It is for similar reasons that Gödel's version of the ontological argument fails, although my comment here has relevance, as with Thomas Aquinas's, more to Gödel's argument-strategy in general terms than to its technical detail, which is discussed in Chapter 14 in this volume; in any case, it is generally accepted that, with some modifications, Gödel's argument is valid. Criticism of his proof therefore relies on challenging his axioms, for if given them, the argument is valid and sound, whereas without them, it is valid but unsound. In this connection, Gödel was, of course, familiar with one of the standard objections to Anselm's first formulation of the argument, the one principally relevant here being that Anselm appears to confuse the subjective *conceivability* of a being "than which nothing greater can be thought" with the objective *possibility* of a being "than which nothing greater can be thought," thinking that the first either is the same as or else entails the second. He appears to conflate the two on the grounds that if the theist and the atheist (the "fool") are to be disagreeing with each other, the one denying what the other affirms, then they must both "have the same idea" of what they are disagreeing about, and so both must be "conceiving" something, namely, what answers to the common definition. That is, if the theist is affirming anything at all under the description "God," then it is *exactly as answering* to that same description that the atheist must deny there is anything, otherwise he fails to pull off an adequate denial of what the theist affirms. For, again, *eadem est scientia oppositorum*. However, suppose that what the theist and the atheist conceive – "that than which nothing greater can be thought" – is not the conception of anything possible. Then neither the theist's affirmation nor the atheist's denial succeed one way or the other, for impossibilities have no truth values, and there is nothing that the theist and the atheist are disagreeing about. However, the first formulation of Anselm's ontological argument relies on the description "that than which nothing greater can be thought" being a possible description, and nothing that Anselm offers shows that it is one.

Gödel's strategy of reformulation acknowledges this difficulty along Leibnizian lines, and he sets out to find a definition of God, something that answers to the description "possibly God exists." That starting point takes the form that possibly there is an individual combining in its definition the set of all "positive" properties, as Gödel calls them, a "positive" property being "positive in the *moral aesthetic* sense (independently of the accidental structure of the world).... It may also mean pure attribution as opposed to privation (or containing privation)." Now while it is far from clear what Gödel means by a property's being "positive" in "the moral aesthetic sense," and though it is somewhat clearer what he means by a property's being "pure attribution as opposed to privation,"[6] it is certainly true, as he goes on to argue, that the property of combining all such positive properties is itself a member of that set, that is to say, is itself a "positive property," there being, tautologically, nothing "privated" therein. That "positive property" of combining all the positive properties is, he says, the property of being "God-like." Now one such positive property, Gödel argues, is "necessary existence." Hence the first series of steps in his reformulation of the ontological argument consists in showing that the "God-like" possession of all the positive properties is consistent, that is to say, is possible, and therefore that possibly a "God-like" being necessarily exists.

At many points on the way thus far, exception can be taken to Gödel's assumptions. Leaving such objections thus far aside, the rest of Gödel's argument depends on the device in modal logic of extracting a necessity from "inside the brackets" so as to stand outside them. His proof consists in showing that from the proposition "possibly a necessary being exists," you can derive the conclusion: necessarily (a necessary being exists) – in short, that the proposition "a necessary being exists" is a necessary truth. Now, as Thomas Aquinas has it, the conclusion is false: "God exists" is not for him a necessary truth, for a necessary truth is one whose negation is a contradiction. Although Thomas thinks that atheism is false, he does not hold that the denial of God's existence is self-contradictory. God's existence, he says, is indeed a necessary existence: *if* God

---

[6] For Gödel's positive properties would seem to have the same logical character as Thomas Aquinas's transcendentals. Terms such as *goodness, oneness,* and, indeed, *existence* are topic-neutral in that they are predicable literally of anything whatsoever, regardless of category – of any kind of thing whatsoever, there is something that literally counts as being a good one of that kind – whereas nontranscendental terms are topic-specific in that their literal predication is limited to a particular category. *Heavy* is predicable literally only of bodies that have mass and so metaphorically of hearts or thoughts. As Thomas maintains, then, *goodness* is predicable literally of both God and creatures, for nothing in the meaning of *good* limits it to creatures – or, as Gödel would say, "contains privation" – whereas "God is the rock of my salvation" is necessarily a metaphor.

exists, then God exists necessarily. It is simply a confusion to suppose that the *de re* necessity of God's existence entails the *de dicto* necessity of the proposition "God exists."

As we know, it is an elementary mistake to confuse the necessity with which the conclusion follows from premises in a valid inference with the necessity of the conclusion itself because *any* conclusion of a valid inference follows necessarily from its premises, including contingent truths. The mistake is common among the Platonists, Thomas notes, who similarly seem to derive from the *de dicto* proposition "necessarily what is known is true" (which is no more than a definition of *to know*) the *de re* conclusion that "only the necessarily true is known." If I know that Socrates is sitting, he says, then necessarily, Socrates is sitting, but of course, there is no necessity *de re* to Socrates' sitting: he has only to stand up and walk away, and the proposition becomes false (*Summa Theologiae* 1a q14 a13 ad3). Now, although of marked Platonist tendencies, Gödel makes no such elementary mistake. The conclusion he derives validly from his premises, namely, that "'a necessary being exists' is a necessary truth," is true if and only if one of the argument's assumptions is true, namely, that "necessary existence" is a positive property of a "God-like" being. It is *that* assumption that Thomas would have rejected, on the grounds that that assumption is crucial to the derivation of the false conclusion that "God exists" is a necessary truth, for "God exists" is deniable without contradiction.

To be clear, Thomas, of course, thinks that *if* you can prove that there is a God, then it is simply a matter of what you mean by "proof" that the conclusion "God exists" necessarily follows from that proof's premises. Of course, Thomas thinks that God is a "necessary being" – he thinks that you can prove that, too. What he denies is that the conclusion "God exists" is a necessary truth because the purported demonstration of that conclusion has to depend on conceiving of "necessary existence" as part of the meaning of "God"; that is to say, it depends on conceiving of "necessary existence" as a *property* because no judgment whether of contingent or necessary existence denotes a property. Existence, necessary or contingent, is expressed in *judgments*, not analyzed out of *concepts*: it is never part of the definition of anything that it exists, necessarily or otherwise. His view is, as one might put it, that there is no kind of thing *such that* it exists. Of course, then, if such a kind of being were possible, then not only would it be a necessary being, *that* it exists would be a necessary truth.

12.7 Aquinas on Proof of God's Existence

If Thomas Aquinas's argument-strategy abandons the deductivism of Anselm's – and by anticipation that of Leibniz – then there seems little prospect thus far of common theological purpose with Thomas being found in Gödel's revision of it. As I say, potential for some such connection worth exploring seems to lie more distantly – although in my view more importantly – in a common but general epistemological strategy, in Gödel's case arithmetical, in Aquinas's theological. Aquinas's conception of proof of God neither begins from nor ends in a *definition of God* but rather from within a strategy of *rational questioning of the world*. His conviction is that if you push such questions, principally causal ones, hard enough to reach the limits of reason, then just there at those limits, reason yields to the conclusion that the world originates in an unknowable cause. Not what there is, but that there is anything at all – *that* is the mystery that, for Thomas, if not for Wittgenstein, calls for the name "God." You could put it in another way: something or other answers to the common noun *God* because it refers to the unknowable answer to a question that arises from our having pushed causal explanations to the very boundary of sense, which we do when we ask, "How is there anything at all rather than nothing?" Thomas, of course, does not put it in quite that Leibnizian way, although what he does say is not unhappily thus paraphrased. For the Leibnizian question has the same formal shape as Aquinas's proofs, and if there is anything at all in the comparison between Gödel's theorems and Aquinas's proofs of God, it lies most plausibly, as I have said, in that general shape. For we may note that the paradoxical result of Gödel's first incompleteness theorem arises in connection with multiplication and not with addition only, and for Thomas, too, proof of God has an essentially *multiplicative* character. Aquinas's causal questions can be said to *arise* from within the bounds of language, for, as Peter Geach (1969, 81–81) puts it, if we have available to us ordinary uses of the noun phrase "every moveable thing" and the relative clause "cause of . . . ," then the question formed from their conjunction, "what is the cause of every mutable thing?" can, contrary to what Kant supposed, have a sense we can grasp. However, the answer to it could not itself fall within the bounds from within which the question arose; that is to say, the conjunction is not merely additive, as if the sense of the word *cause* in "cause of *every* mutable thing" was just the same as that which it possesses in its application *within* the world of mutable things. For the sense of *cause* in "cause of every mutable thing" as it were "multiplies" off the edge of the universe from within which the question arose. Were the answer to be thus contained within the bounds of sense, then it could not be to that question that it is the answer: for then it would be just one of the mutable things within the world that the causal question asks about. Hence no such answer deserves the name "God": to pray to such a being would be the merest idolatry, and your curses would be well merited by such a shibboleth and achieve no blasphemous purpose.

We *could not* know what God is. Of course, as a Christian believer, Thomas does not *need* to prove that God exists. However, as a Christian believer about to embark on a massive structure of theological exposition and argument about the Trinitarian God and the God of Jesus Christ, in which it could seem as if he was, after all, telling us a great deal about what God is, he does need to prove that, on the contrary, anything you go on to say about God can serve no purpose but to reinforce that unknowability of the divine mystery. As he emphasizes, even faith does not somehow fill in a definition of God lacking to reason because although, as he says, through faith we do indeed enter into some sort of communion with God, even by faith, this "communion" is as to something unknown to us, *quasi ei ignoto* (*Summa Theologiae*, 1a q12 a13 ad1). Faith is a life lived, as it were, in the medium of that unknowability.

So that – as, I repeat, in highly general terms – is where things stand for Thomas: argument about the world must, in the end, demonstrate the unknowability of its cause. It is the nature of proof of God itself that generates the metatheological proposition that theology is uncompletable because any conception of theology as a completable system would run it into a self-defeating – because idolatrous – inconsistency.

12.8 Arithmetical and Theological Incompleteness:
Are They the Same?

It was such accounts of theological epistemology – in one way that of Gregory of Nyssa, in another way that of Kant, and in a third, and perhaps more convincing way, that of Thomas – that brought to mind thoughts of those analogies, like-sounding as they seem, with so many parallel notions that today are found in philosophy, particle physics, literary theory, and, most surprising of all, arithmetic. However, if such is the case for the exploitation of analogies between Gödel and at least one Christian theological tradition, as it is called, of "negative theology," it is time now to entertain grounds for doubting whether the analogies hold at any more than at a highly general, and equally risky, level. For I ought at this point to come clean and make it clear that – tempting as it would be in this volume with its cross-disciplinary focus and scope – I have no firm propositions to offer here of comfort to those too casually seeking support for these notions of theological paradox in the paradoxes of Gödel-Rosser about number theory. For my part, at any rate, I simply do not know whether the superficial parallel is genuinely illuminating, if only because I do not know whether the theologian and the mathematician are working with sufficiently similar notions of consistency, completeness, or proof. If anything, I am on balance skeptical. It is in any case clear that Gödel's meta-arithmetical incompleteness theorems are syntactical propositions about proof of unprovability within a calculus and are not without much more ado to be assimilated to the semantical proposition of Thomas concerning the unknowability of a proved result's nature. Moreover, the notion of "proof" within a calculus is inevitably defined in reference to a rigidly deductive system, whereas Aquinas's notion of proof in connection with the existence of God must submit with equal inevitability to different rules of apodicity, as I have shown. The point of difference between deductive proof in arithmetic and Thomas Aquinas's conception of proof as of God is all the more reinforced in Aquinas's case by his refusal to accept the validity in principle of any "ontological" – that is, quasi-deductive – proof of God, as we have also seen. At best, then, any grounds for a more optimistic assimilation between Gödel and Thomas must rely on the most general and formal level of comparability. I suppose at that level, it can be said that something like the Gödelian syntactical proposition of the first theorem about the "demonstrably undecidable" in arithmetic has the role in Thomas of generating the metatheological proposition that theology cannot occupy the ground on which alone its completeness could be known,[7] this theological ignorance being systemic: for it is simply in the nature of theology as a form of knowledge that it is on the grounds of what it *does* know that it knows what it *cannot* know.

That said, my instinct is to conclude that except at such a generalized level of structural similarity, the analogy serves little but the interests of distortion, both of Gödel

---

[7] Incidentally, Thomas Aquinas's conception of the "demonstrably undecidable" is not limited to this general metatheological proposition. It governs the relationship between the oneness of God and God's Trinitarian nature. God's being one is of such a kind, he thinks, that it is demonstrably undecidable whether there are or are not three "persons" in God. In a celebrated polemical work, he argues that it is "demonstrably undecidable" for a rational mind whether the world is eternal, a question that, he argues, is not settled either way by any proof that the world is created "out of nothing"; see *De aeternitate mundi contra murmurantes*, an extract of which is published in translation in McInerney (1998, 710–17).

and of Thomas. Perhaps this will seem a sadly downbeat answer with which to conclude, but not all is lost, there being at least one significant, albeit negative, gain. For it is Gödel's contribution to that series of skeptical questionings, emergent as we have seen in the earlier decades of the twentieth century across a wide range of disciplines, that exploded the myth of a narrowly rationalist and essentially antitheological conception of reason, and it is his distinctive contribution to these questionings that is of particular significance to the theologian. For it was in mathematics that the ideal type of a system of rational thought, demonstrably complete and demonstrably consistent, seemed to be instantiated most incontrovertibly. Gödel demolishes that paradigm of reason precisely in its strongest case, but then, neither does the mirror image of that narrow atheistic rationalism, which consists in a form of fideistic and irrationalist theism, gain any succor from a theological analogy with Gödel. For what Gödel offers in the first of his theorems is a *proof* of incompleteness, and in the second, a *proof* that consistency is unprovable, a paradigmatically *rational* procedure of cracking open the closed circle of reason and exposing it to what lies beyond it. On that ground of reason equivalently mapped, neither rationalist nor irrationalist, the theologian may wish to stand. At any rate, Thomas Aquinas does.

References

Colledge, E., and McGinn, B., eds. and trans. (1981). Sermon 2, Intravit Iesus. In *Meister Eckhart: Essential Sermons, Commentaries, Treatises and Defense.* London: SPCK

Geach, P. (1969). Causality and creation. In *God and the Soul.* London: Routledge and Kegan Paul.

Gödel, K. (1970). Ontological proof. In *Collected Works*, vol. 3 (1995), pp. 403–4. [Introductory note by R. M. Adams, pp. 388–402. Appendix B: Texts Relating to the Ontological Proof, Including Gödel's First Version, 1941, pp. 429–37.]

Gödel, K. (1995). "Ontological Proof". *Collected Works: Unpublished Essays & Lectures, Volume III.* pp. 403–404. Oxford University Press.

Gregory of Nyssa. (1978). *Life of Moses*, I. Translated by A. J. Malherbe and E. Ferguson. New York: Paulist Press.

Kant, I. (1929). *Critique of Pure Reason.* 1st ed. Translated by N. Kemp-Smith. London: Macmillan.

———. (1964). *Critique of Pure Reason.* 2nd ed. Translated by N. Kemp-Smith. London: Macmillan.

———. (2000). *Critique of Judgment*, n. 51, 5:325. Translated by P. Geyer and M. Matthews. Cambridge: Cambridge University Press.

King, A. (2005). Philosophy and salvation: The apophatic in the thought of Arthur Schopenhauer. *Modern Theology*, 2, 253–74.

McCabe, H. (2002). Aquinas on the Trinity. In *God Still Matters*, pp. XX–XX. London: Continuum.

McInerney, R., ed. and trans. (1998). Introduction and notes to *Thomas Aquinas: Selected Writings.* Harmondsworth, UK: Penguin Books.

Milbank, J. (1999). Intensities. *Modern Theology*, 15, 445–97.

Nietzsche, F. (1998). *Twilight of the Idols*, III.6. Translated by D. Large. Oxford: Oxford University Press.

———. (1999). *The Birth of Tragedy.* Translated by R. Speirs. Cambridge: Cambridge University Press.

———. (2001). *The Gay Science*, III. Translated by J. Nauckhoff. Cambridge: Cambridge University Press.

———. (2002). *Beyond Good and Evil* I. Translated by J. Norman. Cambridge: Cambridge University Press.

Rosser, J. B. (1936). Extensions of some theorems of Gödel and Church. *Journal of Symbolic Logic,* 1, 87–91.

Rubenstein, M.-J. (2003). Unknow thyself: Apophaticism, deconstruction and theology after ontotheology. *Modern Theology*, 19, 387–417.

Stone-Davis, F. (2005). Musical perception and the resonance of the material PhD diss., University of Cambridge. [See esp. Part III.]

Turner, D. A. (2004). *Faith, Reason and the Existence of God.* Cambridge: Cambridge University Press.

Wittgenstein, L. (1961). *Tractatus Logico-Philosophicus*, 6, 432–7. Translated by D. F. Pears and B. F. McGuiness. London: Routledge and Kegan Paul.

Gödel and the Mathematics of Philosophy

CHAPTER 13

Gödel's Mathematics
of Philosophy

Piergiorgio Odifreddi

Mathematicians with an interest in philosophy, such as the present author, find an interest in the latter when they see it as a metaphor – or, even better, an inspiration – for the former. Gödel provides a study case here because he is well known to have declared that (part of) his mathematical work was a direct consequence of his philosophical assumptions. If one takes this to mean Gödel's own philosophy in an academic sense, for example, as crystallized in the *Nachlass*, then it is difficult to make precise sense of his remark, as some scholars have experienced. The following observations intend to show that the difficulties disappear if one interprets Gödel's "philosophical assumptions" in a more popular sense, as meaning assumptions of philosophers whose thought he happened to know and find interesting.

I claim that some of Gödel's main results can be interpreted as being mathematically precise formulations of intuitions of Aristotle, Leibniz, and Kant. We know of a direct cause-and-effect connection only in some of the cases, but this is not the point. What we really care about, as nonprofessional readers, is to show that (part of) philosophy can be reinterpreted as asking questions and suggesting answers that mathematics makes precise. To put it in a more general slogan, in intellectual history, everything happens twice: first as philosophy, then as mathematics.

13.1 The Arithmetization Method (1931)

According to Gerald Sacks (1987), Gödel said that he got the idea of arithmetization from Leibniz. Sacks seems to be skeptical about the causal effect and suggests that Gödel may have only thought of it after the fact.

Be that as it may, it cannot be denied that in Leibniz's (1666) *Dissertatio de arte combinatoria*, one indeed finds a detailed exposition of a method to associate numbers with linguistic notions. However, one also finds a quite surprising naïvety: to associate composite numbers with composite notions, Leibniz proposes (in sect. 69) to use multiplication, thus making decomposition uncertain (because a number may have multiple decompositions).

299

Gödel's improvement on Leibniz was the use of prime exponentiation in place of multiplication. This made decomposition unique by the prime factorization theorem.

13.2 The Incompleteness Theorem (1931)

Kant's main point in *Critique of Pure Reason* (Kant, 1781, 1787) and *Prolegomena to Any Future Metaphysics* (Kant, 1783) is to show the limits of pure reason. A quick rendition of the core of his arguments in modern logical language will help me make my point.

Kant uses a system of twelve concepts of the understanding, to which correspond twelve types of judgments, constituting the basis for a system of first-order modal logic. More precisely, the list of judgments in Kant's wording (and their self-explanatory, one hopes, logical translations) is the following:

Quantity	Quality
Universal ($\forall$)	Affirmative (propositional)
Particular ($\exists$)	Negative ($\neg$)
Singular ($\exists!$)	Infinite (first order)
Relation	**Modality**
Categorical (atomic)	Problematic ($\Diamond$)
Hypothetical ($\rightarrow$)	Assertoric ($\vdash$)
Disjunctive ($\vee$)	Apodeictic ($\Box$)

Kant claims that the list of categories is complete and attempts a proof called the "metaphysical deduction of the pure concepts of the understanding." The word *completeness* here means "functional completeness," in the sense in which the usual connectives of classical logic are sufficient to generate any Boolean function.

Kant then claims that if reason is complete, in the different sense of being able to deal freely with the concepts of understanding, then it is inconsistent. His proof proceeds in two steps.

He first concentrates on the three concepts belonging to the relation group and claims that completeness allows one to consider a limit version of each of them, called a "transcendental idea." More precisely, the three transcendental ideas are *soul*, *first cause*, and *God*: they are obtained, respectively, by pushing to the limit the categories of atomic predicate, implication, and disjunction.[1]

The second step of Kant's proof is to show that the transcendental ideas lead to contradiction, which he does by means of the four antinomies of pure reason.

Gödel's work on the incompleteness theorem can be seen as a formal analogue of Kant's own. Indeed, the statement of the incompleteness theorem is a contrapositive of Kant's formulation: a sufficiently powerful and sound formal system that is not inconsistent must be incomplete. Gödel's proof neatly separates and axiomatizes various aspects of Kant's proof, as follows.

The request that the formal system be sufficiently powerful expresses a weak form of completeness, and it can be rephrased as saying that it is possible to consider a

---

[1] Thus God is identified with an omnicomprehensive disjunction.

formal version of a particular transcendental idea, associated with (i.e., obtained by pushing to the limit) the category of "nonprovability" and expressing the statement "I am not provable."

The role of antinomies is played here by the contradiction according to which, if "I am not provable" were provable in a sound system, it would be true and hence also not provable.

13.3 The Double-Negation Translation (1933)

In *The Metaphysics* (Γ, especially 1006a), Aristotle isolates two basic axioms of being: (1) the noncontradiction principle, that is, $\neg(A \wedge \neg A)$ (denied by Heraclitus), and (2) the excluded middle, that is, $A \vee \neg A$ (denied by Anaxagoras). He asserts that a proof of these principles should not be attempted: not everything can be proved, and it is a sign of good education to know when to stop.

One would expect that at this point, nothing much would remain to be said, but Aristotle makes an unexpected move: he claims that those principles, although impossible to *prove*, can, however, be shown to be impossible to *disprove*, in the sense that the assumption of their negation leads to contradiction.

The details of Aristotle's attempted proof, by a method known as the *élenchos*, are of no interest to us. But his statement is of interest, being just an assertion of the provability of the double-negation of the axioms. This is precisely the core of the double-negation interpretation of classical propositional logic into intuitionistic logic, proven by Kolmogorv in 1925, Glivenko in 1929, and Gödel in 1933.

More precisely, one way of proving the double-negation interpretation (whose statement asserts that if a propositional formula is classically provable, then its double negation is intuitionistically provable) is by showing that the double negation of any tautology is an intuitionistic consequence of instances of double negations of the excluded middle, and then to prove the latter outright in intuitionistic logic (this final step being the formal analogue of the *élenchos*).

Of course, one should always take Aristotle's (or anybody else's) intuitions with a grain of salt: after all, he also claimed (in 1012a) that if the excluded middle failed for some formula, it should fail for all formulas, and that if there were more than two truth values, there should be infinitely many – two assertions proved incorrect by later developments of logic.

13.4 Sets and Classes (1940)

In his discussion of the ontological proof, Kant rephrases his main point about limitations of reason in a way that ties it more to set theory than to the incompleteness theorem. More precisely, he observes (Kant, 1781, 390–91) that one could distinguish God from beings by saying that the latter are defined by sets of atomic predicates, whereas the former is "only" defined by a (proper) class: the natural illusion of reason here thus takes the form that "it is possible to apply to classes the same properties of sets."

The difference between classes and sets is exploited and made precise in the so-called Von Neumann–Bernays–Gödel system, in which classes are taken as primitive and seen as extensions of arbitrary predicates, whereas sets are defined as classes belonging to some (other) class. One can then prove that proper classes (i.e., classes that are not sets, of which God is an example, according to Kant) are indistinguishable from the whole universe in the sense that they can all be put into a one-one and onto correspondence with it (thus uncovering an unexpected pantheistic flavor in Kant's theology).

13.5 Cosmological Models (1949)

A basic assumption of Kant's (1781, 1787) *Critique of Pure Reason* is that space and time do not exist. Rather they are only illusions: although the objects of the external world do appear to us as having spatial and temporal extension, these are not properties of the objects; they are instead a consequence of the structure of our sensorial and mental apparatus. (Kant's terminology is that they are a priori constitutive of the form of our perception.) In particular, we cannot have any objective knowledge of the world and are bound to subjective pictures determined by our human nature.

Gödel (1995, 274) explicitly states that his work on cosmological solutions of the field equations of general relativity was prompted by the question of whether the latter could be shown to be in agreement (or at least, not in disagreement) with Kant's assumption on time.

On the positive side, he showed that it is indeed consistent with general relativity that there is no notion of absolute and objective time. More precisely, he first constructed a model with rotation of the major mass points, a condition that is sufficient (and necessary) to show that there is no absolute time in the model. He then provided stronger solutions in which there are even closed timelike lines: in the latter models, one could travel in the past by going sufficiently far in the future, and thus even the notion of an objective time for individual observers is ruled out.

On the negative side, Gödel argued that Kant's thesis of the unknowability of things in themselves was a subjectivistic exaggeration and not a logical consequence of his assumptions on the a priori character of the form of human perception. In particular, relativity theory itself shows that properties of space and time (such as the relativity of simultaneousness, the non-Euclidicity of space, or the Lorentz contraction) that flatly contradict our a priori intuition can nevertheless be discovered: thus it is possible that science could "go beyond the appearances and approach the world of things" (Gödel, 1995, 244).

13.6 The Ontological Proof (1970)

According to Hao Wang (1987, 195) Gödel said that he got the idea for his version of the ontological proof from Leibniz. It is clear from the following analysis that Gödel refers to the short paper *Quod Ens perfectissimum existit*, in which Leibniz (1676) defines God as a being having all perfections and proves that God exists because existence is a perfection. The argument is, of course, Descartes's version of Anselmus's own, and Leibniz's improvement consisted in an attempt to show that the definition was not contradictory.

Despite that, according to Leibniz, his argument satisfied Spinoza, it could certainly not satisfy anybody aware of compactness (such as Gödel, who first discovered it). Leibniz indeed makes the following hair-raising inference: he only shows that perfections are compatible *two by two*, and then deduces from this that they are all compatible.

To avoid such an embarrassing mistake, Gödel reformulated the argument as follows (where we have dropped any reference to modalities because it was later proved that they all collapse in the axiom system used by Gödel). He defines God as a being having all positive properties and assumes that the positive properties form an ultrafilter on the universe: thus closure under intersection (expressing that the positive properties are compatible two by two) is simply postulated. Because any ultrafilter of subsets of a finite set is principal, the existence of God (i.e., the intersection of the ultrafilter) follows immediately from the assumption of the finiteness of the universe.

To avoid this unsatisfying hypothesis, it is enough to notice that it is used in the argument only to show that the ultrafilter contains its own intersection: if the latter is assumed directly, the result follows without any finiteness assumption. But the intersection of the ultrafilter is God itself, and to assume that it is in the ultrafilter means to assume that "being God" is a positive property, which is what Gödel indeed does.

13.7 Conclusion

I have shown that certain passages of Gödel's favorite philosophers can be seen as an inspiration for, or reinterpreted in the light of, some formal developments of his mathematical work. Historical evidence showing factual causal connections, which indeed exists in some of the cases, is, however, not relevant to our main point, namely, that both mathematicians and philosophers may profit from a nonacademic reading of philosophy.

References

Aristotle. *The Metaphysics.*

Gödel, K. (1995). *Collected Works. Vol. III: Unpublished Essays and Lectures.* Edited by S. Feferman et al. New York: Oxford University Press.

Kant, I. (1781). *Critique of Pure Reason.* 1st ed.

Kant, I. (1783). *Prolegomena to Any Future Metaphysics.*

Kant, I. (1787). *Critique of Pure Reason.* 2nd ed.

Leibniz, G. W. (1666). *Dissertatio de arte combinatoria* [*Dissertation on the Art of Combinations*].

Leibniz, G. W. (1676). *Quod Ens perfectissimum existit.*

Sacks, G. Pers. comm.

Wang, H. (1987). *Reflections on Kurt Gödel.* Cambridge, MA: MIT Press.

Gödel and Philosophical
Theology

CHAPTER 14

Gödel's Ontological Proof and Its Variants

Petr Hájek

In the early 1970s, we learned that Gödel had produced a proof of the existence of God after he showed it to Dana Scott, who discussed it in a seminar at Princeton. Notes began to circulate, and the first public analysis of the proof was performed by Sobel (1987). Only after Gödel's death, in the third volume of his collected works (Gödel, 1995), was the proof finally published. It is just one page, preceded by an extensive and very informative introduction by Adams (Gödel, 1995). The volume also contains notes from Gödel's *Nachlass*, dated 1940, containing his first drafts of the proof. Presently, there exist several papers on the topic, including two extremely interesting monographs (Sobel, 2004; Fitting, 2002). A very important variant of Gödel's system, resulting from the first criticisms made of it, was generated by Anderson (1990). Anderson's variant will play an important role in this chapter.

Gödel is famous for his completeness and incompleteness theorems as well as for his work with set theory, so his ontological proof has never received the same attention. The proof belongs to the family of *ontological* arguments, that is to say, arguments that try to establish the existence of God by relying only on pure logic. Such arguments were presented by Anselm (1033–1109), Descartes (1598–1650), Leibniz (1646–1716), and others, and Gödel is known to have studied particularly Leibniz's works (the two books mentioned earlier are recommended for information on the old ontological proofs and their relation to Gödel's proof). Here let us only mention that even if the old proofs were clearly not formal proofs – in the sense of formal logic – and did not use modalities, they show a striking similarity in their form: they (try to) argue that God exists (actually, really, necessarily) if God is possible (consistent, present in our mind), and then to prove that God is indeed possible. Details and weaknesses of these proofs are analyzed particularly in Sobel's book. Hartshorne (1962) seems to be the first to formalize the modal substance of these proofs (see later); Gödel's proof has this structure too.

This chapter concentrates on the formal (mathematical) aspects of Gödelian onto-logical proofs; philosophical aspects are completely ignored. However, at the end, we shall comment on the (possible) relevance of ontological proofs to religious faith. Our plan is as follows: in Section 14.1, we specify the formal logic used (first-order modal S5); Section 14.2 discusses Gödel's original version (as presented by Scott), and we analyze the first criticisms of it (made by Sobel and Magari); Section 14.3 focuses on Anderson's variant and its variants; in Section 14.4, we analyze criticism formulated by Oppy; Section 14.5 contains "Miscellanea," for example, the use of different modal logics, the question of proving the devil's existence, a recent variant of Gödel's original proof by Koons, and some other matters; in Section 14.6, we ask what we have learned about Gödel; finally, in Section 14.7, we consider the importance of ontological proofs for religious belief. Section 14.8 concludes.

Let us close this introduction, for the reader's pleasure, by quoting at least a part of the famous Anselm's Proslogion II as an example of an ontological proof (Hopkins and Richardson, 1974):

> Therefore, Lord, Giver of understanding to faith, grant me to understand – to the degree You deem best – that you exist, as we believe, and that you are what we believe You to be. Indeed, we believe You to be something than which nothing greater can be thought ... even the Fool is convinced that something than which nothing greater can be thought exists at least in his understanding; for when he hears of this being, he understands [what he hears], and whatever is understood is in the understanding. But surely that than which a greater cannot be thought cannot be only in the understanding. For if it were only in the understanding, it could be thought to exist also in reality – which is greater [than existing only in the understanding]. Hence, without doubt, something than which a greater cannot be thought exists both in the understanding and in reality.

14.1 The Logic Used

We shall use the best-known modal logic S5 (for other possibilities, see remarks in Section 14.5). The reader is assumed to be familiar with classical propositional and first-order predicate logic as well as with (at least) propositional S5 (for detailed treatment of first-order S5, see Fitting and Mendelsohn, 1990). As usual, we use $\Box$ for the modality of necessity and $\Diamond$ for possibility. Let us recall the propositional axioms of S5. Besides the axioms of classical propositional logic, these are as follows

$$\Box(\varphi \to \psi) \to (\Box\varphi \to \Box\psi)$$

$$\Box\varphi \equiv \Box\Box\varphi$$

$$\Box\varphi \equiv \Diamond\Box\varphi$$

$$\Box\varphi \to \varphi.$$

Possibility is defined by

$$\Diamond\varphi \equiv \neg\Box\neg\varphi.$$

Deduction rules are modus ponens and necessitation: from φ infer $\Box\varphi$. Semantics come into play when considering Kripke models consisting of a nonempty set W of possible worlds and an evaluation w of each propositional variable in each possible world by a truth value 0 or 1. Connectives are evaluated in each possible world by the usual truth tables; $\Box\varphi$ is true in the model if and only if φ is true in each possible world. Propositional S5 is complete with respect to these models; a formula is provable if and only if it is an S5-tautology (which is true in all Kripke models). (In particular, recall that $\Diamond\Box\varphi \equiv \Box\varphi$ is an S5-tautlogy.)

Let us illustrate the use of propositional S5 by presenting Hartshorne's proof. Let q stand for "God exists." Assuming that $\Diamond q$, and $q \rightarrow \Box q$ (Anselm's principle), then

$$\Box(q \rightarrow \Box q),$$
$$\Diamond q \rightarrow \Diamond\Box q,$$
$$\Diamond\Box q \rightarrow \Box q,$$
$$\Diamond q \rightarrow \Box q,$$
$$\Box q.$$

As stated earlier, this is, in some sense, the common structure of ontological proofs discussed by us. The question is how (or from what) to prove the two assumptions. Let's turn to the first-order S5. There are variables for an object's alias beings (lowercase letters) and for properties (uppercase latters), one constant G for the property being godlike, and one predicate P of properties (being a positive property). This can be understood either as a second-order logic or as a first-order, two-sorted logic with one binary predicate $Appl$ with the formula $Appl(Y, x)$, abbreviated as $Y(x)$. (The reader may choose which version he or she likes.) We shall also use the extensional equality predicate $=$ for beings (thus we assume $x = y \rightarrow \Box(x = y)$).

We shall discuss two variants of the logic: the first has Kripke models with fixed domain $(W, M, Prop, Appl, P_K, G_K)$, where W is a nonempty set of possible worlds, M a nonempty set of objects (beings), $Prop$ a nonempty set of properties, $G_K \in Prop$, $P_K \subseteq Prop \times W$, and $Appl \subseteq Prop \times M \times W$. This has the following meaning: $(X, w) \in P_K$ means that X is a positive property in the world w and that $(X, a, w) \in Appl$ means that the being a has the property X in the world w. In particular, $(G_K, a, w) \in Appl$ means that a is a godlike object in the world w. Satisfaction is defined in the usual way, in particular, $\Box\varphi$ (φ being a sentence) is true in the model if and only if φ is true in each possible world.

Note that with these models, the predicate modal logic S5, with its usual axioms (just axioms, rules of classical predicate logic, of propositional modal S5, Barcan formula $\Box(\forall x)\varphi \equiv (\forall x)\Box\varphi$, and the rule of necessitation (from φ infer $\Box\varphi$)) is complete (see Fitting and Mendelsohn, 1990).

The second set of semantics has, in addition to the preceding, among the predicates a constant predicate E of (actual) existence; that $E(u)$ is true in a world, w means that the object u actually exists in this world. An axiom demands that in each world, at least one object actually exists. The formula $(\exists^E u)\varphi(u)$ stands for $(\exists u)(E(u) \wedge \varphi(u))$; similarly, $(\forall^E u)\varphi(u)$ stands for $(\forall u)(E(u) \rightarrow \varphi(u))$ (see, e.g., Fitting, 2002).

Furthermore, we assume the axiom schema with full comprehension. For each formula $\varphi(u, \dots)$ (the dots standing for possible other free variables for beings or properties), and each variable Y not occurring in φ, the axiom says

$$(\forall \dots)(\exists Y)\Box(\forall u)(Y(u) \equiv \varphi(u, \dots))$$

(satisfying that φ is a property). Full comprehension demands restriction to Kripke models such that properties defined by formulas exist in the model. (One can also work with weaker systems not containing the full comprehension; see comments later.)

14.2 Gödel's Proof

We reproduce Gödel's original system (as presented by Scott) and his proof and prove some additional theorems in the system. Recall that $P(X)$ says that X is a positive property, and $G(a)$ says that a is a godlike being. Furthermore, $X \subseteq Y$ (defined later) says that X entails Y, $X\mathbf{Ess}a$ that the property X is an essence of a (Gödel's name), and $NE(a)$ that a necessarily exists. $\neg X$ denotes a property Y satisfying $(\forall a)(Y(a) \equiv \neg X(a))$. The axioms follow:

(G1)	$P(X) \equiv \neg P(\neg X)$
(def)	$X \subseteq Y \equiv (\forall u)(X(u) \rightarrow Y(u))$
(G2)	$(P(X) \wedge \Box(X \subseteq Y)) \rightarrow P(Y)$
(def)	$G(a) \equiv (\forall Y)(P(Y) \rightarrow Y(a))$ (Godlike)
(G3)	$P(G)$
(G4)	$P(Y) \rightarrow \Box P(Y)$
(def)	$X\mathbf{Ess}a \equiv X(a) \wedge (\forall Y)(Y(a) \rightarrow \Box(X \subseteq Y))$
(def)	$NE(a) \equiv (\forall X)(X\mathbf{Ess}a \rightarrow \Box(\exists a)X(a))$
(G5)	$P(NE)$

In words, a property is positive if and only if its negation is not positive; a property necessarily entailed by a positive property is itself positive; a positive property is necessarily positive; and both godlikeness and necessary existence are positive. This theory (over the predicate modal logic S5) will be denoted by $\mathcal{G}O$.

Theorem 14.2.1 $\mathcal{G}O$ *proves* $P(X) \rightarrow \Diamond(\exists a)X(a)$.

PROOF   The following chain of implications is provable:
$$\Box(\forall u)\neg X(u) \rightarrow \Box(\forall u)(X(u) \rightarrow \neg X(u)) \rightarrow (P(X) \rightarrow \neg P(X)) \rightarrow \neg P(X).$$
Thus we have proved $\neg\Diamond(\exists a)X(a) \rightarrow \neg P(X)$, which, in turn, proves the theorem.

Remark $\mathcal{G}O$ proves $G(u) \equiv (\forall Y)(P(Y) \equiv Y(u))$.
Indeed, $\leftarrow$ is trivial; for $\rightarrow$, observe $(G(u) \wedge Y(u) \wedge \neg P(Y)) \rightarrow ((G(u) \wedge P(\neg Y)) \rightarrow \neg Y(u)$, a contradiction.

Theorem 14.2.2 $\mathcal{G}O$ *proves* $G(u) \rightarrow G\mathbf{Ess}u$.

PROOF   The following are provable:
$(G(u) \wedge Y(u)) \rightarrow P(Y) \rightarrow \Box P(Y);$ $P(Y) \rightarrow (\forall x)(G(x) \rightarrow Y(x)),$ thus
$\Box P(Y) \rightarrow \Box(G \subseteq Y)$, hence $(G(u) \wedge Y(u)) \rightarrow \Box(G \subseteq Y)$.

Theorem 14.2.3 $\mathcal{G}O$ *proves* $\Box(\exists x)G(x)$

PROOF $\mathcal{G}O$ proves $\Diamond(\exists x)G(x)$ by Theorem 14.2.1, $G(x) \rightarrow \mathbf{NE}(x)$ because $P(\mathbf{NE})$, thus $G(x) \rightarrow \Box(\exists x)G(x)$ because $G\mathbf{Ess}\,x$. Take q to be $(\exists x)G(x)$ and apply Hartshorne's proof.

This completes Gödel's proof of the necessary existence of a godlike being. We present some further consequences of the axioms.

Theorem 14.2.4 $\mathcal{G}O$ *proves* $(G(x) \wedge Y(x)) \to \Box Y(x)$.

PROOF   Let I_x be the property of being x, that is to say, satisfying $I_x(z) \equiv z = x$. We have to prove that $(G(x) \wedge Y(x)) \to \Box Y(x)$. We already have $\Box(\exists z)G(z)$ (using all the axioms of $\mathcal{G}O$). Now observe the following provable implications:

$(G(x) \wedge Y(x)) \to (P(Y) \wedge \Box P(Y))$,
$\Box(G(x) \to (P(Y) \to (Y(x) \wedge I_x(x))))$,
$\Box P(Y) \to \Box(G(x) \to (Y(x) \wedge I_x(x))))$,
$\Box P(Y) \to \Box(\exists z)(G(z) \wedge Y(z) \wedge I_x(z))$,
$\Box P(Y) \to \Box(G(x) \wedge Y(x))$,
$(G(x) \wedge Y(x)) \to \Box Y(x)$.

Theorem 14.2.5 $\mathcal{G}O$ *proves* $G(u) \equiv (\forall Y)(P(Y) \equiv \Box Y(u))$.

This follows immediately from the preceding remark and theorem. This equivalence says that u is godlike if and only if the properties that u necessarily has are exactly all positive properties. It will play an important role in Anderson's variant of the ontological proof. (It appears that provability of this theorem in $\mathcal{G}O$ has not previously been observed.)

The next theorem is from Sobel (1987) and is called the "theorem on the collapse of modalities in $\mathcal{G}O$."

Theorem 14.2.6   *For each formula* φ, $\mathcal{G}O$ *proves* $(\varphi \to \Box\varphi)$.

PROOF   The following chain of implications is provable:
$\varphi \to [G(x) \to (G(x) \wedge \varphi)] \to [G(x) \to \Box(G(x) \wedge \varphi)] \to [G(x) \to \Box\varphi]$.
Thus $G(x) \to (\varphi \to \Box\varphi)$,   hence $(\exists x)G(x) \to (\varphi \to \Box\varphi)$,   and   therefore
$\varphi \to \Box\varphi$.

Consequently, in $\mathcal{G}O$, for each formula φ, the formulas φ, $\Box\varphi$, and $\Diamond\varphi$ are mutually equivalent, and modalities become superfluous. This appears to be a significant disadvantage of $\mathcal{G}O$ and led Anderson to his emendation of the system (Anderson, 1990; he says there that, unfortunately, too much follows from these (Gödel's) axioms). But surprisingly, in a very recent paper, Sobel (2006) argues that it is possible that Gödel knew that his axioms lead to the collapse of modalities and accepted this for some philosophical reasons (see Adams in Gödel, 1990, 400). Nevertheless, our main attention will be paid to variants of the system not having a collapse of modalities.

Among the first critics of Gödel's proof, let us mention Magari (1988). He says, *"Teofili hanno spesso fornito ingegnosi argomenti … esistono anche teofobi (io lo sono di tutto cuore)"* (he declares himself to be a "theofob") and *"non è più facile ammettere gli assiomi che ammettere direttamente il teorema"* (it is not easier to accept the axioms than to accept the conclusion). This second quotation is interesting with respect to the possible religious significance of Gödel's proof. Magari claims that the sentence $\vdash \Box(\exists x)G(x)$ is provable using only the first three axioms (G1)–(G3) of Gödel, but this is false. (My counterexample in Hájek (1996) satisfies even Gödel's

(G1)–(G4).) Finally, let us mention Fitting's observation (in his monograph) that in $\mathcal{G}O$, the axiom $P(G)$ (godlikeness is positive) can be equivalently replaced by $\Diamond(\exists x)G(x)$.

14.3 Anderson's Variant and Its Variants

Anderson's variant (mentioned at the end of the preceding section) consists first in weakening Gödel's axiom (G1) to $P(X) \to \neg P(\neg X)$. Thus, at most, one of the properties, X or $\neg X$, may be positive, but not both (but possibly neither of them). The second difference in Anderson's variant is made by changing the definitions of godlikeness and essence, keeping the axioms (G2)–(G5) and the definition of necessary existence as they are but using the new notions of godlikeness and necessary existence. Call these axioms now (A2)–(A5) (Anderson's version). The definition of godlikeness is now

(def) $G(u) \equiv (\forall Y)(P(Y) \equiv \Box Y(u))$.

(Recall that we have shown that this formula is provable in $\mathcal{G}O$.) We postpone any discussion of Anderson's notion of essence and necessary existence because we shall show that in Anderson's system (which we may call $\mathcal{A}O$), they are redundant. (This was first shown in Hájek (1996)). For the ontological proof in the logic with fixed domains, we only need the following three axioms (the three first axioms in Anderson's theory; $X \subseteq Y$ stands for $(\forall u)(X(u) \to Y(u))$):

(A1) $P(X) \to \neg P(\neg X)$, where $(\neg X)(u) \equiv \neg X(u)$,
(A2) $(P(X) \wedge \Box(X \subseteq Y)) \to P(Y)$,
(A3) $P(G)$, where $G(u) \equiv (\forall Y)(P(Y) \equiv \Box Y(u))$.

In words, the negation of a positive property is not positive; a property necessarily implied by a positive property is positive; and godlikeness is positive (u is godlike if properties it necessarily has are exactly all positive properties). This theory will be denoted (in accordance with Hájek, 2002b) by $\mathcal{A}O_0$.

Theorem 14.3.1 $\mathcal{A}O_0$ proves $\Box(\exists x)G(x)$, $(\exists x)\Box G(x)$ and also uniqueness:

$$(\forall x, y)((G(x) \wedge G(y)) \to x = y).$$

PROOF $\mathcal{A}O$ proves $P(X) \to \Diamond(\exists u)X(u)$ because assuming X is necessarily empty gives $\Box(X \subseteq \neg X)$; thus $P(X)$ gives $P(\neg X)$, which contradicts (A1). This proves $\Diamond(\exists x)G(x)$. Furthermore, $\mathcal{A}O_0$ proves $G(u) \to \Box G(u)$ using (A3) and the definition of G. From this follows $\Diamond(\exists u)G(u) \to \Box(\exists u)G(u)$ and hence $\Box(\exists u)G(u)$. Thus one gets $(\exists x)G(x)$, $(\exists x)\Box G(x)$. (Indeed, because $G(x) \to \Box G(x)$, we can conclude that $\Box(\exists x)G(x) \to \Box(\exists x)\Box G(x) \to (\exists x)\Box G(x)$.) To prove uniqueness, observe that for each x, there exists a property I_x such that $\Box(\forall y)(I_x(y) \equiv y = x)$ (being x). Clearly $\Box I_x(x)$ for all x; if $G(x)$, then $P(I_x)$; thus if $G(y)$, then $\Box I_x(y)$, hence $\Box(y = x)$.

Theorem 14.3.2 $\mathcal{A}O_0$ proves $P(Y) \to \Box P(Y)$ and $P(Y) \to P(\Box Y)$. (The former formula is Anderson's axiom (A4).)

PROOF   First,

$$G(g) \to (P(Y) \equiv \Box Y(g)),$$
$$\Box G(g) \to (\Box P(Y) \equiv \Box Y(g)),$$
$$\Box G(g) \to (\Box P(Y) \equiv \Box Y(g) \equiv P(Y)),$$
$$P(Y) \equiv \Box P(Y).$$

We prove the second formula. Indeed, $\Box(\Box Y \to Y)$ is a theorem, thus $P(\Box Y)$ implies $P(Y)$. Conversely, assuming $G(x)$, we get

$$P(Y) \to \Box Y(x) \to \Box(\Box Y(x)) \to P(\Box Y).$$

Following Anderson, let us define (individual) essence and necessary existence as follows:

$$X\mathbf{Ess}u \equiv (\forall Y)(\Box Y(u) \equiv \Box(X \subseteq Y))$$
$$\mathbf{NE}(x) \equiv (\forall Y)(Y\mathbf{Ess}x \to \Box(\exists u)Y(u)).$$

Theorem 14.3.3 $\mathcal{AO}_0$ *proves* $(\forall x)I_x\mathbf{Ess}x$, $(\forall x)\mathbf{NE}(x)$ *and hence* $P(\mathbf{NE})$. *(The last formula is Anderson's axiom (A5).)*

PROOF   The first claim is evident from definition, the second from the provability of $\Box I_x(x)$ and from the provability of the formula $(X\mathbf{Ess}u \wedge Y\mathbf{Ess}u) \to (X \subseteq Y \wedge Y \subseteq X)$. The last formula follows from the fact that any property that all objects have is positive.

Summarizing, $\mathcal{AO}_0$ proves all the axioms (A1)–(A5) of Anderson's theory $\mathcal{AO}$.

Now let us turn to the second set of models, which has among the predicates, in addition to the preceding, a constant predicate E of (actual) existence. Recall that $E(u)$ is true in a world w means that the object u actually exists in this world and that the formula $(\exists^E u)\varphi(u)$ stands for $(\exists u)(E(u) \wedge \varphi(u)$; similarly, $(\forall^E u)\varphi(u)$ stands for $(\forall u)(E(u) \to \varphi(u))$ (see, e.g., Fitting, 2002).

Over this logic, we keep the axioms (A1), (A2), and (A3) and only change the definition of godlikeness:

$$G(x) \equiv (E(x) \wedge (\forall Y)(P(Y) \equiv \Box Y(u))).$$

The theory will be called $\mathcal{AO}E_0$.

Theorem 14.3.4 $\mathcal{AO}E_0$ *proves* $(\exists x)\Box G(x)$, *and hence* $\Box(\exists^E x)G(x)$ *(necessary actual existence of a godlike being) and also uniqueness of the godlike being.*

PROOF   Clearly the theory proves $G(x) \to \Box G(x)$ and $\Diamond(\exists x)G(x)$, that is, $(\exists x)\Diamond G(x)$. From $G(x) \to \Box G(x)$, we get $\Diamond G(x) \to \Box G(x)$ and $(\exists x)\Diamond G(x) \to (\exists x)\Box G(x)$; thus we get $(\exists x)\Box G(x)$. This gives $\Box(\exists x)G(x)$ and hence $\Box(\exists^E x)G(x)$ (because $G(x)$ implies $E(x)$). Uniqueness uses the property I_x as earlier.

Define $X\mathbf{Ess}x$ as earlier and prove that I_x is the individual essence of x for each x, but change the definition of necessary existence to

$$\mathbf{NE}(x) \equiv (\forall Y)(Y\mathbf{Ess}x \to \Box(\exists^E y)Y(y)).$$

Theorem 14.3.5 $\mathcal{AO}E_0$ proves $\mathbf{NE}(x) \equiv \Box E(x)$ and also $P(\mathbf{NE})$.

PROOF   We know that each essence of x is necessarily equivalent to I_x; thus $\mathbf{NE}(x) \equiv \Box(\exists^E u)I_x(u) \equiv \Box E(x)$. Furthermore, for each x, $G(x) \to \Box G(x) \to \Box E(x)$, and thus $\Box(G \subseteq \Box E)$ and hence $P(\Box E)$.

In the rest of this section, we give an alternative presentation of the theory $\mathcal{AO}E_0$ in the logic with variable domains. (This part may be omitted by a hurrying reader.) We now will call the theory $\mathcal{AO}E_0$, $\mathcal{AO}E_{01}$ and its predicate of godlikeness G_1. Now define $X \subseteq^E Y$ to be $(\forall^E x)(X(x) \to Y(x))$, equivalently $(\forall x)((E(x) \wedge X(x)) \to Y(x))$ (see Fitting, 2002). Take $\mathcal{AO}E_0$ and change the axiom (A2) to (A2E), saying $(P(X) \wedge \Box(X \subseteq^E Y)) \to P(Y)$. Keep the original definition of godlikeness; that is, let $G_2(x) \equiv (\forall Y)(P(Y) \equiv \Box Y(x))$; the axiom (A3) becomes $P(G_2)$. Call this theory $\mathcal{AO}E_{02}$.

Theorem 14.3.6 $\mathcal{AO}E_{02}$ proves $(\exists x)\Box(G_2(x) \wedge E(x))$ and hence $\Box(\exists^E x) G_2(x)$. It also proves $(\exists! x)G_2(x)$.

PROOF   *Claim 1:* The theory proves $P(X) \to \Diamond(\exists^E u)X(u)$. Indeed, $(\Box(\forall^E u)\neg X(u)) \to (\Box(X \subseteq^E \neg X)) \to (P(X) \to P(\neg X)) \to \neg P(X)$. Consequently, the theory proves $\Diamond(\exists^E x)G_2(x)$ and hence $(\exists x)\Diamond(E(x) \wedge G_2(x))$.
   Claim 2: The theory proves $\Diamond(E(x) \wedge G_2(x)) \to \Box(E(x) \wedge G_2(x))$. Indeed, $\vdash \Box(G_2 \subseteq^E (G_2 \wedge E))$, which gives $\vdash P(G_2 \wedge E)$ and hence $\vdash (G_2(u) \wedge E(u)) \to \Box(G_2(u) \wedge E(u))$. From this, we get $\vdash \Diamond(G_2(u) \wedge E(u)) \to \Box(G_2(u) \wedge E(u))$. The rest is evident. Uniqueness is proved as usual.

Theorem 14.3.7 $\mathcal{AO}E_{01}$ proves $\mathcal{AO}E_{02}$, and $G_1(x) \equiv G_2(x)$.

PROOF $X \subseteq^E (X \wedge E)$, thus $\vdash P(X) \to P(X \wedge E)$ and $(P(X) \wedge \Box(X \subseteq^E Y)) \to P(Y)$. We see that $\mathcal{AO}E_{01}$ proves (A2E). Clearly $\vdash \Box(G_1(x) \to G_2(x))$, thus $\vdash P(G_2)$. Furthermore, $\vdash (\exists! x)G_1(x)$ and, also, $\vdash (\exists! x)G_2(x)$ (because all axioms of $\mathcal{AO}E_{02}$ are provable). This gives $\vdash \Box(\forall x)((G_1(x) \wedge G_2(y)) \to x = y)$.

Theorem 14.3.8 $\mathcal{AO}E_{02}$ proves $\mathcal{AO}E_{01}$ and $G_1(x) \equiv G_2(x)$.

PROOF   Over $\mathcal{AO}E_{02}$, $\vdash (P(X) \wedge \Box(X \subseteq Y)) \to (P(X) \wedge \Box(X \subseteq^E Y)) \to P(Y)$, thus $\vdash$(A2). Moreover, $\vdash P(G_2 \wedge E)$, thus $P(G_1)$. The rest is as in the previous proof.

14.4 Oppy's Criticism

First, let us quickly mention a criticism from Oppy (1996). The idea was to define a god*-like being to be a being necessarily satisfying many but not all positive properties (and no nonpositive properties) and prove its necessary existence. Without going into details, let us prove a theorem showing that this is not possible.

Theorem 14.4.1 *Let AO_0 be the theory over logic with fixed domain, as earlier. Let AO_0' be an Oppy-style extension of AO_0 with a new predicate G^* for god*-like such that T' proves the following:*

(B1) $G^*(x) \to (\forall Y)(\Box Y(x) \to P(Y))$,
(B2) $G^*(x) \to \Box G^*(x)$,
(B3) $(\exists x)G^*(x)$,
(B4) $(G^*(x) \wedge G^*(y)) \to x = y$.

Then AO_0' proves $G^*(x) \to G(x)$. In words, assume that AO_0' proves that god*-likeness of x implies that each property that x necessarily has is positive; god*-likeness implies necessary god*-likeness; and there is exactly one god*-like being. Then AO_0' proves that the unique god*-like being is the same as the unique godlike being.

PROOF First, observe that (B2) and (B1) imply that G^* is positive. Second, introduce a constant g^* for the unique god*-like being (and g for the unique godlike being). Owing to uniqueness, for each property Y, $Y(g^*)$ is equivalent to $(\forall x)(G^*(x) \to Y(x))$. Let Y be an arbitrary positive property; we claim $\Box Y(g^*)$ (proving this we shall be ready). Indeed,

$\Diamond \neg Y(g^*) \to (\forall x)(G^*(x) \to \Diamond \neg Y(x))$ (see earlier), hence
$\Diamond \neg Y(g^*) \to \Box(\forall x)(G^*(x) \to \Diamond \neg Y(x))$ (by necessitation, $\Diamond$ being equivalent to $\Box \Diamond$),

thus $\Diamond \neg Y(g^*) \to P(\Diamond \neg Y)$ (because $\Diamond \neg Y$ necessarily follows from G^*, as we have just proved). But $\Diamond \neg Y$ is the same as $\neg \Box Y$, and $\Box Y$ is positive, as proved earlier. Thus we get a contradiction with the axiom (A1). We have proved $\Box Y(g^*)$. Thus g^* satisfies the formula defining the godlike being, and hence $g^* = g$. This completes the proof.

This is practically the same reasoning as employed by Gettings (1999). Oppy (2000) also suggests another "Gaunilian parody" of the proof of a godlike being. We slightly simplify (owing to the fact that we need fewer axioms than Anderson's original system), and we systematically speak of theories, not only of their models. Let P^* be a new predicate of properties (possibly definable in AO_0, or just extending its language); we postulate the following (still over the logic with fixed domains):

(A1*) $P^*(X) \to \neg P^*(\neg X)$, where $(\neg X)(u) \equiv \neg X(u)$,
(A2*) $(P^*(X) \wedge \Box(X \subseteq Y) \to P^*(Y))$,
(A3*) $P^*(G^*)$, where $G^*(u) \equiv (\forall Y)(P^*(Y) \equiv \Box Y(u))$.

These are the same axioms as earlier, only with P^*, G^* instead of P, G. Clearly the extended system proves $\Box(\exists u)G^*(u)$ (plus uniqueness), but assuming that P is not equivalent to P^*, we clearly get that the unique godlike being is different from the unique god*-like being. Can we assume this? Is it consistent? The answer is yes, and using the logic of a fixed universe, we can define such systems under the assumption that there are at least two objects. Call a predicate P^* on properties definable in $\mathcal{A}O_0$ *Gaunilian* if $\mathcal{A}O_0$ proves (A1*)–(A3*).

Theorem 14.4.2 *Gaunilian predicates are in one-one correspondence with objects: for each Gaunilian predicate P^*, let g^* denote the corresponding god*-like being, and for each object constant a, let $P^a(Y)$ be defined as $\Box Y(a)$. Then P^a is a Gaunilian predicate, and its god^a like being is a. From $a \neq b$, it follows that $G^a(a) \wedge \neg G^b(a)$ (different elements define nonequivalent Gaunilian predicates).*

PROOF   For P^a, (A1*) and (A2*) are evident. $G^a(u)$ becomes $(\forall Y)(\Box Y(a) \equiv \Box Y(u))$, which implies $\Box G^a(a)$. Thus $P^a(G^a)$, and we have (A3*). The rest is evident. In particular, if $a \neq b$, then $\Box I^a(a) \wedge \neg I^b(a)$, hence $\neg G^b(a)$.

Thus, in each model, each object determines a system of Gaunilian properties whose god*-like object is this object; systems of Gaunilian properties are in one-one correspondence with objects of the model.

Now let us see what happens when we turn to the logic with variable domains of actually existing objects. Having a Gaunilian predicate P^* with the modified definition of G^* (i.e., $G^*(u) \equiv (E(x) \wedge (\forall Y)(P^*(Y) \equiv \Box Y(u))))$, we prove $\Box(\exists^E x)G^*(x)$, that is to say, that the unique object g^* satisfying G^* satisfies $\Box E(g^*)$. (The theory may be called $\mathcal{A}OE'_0$.) Then we get the following.

Theorem 14.4.3 *In $\mathcal{A}OE'_0$, Gaunilian predicates are in one-one correspondence with necessarily actually-existing objects; one of them is the (starting) predicate P of positivity and the corresponding godlike object.*

PROOF   By obvious modification of the preceding proof, we can arrive at this one. In particular, $G^a(u)$ is now equivalent to $E(u) \wedge (\forall Y)(\Box Y(u) \equiv \Box Y(a))$, which gives $\neg E(a) \to \neg G^a(a)$ and $E(a) \to [E(a) \wedge (\forall Y)(\Box Y(a) \equiv \Box Y(a))] \to G^a(a)$.

But we cannot prove existence of objects different from the godlike object, combined with satisfying necessary actual existence from any assumption of (plain) existence of many objects. One easily constructs a model of our theory over the logic with variable domains in which there are as many objects as you want, but the godlike object is the only object necessarily actually existing; that is to say, the model satisfies $(\forall x)(\Box E(x) \to G(x))$. Let us give a trivial example:

	G	M	E
g	1	0	1
m	0	1	0
g	1	0	1
m	0	1	1

with G being godlike; M being me; g being godlike being; m being me; E being actual existence; and on upper world (heaven) and lower world (earth).

14.5 Miscellanea

We shall discuss possible variants of the underlying logic, further variants of the axioms on godlikeness, and the question of proving the existence of the devil. First, let us mention that some authors prefer to present Gödel's system as a system in second-order modal logic, whereas here we prefer to use two-sorted, first-order modal logic. This is unimportant because the second-order logic (and second-order modal logic) with its general Henkin-style (Henkin-Kripke) semantics is equivalent to the corresponding two-sorted, first-order logic (and there is no reason to insist on the standard semantics of second-order logic in analyzing Gödel's ontological proof).

However, what can be discussed are modifications of the underlying modal logic and of the comprehension schema. I argue that the proof is made by using the logic KD45, called the "logic of belief," which is just S5 without the axiom $\Box\varphi \to \varphi$ (Hájek, 1996). Then $\Diamond\varphi$ can be read "φ is admitted" and $\Box\varphi$ read "φ is believed." This reading seems to be of some interest. In the same paper, a schema of "cautious comprehension" is suggested, and it is shown that the system $(\mathcal{G}O)_{\text{caut}}$ (i.e., $\mathcal{G}O$ with only cautious comprehension) is faithfully interpretable in $\mathcal{A}O_0$ (with its full comprehension) (see also Hájek, 2002a).

It is even possible to assume only existence of properties used in the proof and its discussion (as G, I_x, NE) and closedness of properties under negation. One such example is the work of Szatkowski (2005), where the author offers twenty variants of modal logic (second order but with Henkin-Kripke semantics) and shows them being complete with respect to a natural axiomatization. In the appendix, he shows that in all of them, Gödel's ontological proof (of $\Box(\exists x)G(x)$) works.

Kovač (2003) restricts the comprehension schema to prevent the collapse of modalities but keeps it reasonably strong. In particular, it is assumed that the formula $\varphi(x)$ in the comprehension schema does not contain any closed subformula. Axioms on modalities are not uniquely fixed; rather any modal system making the proof work is admitted.

I argue for a further weakening of Anderson's (A1) and (A2), motivated by the observation that (A2) would force us, if we had a predicate "devillike," to infer that *being godlike or devillike* is a positive property (Hájek, 2002b). This may be considered counterintuitive; if so, then the suggestion is to replace (A1) and (A2) by the axiom (A12), saying $(P(X) \wedge \Box(X \subseteq Y)) \to \neg P(\neg Y)$. In words, if X is positive, and X necessarily entails Y, then the negation of Y is not positive. Also, this weakened system proves $\Box(\exists x)G(x)$ (and has no collapse of modalities). Furthermore, a version exists with varying domains.

Koons (2005) extensively discusses possible restrictions of the comprehension schema and modifies the original system, $\mathcal{G}O$, by replacing the axiom (G5) of Gödel (positivity of NE) by (G6) (he names Gödel's axioms (A1)–(A5), and thus he names his axiom (A6)). Koon's axiom is just $P(X) \to P(\Box X)$. He shows that the modified system proves $\Box(\exists x)G(x)$.

Both the papers by Kovač and by Koons pay much attention to philosophical aspects; moreover, Sobel (2006) further analyzes these aspects of Kovač' and Koons' papers. These aspects are not discussed here.

Possibilities of changing Gödelian proofs to a proof of existence of the devil were discussed by Cook (2003) and Pfeiffer (n.d.). Trivially, if you replace the predicate P by N (negative), and G by D (devillike), for instance, in $\mathcal{AO}$, you prove $\Box(\exists x)D(x)$. But you cannot have both the original $\mathcal{AO}$ and its devillike modification and assume $P(X) \rightarrow \neg N(X)$ (that no property is both positive and negative); this is obviously an inconsistent theory. What you can aruge is the variant with (A12) just described earlier, in addition to its devillike variant, and $P(X) \rightarrow \neg N(X)$, but this theory is rather weak. See the papers just cited for more information.

14.6 What Do We Learn about Gödel?

Gödel is reported to have said to Oskar Morgenstern that he did not want to publish his proof, lest he be thought to actually believe in God, whereas he was only engaged in a logical investigation. But Wang reports that Gödel's wife, Adele, revealed that even if Gödel did not go to church, he was religious and read the Bible in bed every Sunday morning (Wang, 1987, 1996). In the so-called Grandjean questionnaire, Gödel stated that his belief was theistic, not pantheistic. From his letters to his mother, it is clear that he believed in a life after death. Wang formulated Gödel's argument as follows: "Science shows that the world is rationally arranged; but without a next life, the potentialities of each person and the preparations in this life make no sense." Concerning religion, Gödel wrote (Hao Wang's translation; German original in Schimanowich-Galidescu, 2002), "I believe that there is much more reason in religion, though not in the churches, than one commonly believes, but we were brought up from early youth to a prejudgment against it. We are, of course, far from being able to confirm scientifically the theological world picture, but it might, I believe, already be possible to perceive by pure reason (without appealing to the faith in any religion) that the theological worldview is thoroughly compatible with all known data (including the conditions that prevail on our earth). The famous philosopher and mathematician Leibniz already tried to do this 250 years ago, and this is also what I tried in my previous letters."

I think that these and similar pieces of information on Gödel's opinions and beliefs show that the ontological proof was, for him, not a pure logical game. The proof helps us to better understand what sort of person he was. See also Köhler (2002).

14.7 Meaning for Religion?

Gaunilian examples discussed earlier were built from the assumption that there are some necessarily actually existing objects; they just show that there may be various systems of properties satisfying the axioms of the "ontological" theory (say, $\mathcal{AO}_0$). It should be explicitly stated that this does not destroy the ontological proof, proving, not assuming, necessary actual existence of a godlike being in the sense of the definition of godlikeness. But to make this notion of godlikeness really interesting from the religious (and philosophical) point of view, one should have an elaborated theory of positiveness (and of actual existence). A theologian should analyze the meaning of axioms (A1)–(A3) and the notion of positive properties, especially in the context of the Bible. Muck (1992) seems to be one of first attempts. Among other things, would the axiom (not used in Gödel's proof) "nobody except (possibly) God has necessary existence" be a theologically acceptable assumption? I believe that Gödel's proof would deserve an analysis similar to the one the famous protestant theologian K. Barth (1958, 622) presented for Anselm's proof. But it must be kept in mind that religious belief is not first a matter of accepting some axioms but accepting a kind of life (following an invitation). The religious notion of God is not the same as the philosophical one; recall Pascal's famous "Fire. God of Abraham, of Isaac, of Jacob, not of philosophers and scientists." On the other hand, let us quote another famous theologian (Tillich, 1955): "Is there any possibility of uniting ontology with biblical religion, if ontology could not accept the central assertion of biblical religion that Jesus is the Christ? . . . To ask the ontological question is a necessary task. *Against* Pascal I say: The God of Abraham, Isaac and Jacob, and the God of the philosophers is the same God. He is a person and the negation of himself as a person."

Finally, the famous Catholic theologian Hans Küng (1978, 626), in his long book, says (shortened):

If God existed, then the substantiating reality itself would not ultimately be unsubstantiated. God would be the *primal reason* of all reality.

If God existed, then the supporting reality itself would not ultimately be unsupported. God would be the *primal support* of all reality.

If God existed, then evolving reality would not ultimately be without aim. God would be the *primal goal* of all reality.

If God existed, then there would be no suspicion that reality, suspended between being and not being, might ultimately be void. God would be the *being itself* of all reality.

If God existed,

then, against all threats of fate and death, I could justifiably affirm the unity and identity of my human existence: for God would be the first ground also of my life;

then, against all threats of emptiness and meaninglessness, I could justifiably affirm the truth and significance of my existence: for God would be the ultimate meaning also of my life;

If God is, he is the answer to the radical uncertainty of reality.

That God is, can however be accepted neither stringently in virtue of a proof or demonstration of pure reason nor absolutely in virtue of a moral postulate of practical reason, still less solely in virtue of the biblical testimony.

That God is, can ultimately be accepted only in a confidence founded on reality itself.

14.8 Conclusion

Gödel's ontological proof is certainly not of central importance in his scientific achievement. However, "the proof and criticisms of it have inspired interesting work" (Fitting, 2002). In particular, in modal logic it gives a nontrivial exercise in modal axiomatic theories and their double semantics (fixed universe, varying universe). Furthermore, we have learned more about Gödel as a personality. Finally, concerning religion, "Even if one concludes that even this form of an ontological argument is no sufficient proof of God, it is a help for clarifying the notion of a property as used in the context of properties of God" (Muck, 1992; my translation). For all of the other incredibly Gödelian accomplishments, Gödel's ontological argument, while often overlooked, is one of his most fascinating legacies.

Acknowledgments

A version of this chapter was presented during the conference "Horizons of Truth," celebrating the one-hundredth anniversary of the birth of Kurt Gödel. The work of the author was partly supported by grant A100300503 of the Grant Agency of the Academy of Sciences of the Czech Republic and partly by the Institutional Research Plan AV0Z10300504.

References

Anderson, C. A. (1990). Some emendations of Gödel's ontological proof. *Faith and Philosophy*, **7**, 291–303.

Barth, K. (1958). *Fides Quaerens Intellectum*. Zollikon, Switzerland: Evangelischer.

Buldt, B., Köhler, E., Stöltzner, M., et al., eds. (2002). *Kurt Gödel: Wahrheit und Beweisbarkeit, Band 2: Kompendium zum Werk*. Wien: öbv & hpt.

Cook, R. T. (2003). God, the Devil and Gödel's other proof. In *Logica Yearbook 2003*, ed. L. Behounek, pp. 97–109. Prague: Filosofia.

Fitting, M. (2002). *Types, Tableaux and Gödel's God*. Dordrecht, Netherlands: Kluwer.

Fitting, M., and Mendelsohn, R. L. (1990). *First-Order Modal Logic*. Dordrecht, Netherlands: Kluwer.

Gettings, M. (1999). Gödel's ontological argument: A response to Oppy. *Analysis*, **59**, 309–13.

Gödel, K. (1970). Ontological proof. In *Collected Works*, vol. 3 (1995), pp. 403–4. [Introductory note by R. M. Adams, pp. 388–402. Appendix B: Texts relating to the ontological proof, including Gödel's first version, 1941, pp. 429–37.]

Hájek, P. (1996). Magari and others on Gödel's ontological proof. In *Logic and Algebra*, ed. A. Ursini and P. Agliano, pp. 125–35. New York: Marcel Dekker.

———. (2002a). *Der Mathematiker und die Frage der Existenz Gottes*. In Buldt et al., 2002, 325–36.

———. (2002b). A new small emendation of Gödel's ontological proof. *Studia Logica*, **71**, 149–64.

Hartshorne, C. (1962). *The Logic of Perfection and Other Essays in Neoclassical Metaphysics*. LaSalle, IL: Open Court.

Hopkins, J., and Richardson, H., eds. (1974). Proslogion by Anselm of Canterbury. In *Anselm of Cantebury, Vol. 1 (Monologion, Proslogion, Debate with Gaunilo and a Meditation on Human Redemption)*. Toronto: Edwin Mellen Press.

Köhler, E. (2002). *Gödels Platonismus*. In *Kurt Gödel: Wahrheit und Beweisbarkeit, Band 2: Kompendium zum Werk*. Ed. B. Buldt et al., pp. 341–86. Viennea: öbv & hpt.

Koons, R. C. (2005). Sobel on Gödel's ontological proof. http://www.scar.utoronto.ca/~sobel/OnL_T/Koons_SobelonGoedel.pdf.

Kovač, S. (2003). Some weakened Gödelian ontological arguments. *Journal of Philosophical Logic*, **32**, 565–88.

Küng, H. (1978). *Existiert Gott?* Munich, Germany: Piper.

Magari, R. (1988). Logica e teofilia. *Notizie di Logica*, 7.

Muck, O. (1992). Religiöser Glaube und Gödels ontologischer Gottesbeweis. *Theologie und Philosophie*, **67**, 263–67.

Oppy, G. (1996). Gödelian ontological arguments. *Analysis*, **56**, 226–30. [See also http://plato.stanford.edu/entries/ontological-arguments, Sect. 6.]

———. (2000). Response to Gettings. *Analysis*, **60**, 363–67.

Pfeiffer, H. (n.d.). Die Existenz Gottes und des Teufels. Unpublished manuscript.

Schimanowich, W. de P. (2002). *Gödels Briefe an seine Mutter*. In *Kurt Gödel: Wahrheit und Beweisbarkeit, Bandl, Dokumente und historische Analysen*, ed. Köhler et al., pp. 185–208. Vienna: öbv & hpt.

Sobel, J. H. (1987). Gödel's ontological proof. In *On Being and Saying: Essays in Honor of Richard Cartwright*, ed. J. J. Thompson, pp. 24–61. Cambridge, MA: MIT Press.

———. (2004). *Logic and Theism: Arguments for and against Beliefs in God*. Cambridge: Cambridge University Press.

———. (2006). On Gödel's ontological proof. In *Modality Matters: Twenty-Five Essays in Honour of Krister Segerberg*, ed. H. Lagerund et al., pp. 397–421. Uppsala, Sweden: Uppsala Philosophical Studies.

Szatkowski, M. (2005). Semantic analysis of some variants of Anderson-like ontological proofs. *Studia Logica*, **79**, 317–55.

Tillich, P. (1955). *Biblical Religion and the Search for Ultimate Reality*. Chicago: University of Chicago Press.

Wang, H. (1987). *Reflections on Kurt Gödel*. Cambridge, MA: MIT Press.

———. (1996). *A Logical Journey: From Gödel to Philosophy*. Cambridge, MA: MIT Press.

Gödel and the Human Mind

CHAPTER 15

The Gödel Theorem and Human Nature

Hilary W. Putnam

In the *Encyclopedia of Philosophy* (Edwards, 1967, 348–49), the article by Jean van Heijenoort titled "Gödel's Theorem" begins with the following terse paragraphs:

> By Gödel's theorem the following statement is generally meant:
>
> In any formal system adequate for number theory there exists an undecidable formula, that is, a formula that is not provable and whose negation is not provable. (This statement is occasionally referred to as Gödel's first theorem.)
>
> A corollary to the theorem is that the consistency of a formal system adequate for number theory cannot be proved within the system. (Sometimes it is this corollary that is referred to as Gödel's theorem; it is also referred to as Gödel's second theorem.)
>
> These statements are somewhat vaguely formulated generalizations of results published in 1931 by Kurt Gödel, then in Vienna.

Despite its forbidding technicality, the Gödel (1931) theorem has never stopped generating enormous interest. Much of that interest has been piqued because with the proof of the Gödel theorem, the human mind has succeeded in proving that – at least in any fixed, consistent system with a fixed, finite set of axioms that are at least minimally adequate for number theory and in a system that has the usual logic as the instrument with which deductions are to be made from those axioms – there has to be a mathematical statement the human mind cannot prove. (In fact, the theorem is much stronger than this: even if we allow the system in question to contain an infinite list of axioms and to have an infinite list of additional rules of inference, the theorem still applies, provided the two lists can be generated by a computer that is allowed to go on running forever.)

Moreover, instead of speaking of formal systems adequate for (at least) number theory, one can speak of computers. In such a version, the theorem says that if a computer is allowed to write sentences of number theory forever, subject to the constraints that (1) the list contains a certain subset of the axioms of Peano arithmetic, (2) the list is not inconsistent (i.e., the list does not contain both a sentence and the negation of that sentence), and (3) any sentence that can be deduced from finitely many sentences in the list via first-order logic is also in the list, then there is a true sentence of number theory – in fact, the sentence that says that the output of the machine is consistent is an example – that is not included in the list of sentences listed by the computer. If we speak of the sentences listed by the computer as "proved" by the computer, and of a computer of the kind just mentioned as an "acceptable" computer, we can phrase what I describe in the preceding succinctly as follows:

> There exists a procedure by which, given any acceptable computer, a human mathematician can write down a sentence of number theory that the icomputer cannot prove or disprove.

Does this or does this not tell us something about the human mind? That is what I shall discuss in this chapter.

15.1 Noam Chomsky, Scientific Competence, and Turing Machines

What provoked the research described in this chapter is a conversation I had with Noam Chomsky more than twenty years ago. I asked Chomsky whether he thought our total competence – not just the competence of his postulated "language organ" but also the competence of the "scientific faculty" that he also postulates – can be represented by a Turing machine (where the notion of competence is supposed to distinguish between our true ability and our so-called performance errors). He said yes. I at once thought that a Gödelian argument could be constructed to show that if that were correct, then we could never know – not just never prove mathematically, which is obvious after the Gödel incompleteness theorems, but never know even with the help of empirical investigation – which Turing machine it is that simulates our scientific competence. That Gödelian argument was given in an article I published more than twenty years ago (Putnam, 1985), but I have always been dissatisfied with one of the assumptions I used in that proof (I called it a "criterion of adequacy" for formalizations of the notion of justification), namely, the assumption that no empirical evidence can justify believing p if p is mathematically false. (Although many would regard this assumption as self-evident, those who would allow quasi-empirical methods in mathematics have good reason to reject it.) Today, I am ready to demonstrate a proof whose assumptions seem to me unproblematic. (Or at least they would be unproblematic if the notion of a formalizable scientific competence made sense. Why I don't think it makes sense is the subject of the final part of this chapter.)

15.2 Gödel's Incompleteness Theorems

Gödel's second incompleteness theorem shows that if the set of mathematical truths that is within human competence to prove were the output of an acceptable computer (as defined earlier), then the consistency of the computer's output could not be one of those mathematical truths. For reasons I will go into in the final part of this chapter, this doesn't show that the human mind (or the competence of a human mathematician) can't be perfectly simulated by a Turing machine. (This remark does not mean, by the way, that I want to defend the view that it can be simulated by a Turing machine; in the final part of this chapter, I argue that it is a mistake to think of us as having a mathematical competence that is so well defined that that question makes sense is a mistake.) Postponing that point for the time being, though, the Gödel incompleteness theorems don't even prove that we can't know which Turing machine simulates our mathematical competence, assuming such a machine exists, for to know that would require empirical, not mathematical, research.

To see why, note that the proposition that we are consistent is ambiguous. For example, this proposition may mean that we know the English sentence "our mathematical competence is consistent" to be true; this is very different from knowing an arithmetized version of the claim that it is consistent. This latter requires knowing which Turing machine we are. Even if, in some sense, we know that we are consistent, to go from that fact to the statement that "Turing machine T_k has a consistent output," where T_k is a machine that simulates our mathematical competence, would require the empirical premise that T_k is such a machine; however, then "T_k is consistent" would be a quasi-empirical mathematical statement, that is, a statement of mathematics that we believe on the basis of an argument some of whose premises were empirical, and nothing in the Gödel theorems rules out our knowing the truth of such a quasi-empirical statement, in addition to knowing the mathematical theorems that we are able to prove in the strict sense. However, I do want to say that reflecting on Gödel's methods and their implications can lead us to a deeper philosophical understanding of the issues raised by Lucas, Penrose, and (on the other side) Chomsky.

15.3 Thinking about Chomsky's Conjecture in a Rigorous Gödelian Way

To think about Chomsky's conjecture in a Gödelian way, let COMPETENCE abbreviate the empirical hypothesis that a particular Turing machine T_k (the kth machine in the standard Kleene enumeration) perfectly simulates the competence of our scientific faculty, in Chomsky's sense. To make this concrete, we may take this to assert that for any proposition p, and for any evidence u expressible in the language that T_k uses for expressing propositions and evidence (and if Chomsky is correct, that language is innate), T_k sooner or later prints out Justified$_u(p)$ if and only if it is justified to accept p when u is our total relevant evidence. (I identify sentences with their Gödel numbers in what follows. Thus "T_k sooner or later prints out 'Justified$_u(p)$'" means that T_k sooner or later prints out the word *Justified* followed by the symbols for the Gödel number of the sentence u (subscripted), followed by a left parenthesis, followed by the symbols for the Gödel number of the sentence p, followed by a right parenthesis.)

In addition, COMPETENCE asserts that the hypotheses and evidence we can describe in an arbitrary natural language can also can be represented in T_k's language, that is, that that language has the full expressive power of Chomskian "Mentalese." I am going to sketch a proof (given in full in the appendix) that if this is true, then COMPETENCE cannot be both true and justified. In other words, if some particular Turing machine T_k correctly enumerates the sentences that our competence would take to be justified on arbitrary evidence, then that statement is one that cannot be empirically justified on any evidence.

The proof proceeds, of course, via reductio ad absurdum. The idea is to assume that COMPETENCE is true and that some sentence e states evidence that justifies accepting COMPETENCE. (Note that e and k are constants throughout the following argument.) Using Gödel's technique, we can construct a sentence GÖDEL in the language of Peano arithmetic, and hence in the language of T_k (because that language is capable of expressing all humanly accessible science), that says that T_k never prints out Justified$_e$(*GÖDEL*). Therefore GÖDEL has the following property:

> It is justified to believe GÖDEL on evidence e if and only if it is justified to believe (on that same evidence) that the machine T_k never says (even if it is allowed to run forever) that it is justified to believe GÖDEL on evidence e.

We are now ready to describe the idea of the proof of the following theorem (the actual proof is given in the appendix).

An anti-Chomskian incompleteness theorem is that

COMPETENCE can't be both true and justified.

It is well known that Gödel's proof mimics the liar's paradox. The proof of the anti-Chomskian incompleteness theorem given in the appendix similarly mimics the following paradox, which we might call the *justification paradox*:

> By the Gödel technique for constructing self-referring sentences (or the technique of Quine, 1966, 9), it is easy to construct a sentence that says of itself that belief in it is not justified. For example (using Quine's technique), (B) "When appended to its own quotation yields a sentence such that believing it is not justified" when appended to its own quotation yields a sentence such that believing it is not justified.

Now reflect on the following: is believing (B) justified? Well, if belief in (B) is justified, then I am justified in believing that belief in (B) is not justified because I see that that is what (B) says. However, if belief in (B) is justified, then believing that belief in (B) is justified is justified, so believing that belief in (B) is justified is both justified and not justified, which is impossible. In this way, I now have justified believing that belief in (B) is not justified, but what (B) says is precisely that belief in (B) is not justified, so I have justified believing (B). Therefore believing (B) is justified, which, we just proved, leads to a contradiction!

Just as Gödel's proof turned the liar's paradox from a contradiction into a proof that the arithmetized version of the sentence "I am not provable" is not provable (unless the system in question is inconsistent), so the proof we give in the appendix turns the justification paradox into a proof that the arithmetized version of the sentence "believing me is not justified" is not printed out by the machine T_k. However, if COMPETENCE were true and also verified (by evidence e) to be true, we could know this proof and so justify the sentence "believing me is not justified" – which would lead to a real contradiction.

15.4 Lucas and Penrose

As mentioned earlier, the Oxford philosopher John Lucas (1961) claimed that the Gödel theorem shows that noncomputational processes (processes that cannot in principle be carried out by a digital computer, even if its memory space is unlimited) go on in our minds. He concluded that our minds cannot be identical with our brains or with any material system (because assuming standard physics, the latter cannot carry out noncomputational processes) and hence that our minds are immaterial. More recently, a thinker whose scientific work I very much admire, Roger Penrose, used a similar but more elaborate argument to claim that we need to make fundamental changes in the way we view both the mind and the physical world. He, too, argued (Penrose, 1994) that the Gödel theorem shows that noncomputational processes must go on in the human mind, but instead of positing an immaterial soul in which they can go on, he concluded that these noncomputational processes must be physical processes in the brain and that our physics needs to change to account for them.

Although neither Lucas nor Penrose is a Chomskian, an important similarity exists between their notion of our mathematical competence and Chomsky's notion of competence. Chomsky (1957) has insisted that our competence in his sense includes the ability to generate arbitrarily long grammatical sentences, together with the proofs of their grammaticality (which had the form of so-called trees in *Syntactic Structures*). It is well known (see, e.g., Hopcroft and Ullman, 1979) that this ability (in the case of unrestricted languages in the Chomsky-Schützenberger hierarchy) is equivalent to the ability to simulate the computations of an arbitrary Turing machine. Thus our competence is supposed to include the ability to verify all truths of the form "machine T_i eventually prints out sentence p." However, Lucas and Penrose must also be making this assumption because they assume that our competence is strictly greater than the ability to prove all the theorems of Peano arithmetic, and all truths of the form "machine T_i eventually prints out sentence p" are (on suitable arithmetization) theorems of Peano arithmetic.

I reviewed Penrose's (1994) argument, and like many others who studied it, I found many gaps (Putnam 1994). Relax, though: I am not going to ask you to think about the details of Penrose's argument, and I am not going to provide a listing of the places at which I believe it goes wrong. It may be that Penrose himself recognized that his argument has gaps (even though, at the beginning, he promised a "clear and simple proof") because he offered a number of arguments that are not mathematical at all but rather philosophical. (In any case, see Penrose's Chapter 16.)

The heart of his reasoning goes as follows: let us suppose that the brain of an ideally competent mathematician can be represented by the program of an acceptable computer (as defined earlier), a program that generates all and only correct and convincing proofs. The statement that the program is consistent is not one that the program generates, by Gödel's second incompleteness theorem. Therefore, if an ideal human mathematician could prove that statement, we would have a contradiction. This part of Penrose's reasoning is incontrovertibly correct.

It is possible, however, that the program is too long to be consciously comprehended at all – as long as the London telephone book or even orders of magnitude longer. In that case, Penrose asks, how could evolution have produced such a program? It would have to have evolved in parts, and there would be no evolutionary advantage to the parts.

As I shall explain shortly, I find some things wrong with the notion of "encapsulating all humanly accessible methods of proof" and hence with the question of whether a computer could do such a thing. However, even if we accept that notion, it is important to recall that just this objection has been made to almost every evolutionary explanation (how could wings have evolved in parts? how could eyes have evolved in parts?). In case after case, an answer has been found (see Carroll, 2006), but the answer typically involves generations of researchers, and we are hardly at a stage at which study of the aspects of the brain that subserve our mathematical abilities are well enough understood even to begin evolutionary study. Still, if you will permit me to offer what are obviously speculations, following are two ways in which our mathematical abilities have parts that could have utility apart from being combined in their present package:

1. As Wittgenstein suggested, the basic laws of arithmetic and geometry could, for example, have first arisen as empirical generalizations that were later hardened into conceptual truths(for an account of this, see Steiner, 2000). Most of this process would belong to the history of our cultural rather than our biological evolution, although it would presuppose the evolution of language and of the skills that underlie, for example, counting.

2. Reflection on their own practices (in any area) is historically an important way in which humans arrive at new practices, but the biological capacity for such reflection could certainly have evolved independently of mathematics itself. Metamathematical reflection, which is what fascinates Penrose, in fact came very late on the mathematical scene. For several millennia, mathematics was exhausted by the two subjects of arithmetic (or number theory) and (Euclidean) geometry, and the explosion in what we today consider to be mathematics, including set theory and its less well known (by nonmathematicians) rival, category theory, was not driven by either Gödelian reflection or new a priori insights; instead, new axioms, such as the axiom of choice and the axiom of replacement, were added on strongly pragmatic grounds (again, a matter of cultural rather than biological evolution). Only with Gödel's incompleteness theorems does reflection on mathematical practice become a way of proving number theoretic propositions that could not have been proved previously. (The most elementary form of metamathematical reflection is semantic reflection, illustrated by the argument that all the axioms of Peano arithmetic are true, and the rules of inference preserve truth; therefore all the theorems of Peano arithmetic are true, but $1 = 0$ is not true and therefore is not a theorem of Peano arithmetic, i.e., Peano arithmetic is consistent. Tarski's work shows how this argument can be formalized in a weak second-order extension of Peano arithmetic, and the Gödel theorem shows that the result cannot be proved in Peano arithmetic itself.) Nothing about this so far, though, suggests any problem for standard evolutionary theory.

What about iterated reflection? Let L_0 be Peano arithmetic. Because we can see, by the kind of metamathematical reflection just described, that the soundness of L_0 (which we accept) entails the truth of $Con(L_0)$, we can accept the soundness of L_1, where L_1 is the result of adding $Con(L_0)$ as an additional axiom to L_0. Continuing this procedure, we generate a hierarchy of stronger and stronger intuitively acceptable systems, L_0, $L_1, L_2, L_3, \ldots$, and so on, ad infinitum.

Turing (1939) proposed to extend this series into the transfinite, using notations for computable ordinals. At limit ordinals λ, the associated L_λ is the union of all the earlier L_λ"ccpms. (The notations for the ordinals are indispensable in defining the sentence Con(L) precisely.) What if our idealized mathematical competencies were represented by the totality of such L_λ with $\lambda < \varphi$, for some computable ordinal φ? (φ would have to be a computable ordinal such that no recursive well ordering of that order type could be seen by a human mathematician to be a well ordering or proved to be a well ordering from below.) Such a totality might well be longer than the London telephone book, but it would be constructed from L_0, a system that is intuitively simple, using a method of extension that is also intuitively simple. (As I just noted, the ability to reflect on our own practices, which is what enabled us to arrive at that method of extension and also enables us to see that various computable orderings are well orderings – normally, by seeing that all their initial segments are well orderings – could well have developed independently of L_0.)

However, what if the ordinal logics procedure Turing described didn't break off at a computable ordinal? Feferman (1962) shows that with stronger reflection principles, one can obtain all true number-theoretical propositions on a suitable path through the computable ordinal notations. In that case, Penrose would be right: our mathematical competence would exceed the powers of any Turing machine. My point is that Penrose cannot exclude the possibility that our powers do break off at some computable ordinal.

Ah, but Penrose will certainly ask, what about us determines the limits of metamathematical reflection? What determines where we break off? Here I have to say that I don't think there is a precise limit to what human mathematicians can do via reflection. The very notion of an ideal mathematician is too problematic for us to speak of a determinate limit here – a point to which I shall return shortly. Different individuals have different limits to what they can do via metamathematical reflection (and the possibilities of metamathematical reflection have barely been discovered to date). I would say that what the limits are in the case of any given individual is obviously an empirical question, and not necessarily a very interesting or important one.

One thing I will grant Penrose, however, is that I do find it mysterious that we have the mathematical abilities we do. The simple fact that we can intuitively understand the notion that the natural number sequence has no end amazes me more than that we can Gödelize any given consistent formal system. I also find it mysterious that we have the natural-scientific abilities, the linguistic abilities, the aesthetic abilities, and so on, that we do. If a change in our physical, biological, and neurological theories makes it possible to explain them, then, of course, I would welcome it – we all would – but I do not believe that Penrose has given compelling reasons to think that our brains must have noncomputable powers and that we must seek a scientific theory to explain them.

To see what we can learn about the human mind, or at least how we should (or rather shouldn't) think about it, from the Gödel theorem, I want now to raise two objections against the very question that Penrose asks – the question of whether the set of theorems that an ideal mathematician can prove could be generated by a computer.

My first objection – and this is a point that I want to emphasize – is that the notion of simulating the performance of a mathematician is highly unclear. Whether it is possible to build a machine that behaves as a typical human mathematician is a meaningful, empirical question, but a typical human mathematician makes mistakes. The output of an actual mathematician contains inconsistencies (especially if we are to imagine that she goes on proving theorems forever, as the application of the Gödel theorem requires), so the question of proving that the whole of this output is consistent does not even arise. To this, Penrose's reply seems to be that the mathematician may make errors, but she corrects them on reflection.

As I pointed out before, even if (via this or some other philosophical argument) we know that the English sentence "our mathematical competence is consistent" is true, this is very different from knowing an arithmetized version of the claim that this is the case. This latter would require knowing which Turing machine simulates our mathematical competence. (That we cannot know a similar fact about our natural-scientific competence by empirical investigation was what our anti-Chomskian incompleteness theorem showed.)

My second objection is that to confuse these questions is to miss the normativity of the notion of ideal mathematics. The notion of ideal mathematics is the notion of a human mathematician who always behaves (in his mathematical work) as a mathematician should (or ideally should). To the extent that we can succeed in describing human beings as physical systems, physics can say (at best) how they will behave, competence errors and all. However, physics is not in the business of telling us how human beings should behave.

15.5 A Caution against a Widespread but Naive Argument for Algorithms in the Brain

It is not my aim merely to criticize those who make the mistake of thinking that the Gödel theorem proves that the human mind or brain can carry out noncomputational processes. That Lucas and Penrose have failed to prove their claims about what the Gödel theorem "shows about the human mind" is widely recognized. At the same time, however, a view that may seem to be at the opposite end of the philosophical spectrum from the Lucas-Penrose view seems to me to be vulnerable to the two objections I just made. I refer to the widespread use of the argument, in philosophy as well as in cognitive science, that whenever human beings are able to recognize that a property applies to a sentence in a "potentially infinite" (i.e., a fantastically large) set of cases, an "algorithm" must be in their brains that accounts for this, and the task of cognitive science must be to describe that algorithm. I think that the case of mathematics shows that that cannot be a universal truth, and my criticism of Penrose may enable us to see why we shouldn't draw any mystical conclusions about either the mind or the brain from the fact that it isn't a universal truth.

First, though, a terribly important qualification is in order. I have no doubt that human beings have the ability to employ *recursions* (another word for *algorithms*) both consciously and unconsciously and that cognitive science should explain (and in many cases does successfully explain) recognition abilities by appealing to the notion of an algorithm. The work of Chomsky and his school, in particular, shows how our ability to recognize the property of grammaticality might be explainable in this way. Why, then, do I say that the case of mathematics shows that it cannot be a universal truth that whenever human beings are able to recognize that a property applies to a sentence in a potentially infinite (i.e., a fantastically large) set of cases, there must be an algorithm in their brains that accounts for it?

Well, just as it is true that we can recognize grammaticality in a fantastically large set of cases, it is true that we can recognize its being proved in a fantastically large set of cases. As we saw (via the anti-Chomskian incompleteness theorem), Gödel's techniques can be used to show – via a very different argument from Penrose's – that if an algorithm accounts for this ability, we are not going to be able to verify this. However, I think it would be wrong just to stop with this conclusion. It would be wrong because talk about a potentially infinite set of cases makes sense only when we can verify the presence of an algorithm. Why do I say this?

Think for a minute about the mathematician about whom Penrose imagines he is talking. If the mathematician could really recognize a potential infinity of theorems, then he or she would be able to recognize theorems of arbitrary finite length, for example, theorems too long to be written down before all the stars cool or are sucked into black holes. However, any adequate physics of the human brain will certainly entail that the brain will disintegrate long before it gets through trying to prove any such theorem. In short, in fact, the set of theorems any physically possible human mathematician can prove is finite.

Moreover, it is not really a set because sets, by definition, are two valued: an item is either in or out. However, the predicate *prove* is vague. We can, of course, make it precise by specifying a fixed list of axioms and rules to be used, but then the Gödel theorem does show that in a perfectly natural sense, we can prove a statement that is not derivable from those axioms by those rules. This may drive us to say that "by 'prove' we mean prove by any means we can 'see' to be correct." However, this uses the essentially vague notion "see to be correct." In short, there isn't a potentially infinite set of theorems a mathematician can prove concerning which we can ask, is it recursive or nonrecursive? What we do have is a vague, finite collection of theorems that a real (flesh-and-blood) mathematician can prove. Vague, finite collections are neither recursive nor nonrecursive; those concepts apply only to well-defined sets.

It may be objected, though, that Chomsky taught us that we can – and should – idealize language users by imagining that (like ideal computers) they can go on generating grammatical sentences forever, sentences of arbitrary length. Yes, he did, but the idealization he showed us how to make is only precise because it corresponds to a well-defined algorithm. If we wish the question "what can an ideal mathematician prove?" to be a precise one, we must specify just how the mathematician's all-too-finite brain is to be idealized, just as Chomsky did. If we do this by specifying an algorithm, then, of course, the result will be computational.

If this is right, then we should not allow ourselves to say that the set of statements that a mathematician can prove is a potentially infinite set. We should say only that there are fantastically many statements a mathematician – an actual mathematician – can prove and let it go at that. It does not follow from this that there is an algorithm such that everything an ideally competent mathematician would (should?) count as proving a mathematical statement is an instance of that algorithm, anymore than it follows from the fact that there are fantastically many jokes that a person with a good sense of humor can see to be funny that there is an algorithm such that everything a person with an ideal sense of humor would see to be funny is an instance of that algorithm.

Appendix

As earlier, we assume that T_k is a machine that perfectly represents our scientific competence in the sense that for any proposition p, and for any evidence u (expressible in the language that T_k uses for expressing propositions and expressing evidence), T_k sooner or later prints out Justified$_u(p)$ if and only if it is justified to accept p when u is our total relevant evidence. Let e be the Gödel number of a sentence that describes evidence that justifies accepting COMPETENCE. Note that k and e are constants in the following argument. As before, COMPETENCE also asserts that whatever hypotheses and evidence we can describe in an arbitrary natural language can also can be represented in T_k's language, that is, that that language has the full of expressive power of the Mentalese that Chomskians mention. Using Gödel's technique, we can construct a sentence GÖDEL in the language of Peano arithmetic, and hence in the language of T_k (because that language is capable of expressing all humanly accessible science), that says that T_k never prints out Justified$_e(GÖDEL)$. Therefore GÖDEL has the following property:

> It is justified to believe GÖDEL on evidence e if and only if it is justified to believe (on that same evidence) that the machine T_k never says (even if it is allowed to run forever) that it is justified to believe GÖDEL on evidence e.

We assume two axioms concerning the notion of justification:

Axiom 1:   It is never both the case that it is justified to believe p on evidence e and justified to believe ¬p on evidence e.

Remark We know various statements of science that are, at a certain stage, justified on all the available relevant evidence but are subsequently rejected on the basis of new relevant evidence, possibly within the framework of new theories. Thus a Turing machine that simulates our scientific competence as it proceeds chronologically may list inconsistent statements because our evidence is not the same at different times. However, T_k is not supposed to mimic the diachronic behavior of our brains but is intended simply to represent a recursive enumeration of the sentences that it is within the competence of the "scientific faculty" – which Chomsky claims to be computational – to accept as justified on any evidence expressible in Mentalese. Thus axiom 1 represents the requirement that the scientific faculty, when acting in accordance with its competence, not regard both p and ¬p as justified on the same evidence. Even this has been questioned by a friend, who writes:

> I suppose that if an agent had purported justifications of a sentence and its negation, he would realize that at least one of these justifications is bogus and should be withdrawn. But the subject may not realize that he has justified a given sentence and its negation. To give our agents the ability to tell whether their (justified) beliefs are consistent is to give them a nonrecursive ability.

My response is that it is part of the notion of justification that justified beliefs are not flatly contradictory (i.e., of the form p&¬p). I believe that this is something on which all epistemologists would agree. If there is some other notion of justification with which both p and ¬p can sometimes be justified on the very same evidence, I am sure that is not the notion of justification either Chomsky or Penrose or Lucas or I have in mind. (The ability to tell whether two beliefs are flatly contradictory is a primitive recursive ability.)

Because COMPETENCE is the statement that T_k perfectly represents our competence, we also have the following:

Axiom 2: If Justified$_e$ (*the machine T_k eventually prints out Justified$_e$ (p)*) and Justified$_e$ (*COMPETENCE*), then Justified$_e$(p). ("Justified$_e$(p)" is, of course, simply the sentence in the language of T_k – which was assumed to be a formalized version of Chomsky's Mentalese – that expresses the statement that "it is justified to believe the sentence p on evidence e." Note that by well-known results of Gödel and Turing, "the machine T_k eventually prints out Justified$_e$ (p))" is expressible by a sentence of Peano arithmetic.)

The anti-Chomskian incompleteness theorem is as follows:

"COMPETENCE" can't be both true and justified.

To guide the reader, following is an outline of the proof I shall give (in both parts of the proof, we will assume that COMPETENCE is both true and justified by the evidence e).

Outline of Part I We will prove (by assuming the opposite and deriving a contradiction) that it is not the case that the machine T_k eventually prints out Justified$_e$ (*GÖDEL*) (i.e., the machine T_k, which we assumed to enumerate correctly the sentences that our epistemic competence would tell us are justified on arbitrary evidence, does not generate the Gödelian sentence on the evidence that justifies our acceptance of COMPETENCE).

Outline of Part II The conclusion of part I was as follows:

(a) It is not the case that the machine T_k eventually prints out Justified$_e$ (*GÖDEL*). Because we can know this proof if we have empirically justified COMPETENCE, and a proof is a justification, it follows immediately that

(b) Justified$_e$ (*the machine T_k never prints out Justified$_e$(GÖDEL)*).

Then the rest of part II will derive a contradiction from (b), thus showing that the assumption that COMPETENCE is both true and justified must be false. The proof in full follows.

Part I of the Proof Assume: The machine T_k prints out Justified$_e$(*GÖDEL*)

(i) Justified$_e$(*GÖDEL*). Reason: For the reductio proof of our incompleteness theorem, we are assuming COMPETENCE. By assumption, the machine T_k prints out Justified$_e$(*GÖDEL*). So by COMPETENCE, Justified$_e$(*GÖDEL*).

(ii) It is justified$_e$ to believe that "the machine T_k eventually prints out Justified$_e$(*GÖDEL*)." Reason: We just showed that Justified$_e$(*GÖDEL*). Then, by COMPETENCE, Justified$_e$ (*the machine T_k eventually prints out Justified$_e$ (GÖDEL)*); that is, we are justified$_e$ in believing that the machine T_k eventually prints out Justified$_e$(*GÖDEL*).

(iii) We are justified in believing "the machine T_k does not eventually print out Justified$_e$(*GÖDEL*)." Reason: Because, as pointed out earlier, on Chomsky's idealized notion of competence (and also on the notion of our mathematical competence assumed by Lucas and Penrose), it is within our competence to recognize all truths of the form "machine T_i eventually prints out sentence p," and, by assumption, "the machine T_k prints out Justified$_e$(GÖDEL)," this is a truth that our scientific faculty is able to prove. Because a proof is a justification, we have the following: Justified$_e$ (*the machine T_k eventually prints out Justified$_e$(GÖDEL)*). However . . .

> It is justified to believe GÖDEL on evidence e if and only if it is justified to believe (on that same evidence) that the machine T_k never says (even if it is allowed to run forever) that it is justified to believe GÖDEL on evidence e, and because this equivalence is known to us, and by (i), we are justified$_e$ in believing the left side, we are justified$_e$ in believing the right side; that is, we are justified$_e$ in believing "the machine T_k does not eventually print out Justified$_e$(GÖDEL))."

However, (ii) and (iii) violate our consistency axiom (axiom 1). Thus (still assuming that COMPETENCE is both true and justified on evidence e) we conclude that the assumption of our subproof is false: it is not the case that the machine T_k prints out Justified$_e$(GÖDEL). However, this is reasoning that we can easily go through if we have discovered COMPETENCE to be true, so without any additional empirical evidence, we have justified$_e$ that "the machine T_k does not eventually print out Justified$_e$(GÖDEL))."

To complete the proof of our incompleteness theorem, we therefore now need a proof that this, too, leads to a contradiction. It follows:

Part II of the Proof We have, in part I, proved the following:

(i) T_k does not eventually print Justified$_e$(GÖDEL). From this, by COMPETENCE, we have

(ii) Justified$_e$(GÖDEL). However, we have from before (B) Justified$_e$(GÖDEL) if and only if (Justified$_e$ (T_k does not eventually print Justified$_e$(GÖDEL)). Therefore

(iii) Justified$_e$ (T_k does not eventually print Justified$_e$(GÖDEL)). However, from (i) and the obvious rule of inference that proving X allows writing Justified$_e$ (X), we have

(iv) Justified$_e$ (T_k does not eventually print Justified$_e$(GÖDEL)).

Now (iii) and (iv) are a contradiction. This completes the proof of our anti-Chomskian incompleteness theorem.

References

Carroll, S. B. (2006). *The Making of the Fittest: DNA and the Ultimate Forensic Record of Evolution.* New York: W. W. Norton.

Chomsky, N. (1957). *Syntactic Structures.* 1st ed. Berlin: Mouton de Gruyter. [2nd ed. 2002.]

Edwards, P., ed. (1967). *Encyclopedia of Philosophy.* Vol. 3. New York: Macmillan.

Feferman, S. (1962). Transfinite recursive progressions of axiomatic theories. *Journal of Symbolic Logic,* **27**, 259–316.

Gödel, K. (1931). Über formal unentscheidbare Sätze der Principia Mathematica und verwandter Systeme I. *Monatshefte für Mathematik und Physik,* **38**, 173–98. [English trans. J. van Heijenoort, ed. (1967). *Frege to Gödel: A Source Book on Mathematical Logic.* Cambridge, MA: Harvard University Press, pp. 596–616. Repr. with facing English trans. On formally undecidable propositions of Principia Mathematica and Related Systems. I. In *Collected Works*, vol. 1 (1986), pp. 145–95.]

Hopcroft, J., and Ullman, J. D. (1979). *Introduction to Automata Theory, Languages and Computation*. Boston: Addison-Wesley.

Lucas, J. R. (1961). Minds, machines, and Gödel. *Philosophy*, **36**, 112–27. [Repr. A. R. Anderson (1964). *Minds and Machines*. Englewood Cliffs, NJ: Prentice Hall.]

Penrose, R. (1994). *Shadows of the Mind: An Approach to the Missing Science of Consciousness*. Oxford: Oxford University Press.

Putnam, H. (1985). Reflexive reflections. *Erkenntnis*, **22**(1), 143–54. [Collected in H. Putnam (1994). *Words and Life*. Cambridge, MA: Harvard University Press, pp. 416–27.]

_____. (1994). Review of *Shadows of the Mind*. *New York Times Book Review*, November 20. [Repr. H. Putnam (1995). Review of *Shadows of the Mind*. *Bulletin of the American Mathematical Society*, **32** (3), 370–73.]

Quine, W. v. O. (1966). *The Ways of Paradox and Other Essays*. New York: Random House.

Steiner, M. (2000). Mathematical intuition and physical intuition in Wittgenstein's later philosophy. *Synthese*, **125**(3), 333–40.

Tarksi, A. (1936). *The Concept of Truth in Formalized Languages*. In Polish; the standard English translation is collected in Tarski's Logic, Semantics, Metamathematics, 2nd. edition (edited by J. Corcoran) (Indianapolis: Hackett, 1983) (1st edition edited and translated by J. H. Woodger, Oxford 1956).

Turing, A. (1939). Systems of logic defined by ordinals. *Proceedings of the London Mathematical Society, Series II*, **45**, 161–228.

CHAPTER 16

Gödel, the Mind, and the Laws of Physics

Roger Penrose

Gödel appears to have believed strongly that the human mind cannot be explained in terms of any kind of computational physics, but he remained cautious in formulating this belief as a rigorous consequence of his incompleteness theorems. In this chapter, I discuss a modification of standard Gödel-type logical arguments, these appearing to strengthen Gödel's conclusions, and attempt to provide a persuasive case in support of his standpoint that the actions of the mind must transcend computation.

It appears that Gödel did not consider the possibility that the laws of physics might themselves involve noncomputational procedures; accordingly, he found himself driven to the conclusion that mentality must lie beyond the actions of the physical brain. My own arguments, on the other hand, are from the scientific standpoint that the mind is a product of the brain's physical activity. Accordingly, there must be something in the physical actions of the world that itself transcends computation.

We do not appear to find such noncomputational action in the known laws of physics, however, so we must seek it in currently undiscovered laws going beyond presently accepted physical theory. I argue that the only plausibly relevant gap in current understanding lies in a fundamental incompleteness in quantum theory, which reveals itself only with significant mass displacements between quantum states ("Schrödinger's cats"). I contend that the need for new physics enters when gravitational effects just begin to play a role. In a scheme developed jointly with Stuart Hameroff, this has direct relevance within neuronal microtubules, and I describe this (still speculative) scheme in the following.

16.1 Three Worlds and the Mysteries That Connect Them

When discussing matters of the mind and of the possibility that the phenomenon of consciousness may be essentially dependent on the particular form of those (mathematical) laws that seem to govern our physical universe – especially in relation to the issue of computation – I have found it useful to describe things with reference to a diagram such as that depicted in Figure 16.1. In this figure, the three spheres represent,

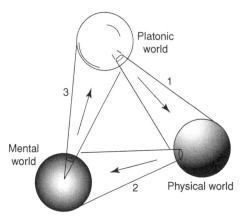

Figure 16.1. Three "worlds" – the Platonic mathematical, the physical, and the mental – and the three profound mysteries in the connections between them.

respectively (clockwise, starting from the far right), the physical world of material objects (including the very space-time that these objects inhabit), the world of mental experience, and the (Platonic) world of mathematical absolutes.

Some may object to the assignment of a reality to all three of these worlds. Mental experiences, for example, might be regarded as merely things that are, in some way, emergent from the activities of physically real brains and therefore have no separate reality from that of the physical brains on which they seem to depend. Others might regard mentality as primary and would take what we call the "physical world" to be some kind of abstraction from conscious experience. Mathematical entities, moreover, are often regarded as being mere constructs of mentality, and accordingly, mathematical entities would have no separate reality beyond that of the minds from which they are conjured forth. Yet I believe that it is useful to separate out these three aspects of reality if we are to discuss issues of mentality and of physical laws in depth. This is made particularly evident, I maintain, because if one takes the view that the very existence of mathematical concepts is dependent on a preexisting mentality (with that being, according to a viewpoint just expressed, dependent on a preexisting physical reality), one is presented with the paradoxical issue of how it could be that the physical universe had already been acting in accordance with very precise and sophisticated mathematical principles long before there were any sentient beings able to conjure up the needed mathematics!

Nor is this issue resolved if we regard the precise mathematical structure of the laws governing the physical universe as something merely apparent, something that our minds simply construct so that the workings of the physical world can become comprehensible to us. The more profoundly we probe these physical laws, the more we discover that sophisticated mathematics is just there in the workings of the world, at the deepest levels that we can yet discern, with a precision that becomes only gradually revealed to us and is found to be far, far greater than the precision that was available to the original discoverers of these laws.[1]

---

[1] For further discussion of these points, see Penrose (2004, sect. 1.3, 1.4, 3.3, 19.8, 34.7).

The separating out of these three aspects of reality, as depicted in Figure 16.1, does not resolve these issues, but it gives us a better framework within which to discuss them. It also brings into focus the three mysteries that are involved in the relationships between these three different aspects of reality, as shown in the figure. It is, indeed, a mysterious fact that the workings of the physical universe accord, with such extraordinary precision, with highly sophisticated, elegant mathematical laws (mystery 1).[2] It is also deeply mysterious how conscious mentality seems to arise when physical structures are exquisitely organized in an appropriate way, as with wakeful, healthy human brains (mystery 2). Moreover, there is even something profoundly mysterious about how conscious mentality, when focused with due concentration and understanding, can have access to deep mathematical truths (mystery 3). The contention that this access cannot be something of a purely computational nature, as I am indeed claiming, and as Gödel himself believed, is one of the main themes of this chapter. Being unable to be fully imitated by a computer, this mental access is, accordingly, something whose precise nature is distinctly mysterious. In this account, I shall attempt to shed a little light on this issue.[3]

A further mystery is inherent in the structure revealed in Figure 16.1, represented in the cyclic nature of the figure. Whereas the actions of the entire physical world are taken to be fully dependent on mathematical laws, as depicted, it is only a small fraction of the Platonic mathematical world that has direct relevance to this physical behavior. (Most mathematical journals, for example, are filled with articles whose concerns are very remote from any physical application.) Also, those parts of the physical world whose actions are relevant to the production of conscious mentality form a very tiny proportion of the whole. (The mass of rocks in the earth, for example, enormously exceeds its collective brain mass.) Also, it is clear that only a very minor part of mental activity is concerned with the ascertaining of mathematical truth. (Even mathematicians – at least, most of them – spend the greater part of their time thinking of other matters!) Yet, if we follow the diagram of Figure 16.1 in a clockwise direction, we find that the relevant small portion of each world encompasses, in principle, the entirety of the next world and hence of all the worlds. This, in itself, is not quite a contradiction, but there is something of a fourth mystery exhibited by this cyclic feature, which I have tried to emphasize by invoking an element of the "impossible figure" in my illustration.[4]

I should point out, in this context, that this seemingly paradoxical feature is made particularly stark because of three prejudices of mine that are implicitly involved in how I have represented the connections between the three worlds in Figure 16.1. I have taken it that these mysterious connections between each world and the next (read clockwise) are such that in each case, the entire following world is encompassed by but a small part of the one preceding it. Yet it is perfectly conceivable that in actuality, this is not the case. Indeed, some physical actions may not be describable in terms of mathematical laws; mentality may exist that is not grounded in physical structures; mathematical truths may exist that can be precisely formulated but that are not accessible in principle

---

[2] See, e.g., Wigner (1960); see also Penrose (2004, sect. 34.8, n. 34.34).

[3] For more detailed accounts of my point of view, see Penrose (1989, 1994, 1997a, 1997b, 2004, sect. 34.7, n. 34.34).

[4] See, e.g., Ernst (1986) and Penrose and Penrose (1958).

by conscious insight and understanding. All these issues have relevance and should be considered seriously. However, for the purposes of this particular chapter, this would just complicate the discussion unnecessarily. To allow a broadening of the picture by including them would not appear to help us resolve the issues with which I am concerned here.

Some points concerning our access to mathematical truth (mystery 3) are worthy of further mention. It appears to be a common view that Gödel's incompleteness theorems have established the existence of well-defined mathematical statements that cannot be resolved ("proved") by use of purely mathematical reasoning. As should become clear in the following discussion, this is very far from my own (or indeed, Gödel's) interpretation of the incompleteness theorems, which can be regarded as yielding only humanly accessible truths, and I shall discuss this point further later. However, it would seem to be an unresolved issue whether absolutely unsolvable mathematical problems exist, as opposed to those that cannot be resolved within some particular formal system of proof procedures.[5] Moreover, I should make it clear that the arrow of mystery 3 in Figure 16.1 is meant to refer to access *in principle*. It is clear, for example, that some multiplication sums between integers are so stupendously huge that no human (or even computer) can ever check the correctness of (or even state) such a product in detail. Yet nothing of principle is involved in such a calculation that is beyond human reason, so the arrow of mystery 3 is intended to include such things, where the only obstruction to human access is sheer complication rather than any matter of principle.

Of course, the notion of the multiplication of a pair of integers is something mathematically very elementary, no matter how large those integers may be, and the Platonic mathematical world contains truths that are far more sophisticated than the correctness of such a calculation. Indeed, mathematical logicians might take differing views of many statements in set theory as to whether the so-called truth of certain mathematical assertions is, or is not, an objective matter. One of the best known of such assertions is Cantor's continuum hypothesis, which asserts that no infinity (i.e., infinite cardinal number) is greater than the number of integers yet, at the same time, is smaller than the number of real numbers. Some, such as Gödel himself, would regard such a mathematical assertion as being necessarily either objectively true or objectively false (where as yet the truth or falsehood of the continuum hypothesis would be unknown); others might regard such a mathematical assertion as being a matter of subjective opinion that might depend on which of a number of different alternative mathematical standpoints one chooses to adopt. The issue has to do with how strong a Platonist a mathematician might be considered to be. Gödel himself was a very strong Platonist, taking the view that all such assertions as the continuum hypothesis should be objectively either true or false, and the unsettled nature of this particular assertion is merely a feature of the present state of mathematical knowledge.[6] Other mathematical logicians might be considered to be much weaker Platonists, taking the view that the truth of a good many apparently well-defined mathematical statements might be a matter of subjective

---

[5] See, e.g., Mostowski (1957), and, for a more directly relevant modern discussion, Feferman (2006).

[6] But see Chapter 20 in this volume. It should be made clear that the Gödel-Cohen demonstration of the independence of the continuum hypothesis of the standard axioms of set theory (see Cohen, 1966) is in accordance with this "unsettled nature."

opinion. In my view, such differences of opinion are of small concern to the issues that will be relevant to us here. The inhabitants of the Platonic mathematical world are to be simply those mathematical entities or assertions whose meanings and validity are indeed objective and unambiguous. Moreover, as we shall see, the arguments that I give will, on the whole, require such objectivity only for statements of a particularly straightforward and unambiguous type, referred to as Π_1-sentences.

16.2 The Gödel-Turing Argument

The Hilbert program of the late nineteenth and early twentieth centuries attempted to provide a basis for the notion of rigorous mathematical proof that could be formulated in terms of what are referred to as formal systems. For the purpose of the discussion here, it is not at all necessary to know what a formal system actually is, in any kind of detail. All that we need to know is that any such system $\mathbb{F}$ provides us with a computational procedure F for checking whether the methods of mathematical proof that $\mathbb{F}$ provides have been correctly applied in any particular case.

Here I depend on the Turing(-Post) notion of a computation, or algorithm. What is an algorithm? Rather than providing a precise mathematical definition of an algorithm in terms of, say, a Turing machine,[7] it will be easier for our present purposes (now that computers constitute such a commonplace part of our lives) simply to define an algorithm to be the action of a digital computer, where that computer is idealized so that it never makes mistakes, has an unlimited storage capacity, and can run indefinitely without ever wearing out. I shall be concerned here (up until Section 16.4) only with the standard *discrete* notion of computation, where without loss of generality, we may consider that the actions of our algorithms are on *natural numbers* (the ordinary finite counting numbers $0, 1, 2, 3, 4, 5, 6, \ldots$).

Note the fact that algorithms need not simply be composed of arithmetical operations but also can involve logical operations. Let us take an example that will be of value to us later. If we are given the pair of natural numbers (a, b), then we can construct the natural number $n = ((a + b)^2 + 3a + b)/2$ simply by means of standard arithmetical operations. However, in this case, we can also *reconstruct* the pair (a, b) from the natural number n by an algorithmic procedure that also uses logical operations. To do this, we first obtain the largest triangular number t that is no larger than n. Recall that the triangular numbers are $0, 1, 3 = 1 + 2, 6 = 1 + 2 + 3, 10 = 1 + 2 + 3 + 4, \ldots, t = 1 + 2 + 3 + \cdots + r = r(r + 1)/2$. Then we find that $t = r(r + 1)/2$ with $r = a + b$, from which it follows that $a = n - t$ and $b = r - a$. Thus, once we have found t (i.e., r), we need only standard arithmetical operations for finding a and b; but how do we obtain t from n? One procedure would be to subtract 1 from n, then 2 from the result, then 3 from the result, then 4, and so on, until we find the first number $r + 1$ for which this subtraction yields a negative number. This gives us r, as required, but we see that logical operations are required as well as the standard arithmetical ones.

---

[7] For the mathematical definition of a Turing machine, see, e.g., Davis (1978) or Penrose (1989, chap. 2).

Note that this gives a way of encoding the *pairs* of natural numbers (a, b) as *single* natural numbers, and vice versa, in an algorithmic way.

Another example of an algorithmic operation is to find the smallest natural number that is not the sum of three square natural numbers. Here we could proceed as follows: start with 0, finding $0 = 0^2 + 0^2 + 0^2$. Then try 1, finding $1 = 1^2 + 0^2 + 0^2$; then 2, finding $2 = 1^2 + 1^2 + 0^2$; then $3 = 1^2 + 1^2 + 1^2$; then $4 = 2^2 + 0^2 + 0^2$. In each case, we can simply try all possibilities of sums of three square numbers that are each no larger than the number to which we are trying to sum. We find that the operation stops at 7, telling us that 7 is the smallest number that is not the sum of three squares. This procedure is entirely algorithmic, but we take note that again, logical operations were employed.

An important feature of algorithmic operations, however, is that they sometimes do not terminate. For a simple example, we can replace 3 by 4 in the preceding example and attempt to find the smallest natural number that is not the sum of four squares. Noting that $7 = 2^2 + 1^2 + 1^2 + 1^2$, we are forced to try larger numbers, but we find that the procedure in fact continues indefinitely forever, without stopping. This is the case because of a famous eighteenth-century theorem, from Lagrange, that every natural number is the sum of four squares.

Lagrange's theorem is a good example of a Π_1-sentence, which is the simple type of mathematical assertion that was alluded to at the end of the introduction. Generally, a Π_1-sentence is an assertion of the type "such-and-such an algorithmic procedure does not terminate." Another well-known Π_1-sentence is the assertion involved in Fermat's last theorem, which states that there are no positive integers x, y, z, m, with $m > 2$ such that $x^m + y^m = z^m$, known to be true since the work of Andrew Wiles. Here the algorithmic procedure could be phrased as running over all quadruplets (x, y, z, m) of natural numbers (which can be done in a way that will be described shortly), where x, y, z, and $m - 2$ are all positive, and then checking that in each case, the Fermat equation is *not* satisfied, whereas the procedure would have had to come to a halt as soon as a case was encountered in which the equation *was* satisfied. Another famous example of a Π_1-sentence is Goldbach's conjecture, which asserts that every even number greater than 2 is the sum of two prime numbers. This Π_1-sentence is somewhat similar in form to the previous example, but with the additional complication that the numbers in the sum have to be checked to be prime in each case, which is an algorithmic (but lengthy) process.

Recall that for the Fermat example, we need to run through the quadruplets (x, y, z, m). This can be done simply by employing a slight elaboration of what we did earlier, according to which we could run through all the pairs (a, b) of natural numbers simply by running through all the natural numbers $n = ((a + b)^2 + 3a + b)/2$. One way of running through all the quadruples of natural numbers would be first to apply this procedure to encode each of the pairs (x, y) and (z, m) as single numbers and then to apply it again to the resulting two numbers to encode the entire quadruple as a single natural number.

An important issue now arises: how are we to tell whether a given Π_1-sentence P is true or false? A *false* Π_1-sentence can, in principle, always be ascertained to be false, simply by running the said algorithm until it terminates. (The "in principle" is important here because the actual running of the algorithm as far as is needed could well be prohibitive. This holds even for very elementary algorithms that obviously terminate such as "start with 10^{999999} and then subtract 1 repeatedly, halting if the number becomes negative.") But it can sometimes be very difficult to ascertain the truth of a Π_1-sentence that happens to be true, as we have seen in the case of Lagrange's theorem, or extremely difficult, as in the case of Fermat's last theorem, or apparently even more difficult than that, in the case of Goldbach's conjecture – if, indeed, it turns out to be true – because it remains unresolved at present. Is there, in fact, any true Π_1-sentence whose truth is in principle inaccessible to human reason and insight?

To see what Gödel's incompleteness theorems have to say on the matter, let us appeal to Turing's more direct and very general version of Gödel's essential result. Let us suppose that we have settled on some formal system $\mathbb{F}$ with which we are happy so that we are prepared to have complete trust in the truth of any Π_1-sentence that can be actually proved by following the rules of $\mathbb{F}$. The checking of such a proof is to be an algorithmic process, so we have some particular algorithm F that decides this.

More specifically, for the purposes of the Gödel-Turing argument, we shall need to consider families of Π_1-sentences, the members of which are computably labeled by a natural number r. Thus each value of r (computably) provides us with a computation, and we are concerned with whether this computation terminates. Examples of such families would be assertions claiming the nontermination of computations of the form "find a natural number that is not the sum of r square numbers," or "find a natural number greater than 2 that is not the sum of up to r prime numbers," or some such. The first of these would be true, and proved, for $r \geq 4$ and false otherwise; the second would only be proved to be true for some very large r but conjectured to be true for $r = 2$ and false for $r = 0$ or 1. Accordingly, each Π_1-sentence of our family would be defined as an assertion that some computation does not terminate, the rth Π_1-sentence of the family being defined in terms of some r-dependent computation that we could write as $\mathbf{C}(r)$, where the computation is considered to act algorithmically on the natural number r. (This, indeed, is the kind of action that a Turing machine effects.) All such (r-dependent) computations can be (computably) listed:

$$C_0, C_1, C_2, C_3, C_4, \ldots,$$

that is, as computer programs listed in an appropriate numerical ordering. We require that this listing be computable in the sense that $\mathbf{C}_q(r)$ is a computable function of the pair (q, r).

Now, suppose that we have some formal system $\mathbb{F}$ that has a certain capacity to prove certain Π_1-sentences, where we fully trust that $\mathbb{F}$ is sound in this respect. Thus $\mathbb{F}$ provides us with an algorithm A that, when presented with the pair of natural numbers (q, r), will sometimes terminate, this termination signaling that $\mathbb{F}$ has indeed ascertained that the computation $\mathbf{C}_q(r)$ does not terminate (i.e., $\mathbb{F}$ has provided us with a proof of this, as confirmed by its proof-checking algorithm F). What we demand of the algorithm A, as the Π_1-sentence truth ascertainer, is that whenever it comes to a halt, this signals that such a proof has been achieved, according to the rules of $\mathbb{F}$, of the nonstopping of the algorithm $\mathbf{C}_q(r)$; so

$$\text{if } A(q, r) \text{ stops, then } \mathbf{C}_q(r) \text{ does not stop.}$$

We do not require that $A(q, r)$ can settle the truth of every Π_1-sentence, but we are demanding that A does not make mistakes so that the formal system $\mathbb{F}$ is at least *sound* with regard to its assessments of Π_1-sentences.

Putting q equal to r in the preceding displayed statement, we obtain the following:

$$\text{if } A(r, r) \text{ stops, then } C_r(r) \text{ does not stop.}$$

However, $A(r, r)$ is an algorithm that depends only on the single natural number r, so it must be one of the C_q acting on r, say, C_K, so

$$A(r, r) = C_K(r),$$

and taking the particular value $r = K$ in the previous displayed equation, we obtain the following:

$$\text{if } A(K, K) \text{ stops, then } C_K(K) \text{ does not stop.}$$

However (from the preceding equation with $r = K$),

$$A(K,K) = C_K(K),$$

whence

$$\text{if } C_K(K) \text{ stops, then } C_K(K) \text{ does not stop,}$$

so $C_K(K)$ does *not* in fact stop, and being the *same* as $A(K, K)$, we have that $A(K, K)$ does not stop either.

The upshot of this is that we have found (in fact, in a manifestly computable way, starting from the formal system $\mathbb{F}$) a computation $C_K(K)$ that does not terminate, giving us a valid Π_1-sentence, yet our particular Π_1-sentence truth ascertainer A is not itself powerful enough to be able to establish this Π_1-sentence because A does not terminate in this case. I shall refer to this Π_1-sentence as $G(\mathbb{F})$, the Gödel assertion for the system $\mathbb{F}$. What we have established is that the Π_1-sentence $G(\mathbb{F})$ is true provided that $\mathbb{F}$ is sound with regard to Π_1-sentences, that is, that no false Π_1-sentences can be proved by the methods allowed by $\mathbb{F}$. Yet, under this assumption, $G(\mathbb{F})$ cannot itself be proved using these methods. We can paraphrase this remarkable conclusion more colloquially as follows: for any computationally checkable system of proof procedures (for Π_1-sentences) $\mathbb{F}$, we can construct a Π_1-sentence $G(\mathbb{F})$ that, on the assumption that $\mathbb{F}$ is sound, is both true and unprovable by the methods of $\mathbb{F}$.

16.3 The Noncomputability of Mathematical Conviction

We may wonder how it can be possible to deduce the truth of a particular Π_1-sentence, namely, $G(\mathbb{F})$, from the *same* belief system – that is, trust in the soundness of $\mathbb{F}$ – as is needed for trusting the direct proof procedures that $\mathbb{F}$ actually provides, while at the same time we find that $G(\mathbb{F})$ is not itself accessible using these very proof procedures.[8]

---

[8] In the context of Gödel's second incompleteness theorem, belief in the soundness of $\mathbb{F}$ entails a belief in its consistency (otherwise we would have to accept absurd assertions such as "2 = 3"), so we must also believe in the Gödel sentence that asserts this consistency.

Indeed, this is what, to me, is the most remarkable thing about Gödel incompleteness, for it strikes at the very heart of the formalist philosophy of mathematics. That philosophy would contend that formal proof is the bedrock of mathematics. In any particular mathematical area, one would settle on some formal system $\mathbb{F}$ and take the view that the establishment of mathematical truth within that area is, by definition, reduction to "proof" within $\mathbb{F}$. However, we have seen that if the area of mathematics in question is the very basic one of "establishment of Π_1-sentences," then this point of view will not do – whatever system $\mathbb{F}$ we choose, our trust in its soundness, with regard to Π_1-sentences, also gives us a trust that extends beyond the scope of $\mathbb{F}$ as a provider of proof procedures, namely, to $G(\mathbb{F})$ and also to many other Π_1-sentences such as those obtained by iterating the Gödel procedure.

All this is beyond contention, as far as I can make out. However, in my opinion, we can go a good deal further than this and gain some glimmer of a deep relevance to the nature of human consciousness. Why do I believe that Gödel's theorems have something deep to say about human consciousness? Basically, I believe so because the incompleteness theorems show, in my view, that human *understanding* – specifically in the area of *mathematical* understanding – is something that cannot be reduced to computation, the quality of understanding being one manifestation of human consciousness.

I should remark that whereas it is fully accepted that Gödel's incompleteness theorems have had a profound importance for the foundations of mathematics, and also that they have significant and deep implications for philosophy in general, the issue of the bearing of Gödel's results on the nature of the human mind appears to remain highly controversial. It seems clear that Gödel himself strongly held the view that human understanding – specifically human insight (or "intuition," as Gödel would more usually refer to it) – must indeed lie outside the scope of purely computational actions, but he confessed to not having a rigorous demonstration of this contention, for he asserted:

> On the other hand, on the basis of what has been proved so far, it remains possible that there may exist (and even be empirically discoverable) a theorem-proving machine which in fact *is* equivalent to mathematical intuition, but cannot be *proved* to be so, nor even proved to yield only *correct* theorems of finitary number theory.[9]

Others[10] have been somewhat less cautious in citing Gödel's incompleteness theorems as providing a demonstration that human understanding cannot be describable as the action of a Turing machine. Gödel's caution stems from the consideration that an $\mathbb{F}$ (say, with regard to Π_1-sentences) may exist that is, unbeknownst to us, equivalent to what in principle is accessible by understanding and insight, but where $\mathbb{F}$ actually having this status (or even providing only true Π_1-sentences) would in principle be inaccessible to us. If such an $\mathbb{F}$ were to exist, and effectively were to be part of the unknown underlying mechanisms of our "computer-brains," then even if we happened

---

[9] See Gödel (1951) (his Gibbs lecture) as printed in Gödel (1995). This quotation can be found in Wang (1993, 118) and should be compared with what Gödel (1995, 309) says. Also see the introductory note by George Boolos to Gödel (1951, 290–304) as well as Feferman (2006), in which Gödel's cautious antimechanistic argument is critically examined.

[10] See Nagel and Newman (1958), Lucas (1961), and Penrose (1989).

Figure 16.2. For our remote ancestors, a specific ability to do sophisticated mathematics can hardly have been a selective advantage, but a general ability to *understand* could well have been.

to come across it ("empirically") and were to guess that it is sound (with regard to Π_1-sentences), we could never be sure of this. If we were, we could infer with certainty the validity of $G(\mathbb{F})$, which is not within the scope of $\mathbb{F}$ and therefore not supposed to be a knowably true Π_1-sentence.

However, is the existence of such an $\mathbb{F}$ really plausible? Even just on the general grounds of evolution, this seems exceedingly unlikely. We would be asked to believe that, purely by natural selection, an "$\mathbb{F}$" could arise, basic to the underlying processes of our supposedly computer-like brains, of such extraordinary sophistication that it could in principle generate all the Π_1-sentences that are knowably true to us. However, this role of $\mathbb{F}$ could not be knowable to us, and its $G(\mathbb{F})$ would lie just beyond what is in principle accessible. I can see no believable way that natural selection could give rise to such a thing, especially when one takes into account how irrelevant to our ancestors' survival a facility with such sophisticated mathematical ideas could be (see Figure 16.2).

Moreover, it seems to me that one can go a good deal further, on purely logical grounds, toward eliminating Gödel's loophole of a "theorem-proving machine," which might even be empirically discoverable. I am concerned only with Π_1-sentences here, so let us suppose that we have somehow come across the rules of such a machine – that is, some formal system $\mathbb{F}$ – whose Π_1-sentence theorems happen to be precisely the Π_1-sentences that are in principle demonstrably true by human reason and insight but we do not know for sure that $\mathbb{F}$ has this status, and certainly not at the level of any kind of mathematical proof. The way that the following argument proceeds, roughly speaking, is to consider, hypothetically, not just the Π_1-sentences that the system $\mathbb{F}$ can itself directly prove but also the broader class of Π_1-sentences that could be established on the assumption that $\mathbb{F}$ actually has this hypothesized status (of being able to establish precisely all Π_1-sentences accessible in principle to human insight and reason). If we are to take it that we are entirely computational beings, and hence that this broader class of Π_1-sentences[11] can also be generated by some formal system, say, $\mathbb{F}^*$, then we could consider the Π_1-sentence $G(\mathbb{F}^*)$ and would have to regard it as true on the basis of the aforesaid assumption. This line of reasoning follows one that I gave in my article "Beyond the Doubting of a Shadow" (Penrose, 1996a), which was written in response to a number of invited criticisms of my earlier book *Shadows of the Mind* (Penrose, 1994). In that article, I proposed a simplified version of the more detailed argument presented in the book (based on a form of words used by David Chalmers (1995) in his commentary), and I repeat this formulation here.

In what follows, the term *sound* refers only to the soundness of a formal system $\mathbb{F}$ with regard to its assertions (i.e., theorems) that have the form of Π_1-sentences. Also, the assertion "I am $\mathbb{F}$" is just shorthand for "the Π_1-sentences that are theorems of $\mathbb{F}$ are precisely those that are in principle demonstrably true by human reason and insight." Furthermore, I am making the (reasonable) assumption that the Π_1-sentences that are in fact "demonstrably true by human reason and insight" are in fact true.

On being presented with the putative system $\mathbb{F}$, we now reason as follows:

($*$) "Though I do not know that I necessarily am $\mathbb{F}$, I conclude that if I were, then the system $\mathbb{F}$ would have to be sound and, more to the point, $\mathbb{F}^*$ would have to be sound, where $\mathbb{F}^*$ is $\mathbb{F}$ supplemented by the further assertion 'I am $\mathbb{F}$.' I perceive that it follows from the assumption 'I am $\mathbb{F}$' that the Gödel statement $G(\mathbb{F}^*)$ would have to be true and, furthermore, that it would not be a consequence of $\mathbb{F}^*$. However, I have just perceived that 'if I happened to be $\mathbb{F}$, then $G(\mathbb{F}^*)$ would have to be true,' and perceptions of this nature would be precisely what $\mathbb{F}^*$ is supposed to achieve. As I am therefore capable of perceiving something beyond the powers of $\mathbb{F}^*$, I deduce that I cannot be $\mathbb{F}$ after all."

The foregoing reasoning deserves some considerable clarification. In particular, we need to clarify what is meant by the system "$\mathbb{F}$ supplemented by the further assertion 'I am $\mathbb{F}$.'" I am taking it that we are happy with the unassailable nature of the Gödel-Turing argument, as outlined in Section 16.1. We are concerned here with unassailable belief, and such belief in "I am $\mathbb{F}$" would therefore imply a corresponding belief in the particular Π_1-sentence $G(\mathbb{F})$ as well as in all the Π_1-sentence theorems of $\mathbb{F}$. Hence $G(\mathbb{F})$ would have to be a consequence of the system $\mathbb{F}^*$. Moreover, this belief in $G(\mathbb{F})$ would have the implication that the formal system $\mathbb{F}'$, which is $\mathbb{F}$ with $G(\mathbb{F})$ adjoined to it, is also sound with regard to Π_1-sentences so that a belief in the Π_1-sentence $G(\mathbb{F}')$ is also a consequence of $\mathbb{F}^*$. This process can be repeated, and we conclude that the Π_1-sentence $G(\mathbb{F}'')$ is also a consequence of $\mathbb{F}^*$, being a consequence of $\mathbb{F}''$, where $\mathbb{F}''$ is the formal system $\mathbb{F}'$ with $G(\mathbb{F}')$ adjoined. This process can be continued, in various kinds of ways, to give additional consequences of $\mathbb{F}^*$.

However, in the previously displayed argument, we would need to be able to go even further and say that the "Π_1-sentence $G(\mathbb{F}^*)$" is also a deduction from "I am $\mathbb{F}$"; however, this argument has a weakness because, as stated, $\mathbb{F}^*$ is not a formal system in the ordinary sense, for "I am $\mathbb{F}$" is not an ordinary kind of logical assertion that

---

[11] I am grateful to Hilary Putnam for pointing out, in conversation, that because of their conditional dependence on $\mathbb{F}$'s status, the actual assertions of this family do not individually have the form of Π_1-sentences. Critiques concerning various other forms of this argument can be found in Shapiro (2003) and Lindström (2006).

we would know how to express within the system $\mathbb{F}$. However, if we are to assume that our ordinary processes of understanding can be reduced to computation (as is being assumed already in the proposed existence of $\mathbb{F}$, in any case), then the family of Π_1-sentences that can be established from the assumption "I am $\mathbb{F}$," together with $\mathbb{F}$ itself, is indeed the family of Π_1-sentences that are computably generated and that are therefore theorems of some formal system $\mathbb{F}*$. Trouble arises, however, because we may not know what the system $\mathbb{F}*$ would actually be (even though $\mathbb{F}$ is taken as given). Although we can contemplate any suggested "$\mathbb{F}*$" and then construct its Gödel Π_1-sentence $G(\mathbb{F}*)$ without being sure that this particular $\mathbb{F}*$ actually does provide proofs of all the Π_1-sentences that can be established from the assumption "I am $\mathbb{F}$," together with $\mathbb{F}$ itself, we cannot be sure that $\mathbb{F}*$ is actually sound (with respect to Π_1-sentences), so we would not be unassailably convinced of the truth of $G(\mathbb{F}*)$.

For reasons of this kind, the arguments that I gave in *Shadows of the Mind* (1994, chap. 3) were not stated in exactly the way that I have stated them here. Instead, the form of argument that I gave was a kind of reductio ad absurdum based on the hypothesis that a family of mathematics-understanding robots could actually be constructed using standard computer-based artificial intelligence (AI) procedures. When presented with the detailed computational procedures $\mathbb{M}$ that led to their construction (and I shall discuss such computational procedures a little more fully in the next section), they could, on the hypothesis that the procedures $\mathbb{M}$ were indeed those underlying their construction, compute what $\mathbb{F}$ would actually have to have been. Moreover, they could compute what the system $\mathbb{F}^{\#}$ would have to have been, where $\mathbb{F}^{\#}$ (similar to $\mathbb{F}*$, but not constructed in quite the same way) would be able to prove precisely the Π_1-sentences that can be established from the assumption "I have been constructed according to $\mathbb{M}$," together with $\mathbb{F}$ itself. Thus, if they were to accept that $\mathbb{M}$ actually has the role claimed for it, then they would have to accept the Π_1-soundness of both $\mathbb{F}$ and $\mathbb{F}^{\#}$ and hence the truth of $G(\mathbb{F}^{\#})$ and its unprovability from $\mathbb{F}^{\#}$. But this *would* be a contradiction for them because $\mathbb{F}^{\#}$ was supposed to be able to prove any Π_1-sentence that they could establish on the basis that they (the robots) had indeed been constructed according to $\mathbb{M}$.

Admittedly, this weakened form of argument does not have quite the force or simplicity that the version extracted earlier and labeled (*) was intended to have provided, but this particular (weakened) argument is essentially that which I actually gave in *Shadows of the Mind* (where the computational AI procedures that are supposed to be able to result in intelligent mathematics-performing robots are humanly knowable), and I believe that it is, in essence, correct. It lends considerable doubt to the hypothesis that standard computational AI procedures can ever provide a robot that could acquire genuine mathematical understanding. Some of the issues involved in formulating such computational procedures will be raised in the next section.

16.4 Mistakes, Vague Arguments, and Idealizations

People often object to the kind of argument given in the previous section, sometimes claiming that actual human reasoning does not follow the idealized precision that is assumed in the Gödel-type arguments for the noncomputability of human understanding and insight. Indeed, it is certainly true that mathematicians, like everyone else, are capable of vague reasoning and sometimes of making serious mistakes. I want to make the point that though this is true, it does not reduce the significance of the arguments given in Section 16.2. The most usual kind of mistake that a mathematician might make is of no real concern to us here, being something that is correctable[12] by that mathematician on further contemplation or when the error is pointed out by someone else. Our essential concern is for the ideal mathematical notions and arguments for which mathematicians strive. In their preliminary thinking, it is quite usual to fall significantly short of such ideals, and their published works are likely to be much closer to such ideals than is their preliminary thinking.

Such an ideal of correctness is an essential feature of the very discipline of mathematics because it gives objectivity to the fact that erroneous arguments are in fact to be rejected. This is not to say that correct arguments need to be formulated in terms of some particular formal system $\mathbb{F}$ because, as we have seen in Section 16.1, Gödel's theorems tell us that this can never encapsulate all correct mathematical arguments.

However, a different aspect of the issue of mathematical correctness relates to the fact that not all mathematicians would hold to the same viewpoint with regard to arguments involving infinite sets or classes, particularly considering the question of how strong a Platonist a particular mathematician might be considered to be (e.g., is it always acceptable to use the axiom of choice in a mathematical argument?). These issues tend not to come up seriously until large, infinite sets or classes are being considered. I have tried to reduce the impact of such differences of opinion by restricting attention to mathematical statements of a particularly simple kind, namely, Π_1-sentences, whose truth or falsehood would normally be considered to be a completely objective matter, despite continuing arguments concerning the validity of proof methods used to establish such sentences.

Another thing about which people tend to worry concerning the validity of the use of Gödel-type arguments in relation to human thinking is that they would envisage such arguments being applied to the entire (horrendously complicated) detailed activity of an individual human brain, and they might quite reasonably question the validity of such activity being regarded as being equivalent (with regard to Π_1-sentences) to some formal system $\mathbb{F}$. I have tried to avoid such criticism by referring not to the action of the brain of some individual person but to the more abstract notion of mathematical reasoning that is only approximated by the actions of individual mathematicians' brains.

In *Shadows of the Mind* (Penrose, 1994, chap. 3), I needed to go rather further than this, as I have indicated in the previous section, by considering the potential capabilities of a community of mathematics-performing, computationally controlled robots, the issue being how far toward achieving something equivalent to the capabilities of mathematical understanding and insight could such a computational system be able to go. The robots could be deliberately programmed in a top-down way, or perhaps

---

[12] I consider the elimination of such "correctable mistakes" at some length in *Shadows of the Mind* (Penrose, 1994, esp. sect. 2.10, query 13, and sect. 3.17–3.21) as well as in "Beyond the Doubting of a Shadow" (Penrose, 1996a, sect. 6). Curiously, as I have discussed, Turing's own proposed loophole to the Gödel argument seemed to depend on the fact that humans are prone to errors (see Penrose, 1994, sect. 3.1). The issue of the idealization of human mathematical thinking is also discussed in Shapiro (2003).

bottom-up learning processes could be invoked, involving natural selection processes with computationally simulated environments. Eventually, an algorithmic-underlying, consistent mathematical belief system that is entirely computational in nature might be attained by the robots. In this case, we are dealing with a system that is (by definition) clearly computational. My claim, ultimately based on arguments like those given earlier, is that such a system would always fall significantly short of what can be achieved through human understanding.

A related matter is the issue of actual computers or brains being finite so that the idealized notion of a Turing machine may be regarded as being only approximately realized by a physical computer. The finiteness issues are also addressed in *Shadows of the Mind* (Penrose, 1994, sect. 2.6, query 8, and sect. 3.20). I argue there that these issues are only marginally relevant to whether we are computational entities.

16.5 Computability in the Laws of Physics

If it is accepted, in accordance with the discussion given previously, that human understanding, reasoning, and insight lie beyond computational activity, then we may reasonably ask how such noncomputable action might arise through some physical action of the brain. It appears that Gödel, on the other hand, did not consider the possibility that the laws of physics might themselves involve noncomputational procedures; accordingly, he found himself driven to the conclusion that mentality must lie outside the actions of the physical brain. My own position is to take a more physicalist stance than this and to argue (in accordance with mystery 2 of Figure 16.1) that the activities of the human mind are grounded in the physical behavior of the particular physical structures that are human brains. In accordance with this, there must be something in the physical actions of the world that themselves transcend computation.

We do not appear to find such noncomputational action in the known laws of physics, however, although this is a matter that is not beyond contention. It should be borne in mind that the currently understood physical laws are phrased in terms of continuous parameters, whereas the standard notion of computability refers to computation with the (discrete) natural numbers. Thus the computability issues that I have so far been considering are not immediately applicable to the laws of physics as we presently understand them.[13] However, my own position, at least as things stand at the moment, has been to ignore this issue because it seems unlikely to me that any plausible kind of noncomputability exists in these considerations that could be usefully taken advantage of by a physical system. It is certainly my belief that many issues of mathematical interest are still to be explored concerning this question, but I find it hard to see how the issues I have raised concerning the noncomputability of mathematical understanding can be

---

[13] See, e.g., Weihrauch (2000) and Pour-El and Richards (1998). Pour-El and Richards (1982, 2003) have pointed out that some of the equations of physics have the property that computably presented initial data (necessarily not smoothly varying, i.e., not C^2) can sometimes lead to a noncomputable evolution. However, the relevance of this type of result for the situation under consideration here is questionable (see Penrose, 1989). For some alternative approaches to the issue of computability for real numbers, see Blum et al. (1989) and Bridges (1999). See also Stannett (2003).

resolved through subtleties of the distinctions between the discrete and continuous notions of noncomputability. The measurements of continuous parameters in physical situations are never absolutely precise, so these subtle distinctions would appear not to be playing any operational role.

Instead, I believe that we must turn to the question of whether any fundamental gaps remain in our understanding of physical laws that might be relevant to conscious brain action. It is my view that we must seek noncomputability in currently undiscovered laws going beyond currently accepted physical theory. Why do I make this claim? The reason is that the standard equations of physics – taking into account the reservations expressed in the previous paragraph – seem to be essentially computable. In fact, I shall use the phrase "essentially computable" to denote "computable," but where the subtleties of the previous paragraph are indeed being ignored.

What are these equations? In the first place, when it comes to describing macroscopic objects, we have those from classical physics: Newtonian mechanics (Newton's laws), the field theories of electromagnetism (the Maxwell-Lorentz equations), and gravitation (Einstein's field equations). Then, for describing submicroscopic objects, we need Schrödinger's equation for the evolution of the quantum state-vector. The solutions of all these equations are essentially computable. However, when a quantum "measurement" or "observation" is made, we allow the state to jump to an eigenstate of an operator that is defined in terms of the particular measurement that is being performed and with a probability that is determined in terms of scalar products between the state-vectors before and after the jump takes place. This jumping is taken to be entirely probabilistic and certainly is not "noncomputable" in the essential sense to which I just referred. It is not described by Scrödinger's equation. That equation would provide an evolution that is as deterministic and computable as is Newtonian dynamics. One should, of course, bear in mind that determinism does not imply computability, it being a fairly simple matter to provide toy-model universes that are deterministic yet whose evolution is not possible to simulate computationally (see Penrose, 1989, 220–21, 287n6; 1997a, 118–19); that is the kind of "essential noncomputability" being referred to here. The state jumping that occurs in the standard description of the measurement process is quite unlike this, being taken to be a probabilistic action, providing us with something of the nature of a Turing machine with a randomizer.

This last procedure is indeed very odd, being of quite a different form from the continuous evolution equations that had been used up to this stage in physics. It is odd also because it is a probabilistic procedure rather than something deterministic, as had been the case for all evolution procedures previously known in the behavior of physical systems, and it is particularly odd because it is in gross contradiction with the smooth evolution of the quantum state described by the very precise Schrödinger equation.

This contradiction is often described as the *measurement paradox*. This paradox is seen as a gross conflict between the two basic procedures of quantum mechanics. A measurement of a small quantum system, after all, can be understood as the evolution of a larger system consisting of the small quantum system under consideration together with a larger object, namely, the measuring apparatus (and relevant elements of the environment as well). Why does this larger system not simply evolve smoothly according to the Schrödinger equation, rather than jumping in this strange way, as described by the measurement process of quantum mechanics? This discrepancy is not a small thing, as was pointed out dramatically by Schrödinger himself. He considered a hypothetical experiment (which would not be difficult to carry out in practice, if one wanted to do such a thing) in which, according to the Schrödinger evolution, the state of a cat would evolve into one in which the cat would be simultaneously alive and dead. This would be an example of a quantum superposition – a well-established occurrence in the case of individual particles or small quantum systems – in which two alternative realities can occur simultaneously such as an individual quantum particle occupying two different locations at the same time. Although small quantum systems indeed appear to indulge in such strange superpositions, one never sees such things occurring for a large system such as a cat, despite that the Schrödinger equation makes no distinction between small things and large things.

The quantum measurement paradox is, in my view, completely *un*-resolved by standard quantum theory, despite that we have numerous ways to address the issue, with different physicists frequently taking completely different standpoints with regard to it. For example, in one quite common point of view, that the detailed state of the environment is "unobservable" is regarded as providing a resolution of the paradox. In another, one considers that the dead and live cats coexist in parallel universes, and an observer who then looks at the cat then has parallel existences in these two universes, one seeing the dead cat and the other a live one, because somehow the observer's conscious experience splits into two so that each records the experience of just one of the two alternatives. This is not the place for me to go into the reasons for my dissatisfaction with both (and indeed all) these "conventional" resolutions of the measurement paradox.[14] Suffice it to say that in my opinion (in agreement with at least three of the founders of the theory – Einstein, Schrödinger, and Dirac), standard quantum theory is a *provisional* theory that needs to be completed or fundamentally changed in some way in the future.

If we need a basic change in physical theory to accommodate the noncomputability that appears to be a feature of the behavior of human brains when, at least, they are involved with sophisticated mathematical reasoning, then it seems reasonable that this change should be at a place where it seems that present-day theory is, in any case, in need of change for direct physical reasons. Such a place would appear to reside in whatever is needed to resolve the measurement paradox. We have many unknowns in our present-day physical picture, but of these, only the measurement paradox seems to present itself at such a scale that it could have relevance to the workings of a human brain.

With this connection, I should emphasize that the change from Gödel-type reasoning that I am demanding is not really expected to have anything specifically to do with mathematical reasoning. The particular quality that is being brought to bear here on a mathematical problem is *understanding*, and understanding need not specifically have anything to do with mathematics. Moreover, genuine understanding is, at least in the normal use of this term, *conscious*. It makes little sense to say of an entity that it "understands" something if it is not actually aware of this thing, and awareness implies

---

[14] See, in particular, Penrose (2004, chap. 29).

consciousness. The notion of consciousness encompasses many other things than just understanding, of course, such as the perception of a color, such as "blue," or the feeling of an emotion, such as "sadness," or the feeling of pain, or the appreciation of the beauty of a melody, and many other things besides. In addition, it seems unlikely that most of these qualities of conscious experience are restricted to human beings, as a good many of these qualities would seem to be quite widespread through the animal kingdom, at least at the higher end of the animal scale with regard to mental capabilities. There is nothing particular about mathematical understanding and insight, in my view, beyond the fact that in mathematics, we find a use of understanding that is sufficiently defined that it is possible to make precise arguments about the understanding process itself. It is the more general quality of consciousness that itself seems to have the distinctiveness that some very special feature of physical behavior may be at work.

What special feature of physical behavior could this be, where something distinctly different seems to be going on from the normal physical actions that appear to govern unconscious actions and seem to be adequately described by the conventional physics that we now understand? I cannot see anything that has much chance beyond the part of our physical picture that at present demands the strange jumping of the quantum state that is needed to describe quantum measurements. If we indeed need a new theory for that, then it seems to me that this new theory would have to be a major revolution, in which new concepts are introduced that are broad enough to be able to accommodate conscious experience. According to the foregoing discussion, these would have to contain elements that are noncomputable.

Needless to say, I am not in possession of such a theory. Nevertheless, I believe that it is possible to make some well-founded guesses as to some of the ingredients that such a theory would involve. One of these guesses is that it is in the appropriate combination of principles from the two great revolutions of twentieth-century physics – quantum theory and Einstein's general theory of relativity – that we must seek appropriate new ideas. A preliminary investigation of these issues[15] suggests that a resolution of the measurement paradox would imply that a quantum superposition of an object in two different locations should decay to one or the other on a time scale that can be calculated from the distribution of mass in the object. In a proposed experiment partly aimed at eventually testing such a scheme,[16] a tiny mirror, of dimension roughly one-tenth of the thickness of a human hair, would be put into a quantum superposition of two very slightly different locations, resulting in a time scale of something around one second.

16.6 Possible Role in Brain Function

Although some speculation is undoubtedly involved in these physical considerations, inevitably, a good deal more guesswork is involved when it comes to the particular role

---

[15] See Penrose (2004, chap. 30); for earlier work along these lines, see Károlyházy (1966), Diósi (1984, 1989), and Penrose (1996b).

[16] See Marshall et al. (2003).

that this speculative physics might play in the operation of the brain when consciousness is somehow evoked. The conventional picture of brain action, including whatever it is that gives rise to consciousness, is that everything is to be explained in terms of the firing pattern of neurons.[17] What the current proposal requires, on the other hand, is the allowance of a significant role for the proposed new physics that is supposed to replace the present-day quantum-measurement procedure. This new physics would become relevant at the boundary between quantum and classical behavior. Ordinary nerve transmission seems to offer little hope for this because the electric fields involved would continually disturb the environment of the individual nerves, allowing very little prospect of maintaining the isolation needed for quantum coherence (quantum effects) on a large scale (like superconductivity), which would be required over considerable areas of the brain.

On the other hand, the cytoskeletons of neurons might be able to preserve such quantum coherence. In particular, Hameroff has argued[18] that microtubules (an important constituent of the cytoskeleton), cylindrical tubes some twenty-five nanometers in outer diameter and fourteen nanometers in inner diameter, might have a fundamental role in the production of consciousness. In a model suggested jointly by Hameroff and me,[19] we proposed that different conformational changes in coherent collections of microtubules, acting in concert over quite large areas of the brain, might exist in quantum superposition, and conscious experiences could be associated with the reduction of such large-scale superpositions to one or other of their separate components. According to this scheme, the conformational states of the microtubules would indirectly affect synapse strengths and thereby influence the circuitry of the neuron connections, and thence the external behavior of the individual.

Various aspects of cognition that have proved puzzling for the conventional, solely neuronal picture, notably the binding problem, whereby distant and largely unconnected areas of the brain seem to conspire to provide a conscious experience, are more likely to be explicable within the scheme that we are proposing. Moreover, the actual dependence of conscious experience on the putative new physics, according to this scheme, provides possible scope for genuine noncomputational behavior. Clearly a good measure of speculation resides in any such proposal. In any case, consciousness is such an extraordinary phenomenon, quite unlike anything else that we know about in physical behavior, that any suggested explanation for it that is not exotic in some very fundamental way would have very little hope of success.

Acknowledgments

The author is grateful to the U.S. National Science Foundation for support under contract PHY 00-90091 and to Solomon Feferman for helpful comments, particularly with regard to references.

[17] See, e.g., McCulloch and Pitts (1943) and Crick (1994).
[18] See Hameroff (1987).
[19] See Hameroff and Penrose (1996).

References

Blum, L., Shub, M., and Smale, S. (1989). On a theory of computation and complexity over the real numbers: NP completeness, recursive functions and universal machines. *Bulletin of the American Mathematical Society*, **21**, 1–46.

Bridges, D. S. (1999). Can constructive mathematics be applied in physics? *Journal of Philosophical Logic*, **28**, 439–53.

Chalmers, D. J. (1995). Minds, machines, and mathematics. *Psyche*, **2**(9). http://psyche.cs.monash .edu.au/v2/psyche-2–09-chalmers.html. [See also *Scientific American*, June 1995, 117–18.]

Cohen, P. J. (1966). *Set Theory and the Continuum Hypothesis*. New York: W. A. Benjamin.

Crick, F. (1994). *The Astonishing Hypothesis: The Scientific Search for the Soul*. New York: Charles Scribner's Sons.

Davis, M. (1978). What is a computation? In *Mathematics Today: Twelve Informal Essays*, ed. L. A. Steen, pp. 241–267. New York: Springer.

Diósi, L. (1984). Gravitation and quantum mechanical localization of macro-objects. *Physics Letters*, **105A**, 199–202.

————. (1989). Models for universal reduction of macroscopic quantum fluctuations. *Physical Review*, **A40**, 1165–74.

Ernst, B. (1986). Escher's impossible figure prints in a new context. In *M. C. Escher: Art and Science*, ed. H. S. M. Coxeter, M. Emmer, R. Penrose and M. L. Tauber, et al., pp. 125–34. Amsterdam, Netherlands: Elsevier Science.

Feferman, S. (2006). Are there absolutely unsolvable problems? Gödel's dichotomy. *Philosophia Mathematica, Series III*, **14**, 134–52.

Gödel, K. (1951). Some basic theorems on the foundations of mathematics and their implications. [Also in *Collected Works*, vol. 3 (1995), pp. 304–23 (Gibbs lecture) *Unpublished Essays and Lectures*. New York: Oxford University Press.]

Hameroff, S. R. (1987). *Ultimate Computing: Biomolecular Consciousness and Nano-Technology*. Amsterdam, Netherlands: North-Holland.

Hameroff, S. R., and Penrose, R. (1996). Conscious events as orchestrated space-time selections. *Journal of Consciousness Studies*, **3**, 36–63.

Károlyházy, F. (1966). Gravitation and quantum mechanics of macroscopic bodies. *Nuovo Cimento*, **A42**, 390.

Lindström, P. (2006). Remarks on Penrose's "new argument." *Journal of Philosophical Logic*, **35**, 231–37.

Lucas, J. R. (1961). Minds, machines, and Gödel. *Philosophy*, **36**, 112–27. [Repr. A. R. Anderson, ed. (1964). *Minds and Machines*. Englewood Cliffs, NJ: Prentice Hall.]

Marshall, W., Simon, C., Penrose, R., and Bouwmeester, D (2003). Towards quantum superpositions of a mirror. *Physical Review Letters*, **91**, 13–16.

McCulloch, W. S., and Pitts, W. (1943). A logical calculus of the ideas immanent in nervous activity. *Bulletin of Mathematical Biophysics*, **5**, 115–33. [Repr. W. S. McCulloch, ed. (1965). *Embodiments of Mind*. Cambridge, MA: MIT Press.]

Mostowski, A. (1957). *Sentences Undecidable in Formalized Arithmetic: An Exposition of the Work of Kurt Gödel*. Amsterdam, Netherlands: North-Holland.

Nagel, E., and Newman, J. R. (1958). *Gödel's Proof*. London: Routledge and Kegan Paul.

Penrose, L. S., and Penrose, R. (1958). Impossible objects: A special type of visual illusion. *British Journal of Psychiatry*, **49**, 31–33.

Penrose, R. (1989). *The Emperor's New Mind: Concerning Computers, Minds, and the Laws of Physics*. Oxford: Oxford University Press. [Repr. 1999 with new preface.]

———. (1994). *Shadows of the Mind: An Approach to the Missing Science of Consciousness.* Oxford: Oxford University Press.

———. (1996a). Beyond the doubting of a shadow. *Psyche,* **2,** 89–129. http://psyche.cs.monash .edu.au/psyche-index-v2_1.html.

———. (1996b). On gravity's role in quantum state reduction. *General Relativity and Gravitation,* **28,** 581–600.

———. (1997a). *The Large, the Small and the Human Mind.* Cambridge: Cambridge University Press. [*Canto* ed. 2000.]

———. (1997b). On understanding understanding. *International Studies in the Philosophy of Science,* **11,** 7–20.

———. (2004). *The Road to Reality: A Complete Guide to the Laws of the Universe.* London: Jonathan Cape.

Pour-El, M. B., and Richards, I. (1982). Non-computability in models of physical phenomena. *International Journal of Theoretical Physics,* **21,** 553–55.

Pour-El, M. B., and Richards, J. I. (1989). Computability in analysis and physics. In *Perspectives in Mathematical Logic,* vol. 1. Berlin: Springer.

Shapiro, S. (2003). Mechanism, truth, and Penrose's new argument. *Journal of Philosophical Logic,* **32,** 19–42.

Stannett, M. (2003). Computation and hypercomputation. *Minds and Machines,* **13,** 115–53.

Wang, H. (1987). *Reflections on Kurt Gödel.* Cambridge, MA: MIT Press.

———. (1993). On physicalism and algorithmism: Can machines think? *Philosophia Mathematica, Ser. III,* 97–138.

Weihrauch, K. (2000). *Computable Analysis: An Introduction.* Texts in Theoretical Computer Science. Berlin: Springer.

Wigner, E. P. (1960). The unreasonable effectiveness of mathematics in the physical sciences. *Communications in Pure Applied Mathematics,* **13,** 1–14.

PART THREE

New Frontiers: Beyond Gödel's Work in Mathematics and Symbolic Logic

Extending Gödel's Work

CHAPTER 17

Gödel's Functional Interpretation and Its Use in Current Mathematics

Ulrich Kohlenbach

17.1 Introduction: General Remarks on Proof Interpretations

This chapter discusses applied aspects of Gödel's functional ('Dialectica') interpretation, which was originally designed for foundational purposes. The reorientation of proof theory toward applications to concrete proofs in different areas of mathematics, which was begun in the 1950s by G. Kreisel's pioneering work on the 'unwinding of proofs,' also led to a reassessment of possible uses of functional interpretations. Since the 1990s, this has resulted in a systematic development of specially designed versions of functional interpretation and their use in numerical analysis, functional analysis, metric fixed point theory, and geodesic geometry. Whereas [67] presents a comprehensive survey of the new results that were obtained in these areas in the course of this investigation, this chapter focuses on the underlying logical aspects of these developments. We start, however, with a general discussion of so-called proof interpretations (and their role in Gödel's work), of which functional interpretation is a particularly interesting instance, and explain the original motivation behind the latter.

Proof interpretations play an important role in Gödel's work and seem to be first used systematically by him.

Let $\mathcal{T}_1$ and $\mathcal{T}_2$ be theories in languages $\mathcal{L}(\mathcal{T}_1)$ and $\mathcal{L}(\mathcal{T}_2)$. A proof interpretation I of $\mathcal{T}_1$ in $\mathcal{T}_2$ in our sense consists of the following elements:

1. To each formula $A \in \mathcal{L}(\mathcal{T}_1)$, a new formula $A^I \in \mathcal{L}(\mathcal{T}_2)$ is assigned by induction on the logical structure of A.
2. The interpretations of the $\mathcal{T}_1$-axioms can be verified in $\mathcal{T}_2$.
3. The interpretations of the $\mathcal{T}_1$-rules are derivable rules of $\mathcal{T}_2$. The most important special case is that of the *modus ponens* rule:

   Modus Ponens Problem: $\dfrac{A^I, \ (A \to B)^I}{B^I}$.
4. Often the soundness proof conveys additional information on A^I, for example, a closed term of $\mathcal{L}(\mathcal{T}_2)$ realizing A^I in cases where A^I is an existential sentence.

As a consequence of these features of proof interpretations, one obtains the following:

1. We get a soundness theorem for I: $T_1 \vdash A \Rightarrow T_2 \vdash A^I$.
2. A proof p of A can be transformed by I (by recursion on p) into a proof p^I of A^I. In particular, the main overall logical structure of p remains intact, and p^I usually is not much longer than p.
3. If $\mathcal{L}(T_1) = \mathcal{L}(T_2)$, $T_2 \subseteq T_1$ and Γ is a class of sentences A of $\mathcal{L}(T_1)$ such that

$$T_2 \vdash A^I \rightarrow A,$$

then T_1 is conservative over T_2 with respect to sentences Γ. In particular, if $\bot \in \Gamma$, then I yields a relative consistency proof of T_1 relative to T_2.

In Gödel's work, the following examples of proof interpretations can be found:

1. Gödel's 1933 negative translation of classical ('Peano') arithmetic (PA) into intuitionistic ('Heyting') arithmetic (HA) ([32]): $A \mapsto A'$ is defined by induction on the logical structure of A, as follows:

$$P' :\equiv \neg\neg P \text{ for prime formulas } P,$$
$$(A \wedge B)' :\equiv A' \wedge B',$$
$$(A \vee B)' :\equiv \neg(\neg A' \wedge \neg B'),$$
$$(A \rightarrow B)' :\equiv \neg(A' \wedge \neg B'),$$
$$(\forall x\, A)' :\equiv \forall x\, A',$$
$$(\exists x\, A)' :\equiv \neg\forall x \neg A'.$$

In [32], Gödel proved that

$$(*) \quad \text{PA} \vdash A \Rightarrow \text{HA} \vdash A'.$$

This establishes the consistency of classical arithmetic relative to intuitionistic arithmetic and (using that intuitionistically $(A \rightarrow B)' \leftrightarrow (A' \rightarrow B')$) the conservativity of the former over the latter for so-called $\exists$-free sentences, that is, sentences that neither contain $\exists$ nor $\vee$.

The characterization theorem for negative translation is trivial:

$$\text{PA} \vdash A \leftrightarrow A',$$

and so also the converse of $(*)$ holds.

A similar such translation was found independently by Gentzen, and prior to Gödel, for propositional logic, Glivenko [31] had shown that the simple translation $A \mapsto \neg\neg A$ already suffices. Other variants were subsequently developed by Kuroda [86] and Krivine (made explicit in [105] and [104]). Negative translations have been extended to a great variety of other systems (see Troelstra's introductory notes to Gödel's paper in [39]).

2. Gödel's 1933 interpretation of intuitionistic propositional logic into classical modal propositional logic S4 [33]: Let IPC denote intuitionistic propositional logic and S4 the

familiar modal proposition logic defined by Lewis. Now define for $A \in \mathcal{L}(\text{IPC})$ $A^{\square}$ by induction on A

$$P^{\square} :\equiv P \text{ for atomic sentences } P,$$
$$(A \vee B)^{\square} :\equiv (\square A^{\square} \vee \square B^{\square}),$$
$$(A \wedge B)^{\square} :\equiv (A^{\square} \wedge B^{\square}),$$
$$(A \rightarrow B)^{\square} :\equiv \square A^{\square} \rightarrow \square B^{\square},$$
$$(\neg A)^{\square} :\equiv \neg \square A^{\square}.$$

The following is shown in [33]:

$$\text{IPC} \vdash A \implies \text{S4} \vdash A^{\square}.$$

The converse implication (conjectured in [33]) was proved in [92]. The resulting equivalence was extended to predicate logic in [96] and [91] independently (taking $(\forall x \, A)^{\square} :\equiv \forall x \, A^{\square}$ and $(\exists x \, A)^{\square} :\equiv \exists x \square A^{\square}$).

These results give a kind of interpretation of intuitionistic logic (as well as intuitionistic arithmetic and analysis, to which this interpretation was extended subsequently in various ways) in classical terms as a logic of proof obligations. For example, intuitionistically, a sentence $A \vee B$ asks for a proof of either A or B. Here provability of A (i.e., $\square A$) must not be understood with respect to a fixed formal system such as PA because the S4-axiom $\square A \rightarrow A$ would – as a consequence of Gödel's second incompleteness theorem – not be valid under such an interpretation. This, however, can be overcome by refining the interpretation in terms of provability into a logic of proofs, which was sketched by Gödel in 1938 (see [34], published only in 1995 in [41]) and fully elaborated by Artemov [2].

3. Gödel's functional ('Dialectica') interpretation ([37]; see also [36] and [38]) gives an interpretation of HA into a quantifier-free calculus T of Hilbert's [48] primitive recursive functionals of finite type. It is based on a formula assignment

$$A \mapsto \exists \underline{x} \forall \underline{y} \, A_D(\underline{x}, \underline{y}),$$

where A_D is quantifier-free and $\underline{x}, \underline{y}$ are tuples of functionals of finite type, together with a soundness proof that shows by induction on a given HA-proof of A how to construct closed terms $\underline{t}$ of T such that

$$T \vdash A_D(\underline{t}, \underline{y}).$$

Because $0 = 1$ is preserved by this interpretation, the result provides a consistency proof for HA (and so by the negative translation discussed in point 1 point also for PA) relative to T. The foundational significance of this reduction rests on the fact that T contains only (a rule of) quantifier-free induction, though stated in an extended language of functionals of finite types. The discussion of this technique and its applications to mathematics in more recent years will be the main issue of the rest of this chapter.

A common feature of all three proof interpretations, but also of Gödel's inner model construction for $V = L$ in [35], is the treatment of constructive reasoning from a classical standpoint:

• The negative translation shows that (irrespectively of the totally different philosophical concepts behind classical and intuitionistic mathematics) they can rather easily be related to each other in the sense that intuitionistic arithmetic can be viewed as a refinement of classical arithmetic, which contains the latter in its $\exists$-free part but which makes finer distinctions in general, for example, by distinguishing the classical $\exists$->quantifier '$\neg\forall\neg$' from a constructive one.

- The interpretation of IPC into S4 shows that the refinement provided by intuitionistic logic can be accounted for also in a classical setting, provided the latter is extended by a suitable modal operator.

- In the next section, we will argue that Gödel's functional interpretation can be viewed as treating the intended meaning of the intuitionistic connectives as spelled out by the so-called Brouwer-Heyting-Kolmogorov (BHK) interpretation from a classical point of view, in which, also, negatively occurring universal quantifiers count as existential ones, and so only purely universal statements carry complete information (compare this with the 'real statements' of Hilbert). To put it in different terms, Gödel uses a classical concept of '$\exists$-free' within an interpretation of intuitionistic arithmetic.

- In Gödel's work on $V = L$, the concept of predicative definability (developed in the 'semi-intuitionistic' context put forward by H. Weyl, E. Borel, H. Poincaré, and others) is used within impredicative classical set theory. In a letter to H. Wang from March 7, 1968, Gödel ([110], p. 205; reprinted in [43], p. 400) wrote,

> However, as far as, in particular, the continuum hypothesis is concerned, there was a special obstacle which *really* made it *practically impossible* for constructivists to discover my consistency proof. It is the fact that the ramified hierarchy, which had been invented *expressly for constructivistic purposes*, has to be used in an *entirely nonconstructive way*.

This, however, seems to be somewhat too strongly worded: after all, Gödel's consistency proof has a constructive interpretation as a *relative* consistency proof of GB + GCH relative to GB(= Gödel-Bernays set theory; see also Kreisel [83]). In his 'Lecture on the Consistency of the Continuums Hypothesis' at Brown University (published posthumously in [41], pp. 175–185), he explicitly links his consistency proof to the program (for a 'proof' of the continuums hypothesis) outlined in Hilbert ([48]; Gödel [41], p. 175):

> Just recently I have succeeded in giving the proof a new shape which makes it somewhat similar to Hilbert's program presented in his lecture "Über das Unendliche."

In addition to this, it is worth noting that in the letter to Wang cited earlier, Gödel also attributes the fact that neither Skolem nor Herbrand had discovered his completeness theorem to 'that reluctance to use non-finitary concepts and arguments in metamathematics' ([43], pp. 403–404). However, there is evidence (see Goldfarb's introduction to [44], p. 12) that Herbrand was aware of the possibility of using an ineffective argument to derive the completeness theorem from his theorem but did not believe that general 'validity' would make sense and focused on the fact that a *particular set of rules* (having the subformula property by omitting the syllogism rule) was as strong as the usual rules:

> the theorem in question permits us to show that the system of rules of reasoning can be changed profoundly while still remaining equivalent to the original ones, so that the

rule of the syllogism, the basis of Aristotelian logic, is of no use in any mathematical argument. (Herbrand [44], p. 276)

17.2 Functional Interpretation

Functional interpretation was developed by Gödel for HA but – via his negative translation – also applies to PA. We start by briefly recalling the well-known BHK interpretation of intuitionistic logic in terms of 'proofs' which, rather than proofs in specific formal systems, should be viewed as 'verifying constructions':

1. A proof of $A \wedge B$ is a pair (p_A, p_B), where p_A is a proof of A and p_B is a proof of B.
2. A proof of $A \vee B$ is a pair (n, p), where n is a natural number and p is a proof of A if $n = 0$ and a proof of B if $n \neq 0$, respectively.
3. A proof of $A \rightarrow B$ is a construction p that transforms any hypothetical proof q of A into a proof $p(q)$ of B.
4. A proof of $\exists x\, A(x)$ is a pair (c, p), where c is an element (or – more precisely – a term denoting an element) of the underlying domain and p is a proof of $A(c)$.
5. A proof of $\forall x\, A(x)$ is a construction p that transforms any element c of the underlying domain into a proof $p(c)$ of $A(c)$.

Negation is treated as a defined notion $\neg A :\equiv (A \rightarrow \bot)$ with the stipulation that there is no proof p for $\bot$.

This informal interpretation has some ambiguities, in particular, in the clause for implication. Making this precise results in various so-called realizability interpretations. The most important ones are Kleene's realizability [54] and Kreisel's modified realizability [78, 79]. Let us sketch now the main differences between these two interpretations:

- Kleene's realizers are natural numbers viewed as codes of partial recursive functions. A number x realizes an implication $A \rightarrow B$ (short: $x\, r\, (A \rightarrow B)$) if

$$\forall y \in \mathbb{N}(y\, r\, A \rightarrow \{x\}(y) \downarrow \wedge \{x\}(y)\, r\, B);$$

That is,

$$\forall y \in \mathbb{N}(y\, r\, A \rightarrow \exists z \in \mathbb{N}(\{x\}(y) \simeq z \wedge z\, r\, B)).$$

In particular, a Kleene realizer does not have to be defined in general on hypothetical realizers for A but only on actual realizers for A. Referring to the standard formalization of elementary recursion theory in HA, the interpretation $x\, r\, A$ of a formula in the language of HA again is a formula in that language.
- In contrast to the partial (and type-free) Kleene realizability, realizers in the sense of modified realizability are (tuples of) total functionals $\underline{x}$ of finite type (in a suitable domain of functionals) such that (denoting modified realizability by 'mr')

$$\underline{x}\, mr\, (A \rightarrow B) :\equiv \forall \underline{y}(\underline{y}\, mr\, A \rightarrow \underline{x}(\underline{y})\, mr\, B),$$

where for $\underline{x} = x_1, \ldots, x_n$, the expression $\underline{x}(\underline{y})$ denotes the tuple $x_1(\underline{y}), \ldots, x_n(\underline{y})$. To make this precise, we need to define a finite type extension HA^ω of HA such that for A in the language of HA (and even for A in the extended language of HA^ω), the interpretation $\underline{a}$ mr A of A is a formula of the language of HA^ω.

Both the formula $\underline{x}$ r A as well as the formula $\underline{x}$ mr A are in general not quantifier free because in both cases, the clause for implication (as well as for the universal quantifier) contains a universal quantifier. However, '$\underline{x}$ mr A' is a so-called $\exists$-free formula, that is, a formula that does not contain $\vee$ or $\exists$ (note that over HA, $A \vee B$ is equivalent to $\exists n[(n = 0 \to A) \wedge (n \neq 0 \to B)]$ and so is usually counted as an existential quantifier).

Owing to the existential quantifier hidden in $\{x\}(y)\downarrow$, the formula $\underline{x}$ r A is only 'essentially $\exists$-free', that is, it does not contain $\vee$ and contains $\exists$ only in front of prime formulas.

As mentioned already, Gödel's functional interpretation A^D can be viewed as a form of modified realizability where all classical existential quantifiers (i.e., not only positively occurring $\exists$-quantifiers but also negatively occurring $\forall$-quantifiers) are included among the data $\exists\underline{x}$ that need to be realized, leaving only a purely universal formula left, that is, $A^D \equiv \exists\underline{x}\forall\underline{y}\, A_D(\underline{x}, \underline{y})$, where A_D is a quantifier-free formula.[1] This requires a more subtle interpretation of implications: not only is an implication

$$\exists x\, A_0(x) \to \exists y\, B_0(y)$$

(A_0, B_0 quantifier-free) interpreted (following mr) as

$$\exists f \forall x\, (A_0(x) \to B_0(f(x))),$$

but in addition, an implication

$$\forall x\, A_0(x) \to \forall y\, B_0(y)$$

needs to be interpreted as

$$\exists g \forall y\, (A_0(g(y)) \to B_0(y)).$$

In general, the functional interpretation of implications goes as follows: suppose that we have defined already functional interpretations $A^D \equiv \exists\underline{x}\forall\underline{y}A_D(\underline{x}, \underline{y})$ and $B^D \equiv \exists\underline{u}\forall\underline{v}B_D(\underline{u}, \underline{v})$. Then the interpretation $(A \to B)^D$ of $A \to B$ is defined as

$$(*)\ \exists\underline{U}, \underline{Y}\forall\underline{x}, \underline{v}(A_D(\underline{x}, \underline{Y}(\underline{x}\,\underline{v})) \to B_D(\underline{U}(\underline{x}), \underline{v})).$$

This is motivated as follows: among all four possibilities of prenexing

$$\exists\underline{x}\forall\underline{y}A_D(\underline{x}, \underline{y}) \to \exists\underline{u}\forall\underline{v}B_D(\underline{u}, \underline{v}),$$

---

[1] Some clarification of the relationship between modified realizability and functional interpretation has been achieved in [103], where an infinite family of interpretations between these two is introduced (see also [93]). However, in our view, functional interpretation is rather different from any realizability notion. E.g., realizability notions all have a so-called 'with-truth'-variant due to Aczel (see [109]), whereas such a variant of (the Diller-Nahm version [19] of) D is not sound (while a closely related so-called q-variant due to Kleene is), as was shown in [52, 53].

one chooses

$$\forall \underline{x} \exists \underline{u} \forall \underline{v} \exists \underline{y} (A_D(\underline{x}, \underline{y}) \to B_D(\underline{u}, \underline{v})).$$

Now applying the axiom of choice (see later) for functionals of arbitrary type yields $(*)$. If we had chosen any of the three remaining possibilities for the prenexation, the resulting functional interpretation would fail to have computable realizers already for certain instances of $A \to A$ (see [108] (3.5.3)).

As we see, functionals of higher type show up (both in the courses of modified realizability and functional interpretation) even for formulas in the language of HA (i.e., formulas that only involve variables of the lowest type) and, in fact, the realizing terms will be so-called primitive recursive functionals in the extended sense of [48] and [37]: the set **T** of finite types is generated inductively by the following clauses:

$$(i)\, 0 \in \mathbf{T}\;\; (ii)\, \rho, \tau \in \mathbf{T}, \Rightarrow (\rho \to \tau) \in \mathbf{T};$$

0 is the type of natural numbers, and $(\rho \to \tau)$ is the type of objects mapping objects of type ρ to objects of type τ. The degree $deg(\rho)$ of ρ is defined by $deg(0) := 0$, $deg(\rho \to \tau) := \max\{deg(\rho) + 1, deg(\tau)\}$. Usually, outmost parentheses are omitted. The types $0 \to 0$ and $(0 \to 0) \to 0$ usually are denoted by 1 and 2, respectively. The primitive recursive functionals of finite type in the sense of Hilbert [48] and [37] are generated starting from 0^0 (zero), $S^{(0 \to 0)}$ (successor), and variables of arbitrary types by the following two schemas:

- For each term $t^\tau[\underline{u}]$ built up out of 0, S, previously defined primitive recursive functionals and at most the variables $\underline{u} = u_1^{\rho_1}, \dots, u_k^{\rho_k}$, there is a primitive recursive functional φ such that

$$\forall \underline{u}(\varphi(\underline{u}) =_\tau t[\underline{u}]).$$

- For previously defined primitive recursive functionals ψ^τ and $\chi^{\tau \to (0 \to \tau)}$, there is a primitive recursive functional φ such that

$$\varphi(0^0) =_\tau \psi, \;\; \forall u^0(\varphi(S(u)) =_\tau \chi(\varphi(u), u)).$$

A possible domain over which to interpret these functionals is the full set-theoretic type structure $\mathcal{S}^\omega = \langle S_\rho \rangle_{\rho \in \mathbf{T}}$, where $S_0 := \mathbb{N}$, and $S_{\rho \to \tau}$ is the set of all set-theoretic functionals $S_\rho \to S_\tau$. As observed already by Hilbert in [48], the preceding form of primitive recursion in higher types, which allows one to use the whole functional $\varphi(u)$ of type τ in each step of the recursion to define $\varphi(S(u))$, makes it possible to define more functions $f : \mathbb{N} \to \mathbb{N}$ than just the ordinary primitive recursive ones, namely, the well-known Ackermann function and – as shown in [37] (see later) – in fact all provably total recursive functions of PA.

It turns out that the same definition of functional interpretation applies to formulas formulated already in an extension HA^ω of HA to all finite types. However, to generalize Gödel's soundness theorem for the interpretation from HA to HA^ω, one has to observe two subtle points:

1. Because the functional interpretation of the contraction axiom $A \to A \wedge A$ depends on the existence of decision functionals for the prime formulas involved, one must not include higher type equality relations $=_\rho$ as primitive concepts to the language unless

one has an intensional interpretation of $=_\rho$ in mind for which it is reasonable to assume the existence of effective decision functions.

2. If only $=_0$ is taken as a primitive notion (so that the only prime formulas are $s =_0 t$, which are decidable), higher type equality has to be defined either as

   (a) extensional equality $s =_\rho t :\equiv \forall \underline{v}(s(\underline{v}) =_0 t(\underline{v}))$ with new variables $\underline{v}$ so that $s(\underline{v})$ is of type 0

   (b) in a way that is neutral with respect to the issue whether $s =_\rho t$ is interpreted extensionally or intensionally, namely, as a Leibniz identity, that is, indistinguishability in all number contexts

$$r[s/x^\rho] =_0 r[t/x^\rho]$$

   for all terms r^0 of type 0.

Whereas Gödel in [37] apparently had an intensional interpretation of higher type equality in mind, in his earlier treatment of 1941 [36], he argued for the option to include only equality for numbers as a primitive notion (see the introductory remarks to [37] by Troelstra [41]). We will take this option here and treat higher type equality as extensional equality as this is the most natural interpretation for applications in mathematics. Then, however, for the soundness theorem to hold, one must not stipulate as an axiom that all functionals respect this extensionality but only a weaker rule version of this (first considered in [102]):

$$\text{QF-ER}: \quad \frac{A_0 \rightarrow s =_\rho t}{A_0 \rightarrow r[s/x^\rho] =_\tau r[t/x]},$$

where A_0 is quantifier-free. We denote the resulting system (usually called 'WE-HA$^\omega$' in the literature; see, e.g., [108], [3], or [69] for precise definitions) by HA$^\omega$ because in this chapter, there is no danger to confuse this with other variants. qf-HA$^\omega$ is the quantifier-free fragment of HA$^\omega$, formulated with a substitution rule

$$\text{SUB}: \quad \frac{A}{A[t^\rho/x^\rho]}$$

and a quantifier-free rule of induction

$$\text{QF-IR}: \quad \frac{A(0), \; A(x) \rightarrow A(S(x))}{A(x)}$$

instead of the schema of full induction, both for all formulas A in this quantifier-free language. Except for our extensional treatment of higher type equality (now formulated as open formulas without quantifiers), this system is called calculus T in [37].

Definition 17.2.1

1. *The so-called Markov principle in all finite types is the schema*

$$\text{M}^\omega: \quad \neg\neg\exists \underline{x} \, A_0(\underline{x}) \rightarrow \exists \underline{x} \, A_0(\underline{x}),$$

where A_0 is an arbitrary quantifier-free formula of HA$^\omega$ and $\underline{x}$ is a tuple of variables of arbitrary types ($A_0(\underline{x})$ may contain further free variables in addition to $\underline{x}$).

2. *The independence of premise schema* $IP_\forall^\omega$ *for universal premises is the schema*

$$IP_\forall^\omega : (\forall \underline{x} A_0(\underline{x}) \rightarrow \exists y\, B(y)) \rightarrow \exists y (\forall \underline{x} A_0(\underline{x}) \rightarrow B(y)),$$

where A_0 *is quantifier-free,* $y, \underline{x}$ *have arbitrary types, and* y *does not occur free in* $\forall \underline{x}\, A_0(\underline{x})$.

3. *The axiom of choice schema* AC *in all types* ρ, τ *is given by*

$$AC : \forall x^\rho \exists y^\tau\, A(x, y) \rightarrow \exists Y^{\rho \rightarrow \tau} \forall x^\rho\, A(x, Y(x)),$$

where A *is an arbitrary formula (not containing* Y *free).*

4. *The axiom of quantifier-free choice schema* QF-AC *in all types is given by*

$$QF\text{-}AC : \forall \underline{x} \exists \underline{y}\, A_0(\underline{x}, \underline{y}) \rightarrow \exists \underline{Y} \forall \underline{x}\, A_0(\underline{x}, \underline{Y}(\underline{x})),$$

where A_0 *is a quantifier-free formula (not containing* $\underline{Y}$ *free) and* $\underline{x}, \underline{y}$ *are tuples of variables of arbitrary types.*

NOTATION In the following, let $H^\omega := HA^\omega + AC + IP_\forall^\omega + M^\omega$.

Theorem 17.2.2 (soundness of functional interpretation [37], [111], [108])
The following rule holds:

$$H^\omega \vdash A(\underline{a}) \text{ implies that } T \vdash A_D(\underline{t}(\underline{a}), \underline{y}, \underline{a}),$$

where $\underline{t}$ *is a suitable tuple of closed terms of* HA^ω *which can be extracted from a given proof of the assumption and* $\underline{a}$ *contains all free variables of* A.

Combined with Gödel's negative translation $A \mapsto A'$, which extends from PA to the classical variant PA^ω of HA^ω (i.e. HA^ω with the law of excluded middle schema $A \vee \neg A$ added), the following holds (see, e.g., [89] or [69]).

Theorem 17.2.3 *The following rule holds:*

$$PA^\omega + QF\text{-}AC \vdash A(\underline{a}) \text{ implies that } T \vdash (A')_D(\underline{t}(\underline{a}), \underline{y}, \underline{a}),$$

where $\underline{t}$ *is a suitable tuple of closed terms of* HA^ω *that can be extracted from a given proof of the assumption.*

Remark 17.2.4

1. *The extraction of the terms* $\underline{t}$ *in both theorems is carried out by recursion over the given proof. The complexity of this extraction procedure is rather low: the size of the extracted terms is linear in the size of the given proof, the extraction algorithm has a cubic worst-time complexity, and the depth of the verifying proof is linear in the depth of the given proof and the maximal size of formulas occurring in that proof (see [47] for this and much more detailed information). The extraction algorithm has been further optimized in the 'Light Functional Interpretation' of [45] and is implemented in [46].*

2. *In general,* $(A')^D$ *causes unnecessarily high types. More efficient is, for example, to use Kuroda's negative translation [86] instead of Gödel's original interpretation, which subsequently was further optimized by Krivine (see [85, 105]). In fact,*

*the combination of Krivine's negative interpretation with functional interpreta-
tion yields precisely the so-called Shoenfield variant [99] of Gödel's functional
interpretation (see [104]).*

Corollary 17.2.5 $\text{PA} \vdash 0 = 1 \Rightarrow T \vdash 0 =_0 1$.

This corollary, which provides a consistency proof of PA relative to T, is the main
objective in [37]. The significance of the proof-theoretic reduction achieved by this rests
on the fact that induction for arbitrarily complex formulas is reduced to quantifier-free
induction (though formulated in an extended language of primitive recursive functionals
of finite type).

A far-reaching extension of Gödel's result was obtained in 1962 by Spector [102],
who succeeded in giving a functional interpretation of classical analysis $\mathcal{A}^\omega$ axiomatized as the extension of $\text{PA}^\omega + \text{QF-AC}$ by the axiom schema of countable choice

$$\text{AC}^{0,\rho} : \ \forall x^0 \exists y^\rho \, A(x, y) \to \exists f^{0 \to \rho} \forall x^0 \, A(x, f(x)),$$

where A is an arbitrary formula of PA^ω (not containing f free). The functional interpretation is carried out in the extension $T + \text{BR}$ of T by a schema BR of so-called bar
recursion. In fact, the interpretation also works for the axiom schema of dependent
choice DC [49, 89, 69]. This yields a relative consistency result of classical analysis
relative to $T + \text{BR}$. Further extensions have been given in, for example, [29, 24, 15].
Moreover, functional interpretation has been adapted to fragments of PA^ω based on
restricted forms of induction and could be used to calibrate the provably recursive
function(al)s of such fragments (see [95, 17]).

17.3 From Consistency Proofs to the Unwinding of Proofs

For applications of functional interpretation to mathematics with the aim of extracting
new effective data from a given proof, the reduction of full induction to the rule of
quantifier-free induction QF-IR is rather irrelevant, whereas now the emphasis is on
the extraction of terms realizing A^D or $(A')^D$ for *interesting* theorems A. This 'shift
of emphasis' (G. Kreisel) also leads to the following observation, already stressed by
G. Kreisel in the 1950s: whereas in reductive proof theory, the provability of universal
sentences, namely, formalized consistency statements, is a main focus of interest, one
may add *arbitrary* true universal sentences to, for example, PA or PA^ω as axioms
without any effect on the extractable programs from proofs. In particular, this means
that in this connection, proofs of purely universal lemmas do not need to be analyzed at
all. Before we can state the main consequences of the (proof of the) soundness theorem
of functional interpretation for the extractability of programs, we need the following
characterization result for A^D.

Proposition 17.3.1 ([111], [108])   *For all formulas A of $\mathcal{L}(\text{HA}^\omega)$, one has*

$$\text{H}^\omega \vdash A \leftrightarrow A^D.$$

Theorem 17.3.2 (program extraction by D-interpretation)   *Let $\mathcal{P}$ be an arbi-
trary set of purely universal sentences $\forall \underline{a}^{\underline{\sigma}} P_0(\underline{a})$ (P_0 quantifier-free) of $\mathcal{L}(\text{HA}^\omega)$,*

$A_0(x^\rho, u^\delta)$ *be a quantifier-free formula containing only* x, u *free,* $B(x^\rho, y^\tau)$ *an arbitrary formula containing only* x, y *free, and* ρ, δ, τ *arbitrary types. Then the following rule holds:*

$$\begin{cases} H^\omega + \mathcal{P} \vdash \forall x^\rho \left(\forall u^\delta A_0(x, u) \to \exists y^\tau B(x, y) \right) \\ \textit{then one can extract a closed term } t \textit{ of } HA^\omega \textit{ s.t.} \\ H^\omega + \mathcal{P} \vdash \forall x^\rho \left(\forall u^\delta A_0(x, u) \to B(x, t(x)) \right). \end{cases}$$

In particular, if $\mathcal{S}^\omega \models \mathcal{P}$*, then the conclusion holds in* $\mathcal{S}^\omega$*, where* $\mathcal{S}^\omega$ *is the full set-theoretic model of* PA^ω *(and hence – using* AC *on the metalevel – of* H^ω*). The result also holds for tuples of variables* $\underline{x}, \underline{u}, \underline{y}$*, where, then,* t *is a tuple of closed terms.*

PROOF **(see also [108](3.7.5))** Making use of $IP^\omega_\forall$, the assumption yields that

$$H^\omega + \mathcal{P} \vdash \forall x^\rho \exists y^\tau \left(\forall u^\delta A_0(x, u) \to B(x, y) \right).$$

Now define $C(x, y) :\equiv \left(\forall u^\delta A_0(x, u) \to B(x, y) \right)$ and consider $C^D(x, y) \equiv \exists \underline{a} \forall \underline{b} \, C_D(\underline{a}, \underline{b}, x, y)$. Then

$$\left(\forall x^\rho \exists y^\tau \left(\forall u^\delta A_0(x, u) \to B(x, y) \right) \right)^D \equiv \exists Y, \underline{A} \forall x, \underline{b} \, C_D(\underline{A}(x), \underline{b}, x, Y(x)).$$

By (the proof of) Theorem 17.2.2, we obtain closed terms $t, \underline{s}$ such that

$$HA^\omega + \mathcal{P} \vdash \forall x, \underline{b} \, C_D(\underline{s}(x), \underline{b}, x, t(x))$$

and hence

$$HA^\omega + \mathcal{P} \vdash \forall x \exists \underline{a} \forall \underline{b} \, C_D(\underline{a}, \underline{b}, x, t(x)).$$

By Proposition 17.3.1, we get

$$H^\omega \vdash \forall x \left(C(x, t(x)) \leftrightarrow \exists \underline{a} \forall \underline{b} \, C_D(\underline{a}, \underline{b}, x, t(x)) \right)$$

and so

$$H^\omega + \mathcal{P} \vdash \forall x^\rho \left(\forall u^\delta A_0(x, u) \to B(x, t(x)) \right). \qquad \square$$

The combination of negative translation and functional interpretation (ND) yields the following program extraction theorem for classical proofs (see, e.g., [69]).

Theorem 17.3.3 (program extraction by ND-interpretation) *Let* $\mathcal{P}$ *be as before and* $A_0(x^\rho, u^\delta)$*,* $B_0(x^\rho, y^\tau)$ *be quantifier-free formulas containing only* x, u *or* x, y *free, respectively, and* ρ, δ, τ *are arbitrary types. Then the following rule holds:*

$$\begin{cases} PA^\omega + QF\text{-}AC + \mathcal{P} \vdash \forall x^\rho \left(\forall u^\delta A_0(x, u) \to \exists y^\tau B_0(x, y) \right) \\ \textit{then one can extract closed terms } s, t \textit{ of } HA^\omega \textit{ s.t.} \\ HA^\omega + \mathcal{P} \vdash \forall x^\rho \left(A_0(x, s(x)) \to B_0(x, t(x)) \right). \end{cases}$$

Again, we may have tuples of variables $\underline{x}, \underline{y}, \underline{u}$*.*

The previous result has as a corollary the so-called no-counterexample interpretation (NCI) of PA in HA^ω (and even T). This interpretation was developed by Kreisel [76, 77] prior to Gödel's publication of the functional interpretation and involves for its formulation only functionals of type degree ≤ 2. Consider a sentence A in the language of PA in prenex normal form,

$$A \equiv \exists x_1 \forall y_1 \ldots \exists x_n \forall y_n \, A_0(x_1, y_1, \ldots, x_n, y_n),$$

and its Herbrand normal form, (written with function quantifiers rather than new function symbols),

$$A^H \equiv \forall h_1, \ldots, h_n \exists x_1, \ldots, x_n \, A_0(x_1, h_1(x_1), \ldots, x_n, h_n(x_1, \ldots, x_n)).$$

The NCI of A asks for functionals $\Phi_1, \ldots, \Phi_n$ realizing $\exists x_1, \ldots, x_n$ in A^H, that is (writing $\underline{h}$ instead of $h_1, \ldots, h_n$),

$$\forall \underline{h} \, A_0(\Phi_1(\underline{h}), h_1(\Phi_1(\underline{h})), \ldots, \Phi_n(\underline{h}), h_n(\Phi_1(\underline{h}), \ldots, \Phi_n(\underline{h}))).$$

We then write $\underline{\Phi}$ NCI A.

Corollary 17.3.4 *Let A be a prenex sentence provable in* PA; *then one can extract from a given proof closed terms $\underline{\Phi}$ of* HA^ω, *that is,*

$$HA^\omega \vdash \underline{\Phi} \text{ NCI } A.$$

PROOF Modulo the canonical embedding of PA into PA^ω, the assumption implies that $PA^\omega \vdash A$, and hence, a fortiori, $PA^\omega \vdash A^H$. The conclusion now follows from Theorem 17.3.3. □

By coding the tuple of number variables $x_1, \ldots, x_n$ into a single number variable x and then searching for the least such x satisfying A^H, it is clear that the mere truth of A already implies the existence of computable (and hence continuous in the sense of the Baire space) functionals $\underline{\Phi}$ satisfying the NCI of A. However, from a proof of A, the extraction procedure via functional interpretation will produce subrecursive functionals whose complexity depends on the proof principles used. Note also that the program extraction Theorem 17.3.3 applies to y^τ of arbitrary type τ so that (except for $\tau = 0$) such a search is no longer possible.

While the NCI is an easy consequence of Gödel's functional interpretation (combined with negative translation), it is complicated to establish it directly as a proof interpretation in its own right. This is caused by the bad behavior of the NCI with respect to the *modus ponens* rule. In fact, the solution of the *modus ponens* under this interpretation cannot be carried out (uniformly) by functionals definable in T but requires a use of bar recursion (of lowest type). While this treatment of the *modus ponens* pointwise stays within T, the type level of the primitive recursion needed to satisfy the NCI of the conclusion B will in general be higher than that for the input functionals from the NCI of the premises A and $A \to B$ [60]. In particular, this approach does not give optimal results for fragments of PA^ω with restricted induction, whereas the route of proving the NCI via functional interpretation does [60, 95]. In fact, the original proof of the NCI (based on a different description of the type-2 functionals in T as the $\alpha(< \varepsilon_0)$-recursive functionals of type 2) as given in [76, 77] was based on the ε-substitution method developed in [1] and so rather on a form of cut elimination than established as a proof interpretation. A proof by direct use of cut elimination was given subsequently in [97]. In fact, [34] shows that the NCI was clearly anticipated by Gödel's analysis of Gentzen's consistency proof for PA based on cut elimination.

Functional interpretation provides a modular way of constructing functionals satisfying the NCI. As the latter can be viewed as a generalization of Herbrand's theorem, it is natural to look also for an extraction algorithm based on functional interpretation of valid Herbrand disjunctions from proofs in classical logic. This was suggested already by Kreisel in [81] and carried out finally in [26].

17.4 A Comparison of Interpretations of $\Pi_3^0 \to \Pi_2^0$

In this section, we discuss in detail the treatment of a special case of the *modus ponens* by functional interpretation (in particular, when combined with negative translation) in comparison with other interpretations. The first nontrivial instance of the *modus ponens* in a context based on classical logic is the situation where a $\forall \exists$-sentence (say, in the language of first-order arithmetic PA, i.e., a so-called Π_2^0-sentence) B is proved using a lemma A of the slightly more complicated logical structure: $\forall \exists \forall$, that is, $A \in \Pi_3^0$. We will see that only the combination of negative translation with functional interpretation applied to ineffective proofs of A and $A \to B$ produces a satisfying computational realizer for B. In fact, the performance of functional interpretation on the negative fragment is so strong that applying first negative translation to the proof of $A \to B$ improves the situation even in the case where $A \to B$ is constructively proven (say, in HA). Note that relative to intuitionistic arithmetic plus the Markov principle for numbers, $(A \to B)'$ is a *stronger* statement than $A \to B$.

This situation (formulated in suitable extensions of PA and PA$^\omega$ by abstract classes of spaces; see later) already covers some important applications of functional interpretation in metric fixed-point theory: here often the convergence toward 0 of some nonincreasing sequence (t_n) of nonnegative real numbers (defined in various parameters involving the spaces and functions in question) toward 0 is proved using the Cauchy property of this or some related sequence (s_n) of real numbers, that is,

$$\forall k \in \mathbb{N} \exists n \in \mathbb{N} \forall m \in \mathbb{N}(|s_n -_{\mathbb{R}} s_{n+m}| \leq_{\mathbb{R}} 2^{-k})$$

$$\to \forall k \in \mathbb{N} \exists n \in \mathbb{N} \forall m \geq n(t_m <_{\mathbb{R}} 2^{-k}).$$

Using an appropriate representation of real numbers as Cauchy sequences of rational numbers with a fixed rate of convergence, one has $\leq_{\mathbb{R}} \in \Pi_1^0$ and $<_{\mathbb{R}} \in \Sigma_1^0$. Because the monotonicity of (t_n) implies that $\forall m \geq n(t_m <_{\mathbb{R}} 2^{-k})$ is equivalent to $t_n <_{\mathbb{R}} 2^{-k}$, this implication has the form $\Pi_3^0 \to \Pi_2^0$. The problem now is to get an effective bound on $\exists n$ in the conclusion, despite that there is in general none for $\exists n$ in the premise; for example, suppose that (s_n), too, is a nonincreasing sequence of nonnegative real numbers. Then the Cauchy property always holds, but by Specker [101], there are simple primitive recursively computable sequences for which there is no effective rate of convergence.

Consider sentences A, B in the language of PA $\mathcal{L}(\text{PA})$ having the following form:

$$A \equiv \forall x \exists y \forall z \, A_0(x, y, z), \quad B \equiv \forall u \exists v \, B_0(u, v),$$

where A_0, B_0 are quantifier-free formulas (containing only the variables x, y, z and u, v, respectively).

We will compare in which data (and with what complexity) the interpretations provide a witness function φ for B, that is,

$$\forall u\, B_0(u, \varphi(u)).$$

The interpretations we consider are Kleene realizability, modified realizability, and functional interpretation as well as each of these interpretations combined with negative translation as a preprocessing step. The preprocessing step using negative translation in general is unavoidable when A and/or $A \to B$ are given with ineffective proofs because all the interpretations mentioned do not apply to classical proofs. Because modified realizability is trivial on the result of the application of negative translation, we furthermore discuss the use of the so-called Friedman-Dragalin A-translation (see later) as an intermediate step. Finally, we compare the result of applying functional interpretation with the NCI.

17.4.1 Kleene Realizability of $\Pi_3^0 \to \Pi_2^0$

A Kleene realizer $e \in \mathbb{N}$ for $A \to B$ (short $e\, r\, (A \to B)$) is (a code $e \in \mathbb{N}$ of) a partial computable 2-place function mapping any code $f \in \mathbb{N}$ of a hypothetical partial recursive function satisfying the Kleene realizability of A, that is,

$$\forall x(\{f\}(x) \downarrow \wedge \forall z\, A_0(x, \{f\}(x), z)),$$

and any number $u \in \mathbb{N}$ into a witness for '$\exists v\, B_0(u, v)$', that is,

$$\begin{aligned} e\, r\, (A \to B) :\equiv \forall f, u \in \mathbb{N}(&\forall x(\{f\}(x) \downarrow \wedge \forall z\, A_0(x, \{f\}(x), z)) \\ &\to \{e\}(f, u) \downarrow \wedge B_0(u, \{e\}(f, u))). \end{aligned}$$

So a given recursive witnessing function $\{f\}$ for '$\exists y$' in the premise A is translated into a recursive witnessing function $\lambda u.\{e\}(f, u)$ for '$\exists v$' in the conclusion B.

This requirement to have a recursive witness for $A \in \Pi_3^0$ can, in general, only be expected to be satisfiable if A has a constructive proof (e.g., in HA), as Specker's example, discussed earlier, shows. For an even simpler counterexample from recursion theory, consider (here T denotes the standard Kleene T-predicate) the following:

$$A :\equiv \forall x \exists y \forall z(T(x, x, y) \vee \neg T(x, x, z))$$

is provable already in plain (classical) logic, but there is not even a recursive bound on '$\exists y$' (in x) as such a bound would provide a decision procedure for the special halting problem $\{x : \exists y\, T(x, x, y)\}$.

17.4.2 Modified Realizability Interpretation of $\Pi_3^0 \to \Pi_2^0$

The modified realizability interpretation of $A \to B$ is satisfied by any functional $\Phi : \mathbb{N}^{\mathbb{N}} \times \mathbb{N} \to \mathbb{N}$ such that

$$\forall f \in \mathbb{N}^{\mathbb{N}}, u \in \mathbb{N}(\forall x, z\, A_0(x, f(x), z) \to B_0(u, \Phi(f, u))).$$

We then write $\Phi\, mr\, (A \to B)$.

In difference to Kleene realizability, the modified realizability interpretation of a (HA-) proof of $A \to B$ will provide a subrecursive functional Φ that is primitive recursive in the sense of Gödel's T (and of much lower complexity for appropriate fragments of HA). Moreover, Φ applies to any function f and not just to computable ones. However, the weakness of this interpretation for ineffective premises A is similar to that of Kleene realizability: if there is no computable f such that $\forall x, z\, A_0(x, f(x), z)$, then the modified realizability interpretation does not yield an effective witness for the conclusion B.

17.4.3 Functional Interpretation of $\Pi_3^0 \to \Pi_2^0$

The functional interpretation $(A \to B)^D$ of $A \to B$ is given by

$$\exists V, Z, X \forall f, u\, (A_0(X(f, u), f(X(f, u)), Z(f, u)) \to B_0(u, V(f, u))).$$

Therefore a realization of the functional interpretation of $A \to B$ is a triple of functionals $\varphi_1, \varphi_2, \varphi_3$ such that

$$\forall f, u\, (A_0(\varphi_1(f, u), f(\varphi_1(f, u)), \varphi_2(f, u)) \to B_0(u, \varphi_3(f, u))).$$

We then write $\varphi\, D\, (A \to B)$.

As in the case of the modified realizability, any function f realizing A, that is,

$$(*)\ \forall x, z\, A_0(x, f(x), z),$$

can be used to perform the *modus ponens* to obtain $\lambda u.\varphi_3(f, u)$ as a witness function for the conclusion B. This time, however, instead of $(*)$ for each u, the weaker requirement on f,

$$(*)^-\ A_0(\varphi_1(f, u), f(\varphi_1(f, u)), \varphi_2(f, u)),$$

suffices. For possible uses of this in number theory, and also for $B \in \Pi_3^0$, where a similar weakening of the premise A is possible, see [84] (this paper contains an extensive discussion of the treatment of $(\Pi_3^0 \to \Pi_3^0)$-implications by functional interpretation).

Suppose now that we have a (subrecursive) functional φ_0 satisfying

$$(\%)\ \forall x, g\, A_0(x, \varphi_0(x, g), g(\varphi_0(x, g))),$$

that is, the NCI of A (which, in this case, coincides with the functional interpretation $(A')^D$ of the negative translation A' of A using the stability of A_0). If we then are able to solve the system of equations

$$x = \varphi_1(f, u)\,,\ \ \varphi_0(x, g) = f(\varphi_1(f, u))\,,\ \ g(\varphi_0(x, g)) = \varphi_2(f, u)$$

for f, x, g in u, then we are able to solve the *modus ponens* and obtain $B_0(u, \varphi_3(f, u))$ for this f.

Such a solution can indeed by constructed by means of Spector's aforementioned bar recursive functionals. In the case at hand, only bar recursion $B_{0,1}$ at the lowest type is needed, which, applied to arguments in Gödel's T, stays within T (although $T + B_{0,1}$ goes beyond T) but, in general, increases the complexity of the arguments: if the arguments only use recursion of type level n, the result of applying $B_{0,1}$ to them will produce a functional whose T-definition may need recursion of level $n + 1$. For details on all this, see [60] and the literature cited there (in particular, [51] and [98]).

For the special case of the *modus* ponens *problem* just discussed, where instead of

$$\forall x \exists y \forall z \, A_0(x, y, z) \to \forall u \exists v \, B_0(u, v),$$

we can prove the stronger form

$$(+) \; \forall u \, (\exists y \forall z \, A_0(u, y, z) \to \exists v \, B_0(u, v)),$$

the use of bar recursion can be avoided: the functional interpretation $(+)^D$ of $(+)$ provides us with functions t, s such that

$$\forall u, y \, (A_0(u, y, t(u, y)) \to B_0(u, s(u, y))).$$

Then, substituting $\varphi_0(u, \lambda y.t(u, y))$ for y and using $(\%)$ with $g := \lambda y.t(x, y)$ yields

$$p(u) := s(u, \varphi_0(u, \lambda y.t(u, y)))$$

as a witness function for B, that is, $\forall u \, B_0(u, p(u))$.

Note that this special case does not lead to an improvement of the situation for Kleene or modified realizability because both interpretations fail to produce a witness $t(u, y)$ for '$\forall z$' in the premise.

The significance of the improvement achieved by functional interpretation depends on whether a nontrivial effective φ_0 satisfying $(\%)$ can in fact be extracted from a classical proof of A. In Section 17.4.5, we will show that this can be achieved by applying negative translation followed by functional interpretation, to the proof of A.

For related discussions of the weakness of a constructive (BHK-style) interpretation of implications $\Pi_3^0 \to \Pi_3^0$ and $\Pi_3^0 \to \Pi_2^0$ as spelled out by (modified) realizability compared to the strengthened interpretation provided by functional interpretation, see Kreisel-Macintyre [84] and Kreisel [82].

17.4.4 Negative Translation Followed by Kleene Realizability Resp. Modified Realizability of $\Pi_3^0 \to \Pi_2^0$

The negative translations of A, $A \to B$ and B are, over HA, equivalent to

$$\forall x \neg \forall y \neg \forall z \, A_0(x, y, z),$$

$$\forall x \neg \forall y \neg \forall z \, A_0(x, y, z) \to \forall u \neg \forall v \neg B_0(u, v),$$

$$\forall u \neg \forall v \neg B_0(u, v),$$

and hence to so-called ∃-free formulas. However, the Kleene realizability interpretation $x \, r \, F$ of an ∃-free formula F is just the formula itself and does not depend at all on x (e.g., we may take $x := 0$). Similarly, for modified realizability, where now the empty tuple is the realizer.

To obtain from

$$\forall u \neg \forall v \neg B_0(u, v)$$

a formula for which realizability interpretations are nontrivial, we must be able to convert this back to the original conclusion,

$$\forall u \exists v \, B_0(u, v),$$

which amounts to an application of the (rule version of the) Markov principle M^0 for numbers

$$M^0 : \quad \neg\forall x \neg F_0(x) \to \exists x\, F_0(x),$$

where $F_0(x)$ is a quantifier-free formula (which may contain additional parameters). However, the modified realizability interpretation of M^0 has (for suitable F_0) no effective solution ([79]): take $F_0(x) :\equiv T(a, a, x)$, where T is Kleene's T-predicate. Then

$$f \; mr \; \forall a (\neg\forall x \neg T(a, a, x) \to \exists x\, T(a, a, x)) \leftrightarrow \forall a (\neg\forall x \neg T(a, a, x) \to T(a, a, f(a))).$$

By the undecidability of the special halting problem, however, there is no computable f with this property. For Kleene realizability, the situation is different but equally unfortunate: whereas modified realizability asks for a total function f, Kleene realizability only requires a partial recursive function, which exists by trivial unbounded search

$$f(a) := \begin{cases} \text{least } x \text{ with } T(a, a, x), & \text{if } \exists x\, T(a, a, x) \\ \text{undefined,} & \text{otherwise.} \end{cases}$$

But although M^0 is Kleene realizable in this way, this does not help for the *modus ponens* problem at hand because rather than analyzing the proof of B (via the proofs of A and $A \to B$), just an unbounded search is performed, which totally disregards the proof of B (but only uses the truth of B for the termination). In particular, no subrecursive complexity information is obtained.

17.4.5 Negative Translation Followed by Functional Interpretation of $\Pi_3^0 \to \Pi_2^0$

The functional interpretations $(A')^D$ and $((A \to B)')^D$ of the negative translations A' and $(A \to B)'$ of A and $A \to B$ are equivalent over HA^ω to

$$(1) \; \exists Y \forall x, g\, A_0(x, Y(x, g), g(Y(x, g)))$$

$$\exists X, G, V \forall Y, u \big(A_0(X(Y, u), Y(X(Y, u), G(Y, u)), (G(Y, u))(Y(X(Y, u), G(Y,u))) \\ \to B_0(u, V(Y, u))\big),$$

and $(B')^D$ is just (equivalent over HA^ω to)

$$\exists f \forall u\, B_0(u, f(u)).$$

If we throw away the information provided by X, G, the (then partial) functional interpretation of $((A \to B)')^D$ simplifies to

$$(2) \; \exists V \forall Y, u \big(\forall x, g\, A_0(x, Y(x, g), g(Y(x, g))) \to B_0(u, V(Y, u))\big).$$

From a realizer φ_0 for '$\exists Y$' in (1) and a realizer Φ for '$\exists V$' in (2), we now obtain a realizer for B just by application; that is, for $f(u) := \Phi(\varphi_0, u)$, we have

$$\forall u\, B_0(u, f(u)).$$

Let us finally indicate how such a solution φ_0 can be obtained for the ineffective Cauchy property of a nonincreasing sequence of nonnegative and (for simplicity) rational numbers (s_n) : define $\varphi(0, g) := 0$, $\varphi(k + 1, g) := \varphi(k, g) + g(\varphi(k, g))$. Then (for $N \geq s_0$), one can show that

$$(3) \quad \exists n \leq \varphi(N2^k, g) \left(|s_n - s_{n+g(n)}| \leq_{\mathbb{Q}} 2^{-k} \right),$$

and hence $\varphi_0(k, g)$ can be defined as the least such n. In fact, n can be taken as $n := \varphi(i, g)$ for a suitable $i < N \cdot 2^k$. Note that, in contrast to φ_0, the bound φ depends on (s_n) only via N.

17.4.6 Negative Translation Followed by A-Translation and Modified Realizability of $\Pi_3^0 \to \Pi_2^0$

Friedman [23] and Dragalin [20] independently developed a so-called A-translation of a formula F where every prime formula P (including $\bot$) in F gets replaced by $P \vee A$ (here $\neg A$ must be defined as $A \to \bot$). Applying this technique to $A :\equiv \exists v^0 B_0(u, v)$, one can convert an HA-proof of $\neg \forall v \neg B_0(u, v)$ (resulting from negative translation) into an HA-proof of $\exists v\, B_0(u, v)$, to which, then, modified realizability can be applied. This does provide effective realizers from classical proofs of B (via classical proofs of A and $A \to B$), but the complexity usually is not optimal: as we saw in Section 17.4.5, the negative translation of the Cauchy property of (s_n) has a functional interpretation using only primitive recursion at the lowest type 0. In contrast to this, the $\exists v\, B_0(u, v)$-translation of the negative translation apparently requires primitive recursion at type 1 (sufficient to define the Ackermann function) for its modified realizability interpretation (see [69]).[2] Strangely enough, functional interpretation of the $\exists v\, B_0(u, v)$-translation (though that step actually is not necessary, by our preceding discussion) again only uses primitive recursion of type 0. So here we have a statement whose modified realizability seems to require more complicated functionals than its functional interpretation. However, this is not really surprising: because modified realizability gives a much weaker treatment of implications, it is usually harder to satisfy when applied to an implication whose premise, again, is an implication but whose conclusion is simple. It is exactly this situation into which the A-translation applied to the result of the negative translation leads. Refined versions of A-translation that use that the result of applying negative translation can be verified in a context based on minimal logic are even more complicated here: one now only has to replace $\bot$ by $\exists v^0 B_0$ (but not the prime formula $P :\equiv |s_n -_{\mathbb{Q}} s_{n+g(n)}| \leq_{\mathbb{Q}} 2^{-k}$ by $P \vee \exists v^0 B_0$), but one can no longer disregard the double negation in front of P that results from the 'official' negative translation (which we tacitly dropped in our intuitionistic context of HA$^\omega$). One should also note that the approach via any form of A-translation interprets A based on the conclusion B of the proof that uses A, whereas functional interpretation is fully modular in the sense that A is interpreted in a way sufficient for *any* use of it in any proof of any conclusion.

---

[2] An optimized Kripke-style version of the A-translation that sometimes, e.g., in this case, gives a solution of the right complexity was developed in [18].

17.4.7 Discussion of the Results of the Comparison

As the treatment of $A \to B$ with $A \in \Pi_3^0$ and $B \in \Pi_2^0$ shows, the combination of functional interpretation with negative translation gives the most successful interpretation of this *modus ponens*. This is even the case when $A \to B$ is already proved constructively so that negative translation would not be necessary to be applied to the proof of $A \to B$. Only in the special case where one can strengthen $A \to B$ to[3]

$$(+) \ \forall u \ (\exists y \forall z \ A_0(u, y, z) \to \exists v \ B_0(u, v))$$

does the direct functional interpretation of a constructive proof of $A \to B$ provide a similarly strong result, whereas in the general case, bar recursion (at the lowest level) would be needed, which is avoided by the use of negative translation.

The direct functional interpretation of $A \to B$ (i.e., without negative translation as a preprocessing step) actually coincides with the NCI of the following prenex normal form $(A \to B)^{pr}$ of $A \to B$:

$$\forall u \exists x \forall y \exists z, v \ (A_0(x, y, z) \to B_0(u, v)),$$

whereas the NCI of A coincides with the functional interpretation of the negative translation A' of A. Therefore, to perform the *modus ponens* under the NCI, one needs exactly the same use of bar recursion $B_{0,1}$ as discussed in Section 17.4.3 (for a detailed discussion, see [60]). For the special case $(+)$, the NCI yields results as strong as the combination of negative translation with functional interpretation, but to verify the soundness of the former for a given system, one either has to prove the soundness of the latter or apply a suitable form of ε-substitution or cut elimination [76, 77, 97] that destroys the modularity of the interpretation.

17.4.7.1 Comparison between $(A')^D$ and Other (Classical) $\exists\forall$-Normal Forms

We now further illuminate the good behavior of the combination $(A')^D$ of negative translation A' and functional interpretation A^D for proofs based on classical logic: by the application of functional interpretation, clearly $(A')^D$ has the form $\exists \underline{u} \forall \underline{v} \ (A')_D(\underline{u}, \underline{v})$, where $(A')_D$ is quantifier-free and $\underline{u}, \underline{v}$ are tuples of variables for functionals of finite types (where the length and the types only depend on the logical structure of A). Because we work with classical logic, we may assume that A is given in prenex normal form (with quantifier-free A_0):

$$\forall x_1 \exists y_1 \ldots \forall x_n \exists y_n \ A_0(x_1, y_1, \ldots, x_n, y_n).$$

As in the preceding section, we only treat the case where A is arithmetical, that is, where $x_1, y_1, \ldots, x_n, y_n$ have the type 0 of natural numbers. However, things also generalize to the case where these variables may have arbitrary types.

The most well-known $\exists\forall$ normal form of A results by forming the Skolem normal form, which, in the presence of function variables, can be written as

$$A^S :\equiv \exists f_1, \ldots, f_n \forall x_1, \ldots, x_n \ A_0(x_1, f_1(x_1), \ldots, x_n, f_n(x_1, \ldots, x_n)).$$

---

[3] In A_0, one can also allow $r(u)$ instead of u for some term r.

Though the implication $A^S \to A$ is trivial, the converse implication (and hence the equivalence between A and A^S) can be justified using the following arithmetical axiom of choice for numbers[4]

$$AC_{ar}^{0,0} : \forall \underline{x} \exists y \, A_{ar}(\underline{x}, y) \to \exists f \forall \underline{x} \, A_{ar}(\underline{x}, f(\underline{x})),$$

where $\underline{x} = x_1, \dots, x_k$, y are variables for numbers and A_{ar} is an arithmetical formula.

Compared to A^S, the NCI first forms the dual of the Skolem normal form, the so-called Herbrand normal form,

$$A^H :\equiv \forall g_1, \dots, g_n \exists y_1, \dots, y_n \, A_0(g_1, y_1, g_2(y_1), y_2, \dots, g_n(y_1, \dots, y_{n-1}), y_n),$$

and then applies quantifier-free choice

$$QF\text{-}AC : \forall \underline{x} \exists \underline{y} \, A_0(\underline{x}, \underline{y}) \to \exists \underline{Y} \forall \underline{x} \, A_0(\underline{x}, \underline{Y}(\underline{x})) \quad (A_0 \text{ quantifier-free})$$

(though with function arguments $\underline{x}$) to obtain

$$A^{NCI} :\equiv$$
$$\exists Y_1, \dots, Y_n \forall \underline{g} \, A_0(g_1, Y_1(\underline{g}), g_2(Y_1(\underline{g})), Y_2(\underline{g}), \dots, g_n(Y_1(\underline{g}), \dots, Y_{n-1}(\underline{g})), Y_n(\underline{g})).$$

Because the implication $A \to A^H$ is trivial, the implication $A \to A^{NCI}$ only requires quantifier-free choice QF-AC. However, the implication $A^H \to A$ and hence $A^{NCI} \to A$ can again be justified only using arithmetical choice $AC_{ar}^{0,0}$. The use of $AC_{ar}^{0,0}$ to prove $A \to A^S$ and $A^H \to A$ is unavoidable because both schemas (when stated for arbitrary arithmetical A) in fact imply $AC_{ar}^{0,0}$, as is easy to show. In contrast to this, the relationship between A and $(A')^D$ is much closer than the relationship between A and A^S as well as A and A^{NCI}, respectively: instead of arithmetical choice (for numbers), only quantifier-free choice (though for higher types) is needed to establish the equivalence.

Proposition 17.4.1

1. Let A be an arbitrary formula of the language of PA^ω. *Then*

$$PA^\omega + QF\text{-}AC \vdash A \leftrightarrow (A')^D.$$

[78]

2. Let A be in prenex normal form and arithmetical[5]

$$HA^\omega \vdash A^S \to (A')^D \to A^{NCI}.$$

In general, both implications cannot be reversed over $PA^\omega + QF\text{-}AC$.

PROOF   1. See [108] (3.5.13), with a correction in [69]. 2. This is proved in [69]. □

As we saw, the Skolem normal form A^S is too strong to be useful for proofs based on classical logic, as already Π_3^0-lemmas in general will not have a computable Skolem function. The NCI weakens the requirement sufficiently to allow effective (and even

---

[4] Because one can avoid choice by choosing the least number satisfying the property, this schema corresponds, viewed from the perspective of set theory, to (arithmetical) comprehension and is not a proper form of choice.

[5] The result also holds for general prenex A with the definitions of A^S and A^{NCI} extended in the obvious way.

subrecursive) solutions but, when applied to some prenex normal form of a sentence $\Pi_3^0 \to \Pi_2^0$, is too weak for a simple treatment of the *modus ponens* (but has to use some amount of bar recursion). $(A')^D$ provides the right balance between the two extreme interpretations A^S and A^{NCI}. The price to be paid for this is that with an increasing number of quantifier alternations, the degrees of the types of the functionals increase (i.e., higher and higher function spaces are needed), whereas for A^S and A^{NCI}, variables of type degree 1 and 2 are sufficient, respectively (i.e., only the arity but not the degree of the type increases).

In practice, $(A')^D$, in many cases, coincides with the NCI owing to possible strengthenings (+) of implications $\Pi_3^0 \to \Pi_2^0$, as discussed in Section 17.4.3, by which $\Pi_3^0 \to \Pi_2^0$ reduces from Σ_3^0 to Π_2^0.

In general, $(A')^D$ and A^{NCI} coincide as long as A is in Π_3^0 (as the Cauchy property of bounded monotone sequences) but differ from $A \in \Sigma_3^0$ on. Let us discuss this a bit further but first revisit the solution for $(A')^D$ of the Cauchy property A for monotone sequences (s_n) in $[0, \infty) \cap \mathbb{Q}$ given in (3). The bound also holds for nonincreasing sequences of real numbers (s_n) in $[0, \infty)$ and can be rewritten in the following form (where $[n; n+m] := \{n, n+1, n+2, \ldots, n+m\}$):

$$(4) \quad \exists n \leq \varphi(N2^k, g) \forall i, j \in [n; n + g(n)] \left(|s_i - s_j| \leq_{\mathbb{R}} 2^{-k} \right).$$

Again, there actually exists such an n of the form $n = \varphi(i, g)$ for a suitable $i < N \cdot 2^k$.

(4) yields (a quantitative form of) the so-called 'finite convergence principle' formulated recently by Tao [106, 107].

Corollary 17.4.2 *For all $k, N \in \mathbb{N}, g \in \mathbb{N}^{\mathbb{N}}$, there exists an $M \in \mathbb{N}$ such that for all nonincreasing finite sequences $0 \leq s_M \leq \ldots \leq s_0 \leq N$ of length $M + 1$ in $[0, N]$, there exists an $n \in \mathbb{N}$ with*

$$n + g(n) \leq M \wedge \forall i, j \in [n; n + g(n)](|s_i - s_j| \leq 2^{-k}).$$

Moreover, we can compute M as $M := \varphi(N2^k, g)$ with φ, as earlier.

That the bound in (4) is independent from the sequence (s_n) in $[0, N]$ (which is crucially used in the preceding corollary) is an instance of a general phenomenon that can be established via a so-called monotone variant of functional interpretation due to [58] and that is discussed in the next section (note that $[0, N]^\infty$ is a compact metric space with regard to the product metric).

We conclude this section by briefly mentioning another principle A recently discussed by Tao for which A^{NCI} and $(A')^D$ radically differ because it is no longer of the form $\forall \exists \forall$ (unless a finite collection principle is used, which is as strong as the principle A itself): the infinitary pigeonhole principle (IPP) is defined as

$$(\text{IPP}): \forall n \in \mathbb{N} \forall f : \mathbb{N} \to C_n \exists i \leq n \forall k \in \mathbb{N} \exists m \geq k \left(f(m) = i \right),$$

where $C_n := \{0, 1, \ldots, n\}$.

The Herbrand normal form of (IPP) is

$$(\text{IPP})^H \equiv \forall n \in \mathbb{N} \forall f : \mathbb{N} \to C_n \forall F : C_n \to \mathbb{N} \exists i \leq n \exists m \geq F(i) \left(f(m) = i \right),$$

which gives rise to the following computationally almost trivial solution for the NCI of IPP:

$$M(n, f, F) := \max\{F(i) : i \leq n\} \text{ and } I(n, f, F) := f(M(n, f, F)),$$

which are realizers for '$\exists m$' and '$\exists i$' in $(\text{IPP})^H$. By contrast, the $\forall\exists$-form of $(A')^D$ is arrived at as follows:

$(\text{IPP}) \overset{\text{QF-AC}}{\Leftrightarrow}$

$$\forall n \in \mathbb{N} \forall f : \mathbb{N} \to C_n \exists i \leq n \exists g : \mathbb{N} \to \mathbb{N} \forall k \in \mathbb{N} \big(g(k) \geq k \wedge f(g(k)) = i\big) \overset{\text{QF-AC}}{\Leftrightarrow}$$
$$\forall n \in \mathbb{N} \forall f : \mathbb{N} \to C_n \forall K : C_n \times \mathbb{N}^{\mathbb{N}} \to \mathbb{N} \exists i \leq n \exists g : \mathbb{N} \to \mathbb{N}$$
$$\big(g(K(i, g)) \geq K(i, g) \wedge f(g(K(i, g))) = i\big)$$

and requires (highly nontrivial) functionals $I(n, f, K)$ and $G(n, f, K)$, realizing $\exists i$ and $\exists g$ to solve $(A')^D$ (see [94] and [69] for details). In fact, because (IPP) implies (over weak base systems) the induction axiom for Σ_1^0-formulas (whereas (IPP) itself follows from Σ_2^0-induction), the computational contribution of a use of (IPP) in a proof can be of arbitrary primitive recursive (in the sense of Kleene) complexity, which is not properly accounted for by the simple NCI solution but only by the solution of its ND interpretation. By the aforementioned monotone variant of functional interpretation (to be discussed later), one obtains a bound on $\exists g$ that is independent from the coloring f and that – combined with a uniform continuity argument – yields a quantitative version of the 'finitary' version of (IPP) introduced by Tao [106, 107]. It therefore seems to be the case that the program of so-called 'hard analysis' as advocated in [106] is closely related to carrying out the monotone functional interpretation of proofs in analysis. In fact, the results discussed in the next section have recently been applied by Avigad et al. [4] to obtain the type of uniform quantitative analysis in ergodic theory discussed in Tao [106] and have been used already in Tao [107].

17.5 Extraction of Effective Uniform Bounds in Analysis

Since the 1990s proof-theoretic methods based on specially designed variants and extensions of functional interpretations have been used extensively for the 'unwinding' (G. Kreisel) of prima facie ineffective proofs in analysis, functional analysis, and, most recently, geodesic geometry, this approach, also called 'proof mining', has led to a number of new, effective quantitative results but also to new qualitative results on the independence of solutions from certain parameters (uniformity results). The following papers in analysis use such techniques directly, use results obtained by these techniques, or have been guided by general logical metatheorems that were established using functional interpretations [4, 10, 11, 12, 13, 14, 25, 57, 61, 62, 63, 65, 70, 71, 72, 73, 74, 88]. For surveys, see [67, 68] and – though covering only results up to 2002 – [74], which explains in detail general aspects of applying functional interpretation to analysis. Much more information can be found in the book [69]. [30] uses functional interpretation (though mostly in the form of NCI) to analyze the proof of van der Waerden's theorem given by Furstenberg and Weiss based on topological dynamics.

In recent years, general logical metatheorems based on functional interpretations have been proved that, for large classes of proofs and theorems, guarantee the extractability of effective and strongly uniform bounds [28, 56, 64, 66, 87]. This shows that the concrete applications are not 'ad hoc' and so meet the critique expressed in [21] of early stages of the unwinding program (see also [90] for a discussion of 'unwinding' in general).

In this section, we sketch some of these metatheorems, but for details, refer to [56, 64, 28]. We state one concrete application in analysis but otherwise point to the papers listed earlier.

The most important variant of functional interpretation for these applications is the *monotone functional interpretation* (MD), which was introduced in [58]. It is like ordinary functional interpretation, except that the statement in the soundness theorem for functional interpretation is replaced by

$$(+) \; \exists \underline{x} \big(\underline{t}^* \gtrsim \underline{x} \land \forall \underline{a}, \underline{y} \, A_D(\underline{x}(\underline{a}), \underline{y}, \underline{a}) \big)$$

for suitable closed terms $\underline{t}^*$, where $\gtrsim$ is a suitable hereditarily defined majorization relation between functionals of type ρ ($\underline{t}^* \gtrsim \underline{x}$ is to be understood coordinatewise). For example, we may take Howard's [50] relation of majorizability, or (which sometimes is more suitable) the following variant ('strong majorizability') owing to Bezem [5],

$$\begin{cases} x^* \gtrsim_0 x :\equiv x^* \geq x \\ x^* \gtrsim_{\rho \to \tau} x :\equiv \forall y^*, y (y^* \gtrsim_\rho y \to x^* y^* \gtrsim_\tau x^* y \land x^* y^* \gtrsim_\tau xy). \end{cases}$$

We then say that $\underline{t}^*$ satisfies the MD of A. The soundness proof for the MD proceeds by establishing $(+)$ by induction on the proof. It is similar to the usual soundness proof combined with some easy majorization arguments. In fact, the construction of the terms $\underline{t}^*$ is much simpler than the construction of $\underline{t}$ in the usual functional interpretation [58], which can be bypassed.

In the following, Δ is any set of sentences

$$\forall a \exists b \leq r(a) \forall c \, F_0(a, b, c),$$

where a, b, c may have arbitrary types, r is a closed term, $\leq$ is defined pointwise, and F_0 is a quantifier-free formula. $\tilde{\Delta}$ consists of the Skolem normal forms

$$\exists B \leq r \forall a, c \, F_0(a, B(a), c)$$

of these sentences. 'NMD' denotes the combination of negative translation with MD.

Theorem 17.5.1 (uniform bound extraction [55, 58])
Let $A_0(x^1, y^\rho, z^\tau)$ be a quantifier-free formula of $\mathcal{L}(PA^\omega)$ containing only x, y, z as free variables and let $deg(\tau) \leq 2$ and s be a closed term. Then

$$\begin{cases} PA^\omega + QF\text{-}AC + \Delta \vdash \forall x^1 \forall y \leq_\rho sx \exists z^\tau A_0(x, y, z) \\ \Rightarrow NMD \text{ extracts a closed term } t \text{ of } HA^\omega \text{ such that} \\ HA^\omega + \tilde{\Delta} \vdash \forall x^1 \forall y \leq_\rho sx \exists z \leq_\tau tx \, A_0(x, y, z). \end{cases}$$

In particular, if $\mathcal{S}^\omega \models \Delta$, then the conclusion holds in $\mathcal{S}^\omega$. The result also applies to tuples of variables.

Remark 17.5.2 *The preceding theorem also has a version that is more in the spirit of the original foundational aims of functional interpretation for consistency proofs: if the premise is provable in* $PA^\omega+QF\text{-}AC$ *in the form*[6]

$$\Delta \to \forall x^1 \forall y \leq_\rho sx \exists z^\tau A_0(x, y, z),$$

and Δ *is of the (with regard to the types) more restricted form:*

$$\forall a^1 \exists b \leq_1 r(a) \forall c^0 F_0(a, b, c),$$

then the verification of the extracted bound can carried out using only the following approximate (ε)*-form:*

$$\Delta_\varepsilon := \forall a^1, c^0 \exists b \leq_1 r(a) \forall \tilde{c} \leq c\, F_0(a, b, c)$$

of Δ*, which for many ineffective principles* Δ *has a simple constructive proof. For example, the well-known binary ('weak') König's lemma WKL, which allows one to carry out many ineffective proofs in analysis and algebra [100], can be written as such a special axiom* Δ*, whose* ε*-version is trivial. This yields a relative consistency proof of the part of mathematics that follows over* $PA^\omega+QF\text{-}AC$ *from WKL relative to* HA^ω *and so – by Gödel [37] – relative to T. This has been carried out not only for* PA^ω *but also for numerous fragments in [55, 59]. For a related so-called bounded functional interpretation, see [22].*

Using the so-called standard representation of complete separable ('Polish') metric spaces X and compact metric spaces K (represented as totally bounded complete metric spaces), Theorem 17.5.1 yields (taking quantification over K and X as a kind of 'macro') the following 'applied' version: from proofs (say, in $PA^\omega+QF\text{-}AC+WKL$) of theorems

$$(1)\ \forall k \in \mathbb{N} \forall x \in X \forall y \in K \exists n \in \mathbb{N}\, A_1(k, x, y, n),$$

where A_1 is, modulo the representation of X, K, a purely existential (Σ_1^0)-formula, one can extract bounds $\Phi \in T$ on $\exists n$ that are independent from $y \in K$ but only depend on k and a representative $f_x \in \mathbb{N}^\mathbb{N}$ of the element $x \in X$, that is,

$$(2)\ \forall k \in \mathbb{N} \forall x \in X \forall y \in K \exists n \leq \Phi(k, f_x)\, A_1(k, x, y, n),$$

where f_x is any representative of $x \in X$ (see [56]).

Remark 17.5.3 (enrichment of data) *As stressed by G. Kreisel since the 1950s in lectures at Stanford (see also [84]), functional interpretation is a systematic tool to enrich the data of a problem in the numerical appropriate form to make an effective solution possible. For example, a strictly positive real number x gets enriched by a witness* $n \in \mathbb{N}$ *such that* $x \geq 2^{-n}$*, and a Cauchy sequence gets enriched by a rate of convergence. Monotone functional interpretation (which,*

---

[6] In particular, this is the case if we can allow the full extensionality rule resulting in a system satisfying the deduction theorem. Full extensionality may be added, e.g., if the types in Δ, QF-AC, and ρ are ≤ 1 because then, the elimination of extensionality procedure from [89] applies. One can also permit that the premise $\Delta(x, y)$ depends on the parameters x and $y \leq sx$.

*in these simple cases, produces the same enrichments as functional interpreta-
tion), moreover, enriches a continuous function, say, on $[0, 1]$, by a modulus of
uniform continuity. Here and in many other cases, it automatically creates the
type of enrichments used, for example, in E. Bishop's treatment of constructive
analysis [6, 8]. Whereas Bishop himself considered the functional interpretation
of implications as 'numerical implication' [7], it is argued in [75] that actually,
the monotone version is the more natural one. We will come back to this issue at
the end of this chapter.*

Often theorems (1) result as prenex normal forms of theorems of the form

$$(3) \ \forall x \in X \forall y \in K (F(x, y) =_{\mathbb{R}} 0 \rightarrow G(x, y) =_{\mathbb{R}} 0)$$

for suitable T-definable functions $F, G : X \times K \rightarrow \mathbb{R}$. This can be rewritten as

$$(4) \ \forall k \in \mathbb{N} \forall x \in X \forall y \in K \exists n \in \mathbb{N}(|F(x, y)| \leq_{\mathbb{R}} 2^{-n} \rightarrow |G(x, y)| <_{\mathbb{R}} 2^{-k}),$$

where the matrix in (4) is logically equivalent to a Σ_1^0-formula.

Even for constructively proven theorems (3), modified realizability could not be
used because it produces the empty realizer for $\exists$-free sentences such as (3). Only a
technique supporting in a nontrivial way (i.e., not by unbounded search, as in Kleene
realizability) the (rule version of the) Markov principle, at least for numbers (needed
to perform the transition from (3) to (4)), such as functional interpretation, is of any
use here.

A particularly important class of theorems having the form (3) are uniqueness
theorems:

$$(5) \ \forall x \in X \forall y_1, y_2 \in K \left(\bigwedge_{i=1}^{2} (F(x, y_i) =_{\mathbb{R}} 0) \right) \rightarrow y_1 =_K y_2).$$

Then (2) provides a *modulus of uniqueness* [56] Φ such that

$$\forall k \in \mathbb{N} \forall x \in X \forall y_1, y_2 \in K \left(\bigwedge_{i=1}^{2} (|F(x, y_i)| \leq_{\mathbb{R}} 2^{-\Phi(k, f_x)}) \right) \rightarrow d_K(y_1, y_2) <_{\mathbb{R}} 2^{-k}).$$

The crucial property of Φ is that it does not depend on $y_1, y_2 \in K$ and hence can be
used to compute the unique root $\widehat{y} \in K$ of $F(x, \cdot)$ in cases where it exists: let $\Psi(n, f_x)$
be any algorithm for computing a 2^{-n}-root in K, that is,

$$\forall n \in \mathbb{N}(|F(x, \Psi(n, f_x))| \leq_{\mathbb{R}} 2^{-n}).$$

Then $d_K(\widehat{y}, \Psi(\Phi(k, f_x), f_x)) < 2^{-k}$. Because of this and several other numerically
significant properties, such moduli Φ feature prominently in numerical analysis under
the name of 'strong unicity'. This, in particular, is the case for best approximation
theory, where, based on functional interpretation, new results on the best Chebycheff
as well as L_1-approximation and of functions in $C[0, 1]$ by polynomials $p \in P_n$ of
degree $\leq n$ (for the case of Chebycheff approximation, also for more general so-called
Haar spaces instead of polynomials) have been obtained ([56, 57, 74]).

Whereas the difficult cases of Chebycheff and L_1-approximation deal with the special spaces $C[0, 1]$ and P_n, the following much simpler uniqueness result applies to a general class of spaces.

Definition 17.5.4 *A normed linear space* $(X, \|\cdot\|)$ *is called strictly convex if*

$$\forall x_1, x_2 \in B \left(\|\tfrac{1}{2}(x_1 + x_2)\| = 1 \rightarrow x_1 = x_2 \right), \quad where \ B := \{x \in X : \|x\| \le 1\}.$$

Proposition 17.5.5 *Let* $(X, \|\cdot\|)$ *be a strictly convex space and* $C \subseteq X$ *be a convex subset. Then each element* $x \in X$ *has at most one element* $y_b \in C$ *of best approximation in* C, *that is, at most one element* $y_b \in C$ *such that* $\|x - y_b\| = d := \inf_{y \in C} \|x - y\|$.

Definition 17.5.6 *A normed space* $(X, \|\cdot\|)$ *is called uniformly convex with modulus of uniform convexity* $\eta : (0, 2] \rightarrow (0, 1]$ *if*

$$\forall x_1, x_2 \in B \forall \varepsilon \in (0, 2] \left(\|\tfrac{1}{2}(x_1 + x_2)\| \ge 1 - \eta(\varepsilon) \rightarrow \|x_1 - x_2\| \le \varepsilon \right).$$

We may assume that $\eta(\varepsilon) < 1$.

One easily shows that for uniformly convex $(X, \|\cdot\|)$ with modulus of uniform convexity η, convex $C \subseteq X$, $x \in X$, and $D \ge d$, the function

$$\Phi(\varepsilon) := \min \left\{ 1, \frac{\varepsilon}{4}, \frac{\varepsilon}{4} \cdot \frac{\eta(\varepsilon/(D+1))}{1 - \eta(\varepsilon/(D+1))} \right\}$$

is a modulus of uniqueness, that is,

$$\forall y_1, y_2 \in C \forall \varepsilon \in (0, 2] \left(\bigwedge_{i=1}^{2} (\|x - y_i\| \le d + \Phi(\varepsilon)) \rightarrow \|y_1 - y_2\| \le \varepsilon \right).$$

As an immediate corollary to this *uniform uniqueness* result, one gets (for uniformly convex Banach spaces and closed convex $C \subseteq X$) the *existence* of a (then, of course, unique) best approximation, despite that compactness arguments cannot be applied here: by the definition of d, there is a sequence (y_n) in C with

$$\|x - y_n\| \le d + 2^{-n}.$$

Hence $(y_{\Phi(2^{-n})})$ is a Cauchy sequence whose limit clearly is a best approximation.

In fact, the standard well-known existence proof for best approximations in uniformly convex spaces implicitly uses this very reasoning. Note that the plain uniqueness result that follows already from the weaker assumption of strict convexity is not sufficient to conclude this.

This simple example illustrates two things:

1. Despite that we did not assume X to be separable or C to be compact, we obtained a uniform modulus of uniqueness not depending on $y_1, y_2 \in C$, except for some upper bound on $\|x - y_1\|, \|x - y_2\|$ (clearly any such bound also is an upper bound on d and, given a bound D on the latter, $\|x - y_i\| \le d + \Phi(\varepsilon)$ implies $\|x - y_i\| \le D + 1$). For this it was sufficient to 'uniformize' the condition of strict convexity to uniform convexity, which would follow in the very special case of a *compact* unit ball (implying

the space to be finite dimensional) but also holds in much more general contexts. In fact, essentially all strictly convex spaces of interest are also uniformly convex.

2. In the compact case K, and in contexts where all constants of the language are computable, the existence of bounds that are independent from parameters in K can be established using compactness, arguments (even yielding an effective uniform bound): by unbounded search, one gets a computable nonuniform bound. Because computable type-2 functionals are uniformly continuous when restricted to the Cantor space (effectively in further type 1 parameters), the result follows. Hence here it is the concrete (subrecursive) description of the bounds extracted from given proofs that is of interest. However, in the absence of compactness, already the existence of a (qualitative) uniformity result can be of interest, for example, by providing an existence result that even ineffectively could not be obtained otherwise. Moreover, in the presence of abstract metric or normed spaces without any computability structure (see later) one cannot even search for an effective (nonuniform) bound. For a nontrivial new fixed-point theorem that was obtained in this way by removing a compactness assumption, see [11].

For concrete spaces $(X, \| \cdot \|)$, one can easily construct counterexamples to the claim that the observations just mentioned hold in general. In fact, separability (which was used in the previous metatheorem to represent K via the Cantor space) now is the main obstacle as the uniform version of separability on bounded subsets is nothing else but the total boundedness of these sets and so (up to completeness) brings one back to the compact case. The situation, however, changes if a proof applies to a general class of spaces X whose defining axioms have the right uniformity built in, as is the case for uniformly convex normed spaces but also many other structures, including metric spaces, hyperbolic spaces, CAT(0)-spaces, normed spaces, inner product spaces, uniformly convex hyperbolic spaces, $\mathbb{R}$-trees, and δ-hyperbolic spaces (see [64, 28, 87]). In fact, general metatheorems (based on extensions of monotone functional interpretation) have been proved that guarantee the extractability of effective uniform bounds that only depend on certain local upper bounds on metric distances of parameters in X, sequences in X and functions $f : X \rightarrow X$. Because the main interest here is in new qualitative uniformity results, we will work in extensions of the system of classical analysis $\mathcal{A}^\omega :=\mathrm{PA}^\omega+\mathrm{QF\text{-}}\mathrm{AC}+\mathrm{DC}$, as treated by Spector and Howard using bar recursion [102, 49]. Although the latter has a complexity too vast to be of any numerical use, it does yield effective *uniform* bounds when combined with a novel majorization relation (see latter). Moreover, for concrete proofs, usually, only small fragments of the systems will be used so that the bounds actually extracted often do have numerical value (see [67] for a survey).

To formalize proofs dealing with abstract classes of structures X such as general metric or normed spaces, we add such structures as kinds of atoms (*Urelemente*) to the system $\mathcal{A}^\omega$ by extending the set of types **T** to a new base type X (ranging over elements from X) over which we form the set of all finite types (in fact, one can add several such structures simultaneously, but we treat here only the case of one structure). Then, by adding appropriate new constants and axioms, one axiomatizes the class of structures at hand. In such a framework, one can formalize proofs of theorems that hold for all structures in the class being axiomatized (treated as parameters), as long as we only consider theorems that do not quantify over the class of structures.

For proofs in intuitionistic mathematics, this approach has already clearly been anticipated in Gödel's early 1941 presentation of functional interpretation ([36], pp. 195–196), where he writes

> more generally, if you apply intuitionistic logic in any branch of mathematics, you can reduce it to a finitistic system of this kind under the sole hypothesis that the primitive functions and primitive recursive relations of this branch of mathematics are calculable, respectively, decidable. . . . This finitistic system to which intuitionistic logic, applied in the branch of mathematics under consideration, can be reduced is always obtained by introducing functions of higher types analogous to these, with the only difference that the individuals upon which the hierarchy of functions is built up are no longer the integers but the primitive objects of the branch of mathematics under consideration.

Definition 17.5.7   *The set* $\mathbf{T}^X$ *of all finite types over the two ground types* 0 *and* X *is defined by*

$$(i)\ 0,\ X \in \mathbf{T}^X\ (ii)\ \rho, \tau \in \mathbf{T}^X \Rightarrow (\rho \to \tau) \in \mathbf{T}^X.$$

A type is called small if it is of degree ≤ 1 (i.e., $0 \to \cdots \to 0 \to 0$) or the form $\rho_1 \to \cdots \to \rho_k \to X$, where the ρ_is are 0 or X.

The theories $\mathcal{A}^\omega[X, d]_{-b}$ and $\mathcal{A}^\omega[X, d, W]_{-b}$ result[7] by extending $\mathcal{A}^\omega$ to all types in $\mathbf{T}^X$ and adding axioms for an abstract metric (in the case of $\mathcal{A}^\omega[X, d]_{-b}$) or hyperbolic (in the case of $\mathcal{A}^\omega[X, d, W]_{-b}$) space.

$\mathcal{A}^\omega[X, d, W, \text{CAT}(0)]_{-b}$ is the extension by an abstract CAT(0)-space. Analogously, one has theories $\mathcal{A}^\omega[X, \| \cdot \|]$ with an abstract nontrivial real normed space added (as well as further extensions $\mathcal{A}^\omega[X, \| \cdot \|, C]$ and $\mathcal{A}^\omega[X, \| \cdot \|, C]_{-b}$ with bounded and general convex subsets $C \subseteq X$, respectively, which we will, however, owing to lack of space, not treat here). Our theories also contain a constant 0_X of type X, which, in the normed case, represents the zero vector and, in the other cases, stands for an arbitrary element of the metric space. For details on all this, see [64, 28].

Equality $=_X$ for the new type X is a defined notion

$$x =_X y :\equiv (d_X(x, y) =_{\mathbb{R}} 0_{\mathbb{R}}),$$

and so we still only have decidable prime formulas $s =_0 t$. Because we have to work in the weakly extensional setting of Spector's quantifier-free extensionality, we do not have

$$x =_X y \to f^{X \to X}(x) =_X f(y)$$

but only from a proof of $A_0 \to s =_X t$ can we infer that $A_0 \to f(s) =_X f(t)$ (A_0 quantifier-free). This is of crucial importance for our metatheorems to hold. Fortunately, we can, in most cases, prove the extensionality of f for those functions we consider,

---

[7] The index '$-b$' indicates that in contrast to the corresponding theories in [64], we (following [28]) do not require the metric space to be bounded.

for example, for nonexpansive functions, so that this only causes some need for extra care in a few cases (for an extensive discussion of this point, see [64]).

Definition 17.5.8   *For $\rho \in \mathbf{T}^X$, we define $\widehat{\rho} \in \mathbf{T}$ inductively as follows:*

$$\widehat{0} := 0, \quad \widehat{X} := 0, \quad \widehat{(\rho \to \tau)} := (\widehat{\rho} \to \widehat{\tau});$$

that is, $\widehat{\rho}$ is the result of replacing all occurrences of the type X in ρ by the type 0.

We now introduce an extension of the majorization relation to objects with types $\rho \in \mathbf{T}^X$ where, however, the majorants always have types $\widehat{\rho} \in \mathbf{T}$. This relation is parametrized by an arbitrary reference point $a \in X$.

Definition 17.5.9 ([28])   *We define a ternary majorization relation $\gtrsim_{\rho}^{a}$ between objects x, y, and "a" of type $\widehat{\rho}$, ρ, and X, respectively, by induction on (the depth of) ρ, as follows[8]:*

- $x^0 \gtrsim_0^a y^0 :\equiv x \geq y$,
- $x^0 \gtrsim_X^a y^X :\equiv (x)_{\mathbb{R}} \geq_{\mathbb{R}} d_X(y, a)$,
- $x \gtrsim_{\rho \to \tau}^a y :\equiv \forall z', z(z' \gtrsim_{\rho}^a z \to x(z') \gtrsim_{\tau}^a y(z)) \wedge \forall z', z(z' \gtrsim_{\rho}^a z \to x(z') \gtrsim_{\tau}^a x(z)).$

For normed linear spaces, we choose $a = 0_X$.

Definition 17.5.10   *A formula F in $\mathcal{L}(\mathcal{A}^{\omega}[X, \ldots]_{-b})$ is called a $\forall$-formula ($\exists$-formula) if it has the form $F \equiv \forall \underline{a}^{\underline{\sigma}} F_{qf}(\underline{a})$ ($F \equiv \exists \underline{a}^{\underline{\sigma}} F_{qf}(\underline{a})$), where F_{qf} does not contain any quantifier and the types in $\underline{\sigma}$ are small.*

Theorem 17.5.11 ([28])

1. Let ρ be a small type and let $A_{\forall}(x, u)$ and $B_{\exists}(x, v)$ be $\forall$- and $\exists$-formulas that contain only x, u free and x, v free, respectively. Assume that the constant 0_X does not occur in $A_{\forall}, B_{\exists}$ and that

$$\mathcal{A}^{\omega}[X, d]_{-b} \vdash \forall x^{\rho} (\forall u^0 A_{\forall}(x, u) \to \exists v^0 B_{\exists}(x, v)).$$

Then one can extract a computable functional[9] $\Phi : S_{\widehat{\rho}} \to \mathbb{N}$ such that the following holds in all nonempty metric spaces (X, d): for all $x \in S_{\rho}$, $x^* \in S_{\widehat{\rho}}$ if there exists an $a \in X$ such that $x^* \gtrsim^a x$, then

$$\forall u \leq \Phi(x^*) A_{\forall}(x, u) \to \exists v \leq \Phi(x^*) B_{\exists}(x, v).$$

2. The theorem also holds for nonempty hyperbolic spaces $\mathcal{A}^{\omega}[X, d, W]_{-b}$, (X, d, W), and for $\mathcal{A}^{\omega}[X, d, W, CAT(0)]_{-b}$, where (X, d) is a CAT(0)-space.

3. The theorem also holds for nontrivial real normed spaces $\mathcal{A}^{\omega}[X, \|\cdot\|], (X, \|\cdot\|)$, where, then, 'a' has to be interpreted by the zero vector in $(X, \|\cdot\|)$ and 0_X is allowed to occur in $A_{\forall}, B_{\exists}$.

Instead of single variables x, u, v and single premises $\forall u A_{\forall}(x, u)$, we may have tuples of variables and finite conjunctions of premises. In the case of a tuple

---

[8] Here $(x)_{\mathbb{R}}$ refers to the embedding of $\mathbb{N}$ into $\mathbb{R}$ in the sense of our representation of $\mathbb{R}$.
[9] Note that for small types ρ, the type $\widehat{\rho}$ is of degree 1. So Φ, essentially, is a type 2 functional : $\mathbb{N}^{\mathbb{N}} \to \mathbb{N}$.

x, we then have to require that we have a tuple $\underline{x}^*$ of a-majorants for a common $a \in X$ for all the components of the tuple $\underline{x}$.

Remark 17.5.12 *From the proof of Theorem 17.5.11, it follows that the theorem also holds with additional purely universal axioms and majorizable constants of sufficiently small types added where, then, the bound depends on those majorants. Based on this, the preceding theorem has been adapted to other structures such uniformly convex normed spaces or inner product spaces [28] as well as uniformly convex hyperbolic spaces, δ-hyperbolic spaces (in the sense of M. Gromov), and $\mathbb{R}$-trees in the sense of Tits (see [87]).*

Because the bound Φ operates on objects of degree ≤ 1, that is, natural numbers or n-ary number theoretic functions, rather than $x \in X$ or $f : X \to X$, the usual type 2 computability theory as well as well-known subrecursive classes of such functionals apply here irrespective of whether the metric or normed spaces to which the bounds are applied come with any notion of computability.

The proof of Theorem 17.5.11 provides an algorithm based on (monotone) functional interpretation for the extraction of Φ.

In the concrete applications, Theorem 17.5.11 is used via various applied corollaries, of which we give an example now.

Definition 17.5.13 *Let (X, d) be a metric space. A mapping $f : X \to X$ is called* nonexpansive *if*

$$\forall x, y \in X(d(f(x), f(y)) \leq d(x, y)).$$

Corollary 17.5.14 ([28]) *Let $A_\exists$ be an $\exists$-formula and P, K Polish and compact metric spaces respectively in standard representation by $\mathcal{A}^\omega$-definable terms (see [56] for a precise definition). If $\mathcal{A}^\omega[X, d, W]_{-b}$ proves a sentence*

$$\forall x \in P \forall y \in K \forall z^X, \tilde{z}^X, f^{X \to X} \big(f \text{ nonexpansive} \to \exists v^{\mathbb{N}} A_\exists \big),$$

then one can extract a computable functional $\Phi(g_x, b)$ such that for all $x \in P, g_x \in \mathbb{N}^{\mathbb{N}}$ representative of $x, b \in \mathbb{N}$,

$$\forall y \in K \forall z, \tilde{z} \in X \forall f : X \to X \big(f \text{ n.e.} \wedge d(z, f(z)), d(z, \tilde{z}) \leq b \to \exists v \leq \Phi(g_x, b) A_\exists \big)$$

holds in any nonempty hyperbolic space (X, d, W).

PROOF (sketch) The fact that P, K have a standard representation by $\mathcal{A}^\omega$-terms essentially means that $\forall$-quantification over P and K can be expressed as quantification $\forall x^1$ and $\forall y \leq_1 N$, respectively, where N is a fixed simple (primitive recursive) function depending on K. Here the number-theoretic functions encode Cauchy sequences (with a fixed rate of convergence) of elements from the countable dense subset of P and K, respectively, on which the standard representations are based. We now apply Theorem 17.5.11 with $a := z$. For this, we have to construct $\gtrsim^z$-majorants for $x^1, y^1, z^X, \tilde{z}^X$, and $f^{X \to X}$:

$$x^* := x^M := \lambda n. \max\{x(i) : i \leq n\}, y^* := N^M, z^* := 0^0, \tilde{z}^* := b, f^* := \lambda n^0.n + b.$$

For f^*, we use that

$$d(x, z) \leq n \rightarrow d(f(x), z) \leq d(f(x), f(z)) + d(f(z), z)$$
$$\leq d(x, z) + d(f(z), z) \leq n + b.$$

Note that the majorants only depend on x, b. $\qquad\qquad\square$

Remark 17.5.15 *The preceding corollary holds similarly for the case of normed and inner product spaces (as well as their completions; see [69]) but – owing to the fact that, then, 'a' must be fixed as the zero vector 0_X – one has to add the assumption $\|z\| \leq b$ in the conclusion.*

We conclude this section with one concrete application of corollary 17.5.14: let (X, d, W) be a hyperbolic space, $f : X \rightarrow X$ be nonexpansive, and (λ_n) be a sequence in $[0, 1]$ that is bounded away from 1 and divergent in sum. Let (x_n) be the so-called Krasnoselski-Mann iteration starting from $x_0 := x \in X$:

$$x_{n+1} := (1 - \lambda_n)x_n \oplus \lambda_n f(x_n), \text{ where } (1 - \lambda)x \oplus \lambda y \text{ denotes } W(x, y, \lambda).$$

Theorem 17.5.16 (Borwein-Reich-Shafrir [9])

$$\forall x \in X(d(x_n, f(x_n)) \overset{n \to \infty}{\to} r(f) := \inf_{y \in X} d(y, f(y))).$$

As shown in [28], Corollary 17.5.14 implies that there exists a computable function $\Phi : \mathbb{N}^{\mathbb{N}} \times \mathbb{N}^3 \rightarrow \mathbb{N}$ such that (using that $(d(x_n, f(x_n)))_n$ is nonincreasing) for all hyperbolic spaces X, all nonexpansive functions $f : X \rightarrow X$, all (λ_n) in $[0,1], k \in \mathbb{N}$,

$$\alpha : \mathbb{N} \rightarrow \mathbb{N}, \text{ with } \forall n \left(\lambda_n \leq 1 - \frac{1}{k} \wedge n \leq \sum_{i=0}^{\alpha(n)} \lambda_i \right),$$

and all $x, x^* \in X, b \in \mathbb{N}$ with $b \geq d(x, x^*), d(x, f(x))$:

$$\forall l \in \mathbb{N} \forall n \geq \Phi(\alpha, k, b, l) \, (d(x_n, f(x_n)) < d(x^*, f(x^*)) + 2^{-l}).$$

Such a Φ has been extracted in [71] (for the normed case already in [62]) from the original ineffective proof in [9]:

$$\Phi(\alpha, k, b, l) := \widehat{\alpha}(\lceil 2b \cdot \exp(k(M + 1)) \rceil - 1, M), \text{ where}$$

$$M := (1 + 2b) \cdot 2^l, \, \widehat{\alpha}(0, n) := \tilde{\alpha}(0, n), \widehat{\alpha}(i + 1, n) := \tilde{\alpha}(\widehat{\alpha}(i, n), n), \text{ with}$$

$$\tilde{\alpha}(i, n) := i + \alpha^+(i, n), \text{ where } \alpha^+(i, n) := \max_{j \leq i}[\alpha(n + j) - j + 1].$$

For various applications of this result (also, for new qualitative results that only use the uniformity of the bound but not its numerical value), see [71] and [72]. For uniformly convex hyperbolic spaces, often special arguments yielding better bounds apply. This, in particular, covers the important class of CAT(0)-spaces in the sense of Gromov and in many cases provides even quadratic bounds. Here is one example: let $f : C \rightarrow C$ be a self-mapping of a convex subset C of some CAT(0)-space (X, d); f is called *asymptotically nonexpansive* if, for some sequence (k_n) in $[0, \infty)$ with $\lim_{n \to \infty} k_n = 0$, one has

$$d(f^n x, f^n y) \le (1 + k_n) d(x, y), \quad \forall n \in \mathbb{N}, \forall x, y \in C.$$

In this case, the Krasnoselski-Mann iteration of f starting from $x \in C$ is defined by

$$x_0 := x, \quad x_{n+1} := (1 - \lambda_n) x_n \oplus \lambda_n f^n(x_n).$$

Based on a suitable variant of Corollary 17.5.14, the following quadratic (in ε) bound has been extracted recently.

> **Theorem 17.5.17 ([73])**   *Let (X, d) be a CAT(0)-space, C be a nonempty convex bounded subset of X whose diameter is bounded by d_C, and $f : C \to C$ be asymptotically nonexpansive with sequence (k_n).*
>
> *Assume that $K \ge 0$ is such that $\sum_{n=0}^{\infty} k_n \le K$ and that $L \in \mathbb{N}, L \ge 2$ is such that $\frac{1}{L} \le \lambda_n \le 1 - \frac{1}{L}$ for all $n \in \mathbb{N}$. Then the following holds for all $x \in C$:*
>
> $$\forall \varepsilon \in (0, 1] \exists n \le \Phi(K, L, d_C, \varepsilon) \, (d(x_n, f(x_n)) < \varepsilon),$$
>
> *where*
>
> $$\Phi(K, L, d_C, \varepsilon) := 2M,$$
> $$M := \left\lceil \frac{1}{\varepsilon^2} \cdot 24 L^2 \left(5KD + D + \frac{11}{2} \right) (h(K))^3 ((1 + K)D + 1)^2 \right\rceil,$$
> $$D := e^K (d_C + 2), \quad h(K) := 2(1 + (1 + K)^2 (2 + K)).$$

As mentioned earlier, the extraction technique of monotone functional interpretation underlying the proof of Corollary 17.5.14 (in the form for Hilbert spaces) has recently been used in [4] to extract a uniform bound on the Herbrand normal form of the von Neumann mean ergodic theorem (see also [107]).

17.6 Concluding Remarks

17.6.1 Foundational Reductions Revisited

As discussed earlier, Gödel's aim in developing functional interpretation (and negative translation) was to give a consistency proof for PA by reducing it to T, which Gödel considered as a legitimate extension of strict finitism in the narrow sense. The foundational relevance of this reduction (and other consistency proofs) remains debatable (see Troelstra's introductory remarks to [37] in [40]). The shift of emphasis towards applications in mathematics deviates from this original motivation and replaces the issue of foundational reductions with concrete mathematical applications. However, there is one aspect of the original preoccupation with consistency proofs that has shown up again in the course of this applied reorientation: monotone functional interpretation can be used to prove a useful elimination result for a *classically false* strong uniform boundedness principle $\exists$-UB^X ([66]) over, for example, $\mathcal{A}^\omega[X, d, W]$ (i.e., $\mathcal{A}^\omega[X, d, W]_{-b}$, plus an axiom stating the boundedness of (X, d)). This principle allows one (among many other things) to prove (over $\mathcal{A}^\omega[X, d, W]$) that every nonexpansive mapping $f : X \to X$ has a fixed point which is known to be false already for bounded closed convex subsets of

Banach spaces such as c_0. Nevertheless, for a large class of sentences A provable using $\exists$-UB^X (including so-called asymptotic regularity statements), one can show that they are classically correct (see [66] and [63, 70] for concrete instances of this). Because in metric fixed point theory many proofs of asymptotic regularity exist that use as an assumption that f has fixed points, this can (and has been) applied for

- removing the need for complicated fixed-point theorems used to cancel this assumption (resulting in elementary proofs)
- at the same time, dropping assumptions only needed to have these fixed-point theorems available (thereby generalizing proofs).

17.6.2 Enrichment of Data Revisited

Whereas monotone functional interpretation over the original types $\mathbf{T}$ over $\mathbf{N}$ creates (irrespective of whether negative translation is used first) constructive enrichments of data that classically are redundant (for statements involving continuous functions only), our extension to the types for abstract classes of spaces based on $\succsim^a$ creates enrichments that even classically are a proper strengthening of the original assumptions (owing to the lack of compactness). Using the uniform boundedness principle $\exists$-UB^X just mentioned, these enrichments become provable. For example, we translate for *bounded* metric or hyperbolic spaces

- separability $\Rightarrow$ total boundedness (with modulus)
- extensionality $\Rightarrow$ uniform continuity (with modulus)[10]
- pointwise monotone convergence $\Rightarrow$ uniform convergence (with modulus)
- existence of approximate solutions $\Rightarrow$ existence of solutions
- strict convexity $\Rightarrow$ uniform convexity (with modulus) in the hyperbolic or normed case
- contractive functions (Edelstein) $\Rightarrow$ uniformly contractive functions (with modulus, Rakotch)
- uniqueness $\Rightarrow$ uniform uniqueness (with modulus).

What essentially is shown by theorems such as Theorem 17.5.11 is that provided we restrict ourselves to input data having these uniformity features, this uniformity prevails throughout even prima facie highly ineffective proofs formalizable in $\mathcal{A}^\omega[X, d, W]$ and yields an effective uniform version of the theorem in question. In many concrete applications, even ineffectively, the uniform version was not known before.

Acknowledgments

The author is grateful to Professor G. Kreisel for numerous comments on an earlier version of this chapter.

---

[10] This illustrates the need to be restrictive with respect to extensionality.

References

[1] Ackermann, W., Zur Widerspruchsfreiheit der reinen Zahlentheorie. Math. Ann. **117**, pp. 162–194 (1940).

[2] Artemov, S., Explicit provability and constructive semantics. Bull. Symbolic Logic **7**, pp. 1–36 (2001).

[3] Avigad, J., Feferman, S., Gödel's functional ('Dialectica') interpretation. In: [16], pp. 337–405 (1998).

[4] Avigad, J., Gerhardy, P., Towsner, H., Local stability of ergodic averages. Trans. Amer. Math. Soc. **362**, pp. 261–288 (2010).

[5] Bezem, M., Strongly majorizable functionals of finite type: A model for bar recursion containing discontinuous functionals. J. Symbolic Logic **50** pp. 652–660 (1985).

[6] Bishop, E., *Foundations of Constructive Analysis*. McGraw-Hill, New York (1967).

[7] Bishop, E., Mathematics as a numerical language. In: Kino, Myhill, Vesley (eds.), *Intuitionism and Proof Theory*, North-Holland, Amsterdam, pp. 53–71 (1970).

[8] Bishop, E., Bridges, D., *Constructive Analysis*. Springer-Verlag, Berlin (1985).

[9] Borwein, J., Reich, S., Shafrir, I., Krasnoselski-Mann iterations in normed spaces. Canad. Math. Bull. **35**, pp. 21–28 (1992).

[10] Briseid, E. M., Proof mining applied to fixed point theorems for mappings of contractive type. Master's thesis, University of Oslo, Oslo (2005).

[11] Briseid, E. M., Fixed points of generalized contractive mappings. J. Nonlinear Convex Analysis **9**, pp. 181–204 (2008).

[12] Briseid, E. M., A rate of convergence for asymptotic contractions. J. Math. Anal. Appl. **330**, pp. 364–376 (2007).

[13] Briseid, E. M., Some results on Kirk's asymptotic contractions. Fixed Point Theory **8**, pp. 17–27 (2007).

[14] Briseid, E. M., Logical aspects of rates of convergence in metric spaces. J. Symbolic Logic **74**, pp. 1401–1428 (2009).

[15] Burr, W., Functional interpretation of Aczel's constructive set theory. Ann. Pure Applied Logic **104**, pp. 31–73 (2000).

[16] Buss, S. R. (ed.), *Handbook of Proof Theory*. Studies in Logic and the Foundations of Mathematics, vol. 137, Elsevier, New York (1998).

[17] Cook, S., Urquhart, A., Functional interpretations of feasibly constructive arithmetic. Ann. Pure Applied Logic **63**, pp. 103–200 (1993).

[18] Coquand, T., Hofmann, M., A new method for establishing conservativity of classical systems over their intuitionistic version: Lambda-calculus and logic. Math. Structures Comput. Sci. **9**, pp. 323–333 (1999).

[19] Diller, J., Nahm, W., Eine Variante zur Dialectica-Interpretation der Heyting-Arithmetik endlicher Typen. Arch. Math. Logik Grundlagenforsch. **16**, pp. 49–66 (1974).

[20] Dragalin, A. G., New kinds of realizability and the Markov rule (in Russian). Dokl. Akad. Nauk. SSSR **251**, pp. 534–537 (1980). English translation: Soviet Math. Dokl. **21**, pp. 461–464 (1980).

[21] Feferman, S., Kreisel's 'unwinding program'. In: P. Odifreddi (ed.), *Kreiseliana: About and Around Georg Kreisel*, A. K. Peters, Wellesley, MA, pp. 247–273 (1996).

[22] Ferreira, F., Oliva, P., Bounded functional interpretation. Ann. Pure Applied Logic **135**, pp. 73–112 (2005) .

[23] Friedman, H., Classical and intuitionistically provably recursive functions. In: G. H. Müller, D. S. Scott (eds.), *Higher Set Theory*, LNM, vol. 669, Springer, New York, pp. 21–27 (1978).

[24] Friedrich, W., Gödelsche Funktionalinterpretation für eine Erweiterung der klassischen Analysis. Zeitschr. Math. Logik Grundl. Math. **31**, pp. 3–29 (1985).

[25] Gerhardy, P., A quantitative version of Kirk's fixed point theorem for asymptotic contractions. J. Math. Anal. Applied **316**, pp. 339–345 (2006).

[26] Gerhardy, P., Kohlenbach, U., Extracting Herbrand disjunctions by functional interpretation. Arch. Math. Logic **44**, pp. 633–644 (2005).

[27] Gerhardy, P., Kohlenbach, U., Strongly uniform bounds from semi-constructive proofs. Ann. Pure Applied. Logic **141**, 89–107 (2006).

[28] Gerhardy, P., Kohlenbach, U., General logical metatheorems for functional analysis. Trans. Amer. Math. Soc. **360**, pp. 2615–2660 (2008).

[29] Girard, J.-Y., Une extension de l'interpretation de Gödel à l'analyse, et son application à l' élimination des coupures dans l'analyse et dans la théorie des types. In: J. E. Fenstad (ed.), *Proceedings of the Second Scandinavian Logic Symposium*, North-Holland, Amsterdam, pp. 63–92 (1971).

[30] Girard, J.-Y., *Proof Theory and Logical Complexity*, vol. 1, *Studies in Proof Theory*. Bibliopolis, Napoli (1987).

[31] Glivenko, V. I., Sur quelques points de la logique de M. Brouwer. Bull. Soc. Math. Belg. **15**, pp. 183–188 (1929).

[32] Gödel, K., Zur intuitionistischen Arithmetik und Zahlentheorie. Ergebnisse Math. Kolloquiums **4**, pp. 34–38 (1933). Reprinted together with an English translation in: [39].

[33] Gödel, K., Eine Interpretation des intuitionistischen Aussagenkalküls. Ergebnisse Math. Kolloquiums **4**, pp. 39–40 (1933). Reprinted together with an English translation in: [39].

[34] Gödel, K., *Vortrag bei Zilsel* (1938). First published together with an English translation in: [41], pp. 86–113 (1995).

[35] Gödel, K., The consistency of the axiom of choice and of the generalized continuum hypothesis with the axioms of set theory. *Annals of mathematics studies*, **3**, Princeton University Press (1940).

[36] Gödel, K., In what sense is intuitionistic logic constructive? Lecture at Yale (1941). First published in: [41], pp. 189–200 (1995).

[37] Gödel, K., Über eine bisher noch nicht benützte Erweiterung des finiten Standpunktes. Dialectica **12**, pp. 280–287 (1958). Reprinted together with an English translation in: [40].

[38] Gödel, K., On an extension of finitary mathematics which has not yet been used (1972). First published in: [40], pp. 271–280 (1990).

[39] Gödel, K., *Collected Work*, vol. 1, S. Feferman et al. (eds.), Oxford University Press, New York (1986).

[40] Gödel, K., *Collected Work*, vol. 2, S. Feferman et al. (eds.), Oxford University Press, New York (1990).

[41] Gödel, K., *Collected Work*, vol. 3, S. Feferman et al. (eds.), Oxford University Press, New York (1995).

[42] Gödel, K., *Collected Work*, vol. 4, S. Feferman et al. (eds.), Oxford University Press, New York (2003).

[43] Gödel, K., *Collected Work*, vol. 5, S. Feferman et al. (eds.), Oxford University Press, New York (2003).

[44] Herbrand, J., *Logic Writings*, W. D. Goldfarb (ed.), D. Reidel, Dordrecht-Holland (1971).

[45] Hernest, M.-D., Light functional interpretation: An optimization of Gödel's technique twoards the extraction of (more) efficient programs from (classical) proofs. In: L. Ong (ed.), *CSL 2005*, LNCS, vol. 3634, Springer, New York, pp. 477–492 (2005).

[46] Hernest, M.-D., The MinLog proof-system for Dialectica program-extraction. Free software available at http://www.brics.dk/~danher/MinLogForDialectica.

[47] Hernest, M.-D., Kohlenbach, U., A complexity analysis of functional interpretations. Theoretical Computer Science **338**, pp. 200–246 (2005).

[48] Hilbert, D., Über das Unendliche. Math. Ann. **95**, pp. 161–190 (1926).

[49] Howard, W. A., Functional interpretation of bar induction by bar recursion. Compositio Mathematica **20**, pp. 107–124 (1968).

[50] Howard, W. A., Hereditarily majorizable functionals of finite type. In: Troelstra (ed.), *Metamathematical Investigation of Intuitionistic Arithmetic and Analysis*, LNM, vol. 344, Springer, New York, pp. 454–461 (1973).

[51] Howard, W. A., Ordinal analysis of simple cases of bar recursion. J. Symbolic Logic **46**, pp. 17–30 (1981).

[52] Jørgensen, K. F., Finite type arithmetic: Computable existence analysed by modified realizability and functional interpretation. Master's thesis, University of Roskilde (2001).

[53] Jørgensen, K. F., Functional interpretation and the existence property. Math. Logic Quart. **50**, pp. 573–576 (2004).

[54] Kleene, S. C., On the interpretation of intuitionistic number theory. J. Symbolic Logic **10**, pp. 109–124 (1945).

[55] Kohlenbach, U., Effective bounds from ineffective proofs in analysis: An application of functional interpretation and majorization. J. Symbolic Logic **57**, pp. 1239–1273 (1992) .

[56] Kohlenbach, U., Effective moduli from ineffective uniqueness proofs: An unwinding of de La Vallée Poussin's proof for Chebycheff approximation. Ann. Pure Appl. Logic **64**, pp. 27–94 (1993).

[57] Kohlenbach, U., New effective moduli of uniqueness and uniform a-priori estimates for constants of strong unicity by logical analysis of known proofs in best approximation theory. Numer. Funct. Anal. Optimiz. **14**, pp. 581–606 (1993).

[58] Kohlenbach, U., Analysing proofs in analysis. In: W. Hodges, M. Hyland, C. Steinhorn, J. Truss (eds.), *Logic: From Foundations to Applications. European Logic Colloquium* (Keele, 1993), Oxford University Press, Oxford, pp. 225–260 (1996).

[59] Kohlenbach, U., Mathematically strong subsystems of analysis with low rate of growth of provably recursive functionals. Arch. Math. Logic **36**, pp. 31–71 (1996).

[60] Kohlenbach, U., On the no-counterexample interpretation. J. Symbolic Logic **64**, pp. 1491–1511 (1999).

[61] Kohlenbach, U., On the computational content of the Krasnoselski and Ishikawa fixed point theorems. In: J. Blanck, V. Brattka, P. Hertling (eds.), *Proceedings of the Fourth Workshop on Computability and Complexity in Analysis*, LNCS, vol. 2064, Springer, New York, pp. 119–145 (2001).

[62] Kohlenbach, U., A quantitative version of a theorem due to Borwein-Reich-Shafrir. Numer. Funct. Anal. Optimiz. **22**, pp. 641–656 (2001).

[63] Kohlenbach, U., Uniform asymptotic regularity for Mann iterates. J. Math. Anal. Appl. **279**, pp. 531–544 (2003).

[64] Kohlenbach, U., Some logical metatheorems with applications in functional analysis. Trans. Amer. Math. Soc. **357**, pp. 89–128 (2005) [Some minor errata are corrected at the end of [27]].

[65] Kohlenbach, U., Some computational aspects of metric fixed point theory. Nonlinear Analysis **61**, pp. 823–837 (2005).

[66] Kohlenbach, U., A logical uniform boundedness principle for abstract metric and hyperbolic spaces. Electronic Notes Theoretical Computer Science (Proc. WoLLIC 2006) **165**, pp. 81–93 (2006).

[67] Kohlenbach, U., Effective uniform bounds from proofs in abstract functional analysis. In: Cooper, B., Loewe, B., Sorbi, A. (eds.), *New Computational Paradigms: Changing Conceptions of What Is Computable*, Springer, New York, pp. 223–258 (2008).

[68] Kohlenbach, U., Proof interpretations and the computational content of proofs in mathematics. Bulletin EATCS **93**, pp. 143–173 (2007).

[69] Kohlenbach, U., *Applied Proof Theory: Proof Interpretations and their Use in Mathematics.* Springer Monograph in Mathematics, Springer, New York (2008).

[70] Kohlenbach, U., Lambov, B., Bounds on iterations of asymptotically quasi-nonexpansive mappings. In: J. G. Falset, E. L-Fuster, B. Sims (eds.), *Proceedings International Conference on Fixed Point Theory*, Valencia 2003, pp. 143–172, Yokohama Press, Yokohama, Japan (2004).

[71] Kohlenbach, U., Leuştean, L., Mann iterates of directionally nonexpansive mappings in hyperbolic spaces. Abstract Applied Analysis **2003**, pp. 449–477 (2003).

[72] Kohlenbach, U., Leuştean, L., The approximate fixed point property in product spaces. Nonlinear Analysis **66**, pp. 806–818 (2007).

[73] Kohlenbach, U., Leuştean, L., Asymptotically nonexpansive mappings in uniformly convex hyperbolic spaces. J. European Math. Soc. **12**, pp. 71–92 (2010).

[74] Kohlenbach, U., Oliva, P., Proof mining in L_1-approximation. Ann. Pure Appl. Logic **121**, pp. 1–38 (2003).

[75] Kohlenbach, U., Oliva, P., Proof mining: A systematic way of analysing proofs in mathematics. Proc. Steklov Inst. Math. **242**, pp. 136–164 (2003).

[76] Kreisel, G., On the interpretation of non-finitist proofs, part I. J. Symbolic Logic **16**, pp. 241–267 (1951).

[77] Kreisel, G., On the interpretation of non-finitist proofs, part II: Interpretation of number theory, applications. J. Symbolic Logic **17**, pp. 43–58 (1952).

[78] Kreisel, G., Interpretation of analysis by means of constructive functionals of finite types. In: A. Heyting (ed.), *Constructivity in Mathematics, North-Holland*, Amsterdam, pp. 101–128 (1959).

[79] Kreisel, G., On weak completeness of intuitionistic predicate logic. J. Symbolic Logic **27**, pp. 139–158 (1962).

[80] Kreisel, G., Foundations of intuitionistic logic. In: E. Nagel, P. Suppes, A. Tarski (eds.) *Proceedings Logic Methodology and Philosophy of Science*, Stanford University Press, Stanford, CA, pp. 198–210 (1962).

[81] Kreisel, G., Review of [99]. In: Math. Rev. **37** # 1224 (1967).

[82] Kreisel, G., Logical aspects of computation: Contributions and distractions. In: P. Odifreddi (ed.), *Logic and Computer Science*, Academic Press, London, pp. 205–278 (1990).

[83] Kreisel, G., Review of [40]. J. Symbolic Logic **56**, pp. 1085–1089 (1991).

[84] Kreisel, G., Macintyre, A., Constructive logic versus algebraization I. In: A. S. Troelstra, D. van Dalen (eds.), *Proceedings L.E.J. Brouwer Centenary Symposium (Noordwijkerhout 1981)*, North-Holland, Amsterdam, pp. 217–260 (1982).

[85] Krivine, J.-L., Opérateurs de mise en mémoire et traduction de Gödel. Arch. Math. Logic **30**, pp. 241–267 (1990).

[86] Kuroda, S., Intuitionistische Untersuchungen der formalistischen Logik. Nagoya Math. **3**, pp. 35–47 (1951).

[87] Leuştean, L., Proof mining in ℝ-trees and hyperbolic spaces. Electronic Notes Theoretical Computer Science (Proc. WoLLIC 2006) **165**, pp. 95–106 (2006).

[88] Leuştean, L., A quadratic rate of asymptotic regularity for CAT(0)-spaces. J. Math. Anal. Appl. **325**, pp. 386–399 (2007).

[89] Luckhardt, H., *Extensional Gödel Functional Interpretation.* Springer Lecture Notes in Mathematics, vol. 306, Springer, New York (1973).

[90] Macintyre, A., The mathematical significance of proof theory. Phil. Trans. R. Soc. A **363**, pp. 2419–2435 (2005).

[91] Maehara, S., Eine Darstellung der intuitionistischen Logik in der Klassischen. Nagoya Math. J. **7**, pp. 45–64 (1954).

[92] McKinsey, J. C. C., Tarski, A., Some theorems about the sentential calculi of Lewis and Heyting. J. Symbolic Logic **13**, pp. 1–15 (1948).

[93] Oliva, P., Unifying functional interpretations. Notre Dame J. Formal Logic **47**, pp. 263–290 (2006).

[94] Oliva, P., Understanding and using Spector's bar recursive interpretation of classical analysis. In: *Proceedings of CiE 2006*, LNCS, vol. 3988, Springer, New York, pp. 423–434 (2006).

[95] Parsons, C., On *n*-quantifier induction. J. Symbolic Logic **37**, pp. 466–482 (1972).

[96] Rasiowa, H., Sikorski, R., Algebraic treatment of the notion of satisfiability. Fund. Math. **40**, pp. 62–95 (1953).

[97] Schwichtenberg, H., Proof theory: Some aspects of cut-elimination. In: J. Barwise (ed.), *The Handbook of Mathematical Logic*, North-Holland, Amsterdam, pp. 867–895 (1977).

[98] Schwichtenberg, H., On bar recursion of types 0 and 1. J. Symbolic Logic **44**, pp. 325–329 (1979).

[99] Shoenfield, J. S., *Mathematical Logic*. Addison-Wesley, Reading, MA 1967.

[100] Simpson, S. G., *Subsystems of Second Order Arithmetic*. Perspectives in Mathematical Logic. Springer, New York (1999).

[101] Specker, E., Nicht konstruktiv beweisbare Sätze der Analysis. J. Symbolic Logic **14**, pp. 145–158 (1949).

[102] Spector, C., Provably recursive functionals of analysis: A consistency proof of analysis by an extension of principles formulated in current intuitionistic mathematics. In: J. C. E. Dekker (ed.), *Recursive Function Theory, Proceedings of Symposia in Pure Mathematics*, vol. 5, AMS, Providence, RI, pp. 1–27 (1962).

[103] Stein, M., Interpretation der Heyting-Arithmetik endlicher Typen. Archiv Math. Logik Grundlagenforschung **19**, pp. 175–189 (1978).

[104] Streicher, T., Kohlenbach, U., Shoenfield is Gödel after Krivine. Math. Log. Quart. **53**, pp. 176–179 (2007).

[105] Streicher, T., Reus, B., Classical logic: Continuation semantics and abstract machines. J. Functional Programming **8**, pp. 543–572 (1998).

[106] Tao, T., Soft analysis, hard analysis, and the finite convergence principle. Essay posted May 23, 2007. Appeared in: T. Tao, *Structure and Randomness: Pages from Year One of a Mathematical Blog*, AMS, Providence, RI (2008).

[107] Tao, T., Norm convergence of multiple ergodic averages for commuting transformations. Ergodic Theory Dynamical Systems **28**, pp. 657–688 (2008).

[108] Troelstra, A. S. (ed.), *Metamathematical Investigation of Intuitionistic Arithmetic and Analysis*. Springer Lecture Notes in Mathematics, vol. 344, Springer, New York (1973).

[109] Troelstra, A. S., Realizability. In: [16], pp. 407–473 (1998).

[110] Wang, H., *Reflections on Kurt Gödel*, 2nd printing, MIT Press, Cambridge, MA (1988).

[111] Yasugi, M., Intuitionistic analysis and Gödel's interpretation. J. Math. Soc. Japan **15**, pp. 101–112 (1963).

CHAPTER 18

My Forty Years on His Shoulders

Harvey M. Friedman

Gödel's legacy is still very much in evidence. His legacy is overwhelming, particularly in the arena of general mathematical and philosophical inquiry.

The extent of Gödel's impact in the more restricted domain of mathematical practice, however, is more open to question. In fact, Macintyre provides an in-depth assessment of this impact in Chapter 1 of this volume. However, even in this comparatively specialized domain, Gödel's impact is seen to be substantial. As discussed in this chapter, particularly in Section 18.11, I believe that the potential impact of Gödel's work on mathematical practice is also extensive. Although the full realization of this potential impact will have to wait for some new breakthroughs, I have every confidence that these breakthroughs will materialize.

Generally speaking, current mathematical practice has now become very far removed from general mathematical and philosophical inquiry, where Gödel's legacy is most obviously decisive. However, there are signs that some of our most distinguished mathematicians recognize the need for some sort of reconciliation: "Mathematicians took the role of philosophers, but I want to bring the philosophers back in. I hope someday we will be able to explain mathematics in a philosophical way using philosophical methods" (Atiyah, 2008a, 2008b).

I will not attempt to exhaustively discuss the full impact of Gödel's work and all of the ongoing important research programs that it suggests, as this would require a book-length manuscript. Indeed, several books discuss the Gödel legacy from many points of view, including, for example, Wang (1987, 1996), Dawson (2005), and the historically comprehensive five-volume set of Gödel's work (1986, 1990, 1995, 2003a, 2003b).

In Sections 18.1–18.6, I briefly discuss a sample of research projects that are suggested by some of Gödel's most famous contributions. In Sections 18.7–18.10, I discuss a few of the highlights of a main recurrent theme in my own research, which amounts to an expansion of the Gödel incompleteness phenomena in new critical directions. The incompleteness phenomena lie at the heart of the Gödel legacy. Some careful formulations, informed by various post-Gödelian developments, are presented in Sections 18.2–18.4.

One particular issue that arises with regard to incompleteness has been a driving force for a considerable portion of my work over the last forty years. This has been the ongoing search for necessary uses of set-theoretic methods in normal mathematics. By way of background, Gödel's first incompleteness theorem is an existence theorem not intended to provide a mathematically intelligible example of an unprovable sentence. Gödel's second incompleteness theorem does provide an entirely intelligible example of an unprovable sentence – specifically, the crucially important consistency statement. (Remarkably, Gödel demonstrates by a brief semiformal argument that the sentence he constructs for his first incompleteness theorem is demonstrably implied by the consistency statement – hence the consistency statement is not provable. It was later established that the two are in fact demonstrably equivalent.) Nevertheless, the consistency statement is obviously of a logical nature rather than of a mathematical nature. This is a distinction that is readily noticed by the members of the general mathematical community, which naturally resist the notion that the incompleteness theorem will have practical consequences for their own research.

Genuinely mathematical examples of incompleteness from substantial set-theoretic systems had to wait until the well-known work on the axiom of choice and the continuum hypothesis by Kurt Gödel (1940) and Paul Cohen (1963, 1964). Here the statement being shown to be independent of ZFC (the Zermelo-Fraenkel axioms plus the axiom of choice) – the continuum hypothesis (CH) – is of crucial importance for abstract set theory. However, mathematicians generally find it easy to recognize an essential difference between overtly set-theoretic statements, such as CH, and so-called normal mathematical statements. Again, this is a particularly useful observation for mathematicians. Specifically, the reference to unrestricted uncountable sets (of real numbers) in CH readily distinguishes CH from normal mathematics, which relies, almost exclusively, on the essentially countable (e.g., the continuous or piecewise continuous).

A more subtle example of an overtly set-theoretic statement that requires a second look to see its overtly set-theoretic character is Kaplansky's conjecture concerning automatic continuity. In one of its more concrete special forms, it asserts:

(*) Every homomorphism from the Banach algebra c_0 of infinite sequences of reals converging to 0 (under the sup norm) to any separable Banach algebra is continuous.

Now (*) was refuted using the continuum hypothesis (due independently to H. G. Dales and J. Esterle) and was later shown to be not refutable without the continuum hypothesis, that is, not refutable in the usual ZFC axioms (due to R. Solovay). See Dales (2001) for the refutation and Dales and Woodin (1987) for the consistency (nonrefutability) result.

It is, of course, much easier for mathematicians to recognize the overtly set-theoretic character after they learn that there are set-theoretic difficulties. By taking the following negation, it is clear that one is asking about the existence of an object that was well known, even at the time, to necessarily have rather pathological properties:

(**) there exists a discontinuous homomorphism from the Banach algebra c_0 of infinite sequences of reals converging to 0 (under the sup norm) to some separable Banach algebra.

This is the case even for discontinuous group homomorphisms from $\Re$ into $\Re$ (which can be shown to exist without the CH). For instance, it is well known that there are no discontinuous group homomorphisms from $\mathfrak{R}$ into $\mathfrak{R}$ that are Borel measurable.

At the outer limits, normal mathematics is conducted within complete separable metric spaces. (Of course, we grant that it is sometimes convenient to use fluff – as long as it doesn't cause any trouble.) Functions and sets are normally Borel measurable within such so-called Polish spaces. In fact, the sets and functions normally considered in mathematics are substantially nicer than Borel measurable, generally being continuous or at least piecewise continuous – if not outright countable or even finite.[1]

We now know that the incompleteness phenomena do penetrate the barrier into the relatively concrete world of Borel measurability – and even into the countable and finite world – with independence results of a mathematical character. In Sections 18.7–18.10, I discuss my efforts concerning such concrete incompleteness, establishing the necessary use of abstract set-theoretic methods in a number of contexts, some of which go well beyond the ZFC axioms. Yet it must be said that the results to date are very limited in scope and demand considerable improvement. We are only at the very beginning of being able to assess the full impact of the Gödel incompleteness phenomena. In particular, it is not yet clear how strongly and in what way the Gödel incompleteness phenomena will penetrate normal mathematical activity. Progress along these lines is steady but painfully slow. I am confident that a much clearer assessment will be possible by the end of this century – and perhaps not much earlier. In Section 18.11, I take the opportunity to speculate far into the future.

18.1 The Completeness Theorem

In his PhD dissertation, Gödel (1929) proved his celebrated completeness theorem for a standard version of the axioms and rules of first-order predicate calculus with equality. This result of Gödel was anticipated, in various senses, by the earlier work of T. Skolem, as discussed in detail in the introductory notes of Gödel (1986, 44–59), which were written by B. Dreben and J. van Heijenoort.

Consider the following passage (Gödel, 1986, 52) from a letter Gödel wrote to Hao Wang (December 7, 1967):

> The completeness theorem, mathematically, is indeed an almost trivial consequence of Skolem 1923a.[2] However, the fact is that, at the time, nobody (including Skolem himself) drew this conclusion (neither from Skolem 1923 nor, as I did, from similar considerations).

According to these introductory notes, the situation is properly summarized as follows: thus, Gödel says, the only significant difference between Skolem 1923a and Gödel

---

[1] Apparently, nonseparable arguments are being used in the proofs of certain number-theoretic results such as Fermat's last theorem. I have been suggesting strongly that this is an area where logicians and number theorists should collaborate to see just how necessary such appeals to nonseparable arguments are. I have conjectured that they are not necessary and that EFA = $I\Sigma_0(\exp)$ = exponential function arithmetic suffices. See Avigad (2003), Angus MacIntyre's contribution to this volume, and McClarty (2010).

[2] "Skolem 1923a" in this quote is Skolem (1922) in the list of references.

(1929, 1930) lies in the replacement of an informal notion of "provable" by a formal one – and the explicit recognition that there is a question to be answered.

To this, I would add that Gödel himself relied on a semiformal notion of "valid" or "valid in all set theoretic structures." The appropriate fully formal treatment of the semantics of first-order predicate calculus with equality is credited to A. Tarski. However, as discussed in detail in Feferman (2004), surprisingly, the first clear statement in Tarski's work of the formal semantics for predicate calculus did not appear until the 1950s (Tarski, 1952; Tarski and Vaught, 1957).

Let us return to the fundamental setup for the completeness theorem. The notion of structure is taken in the sense most relevant to mathematics and, in particular, general algebra: a nonempty domain, together with a system of constants, relations, and functions, with equality as understood. It is well known that the completeness proof is so robust that no analysis of the notion of structure need be given. The proof requires only that we at least admit the structures whose domain is an initial segment of the natural numbers (finite or infinite). In fact, we need only admit structures whose relations and functions are arithmetically defined, that is, first-order defined in the ring of integers. However, the axioms and rules of logic are meant to be so generally applicable as to transcend their application in mathematics. Accordingly, it is important to interpret logic with structures that may lie outside the realm of ordinary mathematics. A particularly important type of structure is a structure whose domain includes *absolutely everything*. Indeed, it can be argued that the original Fregean conception of logic demands that quantifiers range over absolutely everything. From this viewpoint, quantification over mathematical domains is a special case, as "being in a given mathematical domain" is treated as (the extensions of) a unary predicate on everything.

These general philosophical considerations were sufficient for an applied philosopher like me to begin reworking logic using structures whose domain consists of absolutely everything. The topic of logic in the universal domain has been taken up in the philosophy community, in particular, by T. Williamson (Rayo and Williamson, 2003; Williamson, 2000, 2003, 2006). I have not yet published on this topic, but unpublished reports on my results are available on the Internet, specifically in Friedman (1999, 2002b, 65–99). I plan to publish a monograph on this topic in the not-too-distant future.

18.2 The First Incompleteness Theorem

The first incompleteness theorem is first proved in Gödel (1931). It is proved there in detail for a specific variant of what is now known as the simple theory of types (going back to Bertrand Russell), with natural numbers at the lowest type. This is a rather strong system, nearly as strong as Zermelo set theory. It asserts that there is a sentence that is neither provable nor refutable in this system.

Gödel (1932a) formulates his incompleteness theorems for extensions of a variant of what is now known as Peano arithmetic (PA). Gödel (1934) not only gives another treatment of the results in Gödel (1931) but also, and most important, introduces the notion of recursive functions and relations.

Gödel (1931, 195) writes at the end of his article that "the results will be stated and proved in full generality in a sequel to be published soon." On the same page, we find the following:

> *Note added 28 August 1963.* In consequence of later advances, in particular of the fact that due to A. M. Turing's work a precise and unquestionably adequate definition of the general notion of formal system can now be given, a completely general version of Theorems VI and XI is now possible. That is, it can be proved rigorously that in *every* consistent formal system that contains a certain amount of finitary number theory there exist undecidable arithmetic propositions and that, moreover, the consistency of any such system cannot be proved in the system.

The sequel was never published, at least partly because of the prompt acceptance of Gödel's results after the publication of his article (Gödel, 1931).

Today, Gödel is credited for quite general forms of the first incompleteness theorem. There are already claims of generality in the original paper (Gödel, 1931). In modern terms, in every 1-consistent recursively enumerable formal system containing a small amount of arithmetic, there exist arithmetic sentences that are neither provable nor refutable. Rosser (1936) is credited for significant additional generality, using a clever modification of Gödel's original formal self-referential construction. It is shown there that the hypothesis of 1-consistency can be replaced with the weaker hypothesis of consistency.

Later, methods from recursion theory were used to prove yet more general forms of first incompleteness, where the proof avoids use of formal self-reference – although even in the recursion theory, there is, arguably, a trace of self-reference present in the elementary recursion theory used. The recursion theory approach, in a powerful form, appears in Robinson (1952) and Tarski et al. (1953), with the use of the formal system Q. Q is a single-sorted system based on $0, S, +, \bullet, \leq, =$. In addition to the usual axioms and rules of logic for this language, we have the following nonlogical axioms:

1. $Sx \neq 0$
2. $Sx = Sy \rightarrow x = y$
3. $x \neq 0 \rightarrow (\exists y)(x = Sy)$
4. $x + 0 = x$
5. $x + Sy = S(x + y)$
6. $x \bullet 0 = 0$
7. $x \bullet Sy = (x \bullet y) + x$
8. $x \leq y \leftrightarrow (\exists z)(z + x = y)$

The last axiom is purely definitional and is not needed for present purposes (in fact, we do not need $\leq$).

Theorem 18.2.1 *Let T be a consistent extension of Q of a relational type in many-sorted predicate calculus of arbitrary cardinality. The sets of all existential sentences in L(Q), with bounded universal quantifiers allowed, that are (1) provable in T, (2) refutable in T, and (3) provable or refutable in T are each not recursive.*

For the proof, see Robinson (1952) and Tarski et al. (1953). It uses the construction of recursively inseparable recursively enumerable sets, for example, $\{n: \varphi_n(n) = 0\}$ and $\{n: \varphi_n(n) = 1\}$.

One can obtain the following strong form of first incompleteness as an immediate corollary:

Theorem 18.2.2   *Let* T *be a consistent extension of* Q *in many-sorted predicate calculus whose relational type and axioms are recursively enumerable. There is an existential sentence in* L(Q), *with bounded universal quantifiers allowed, that is neither provable nor refutable in* T.

We can use the negative s solution to Hilbert's tenth problem to obtain other forms of first incompleteness that are stronger in certain respects. In fact, Hilbert's tenth problem is still a great source of very difficult problems on the border between logic and number theory, which I discuss later.

Hilbert asked for a decision procedure for determining whether a given polynomial with integer coefficients in several integer variables has a zero. The problem received a negative answer in 1970 by Y. Matiyasevich, building heavily on earlier work of J. Robinson, M. Davis, and H. Putnam. It is commonly referred to as the MRDP theorem (in reverse historical order; see Davis, 1973; Matiyasevich, 1993). The MRDP theorem was shown to be provable in the weak fragment of arithmetic, $EFA = I\Sigma_0(exp)$, in Dimitracopoulus and Gaifman (1982), which we can use to obtain the following:

Theorem 18.2.3   *Let* T *be a consistent extension of EFA in many-sorted predicate calculus whose relational type and axioms are recursively enumerable. There is a purely existential equation* $(\exists x_1, \ldots, x_n)(s = t)$ *in* L(Q) *that is neither provable nor refutable in* T.

It is not clear whether EFA can be replaced by a weaker system in Theorem 18.2.3 such as Q.

An important issue is whether there is a "reasonable" existential equation $(\exists x_1, \ldots, x_n)(s = t)$ that can be used in Theorem 18.2.3 for, say, $T = PA$ or $T = ZFC$. Note that $(\exists x_1, \ldots, x_n)(s = t)$ corresponds to the Diophantine problem "does the polynomial s-t with integer coefficients have a solution in the nonnegative integers?"

Let us see what can be done on the purely recursion-theoretic side with regard to the complexity of polynomials with integer coefficients. The most obvious criteria are the following:

a. the number of unknowns
b. the degree of the polynomial
c. the number of operations (additions and multiplications)

In 1992, Matiyasevich showed that nine unknowns over the nonnegative integers suffice for recursive unsolvability. One form of the result (not the strongest form) says that the problem of deciding whether a polynomial with integer coefficients in nine unknowns has a zero in the nonnegative integers is recursively unsolvable. A detailed proof of this result (in sharper form) was given in Jones (1982). In addition, Jones (1982) proves that, for example, the problem of deciding whether a polynomial with integer coefficients defined by at most one hundred operations (additions and multiplications with integer constants) has a zero in the nonnegative integers is recursively unsolvable.

It is well known that degree 4 suffices for recursive unsolvability. In Jones (1982), it is shown that degree 4 and 58 nonnegative integer unknowns suffice for recursive unsolvability; that is, the problem is deciding whether a polynomial with integer coefficients, of degree 4 with at most 58 unknowns, has a solution in the nonnegative integers, is recursively unsolvable. In fact, Jones (1982) provides the following sufficient pairs <degree, unknowns>, where all unknowns range over nonnegative integers:

<4,58>, <8,38>, <12,32>, <16,29>,

<20,28>, <24,26>, <28,25>, <36,24>,

<96,21>, <2668,19>, $<2 \times 10^5,14>$,

$<6.6 \times 10^{43},13>$, $<1.3 \times 10^{44},12>$,

$<4.6 \times 10^{44},11>$, $<8.6 \times 10^{44},10>$,

$<1.6 \times 10^{45},9>$

For degree 2 (a single quadratic), we have an algorithm (over the nonnegative integers, the integers, and the rationals) going back to Siegel (1972). See Grunewald and Segal (1981) and Masser (1998). For degree 3, the existence of an algorithm is wide open, even for three variables (over the integers, the nonnegative integers, or the rationals). For degree 3 in two integer variables, an algorithm is known, but it is wide open for degree 3 in two rational variables.

It is clear from this discussion that the gap between what is known and what could be the case is enormous, just in this original context of deciding whether polynomials with integer coefficients have a zero in the (nonnegative) integers. Specifically, <3,3> could conceivably be on this list of pairs.

These upper bounds on the complexity sufficient to obtain recursive unsolvability can be directly imported into Theorem 18.2.3, as the underlying number theory and recursion theory can be done in EFA. Although one obtains upper bounds on pairs (number of variables, degree) in this way, this does not address the question of the size of the coefficients needed in Theorem 18.2.3.

In particular, let us call a polynomial P a Gödel polynomial if

i. P is a polynomial in several variables with integer coefficients
ii. the question of whether P has a solution in nonnegative integers is neither provable nor refutable in PA (we can also use ZFC here instead of PA)

I have never seen an upper bound on the size of a Gödel polynomial in the literature. In particular, I have never seen a Gödel polynomial written fully in base 10 on a small piece of paper. One interesting theoretical issue is whether one can establish any relationship between the size of a Gödel polynomial using PA and the size of a Gödel polynomial using ZFC.

18.3 The Second Incompleteness Theorem

Gödel (1931) only sketches a proof of his second incompleteness theorem, after proving his first incompleteness theorem in detail. His sketch depends on the proof of the first incompleteness theorem, which is conducted in normal semiformal mathematics, being formalized and proven within (systems such as) PA.

Gödel promised a part 2 for his 1931 paper, but this never appeared. There is some difference of opinion whether Gödel planned to provide detailed proofs of his second incompleteness theorem in part 2 or whether he planned to let others carry out the details. In any case, the necessary details were carried out in Hilbert and Bernays (1934, 1939), later in Feferman (1960), and most recently in Boolos (1993). In Hilbert and Bernays (1934, 1939), the so-called Hilbert-Bernays derivability conditions were isolated in connection with a detailed proof of Gödel's second incompleteness theorem. Later, these conditions were streamlined in Jerosolow (1973).

I take the liberty of presenting my own particularly careful and clear version of the Hilbert-Bernays conditions. The starting point is the usual language $L =$ predicate calculus with equality, with infinitely many constant, relation, and function symbols. For specificity, let us use the following:

    (i) variables x_n, $n \geq 1$
  (ii) constant symbols c_n, $n \geq 1$
 (iii) relation symbols R_m^n, $n,m \geq 1$
 (iv) function symbols F_m^n, $n,m \geq 1$
  (v) connectives $\neg$, $\wedge$, $\vee$, $\rightarrow$, $\leftrightarrow$
 (vi) quantifiers $\exists$, $\forall$

Let us start with the following data:

1. a relational type RT of constant symbols, relation symbols, and function symbols
2. a set T of sentences in (the language based on) RT
3. a one-one function number from formulas of RT into closed terms of RT
4. a distinguished unary function symbol NEG in RT, meaning "negation"
5. a distinguished unary function symbol SSUB in RT, meaning "self-substitution"
6. a distinguished unary function symbol PR in RT, meaning "provability statement"
7. a distinguished formula PROV with at most the free variable x_1, expressing "provable in T"

The following is required. Let A be a formula of RT:

  8. $NEG(\#(A)) = \#(\neg A)$ is provable in T.
  9. $SSUB(\#(A)) = \#(A[x_1/\#(A)])$ is provable in T.
10. $PR(\#(A)) = \#(PROV[x_1/\#(A)])$ is provable in T.
11. $PROV[x_1/\#(A)] \rightarrow PROV[x_1/PR(\#(A))]$ is provable in T.
12. If A is provable in T, then $PROV[x_1/\#(A)]$ is provable in T.

Here $\#(A)$ is the Gödel number of the formula A, as a closed term of RT.

> **Theorem 18.3.1 (self-reference lemma)** *Let A be a formula of* RT. *There exists a closed term* t *of* RT *such that* T *proves* $t = \#(A[x_1/t])$.

PROOF Let $s = \#(A[x_1/SSUB(x_1)])$. Write $s = \#B$, where $B = A[x_1/SSUB(x_1)]$.

Note that

$$B[x_1/\#(B)] = B[x_1/s] = A[x_1/SSUB(s)].$$

Let us now apply condition 9 to B. We have that

$$SSUB(\#(B)) = \#(B[x_1/\#(B)])$$

is provable in T. Hence

$$SSUB(s) = \#(A[x_1/SSUB(s)])$$

is provable in T. Thus the closed term $SSUB(s)$ is as required. QED

Lemma 18.3.2 ("I am not provable" lemma) *There exists a closed term* t *such that* T *proves* $t = \#(\neg PROV[x_1/t])$.

PROOF By Theorem 18.3.1, setting $A = \neg PROV$. QED

We fix a closed term t provided by Lemma 18.3.2.

Lemma 18.3.3 *Suppose* T *proves* $\neg PROV[x_1/t]$. *Then* T *is inconsistent.*

PROOF Assume T proves $\neg PROV[x_1/t]$. By condition 12,

$$PROV[x_1/\#(\emptyset PROV[x_1/t])]$$

is provable in T. By Lemma 18.3.2, T proves $PROV[x_1/t]$. Hence T is inconsistent. QED

Lemma 18.3.4 T *proves* $PROV[x_1/t] \rightarrow PROV[x_1/PR(t)]$. T *proves* $PROV[x_1/t] \rightarrow PROV[x_1/NEG(PR(t))]$.

PROOF Let $A = \neg PROV[x_1/t]$. By condition 11,

$$PROV[x_1/\#(A)] \rightarrow PROV[x_1/PR(\#(A))]$$

is provable in T. By Lemma 18.3.2,

$$PROV[x_1/t] \rightarrow PROV[x_1/PR(t)]$$

is provable in T. By condition 10, T proves

$$PR(t) = \#(PROV[x_1/t]).$$

By condition 8, T proves

$$NEG(\#(PROV[x_1/t])) = \#(\neg PROV[x_1/t]).$$

By Lemma 18.3.2, T proves

$$NEG(PR(t)) = t.$$

The second claim follows immediately. QED

We let CON be the sentence

$$(\forall x_1)(\neg(\text{PROV} \wedge \text{PROV}[x_1/\text{NEG}(x_1)])).$$

Theorem 18.3.5 (abstract second incompleteness)   *Let* T *obey conditions 1–12. Suppose* T *proves* CON. *Then* T *is inconsistent.*

PROOF   Suppose T is as given. By Lemma 18.3.4, T proves

$$\text{PROV}[x_1/t] \; \rightarrow \; \text{PROV}[x_1/\text{PR}(t)] \wedge \text{PROV}[x_1/\text{NEG}(\text{PR}(t))].$$

Given that T proves CON, T proves

$$\neg(\text{PROV}[x_1/\text{PR}(t)] \; \wedge \; \text{PROV}[x_1/\text{NEG}(\text{PR}(t))]).$$

Hence T proves $\neg\text{PROV}[x_1/t]$. By Lemma 18.3.3, T is inconsistent. QED

Informal statements of Gödel's second incompleteness theorem are simple and dramatic. However, current versions of the formal second incompleteness are complicated and awkward. Even the abstract form of second incompleteness given earlier using derivability conditions is rather subtle and involved.

I recently addressed this problem in Friedman (2007a), where I present new versions of formal second incompleteness that are simple and informally imply informal second incompleteness. These results rest on the isolation of simple formal properties shared by consistency statements. Here I do not address any issues concerning proofs of second incompleteness.

Let us start with the most commonly quoted form of Gödel's second incompleteness theorem – for the system PA. PA can be formulated in a number of languages. Of these, L(prim) is the most suitable for supporting formalizations of the consistency of PA.

We write L(prim) for the language based on 0,S and all primitive recursive function symbols. Let PA(prim) be the formulation of PA for the language L(prim); that is, the nonlogical axioms of PA(prim) consist of the axioms for successor, primitive recursive defining equations, and the induction scheme applied to all formulas in L(prim).

Informal second incompleteness (PA(prim)) Let A be a sentence in L(prim) that adequately formalizes the consistency of PA(prim), in the informal sense. Then PA(prim) does not prove A.

We discover the following result. Let PRA be the important subsystem of PA(prim), based on the same language L(prim), where it is required that the induction scheme be applied only to quantifier-free formulas of L(prim).

Formal second incompleteness (PA(prim)) Let A be a sentence in L(prim) such that every equation in L(prim) that is provable in PA(prim) is also provable in PRA + A. Then PA(prim) does not prove A.

The informal second incompleteness for PA(prim) can be derived in the usual semiformal way from the preceding formal second incompleteness for PA(prim).

Formal criterion theorem I Let A be a sentence in L(prim) such that every equation in L(prim) that is provable in PA(prim) is also provable in PRA + A. Then for all n, PRA + A proves the consistency of PA(prim)$_n$.

Here PA(prim)$_n$ consists of the axioms of PA(prim) in prenex form with at most n quantifiers.

The preceding development can be appropriately carried out for systems with full induction. However, a more general treatment covers finitely axiomatized theories as well. Let us use the system of exponential arithmetic (EFA) for this more general treatment. EFA is the system of arithmetic based on addition, multiplication, and exponentiation, with induction applied only to formulas all of whose quantifiers are bounded to terms. This is the same as the system $I\Sigma_0(\exp)$ in Hajek and Pudlak (1993, 37).

Informal second incompleteness (general many sorted, EFA) Let L be a fragment of L(many) containing L(EFA). Let T be a consistent extension of EFA in L. Let A be a sentence in L that adequately formalizes the consistency of T, in the informal sense. Then T does not prove A.

Formal second incompleteness (general many sorted, EFA) Let L be a fragment of L(many) containing L(EFA). Let T be a consistent extension of EFA in L. Let A be a sentence in L such that every universalized inequation in L(EFA) with a relativization in T is provable in EFA + A. Then T does not prove A.

Formal criterion theorem II Let L be a fragment of L(many) containing L(EFA). Let T be a consistent extension of EFA in L. Let A be a sentence in L such that every universalized inequation in L(EFA) with a relativization in T is provable in EFA + A. Then EFA proves the consistency of every finite fragment of T.

Here a relativization of a sentence φ of L(EFA), in T, is an interpretation of φ in T that leaves the meaning of all symbols unchanged but in which the domain is allowed to consist of only some of the nonnegative integers from the point of view of T.

Finally, I mention an interesting issue about which I am somewhat unclear but that can be gotten around in a satisfactory way. It can be said that Gödel's second incompleteness theorem has a defect in that one is relying on a formalization of Con(T) within T via the indirect method of Gödel numbers. Not only is the assignment of Gödel numbers to formulas (and the relevant syntactic objects) ad hoc but one is still being indirect and not directly dealing with the objects at hand – which are syntactic and not numerical.

It would be preferable to directly formalize Con(T) within T, without use of any indirection. Thus, in such an approach, one would add new sorts for the relevant syntactic objects and introduce the various relevant relations and function symbols, together with the relevant axioms. Precisely this approach was adopted by Quine (1951, chap. 7).

However, in so doing, one has expanded the language of T. Accordingly, two choices are apparent. The first choice is to make sure that as one adds new sorts and new relevant relations and function symbols and new axioms to T, associated with syntax, one also somehow has already appropriately treated, directly, the new syntactic objects and axioms beyond T that arise when one is performing this addition to T. The second choice is to be content with adding the new sorts and new relevant relations and function symbols and new axioms to T, associated with the syntax of T only – and not to try to deal in this manner with the extended syntax that arises from this very process. This is the choice made in Quine (1951, chap. 7).

I lean toward the opinion that the first choice is impossible to realize in an appropriate way. Some level of indirection will remain. Perhaps the level of indirection can be made rather weak and subtle. Thus I lean toward the opinion that it is impossible to construct extensions of, say, PA that directly and adequately formalize their entire syntax. I have not tried to prove such an impossibility result, but it seems possible to do so.

In any case, the second choice, on reflection, turns out to be wholly adequate for casting what may be called "direct second incompleteness." This formulation asserts that for any suitable theory T, if T' is the (or any) extension of T through the addition of appropriate sorts, relations, functions, and axioms, directly formalizing the syntax of T, including a direct formalization of the consistency of T, then T' does not prove the consistency of T (so expressed).

One can recover the usual second incompleteness theorem for T from the preceding direct second incompleteness by proving that there is an interpretation of T' in T. This was also done in Quine (1951, chap. 7). Thus, under this view of second incompleteness, one does not view Con(T) as a sentence in the language of T but instead as a sentence in the language of an extension T' of T. Con(T) becomes a sentence in the language of T only through an interpretation (in the sense of Tarski) of T' in T. There are many such interpretations, all of which are ad hoc. This view would then eliminate ad hoc features in the formulation of second incompleteness while preserving the foundational implications.

In the Introduction to Friedman (2011), the reader will find new statements of Gödel's second incompleteness theorem based on Gödel's completeness theorem.

18.4 Lengths of Proofs

Gödel (1936) discusses a result that, in modern terminology, asserts the following. Let RTT be Russell's simple theory of types with the axiom of infinity. Let RTT_n be the fragment of RTT using only the first n types. Let $f:N \to N$ be a recursive function. For each $n \geq 0$, there are infinitely many sentences φ such that

$$f(n) < m,$$

where n is the least Gödel number of a proof of φ in RTT_{n+1} and m is the least Gödel number of a proof of φ in RTT_n.

Gödel expressed the result in terms of lengths of proofs rather than in terms of Gödel numbers or the total number of symbols. He did not publish any proofs of this result or results of a similar nature. As can be surmised from the introductory remarks by R. Parikh, it is likely that Gödel had inadvertently used lengths and probably intended Gödel numbers or numbers of symbols. In any case, the analogous result with Gödel numbers was proved in Mostowski (1952). Similar results were also proved in Ehrenfeucht and Mycielski (1971) and Parikh (1971). See also Parikh (1973) for results going in the opposite direction concerning the number of lines in proofs in certain systems.

In an unpublished manuscript, I considered, for any reasonable system T and positive integer n, that the finite consistency statement $Con_n(T)$ expresses that "every inconsistency in T uses at least n symbols." I gave a lower bound of $n^{1/4}$ on the number of symbols required to prove $Con_n(T)$ in T, provided that n is sufficiently large. A more careful version of the argument gives the lower bound of $n^{1/2}$ for sufficiently large n. I called this "finite second incompleteness." A much more careful analysis of finite second incompleteness is in Pudlak (1985), which establishes an $(n(\log(n))^{-1/2})$ lower bound and an $O(n)$ upper bound for systems T satisfying certain reasonable conditions.

It would be very interesting to extend finite second incompleteness in several directions. One direction is to give a treatment of a good lower bound for a proof of $\text{Con}_n(T)$ in T, which is along the lines of the Hilbert-Bernays derivability conditions, adapted carefully for finite second incompleteness. I offer a treatment of the derivability conditions in Section 18.3 as a launching point. A number of issues arise as to the best way to set this up and what level of generality is appropriate. Another direction to take finite second incompleteness is to give some versions that are not asymptotic; that is, they involve specific numbers of symbols that are argued to be related to actual mathematical practice.

Although the very good upper bound of O(n) is given in Pudlak (1985) for a proof of $\text{Con}_n(T)$ in T, at least for some reasonable systems T, the situation seems quite different if we are talking about proofs in S of $\text{Con}_n(T)$, where S is significantly weaker than T. For specificity, consider how many symbols it takes to prove $\text{Con}_n(ZF)$ in PA, where n is large. It seems plausible that there is no subexponential upper bound here.

Obviously, though, if there is some algorithm and polynomial P that PA can prove is an algorithm for testing the satisfiability of Boolean expressions whose run time is bounded by P, then PA proves $\text{Con}_n(ZF)$ using a polynomial number of symbols in n (assuming that Con(ZF) is, in fact, true). Thus, to show that there is no subexponential upper bound here, we will have to refute this strong version of P $\neq$ NP. However, this appears to be as challenging as proving P $\neq$ NP.

Some other aspects of lengths of proofs seem important. One such aspect is the issue of overhead. Gödel (1940) established that any proof of an arithmetic sentence A in Von Neumann–Bernays–Gödel set theory (NBG) + AxC can be converted to a proof of A in NBG.

Gödel used the method of relativization. Thus one obtains constants c,d such that if arithmetic A is provable in NBG + AxC using n symbols, then A is provable in NBG using at most cn+d symbols.

What is not at all clear here is whether c,d can be made reasonably small. Clearly a lot of overhead is involved on two counts. One is in the execution of the actual relativization, which involves relativizing to the constructible sets. The other is that one must insert the proofs of various facts about the constructible sets, including that they form a model of NBG.

The same remarks can be made with regard to NBG + GC + CH and NBG + GC, where GC is the global axiom of choice. Furthermore, these remarks apply to ZFC and ZF and also to ZFC + CH and ZFC. They also apply equally well to the Cohen forcing method (Cohen, 1963, 1964) and to proofs from ZF + ¬AxC and ZFC + ¬CH.

I close with another issue regarding lengths of proofs in a context that is often considered immune to incompleteness phenomena. Finite incompleteness phenomena are very much in evidence here. Tarski and McKinsey (1951) proved the completeness of the usual axioms for real closed fields using quantifier elimination. This also provides a decision procedure for recognizing the first-order sentences in ($\Re$, <, 0, 1, +, −, •). His method applies to the following three fundamental axiom systems:

1. The language is 0, 1, +, −, •. The axioms consist of the usual field axioms, together with −1 is not the sum of squares, x or −x is a square, and every polynomial of odd degree with leading coefficient 1 has a zero.

2. The language is 0, 1, +, −, •, <. The axioms consist of the usual ordered field axioms, together with every positive element has a square root and every polynomial of odd degree with leading coefficient 1 has a zero.

3. The language is 0, 1, +, −, •, <. The axioms consist of the usual ordered field axioms, together with the axiom scheme asserting that if a first-order property holds of something, and there is an upper bound to what it holds of, then there is a least upper bound to what it holds of.

For reworkings of and improvements on Tarski, see Cohen (1969), Renegar (1992a, 1992b, 1992c), and Basu et al. (2006). In terms of computational complexity, the set of true first-order sentences in $(\Re, <, 0, 1, +, −, •)$ is exponential space easy and nondeterministic exponential time hard. The gap has not been filled. Even the first-order theory of $(\Re, +)$ is nondeterministic exponential time hard (see Rabin, 1977). The work just cited concerns mainly the computational complexity of the set of true sentences in the reals (sometimes with only addition). It does not deal directly with the lengths of proofs in systems 1, 2, and 3.

What can I say about the number of symbols in proofs in systems 1, 2, and 3? I conjecture that with the usual axioms and rules of logic, in all three cases, there is a double exponential lower and upper bound on the number of symbols required in a proof of any true sentence in each of 1, 2, and 3.

What is the relationship between sizes of proofs of the same sentence (without <) in 1, 2, and 3? I conjecture that asymptotically, there are infinitely many true sentences without < such that there is a double exponential reduction in the number of symbols needed to prove it when passing from system 1 to system 3.

These issues concerning sizes of proofs are particularly interesting when the quantifier structure of the sentence is restricted. For instance, the cases of purely universal, purely existential are particularly interesting, especially when the matrix is particularly simple. Other cases of clear interest are $\forall\ldots\forall\exists\ldots\exists$ and $\exists\ldots\exists\forall\ldots\forall$, with the obviously related conditions of surjectivity and nonsurjectivity being of particular interest.

Another aspect of sizes of proofs comes out of strong mathematical Π_2^0-sentences. The earliest ones were presented in Goodstein (1944) and Paris and Harrington (1977) and are proved just beyond PA. I discovered many examples in connection with theorems of Kruskal (1960) and Robertson and Seymour (1985, 2004) that are far stronger, with no predicative proofs (see Friedman, 2002a). None of these three references discuss the connection with sizes of proofs. This connection is discussed in Smith (1985, 132–35) and in the unpublished abstracts (Friedman, 2006a, 2006b, 2006c, 2006d, 2006e, 2006f, 2006g) from the Foundations of Mathematics Archives.[3]

The basic idea is this: a number of mathematically natural Π_2^0-sentences $(\forall n)(\exists m)(R(n,m))$ are provably equivalent to the 1-consistency of various systems T. One normally gets, as a consequence, that the Skolem function m of n grows very fast, asymptotically, so that it dominates the provably recursive functions of T.

However, we have observed that in many cases, one can essentially remove the asymptotics. In other words, in many cases, we have verified that we can fix n to

---

[3] See http://cs.nyu.edu/pipermail/fom/.

be very small (numbers like 3 or 9 or 15) and consider the resulting Σ_1^0-sentence $(\exists m)(R(n,m))$. The result is that any proof in T (or certain strong fragments of T) of this Σ_1^0-sentence must have an absurd number of symbols, for example, an exponential stack of one hundred 2s. Yet, if we go a little beyond T, we can prove the full Π_2^0-sentence $(\forall n)(\exists m)(R(n,m))$ in a normal-sized mathematics manuscript, thereby yielding a proof just beyond T of the resulting Σ_1^0-sentence R(n,m) with n fixed to be a small (or remotely reasonable) number. This provides myriad mathematical examples of Gödel's (1936) original length of proof phenomena.

18.5 The Negative Interpretation

Gödel wrote four fundamental papers concerning formal systems based on intuitionistic logic (Gödel, 1932b, 1933a, 1933b, 1958). Gödel (1972) is a revised version of Gödel (1958).

Gödel (1932b) proves that the intuitionistic propositional calculus cannot be viewed as a classical system with finitely many truth values. He shows this by constructing an infinite descending chain of logics intermediate in strength between classical propositional calculus and intuitionistic propositional calculus. (For more on intermediate logics, see Hosoi and Ono (1973) and Minari (1983).)

Gödel (1933b) introduces his negative interpretation in the form of an interpretation of PA in Heyting arithmetic (HA). Here HA is the corresponding version of PA based on intuitionistic logic. It can be axiomatized by taking the usual axioms and rules of intuitionistic predicate logic, together with the axioms of PA as usually given. Of course, one must be careful to present ordinary induction in the usual way and not use the least number principle.

It is natural to isolate Gödel's negative interpretation in the following two ways:

a. an interpretation of classical propositional calculus in intuitionistic propositional calculus
b. an interpretation of classical predicate calculus in intuitionistic predicate calculus

In modern terms, it is convenient to use $\bot, \neg, \vee, \wedge, \rightarrow$. The interpretation for propositional calculus inductively interprets

$\bot$ as $\bot$

$\neg$ as $\neg$

$\wedge$ as $\wedge$

$\rightarrow$ as $\rightarrow$

$\vee$ as $\neg\neg\vee$

For predicate calculus,

$\forall$ as $\forall$

$\exists$ as $\neg\neg\exists$

φ as $\neg\neg\varphi$, where φ is atomic

Now, in HA, we can prove $n = m \lor \neg n = m$. It is then easy to see that the successor axioms and the defining equations of PA are sent to theorems of HA, and also, each induction axiom of PA is sent to a theorem of HA. Furthermore, the axioms of classical predicate calculus become theorems of intuitionistic predicate calculus, and the rules of classical predicate calculus become rules of intuitionistic predicate calculus. Hence, under the negative interpretation, theorems of classical propositional calculus become theorems of intuitionistic propositional calculus, theorems of classical predicate calculus become theorems of intuitionistic predicate calculus, and theorems of PA become theorems of HA.

In addition, any Π_1^0-sentence $(\forall n)(F(n) = 0)$, where F is a primitive recursive function symbol of PA, is sent to a sentence that is provably equivalent to $(\forall n)(F(n) = 0)$. It is then easy to conclude that every Π_1^0 theorem of PA is a theorem of HA.

Gödel's negative interpretation has been extended to many pairs of systems, most of them of the form T,T′, where T,T′ have the same nonlogical axioms and where T is based on classical predicate calculus, whereas T′ is based on intuitionistic predicate calculus (e.g., see Kreisel, 1968a, 344; 1968b, sect. 5; Myhill, 1974; Friedman, 1973; Leivant, 1985).

A much stronger result holds for PA over HA. Every Π_2^0-sentence provable in PA is provable in HA. The first proofs of this result were from the proof theory of PA via Gentzen (1969; Schütte, 1977) and from Gödel's so-called *Dialectica* or functional interpretation (Gödel, 1958, 1972).

However, for other pairs for which the negative interpretation shows that they have the same provable Π_1^0-sentences, say, classical and intuitionistic second-order arithmetic, one does not have the required proof theory. In this case, the *Dialectica* interpretation has been extended by Spector (1962), and that these two systems have the same provable Π_2^0-sentences then follows.

Nevertheless, there are many appropriate pairs for which the negative interpretation works, yet there is no proof theory and no functional interpretation. In Friedman (1978), I broke this impasse by modifying Gödel's negative interpretation via what is now called the "A-translation" (see also Dragalin, 1980). I illustrate the technique for PA over HA, formulated with primitive recursive function symbols.

Let A be any formula in L(HA) = L(PA). We define the A-translation φ_A of the formula φ in L(HA), in case no free variable of A is bound in φ. Take φ^A to be the result of simultaneously replacing every atomic subformula ψ of φ by $(\psi \lor A)$. In particular, $\bot$ gets replaced by what amounts to A.

The A-translation is an interpretation of HA in HA; that is, if φ^A is defined, and HA proves A, then HA proves φ^A. Also, obviously, HA proves $A \to \varphi^A$.

Now suppose $(\exists n)(F(n,m) = 0)$ is provable in PA, where F is a primitive recursive function symbol. By Gödel's negative interpretation, $\neg\neg(\exists n)(F(n,m) = 0)$ is provable in HA. Write this as $((\exists n)(F(n,m) = 0) \to \bot) \to \bot$.

By taking the A-translation, with $A = (\exists n)(F(n,m) = 0)$, we obtain that HA proves $((\exists n)(F(n,m) = 0 \lor (\exists n)(F(n,m) = 0)) \to (\exists n)(F(n,m) = 0)) \to (\exists n)(F(n,m) = 0)$:

$$((\exists n)(F(n,m) = 0) \to (\exists n)(F(n,m) = 0)) \to$$
$$(\exists n)(F(n,m) = 0),$$
$$(\exists n)(F(n,m) = 0).$$

This method applies to a large number of pairs T/T', as indicated in Friedman (1973) and Leivant (1985). Gödel's so-called *Dialectica* interpretation, or functional interpretation, of HA is presented in Gödel (1958, 1972).

In Gödel's *Dialectica* interpretation, theorems of HA are interpreted as derivations in a quantifier-free system T of primitive recursive functionals of finite type that is based on quantifier-free axioms and rules, including a rule of induction.

The *Dialectica* interpretation has had several applications in different directions. We will not discuss applications to programming languages and category theory. To begin with, the *Dialectica* interpretation can be combined with Gödel's negative interpretation of PA in HA to form an interpretation of PA in Gödel's quantifier-free system T.

One obvious application, and motivation, is philosophical, and Gödel discusses this aspect in both papers, especially the second. The idea is that the quantifiers in HA or PA, ranging over all natural numbers, are not finitary, whereas T is arguably finitary – at least in the sense that T is quantifier-free. However, the objects of T are at least prima facie infinitary, and so there is the difficult question of how to gauge this trade-off. One idea is that the objects of T should not be construed as infinite completed totalities but rather as rules. Interested readers can consult the rather extensive introductory notes to Gödel (1958) in Gödel (1990).

Another application is to extend the interpretation to the two-sorted first-order system known as second-order arithmetic, or Z_2. This was carried out by Spector (1962). Here the idea is that one may construe such a powerful extension of Gödel's *Dialectica* interpretation as some sort of constructive consistency proof for the rather metamathematically strong and highly impredicative system Z_2. However, in various communications, Gödel was not entirely satisfied that the quantifier-free system Spector used was truly constructive.

We believe that the Spector development has not been fully exploited. In particular, it ought to give rather striking mathematically interesting characterizations of the provably recursive functions and provable ordinals of Z_2 and various fragments of Z_2.

Another fairly recent application is to use the *Dialectica* interpretation, and extensions of it to systems involving functions and real numbers, to obtain sharper uniformities in certain areas of functional analysis that had been obtained before by the specialists. This work has been pioneered by Kohlenbach (see Kohlenbach, 2005; Kohlenbach and Gerhardy, 2008; Kohlenbach and Oliva, 2003; see also Chapter 17 in this volume).

18.6 The Axiom of Choice and the Continuum Hypothesis

Gödel wrote six manuscripts directly concerned with the continuum hypothesis: two abstracts (Gödel, 1938, 1939a), one paper with sketches of proofs (Gödel, 1939b), one research monograph with fully detailed proofs (Gödel, 1940), and one philosophical paper, in two versions (Gödel, 1964 [1947]). The normal abbreviation for the axiom of choice is AxC. The normal abbreviation for the continuum hypothesis is CH. A particularly attractive formulation of CH asserts that every set of real numbers is either in one-one correspondence with a set of natural numbers or in one-one correspondence with the set of real numbers.

Normally, one follows Gödel in considering CH only in the presence of AxC. However, note that in this form, CH can be naturally considered without the presence of AxC. In fact, Solovay's model satisfying ZF + DC + "all sets are Lebesgue measurable" also satisfies CH in the strong form that every set of reals is countable or has a perfect subset (this strong form is incompatible with AxC; see Solovay, 1970). Here DC is the axiom of dependent choice.

The statement of CH is from Cantor. Gödel also considers the generalized continuum hypothesis (GCH), the statement of which is credited to Hausdorff. The GCH asserts that for all sets A, every subset of $\wp(A)$ is either in one-one correspondence with a subset of A or in one-one correspondence with $\wp(A)$. Here $\wp$ is the power set operation. Gödel's work establishes an interpretation of ZFC + GCH in ZF. This provides a very explicit way of converting any inconsistency in ZFC + GCH to an inconsistency in ZF.

We can attempt to quantify these results. In particular, it is clear that the interpretation given by Gödel of ZFC + GCH in ZF, by relativizing to the constructible sets, is rather large in the sense that, when fully formalized, it results in a lot of symbols. It also seems to result in a lot of quantifiers. How many?

So far I have been talking about the crudest formulations in primitive notation, without the benefit of abbreviation mechanisms. However, abbreviation mechanisms are essential for the actual conduct of mathematics. In fact, current proof assistants – where humans and computers interact to create verified proofs – necessarily incorporate very substantial abbreviation mechanisms (see, e.g., Barendregt and Wiedijk, 2005; Wiedijk, 2006). Hence the question arises how simple an interpretation of ZFC + GCH in ZF can be, with abbreviations allowed in the presentation of the interpretation. This is far from clear.

P. J. Cohen proved that if ZF is consistent, then so is ZF + ¬AxC and ZFC + ¬CH, thus complementing Gödel's results (see Cohen, 1963, 1964). The proof does not readily give an interpretation of ZF + ¬AxC or of ZFC + ¬CH in ZF. It can be converted into such an interpretation by a general method whereby under certain conditions (met here), if the consistency of every given finite subsystem of one system is provable in another, then the first system is interpretable in the other (see Feferman, 1960).

Again, the question arises, how simple can an interpretation be of ZF + ¬AxC or of ZFC + ¬CH in ZF, with abbreviations allowed in the presentation of the interpretation? Again, this is far from clear. Also, how does this question compare with the previous question?

Another kind of complexity issue associated with the CH is of interest. First, I will provide some background. It is known that every three-quantifier sentence in primitive notation $\in,=$ is decided in a weak fragment of ZF (see Gogol, 1979; Friedman, 2003b). Moreover, the five-quantifier sentence in $\in,=$ is not decided in ZFC (it is equivalent to the existence of a subtle cardinal over ZFC; see Friedman, 2003a). It is also known that AxC can be written with five quantifiers in $\in,=$ over ZFC (see Maes, 2007). The question is, how many quantifiers are needed to express CH over ZFC in $\in,=$? We can also ask this and related questions where abbreviations are allowed.

Most mathematicians instinctively take the view that because CH is neither provable nor refutable from the standard axioms for mathematics (ZFC), the ultimate status of CH has been settled, and nothing is left to ponder. However, many mathematical logicians, particularly those in set theory, take quite a different view, including Gödel. These mathematicians take the view that the CH is a well-defined mathematical assertion with a definite truth value. The problem is to determine just what this truth value is.

The idea here is that a definite system of objects exists independently of human minds and that human minds can no more manipulate the truth value of statements of set theory than they can manipulate the truth value of statements about electrons and stars and galaxies. This is the so-called Platonist point of view that is argued so forcefully and explicitly in Gödel (1947 [1964]).

The late P. J. Cohen led a panel discussion at the Gödel Centenary celebration in Vienna in 2006 called "On Unknowability," in which he conducted a poll roughly along these lines. The question he asked was, "Does the continuum hypothesis have a definite answer?" or "Does the continuum hypothesis have a definite truth value?" The response from the audience appeared quite divided on the issue. Of the panelists, the ones who have expressed very clear views on this topic were most notably Cohen and Hugh Woodin. Cohen took a formalist viewpoint, whereas Woodin took a Platonist one. (See Chapters 19 and 20 – their respective contributions to this volume.)

My own view is that we simply do not know enough in the foundations of mathematics to decide the truth or appropriateness of the formalist versus the Platonist viewpoint – or, for that matter, what mixture of the two is true or appropriate. Then it is reasonable to place the burden on me to explain what kind of additional knowledge could be relevant for this issue.

My ideas are not very well developed, but I will offer at least something for consideration. It may be possible to develop a theory of "fundamental mental pictures" that is so powerful and compelling that it supplants any discussion of formalism-Platonism in anything like its present terms. What may come out is a fundamental mental picture of the axioms of ZFC, even with some large cardinals, along with a theorem to the effect that there is no fundamental mental picture for CH and no fundamental mental picture for $\neg$CH.

18.7 Wqo Theory

Wqo theory is a branch of combinatorics that has proved to be a fertile source of deep metamathematical phenomena. A qo (quasi-order) is a reflexive transitive relation $(A, \leq)$. A wqo (well quasi-order) is a qo $(A, \leq)$ such that for all infinite sequences $x_1, x_2, \ldots$ from A, $\exists\ i < j$ such that $x_i \leq x_j$. The highlights of wqo theory are that certain qos are wqos and certain operations on wqos produce wqos. Kruskal (1960) treats finite trees as finite posets and studies the qo; there exists an inf preserving embedding from T_1 into T_2.

Theorem 18.7.1 (Kruskal, 1960) *The preceding qo of finite trees as posets is a wqo.*

The simplest proof of Theorem 18.7.1 and some extensions is in Nash-Williams (1963), with the introduction of minimal bad sequences. I have observed that the connection between wqos and well orderings can be combined with known proof theory to establish independence results. The standard formalization of "predicative mathematics" is from Feferman and Schutte (FS; see Feferman, 1964, 1968, 1998).

Poincaré, Weyl, and others railed against impredicative mathematics (see Weyl, 1918, 1987; Feferman, 1998, 289–91; Folina, 1992).

Theorem 18.7.2 (Friedman 2002a) *Kruskal's tree theorem* (KT) *cannot be proved in FS.*

KT goes considerably beyond FS, and an exact measure of KT is known (see Rathjen and Weiermann, 1993). J. B. Kruskal actually considered finite trees whose vertices were labeled from a wqo $\leq^*$. The additional requirement on embeddings is that label(v) $\leq^*$ label(h(v)).

Theorem 18.7.3 (Kruskal, 1960) *The qo of finite trees as posets, with vertices labeled from any given wqo, is a wqo.*

Labeled KT is considerably stronger, proof theoretically, than KT, even with only two labels, $0 \leq 1$. We have not seen a metamathematical analysis of labeled KT. Note that KT is a Π_1^1-sentence and that labeled KT is a Π_2^1-sentence.

Theorem 18.7.4 *Labeled* KT *does not hold in the hyperarithmetic sets. In fact,* $RCA_0 + KT$ *implies* ATR_0.

A proof of Theorem 18.7.4 will appear in Friedman et al. (2011).
It is natural to impose a growth rate in KT in terms of the number of vertices of T_i.

Corollary 18.7.5 (linearly bounded KT) *Let* $T_1, T_2, \ldots$ *be a linearly bounded sequence of finite trees.* $\exists\, i < j$ *such that* T_i *is inf preserving embeddable into* T_j.

Corollary 18.7.6 (computational KT) *Let* $T_1, T_2, \ldots$ *be a sequence of finite trees in a given complexity class. There exists* $i < j$ *such that* T_i *is inf preserving embeddable into* T_j.

Note that Corollary 18.7.6 is Π_2^0.

Theorem 18.7.7 *Corollary 18.7.5 cannot be proved in FS. This holds even for linear bounds* $n + k$ *with variable n and constant k.*

Theorem 18.7.8 *Corollary 18.7.6 cannot be proved in FS, even for linear time and logarithmic space.*

By an obvious application of weak Konig's lemma, Corollary 18.7.5 has very strong uniformities.

Theorem 18.7.9 (uniform linearly bounded KT) *Let* $T_1, T_2, \ldots$ *be a linearly bounded sequence of finite trees. There exists* $i < j \leq n$ *such that* T_i *is inf preserving embeddable into* T_j, *where n depends only on the given linear bound and not on* $T_1, T_2, \ldots$.

With this kind of strong uniformity, we can obviously strip Theorem 18.7.9 of infinite sequences of trees. Using the linear bounds $n + k$, with k being fixed, we obtain the following:

Theorem 18.7.10 (finite KT) *Let* $n \gg k$. *For all finite trees* $T_1, \ldots, T_n$ *with each* $|T_i| \leq i + k$, *there exists* $i < j$ *such that* T_i *is inf preserving embeddable into* T_j.

Given that Theorem 18.7.10 → Theorem 18.7.9 → Corollary 18.7.5 (using bounds n + k, with variable n and k constant), we see that Theorem 18.7.10 is not provable in FS. Other Π_2^0 forms of KT involving only the internal structure of a single finite tree can be found in Friedman (2002a).

I proved analogous results for the extended Kruskal theorem (EKT), which involves a finite label set and a gap embedding condition, only here, the strength jumps to that of Π_1^1-CA$_0$. I said that the gap condition was natural (i.e., EKT was natural). Many people were unconvinced. Soon thereafter, EKT became a tool in the proof of the well-known graph minor theorem of Robertson and Seymour (1985, 2004):

Theorem 18.7.11 *Let* $G_1, G_2, \ldots$ *be finite graphs. There exists* i < j *such that* G_i *is minor included in* G_j.

I then asked Robertson and Seymour to prove a form of EKT that I knew implied full EKT, just from graph minor theorem (GMT). They complied, and we wrote a triple paper (Friedman et al., 1987). The upshot is that GMT is not provable in Π_1^1-CA$_0$. Just where GMT is provable is unclear, and recent discussions with Robertson have not stabilized. I disavow remarks in Friedman et al. (1987) about where GMT can be proved. An extremely interesting consequence of GMT is the subcubic graph theorem. A subcubic graph is a graph in which every vertex has valence ≤ 3. (Loops and multiple edges are allowed.)

Theorem 18.7.12 *Let* $G_1, G_2, \ldots$ *be subcubic graphs. There exists* i < j *such that* G_i *is embeddable into* G_j *as topological spaces (with vertices going to vertices).*

Robertson and Seymour also claim to be able to use the subcubic graph theorem for linkage to EKT (see Robertson and Seymour, 1985; Friedman et al., 1987). Therefore the subcubic graph theorem (even in the plane) is not provable in Π_1^1-CA$_0$.

I have discovered lengths of proof phenomena in wqo theory and use Σ_1^0 sentences (see Friedman, 2006a, 2006b, 2006c, 2006d, 2006e, 2006f, 2006g).

(*) Let $T_1, \ldots, T_n$ be a sufficiently long sequence of trees with vertices labeled from {1,2,3}, where each $|T_i| \leq i$. There exists i < j such that T_i is inf and label preserving embeddable into T_j.

(**) Let $T_1, \ldots, T_n$ be a sufficiently long sequence of subcubic graphs, where each $|T_i| \leq i + 13$. There exists i < j such that G_i is homeomorphically embeddable into G_j.

Theorem 18.7.13 *Every proof of (*) in FS uses at least* $2^{[1000]}$ *symbols. Every proof of (**) in* Π_1^1-CA$_0$ *uses at least* $2^{[1000]}$ *symbols.*

18.8 Borel Selection

Let $S \subseteq \Re^2$ and $E \subseteq \Re$. A selection for A on E is a function f:E → $\Re$ whose graph is contained in S. A selection for S is a selection for S on $\Re$. Let us say that S is symmetric if and only if $S(x,y) \leftrightarrow S(y,x)$.

Theorem 18.8.1 *Let* $S \subseteq \Re^2$ *be a symmetric Borel set. Then* S *or* $\Re^2 \backslash S$ *has a Borel selection.*

My proof of Theorem 18.8.1 in Friedman (1981) relied heavily on Borel determinacy, from D. A. Martin (see Martin, 1975, 1985; Kechris, 1994, 137–48).

Theorem 18.8.2 (Friedman, 1981) *Theorem 18.8.1 is provable in ZFC, but not without the axiom scheme of replacement.*

Another kind of Borel selection theorem is implicit in the work of Debs and Saint Raymond of Paris VII. They take the general form that if there is a nice selection for S on compact subsets of E, then there is a nice selection for S on E (see Debs and Saint Raymond, 1996, 1999, 2001, 2004, 2007).

Theorem 18.8.3 *Let $S \subseteq \mathfrak{R}^2$ be Borel and $E \subseteq \mathfrak{R}$ be Borel with empty interior. If there is a continuous selection for S on every compact subset of E, then there is a continuous selection for S on E.*

Theorem 18.8.4 *Let $S \subseteq \mathfrak{R}^2$ be Borel and $E \subseteq \mathfrak{R}$ be Borel. If there is a Borel selection for S on every compact subset of E, then there is a Borel selection for S on E.*

Theorem 18.8.5 (Friedman, 2005) *Theorem 18.8.3 is provable in ZFC, but not without the axiom scheme of replacement. Theorem 18.8.4 is neither provable nor refutable in ZFC.*

We can say more.

Theorem 18.8.6 (Friedman, 2005) *The existence of the cumulative hierarchy up through every countable ordinal is sufficient to prove Theorems 18.8.1 and 18.8.3. However, the existence of the cumulative hierarchy up through any suitably defined countable ordinal is not sufficient to prove Theorem 18.8.1 or Theorem 18.8.3.*

The $f:N \rightarrow N$ constructible in any given $x \subseteq N$ are eventually dominated by some $g:N \rightarrow N$.

Theorem 18.8.7 *ZFC + Theorem 18.8.4 implies DOM (Friedman, 2005). ZFC + DOM implies Theorem 18.8.4 (Debs and Saint Raymond, 2007).*

18.9 Boolean Relation Theory

The principal reference for this section is Friedman (2011). I begin with two examples of statements in Boolean relation theory (BRT) of special importance for the theory.

Thin set theorem Let $k \geq 1$ and $f:N^k \rightarrow N$. There exists an infinite set $A \subseteq N$ such that $f[A^k] \neq N$.

Complementation theorem Let $k \geq 1$ and $f:N^k \rightarrow N$. Suppose that for all $x \in N^k$, $f(x) > \max(x)$. There exists an infinite set $A \subseteq N$ such that $f[A^k] = N \backslash A$.

These two theorems are official statements in BRT. In the complementation theorem, A is unique. We now write them in BRT form.

Let MF be the set of all functions from some N^k into N. Let INF be the family of all infinite subsets of N. We use IBRT for inequational Boolean relation theory. We use EBRT for equational Boolean relation theory.

Thin set theorem For all $f \in$ MF, there exists $A \in$ INF such that $fA \neq N$.

Complementation theorem   For all $f \in$ SD, there exists $A \in$ INF such that fA = N\A.

The thin set theorem lives in IBRT in A,fA. There are only $2^{2^2} = 16$ statements in IBRT in A,fA. These are easily handled.

The complementation theorem lives in EBRT in A,fA. There are only $2^{2^2} = 16$ statements in IBRT in A,fA. These are easily handled.

For EBRT/IBRT in A,B,C,fA,fB,fC,gA,gB,gC, we have $2^{2^9} = 2^{512}$ statements. This is entirely unmanageable. It would take several major new ideas to make this manageable.

Discovery   There is a statement in EBRT in A,B,C,fA,fB,fC,gA,gB,gC that is independent of ZFC. It can be proved in SMAH+ but not in SMAH, even with the axiom of constructibility.

Here SMAH+ = ZFC + $(\forall n)(\exists \kappa)(\kappa$ is a strongly k-Mahlo cardinal). SMAH = ZFC + $\{(\exists \kappa)(\kappa$ is a strongly k-Mahlo cardinal$\}_k$. The particular example is far nicer than any typical statement in EBRT in A,B,C,fA,fB,fC,gA,gB,gC. However, it is not nice enough to be regarded as suitably natural.

Showing that all such statements can be decided in SMAH+ seems to be too hard. What to do? Look for a natural fragment of full EBRT in A,B,C,fA,fB,fC,gA,gB,gC that includes the example, where we can decide all statements in the fragment within SMAH+.

We can also look for a bonus: a striking feature of the classification that is itself independent of ZFC. Then we have a single natural statement independent of ZFC. To carry this off, we need to use the function class ELG of functions of expansive linear growth. These are functions $f:N^k \to N$ such that there exist constants c,d > 1 such that

$$c|x| \leq f(x) \leq d|x|$$

holds for all but finitely many $x \in N^k$.

Template   For all $f,g \in$ ELG, there exist A,B,C $\in$ INF such that

$$X \cup .fY \subseteq V \cup .gW$$
$$P \cup .fR \subseteq S \cup .gT.$$

Here X,Y,V,W,P,R,S,T are among the three letters A, B, C.

Note that there are 6,561 such statements. I have shown that all these statements are provable or refutable in RCA_0, with exactly twelve exceptions. These twelve exceptions are really exactly one exception up to the obvious symmetry: permuting A,B,C, and switching the two clauses. The single exception is the *exotic case*:

Proposition 18.9.1 *A For all* f,g $\in$ ELG, *there exist* A,B,C $\in$ INF *such that*

$$A \cup .fA \subseteq C \cup .gB$$
$$A \cup .fB \subseteq C \cup .gC.$$

This statement is provably equivalent to the 1-consistency of SMAH, over ACA'. If we replace "infinite" with "arbitrarily large finite," then we can carry out this second classification entirely within RCA$_0$.

Inspection shows that all the nonexotic cases come out with the same truth value in the two classifications, and that is, of course, provable in RCA$_0$. Furthermore, the exotic case comes out true in the second classification.

Theorem 18.9.1 *The following is provable in SMAH+ but not in SMAH, even with the axiom of constructibility. An instance of the template holds if and only if, in that instance, "infinite" is replaced with "arbitrarily large finite."*

18.10 Finite Incompleteness

A major direction of current research is to find an explicit Π_1^0-sentence that can be proved only using large cardinals and that arguably represents clear and compelling information in the finite mathematical realm. I have been engaged in this search in recent years and now present a recent version of this work in progress. The reader will find subsequent developments discussed in the Introduction to Friedman (2011).

Kernels in digraphs are intensively studied with a large literature base, including applications to game theory, computer science, and other areas. A digraph is a pair $G = (V,E)$, where $E \subseteq V \times V$. The vertices of G are the elements of V, and the edges of G are the elements of E. A dag is a digraph in which there are no cycles. We say that S is a kernel in (V,E) if and only if

i. No element of S connects to any element of S.
ii. Every element of V\S connects to some element of S.

Theorem 18.10.1 (VM44) *There is a unique kernel in every finite dag.*
Let Q be the set of all rational numbers.
We say that x,y $\in Q^k$ *are order equivalent if and only if, for all* $1 \le i,j \le k$, $x_i < x_j \leftrightarrow y_i < y_j$.
We say that $E \subseteq A^k$ *is order-invariant if and only if, for all order equivalent* x,y $\in A^k$, x $\in$ E $\leftrightarrow$ y $\in$ E.
Let $A \subseteq Q$ *be fixed. The A-digraphs are the digraphs* (A^k,E), *where E is an order invariant subset of* A^{2k}.

A digraph (A^k,E) is downward if and only if x E y $\rightarrow$ max(x) > max(y).

The upper shift of a vector from Q is obtained by adding 1 to all of its nonnegative coordinates.

The upper shift of a set of vectors from Q is the set of upper shifts of its elements.

UPPER SHIFT KERNEL THEOREM. There exists $0 \in A \subseteq Q$ such that every downward A-digraph has a kernel containing its upper shift.

Theorem 18.10.2 USKT is provable in SRP+ but not in SRP. ACA_0 proves USKT if and only if Con(SRP).

Here SRP+ = ZFC + ($\forall$k)(there is a cardinal with the k-SRP). SRP = ZFC + {there is a cardinal with the k-SRP$\}_k$. We say that λ has the k-SRP if and only if every 2 coloring of the unordered k-tuples from λ has a homogeneous set which is stationary in λ.

Theorem 18.10.3 is essentially finitary in the following sense.

Theorem 18.10.3 *There is an algorithm α such that the following is provable in ACA_0. USKT holds if and only if α never terminates.*

We now present an explicitly finite form of USKT. We use the following notion of height of a tuple from Q. This is the sum of the magnitudes of the numerators and denominators in the reduced forms of the coordinates.

Let (V,E) be a digraph, where V is a set of tuples from Q. An n-kernel is a set $S \subseteq V$ such that

i. No element of S connects to any element of S.
i. Every element of V\S of height $p \leq n$ connects to some element of S of height $\leq 8p^2$.

The n-upper shift of S is the set of elements of the upper shift of S of height $\leq n$.

FINITE UPPER SHIFT KERNEL THEOREM Let $n \geq 1$. There exists finite $0 \in A \subseteq Q$ such that every downward A-digraph has an n-kernel containing its n-upper shift. We can choose A and the n-kernels to consist entirely of tuples of height $\leq 8n^2$.

Theorem 18.10.4 *EFA proves FUSKT if and only if Con(SRP).*

Note that FUSKT is explicitly Π_1^0.

18.11 Incompleteness in the Future

The incompleteness phenomena, the centerpiece of Gödel's legacy, have come a long way. The same is true of the related phenomenon of recursive unsolvability, also part of the Gödel legacy. The phenomena are so deep and rich in possibilities that we expect the future to eclipse the past and present. Yet continued substantial progress is expected to be painfully slow, requiring considerably more than the present investment of mathematical and conceptual power devoted to the extension and expansion of the phenomena. This assessment also applies if we consider the P = NP problem as part of the Gödel legacy (as is common today) on the basis of his letter of March 20, 1956, to John von Neumann (Gödel, 2003b, letter 21, 373–77).

Also consider the recursive unsolvability phenomena. Perhaps the most striking example of this for the working mathematician is the recursive unsolvability of Diophantine problems over the integers (Hilbert's tenth problem), as discussed in

Section 18.2. We have, at present, no idea of the boundary between recursive decidability and recursive undecidability in this realm. Yet I conjecture that we will understand this in the future and that we will find, perhaps, that recursive undecidability kicks in already for degree 4 with four variables. However, this would require a complete overhaul of the current solution to Hilbert's tenth problem, replete with new, deep ideas. This would result in a sharp increase in the level of interest for the working mathematician who is not particularly concerned with issues in the foundations of mathematics.

In addition, we still do not know whether an algorithm exists to decide whether a Diophantine problem has a solution over the rationals. I conjecture that this will be answered in the negative and that the solution will involve some clever number-theoretic constructions of independent interest for number theory.

We now come to the future of the incompleteness phenomena. We have seen how this has developed thus far:

i. **First incompleteness:** Some incompleteness in the presence of some arithmetic (Gödel, 1931)

ii. **Second incompleteness:** Incompleteness concerning the most basic metamathematical property: consistency (Gödel, 1931; Hilbert and Bernays, 1934, 1939; Feferman, 1960; Boolos, 1993)

iii. **Consistency of the AxC:** Consistency of the most basic, and once controversial, early candidate for a new axiom of set theory (Gödel, 1940)

iv. **Consistency of the CH:** Consistency of the most basic set-theoretic mathematical problem highlighted by Cantor (Gödel, 1940)

v. ϵ_0**consistency proof:** Consistency proof of PA using quantifier-free reasoning on the fundamental combinatorial structure, ϵ_0 (Gentzen, 1969)

vi. **Functional recursion consistency proof:** Consistency proof of PA using higher-type primitive recursion, without quantifiers (Gödel, 1958, 1972)

vii. **Independence of AxC:** Independence of CH (over AxC); complements iii and iv (Cohen, 1963, 1964); forcing

viii. **Open set-theoretic problems in core areas shown to be independent:** Starting soon after Cohen (1963, 1964), starting dramatically with R. M. Solovay, for example, his work on Lebesgue measurability (Solovay, 1970) and his independence proof of Kaplansky's conjecture (Dales and Woodin, 1987), and continuing with many others; see the rather comprehensive Jech (2006); also see the many set theory papers in Shelah (1969–2007) (core mathematicians have learned to avoid raising new set theoretic problems, and the area is greatly mined)

ix. **Large cardinals necessarily used to prove independent set-theoretic statements:** Starting dramatically with measurable cardinals implies $V \neq L$ (Scott, 1961); continuing with solutions to open problems in the theory of projective sets (using large cardinals), culminating with the proof of projective determinacy (Martin and Steel, 1989)

x. **Large cardinals necessarily used to prove the consistency of set-theoretic statements:** See Jech (2006)

xi. **An uncountable number of iterations of the power set operation necessarily used to prove statements in and around Borel mathematics:** Includes Borel determinacy and some Borel selection theorems of Debs and Saint Raymond (see Section 18.8; see also Friedman, 1971, 2005, 2007b; Martin, 1975)

xii. **Large cardinals necessarily used to prove statements around Borel mathematics:** Includes some Borel selection theorems of Debs and Saint Raymond (see Section 18.8; Debs and Saint Raymond, 1996, 1999, 2001, 2004, 2007; see also Friedman, 1981, 2005, 2007b; Stanley, 1985)

xiii. **Independence of finite statements in or around existing combinatorics from PA and subsystems of second-order arithmetic:** Starting with Goodstein (1944), Paris and Harrington (1977), and most recently, with Friedman (2002a, 2006a, 2006b, 2006c, 2006d, 2006e, 2006f, 2006g); uses extensions of point v from earlier (Gentzen, 1969), from Buchholz et al. (1981), and includes Kruskal's theorem, the graph minor theorem of Robertson and Seymour (1985, 2004), and the trivalent graph theorem of Robertson and Seymour (1985)

xiv. **Large cardinals necessarily used to prove sentences in discrete mathematics, as part of a wider theory:** Boolean relation theory (Friedman, 1998, 2010)

xv. **Large cardinals necessarily used to prove explicit Π_1^0-sentences:** See Section 18.10 for the current state of the art as well as the Introduction to Friedman (2011) for subsequent developments.

Yet this development of the incompleteness phenomena has a long way to go before it realizes its potential to dramatically penetrate core mathematics.

However, I am convinced that this is a matter of a lot of time and resources. The quality man-hours devoted to expansion of the incompleteness phenomena are trivial when compared with other pursuits. Even the creative (and high-quality) study of U.S. tax law dwarfs the effort devoted to expansion of the incompleteness phenomena by orders of magnitude – let alone any major sector of technology, particularly the development of air travel, telecommunications, or computer software and hardware.

Through my efforts over forty years, I can see, touch, and feel a certain combinatorial structure that keeps arising – a demonstrably indelible footprint of large cardinals. I am able to display this combinatorial structure through Borel, discrete and finitary statements that are increasingly compelling mathematically. For subsequent developments in this project, see the Introduction to Friedman (2011).

However, I don't quite have the right way to express it in core mathematics. I likely need some richer context than the completely primitive combinatorial settings that I currently use. This difficulty will definitely be overcome in the future, and that will make a huge difference in the quality, force, and relevance of the results to mathematical practice. In fact, I will go so far as to make the following dramatic conjecture: it's not that the incompleteness phenomena are a freak occurrence; rather they are everywhere. Every interesting substantial mathematical theorem can be recast as one among a natural finite set of statements, all of which can be decided using well-studied extensions of ZFC, but not all of which can be decided within ZFC itself.

Recasting of mathematical theorems as elements of natural finite sets of statements represents an inevitable general expansion of mathematical activity. This, I conjecture, will apply to any standard mathematical context.

This program has been carried out, to a very limited extent, by BRT (as can be seen in Section 18.9). This may seem like a ridiculously ambitious conjecture that goes totally against the current conventional wisdom of mathematicians – who think that they are immune to the incompleteness phenomena. However, I submit that even fundamental features of current mathematics are not likely to bear much resemblance to the mathematics of the future.

Mathematics as a professional activity with serious numbers of workers is quite new, let's say, one hundred years old, although even that is a stretch. Assuming that the human race thrives, what is this compared to, say, a thousand more years? It is probably merely a bunch of simple observations in comparison.

Of course, a thousand years is absolutely nothing in evolutionary or geological time. A more reasonable number is a million years. What does our present mathematics look like compared with mathematics in a million years' time? These considerations should apply to our present understanding of the Gödel phenomena.

We can, of course, take this even further. A million years' time is absolutely nothing in astronomical time. Our sun has several billion good years left (although the sun will cause a lot of global warming). Mathematics in a billion years' time? Who can know what that will be like. However, I am convinced that the Gödel legacy will remain very much alive – at least as long as there is vibrant mathematical activity.

Acknowledgments

This research was partially supported by grant DMS 0245349 from the National Science Foundation and by grant 15400 from the John Templeton Foundation. I also wish to thank Warren Goldfarb and Hilary Putnam for help with several historical points.

References

Atiyah, M. (2008a). Atiyah expounds on "Mind, Matter and Mathematics." http://www.dailystar .com.lb/article.asp?edition_id=1&categ_id=2&article_id=97190.

———. (2008b). Mind, matter and mathematics. Presidential address to the Royal Society of Edinburgh. http://www.rse.org.uk/events/reports/2007-2008/presidential_address.pdf.

Avigad, J. (2003). Number theory and elementary arithmetic. *Philosophia Mathematica*, **11**, 257–84.

Barendregt, H., and Wiedijk, F. (2005). The challenge of computer mathematics. *Transactions of the Royal Society, Series A*, **363**, 2351–75.

Basu, S., Pollack, R., and Roy, M. (2006). *Algorithms in Real Algebraic Geometry*. Berlin: Springer.

Boolos, G. (1993). *The Logic of Provability*. Cambridge: Cambridge University Press.

Buchholz, W., Feferman, S., Pohlers, W., and Seig, W. (1981). *Iterated Inductive Definitions and Subsystems of Analysis: Recent Proof-Theoretical Studies*. Lecture Notes in Mathematics 897. Berlin: Springer.

Cohen, P. J. (1963). The independence of the continuum hypothesis. I. *Proceedings of the National Academy of Sciences of the United States of America*, **50**, 1143–48.

———. (1964). The independence of the continuum hypothesis. II. *Proceedings of the National Academy of Sciences of the United States of America*, **51**, 105–10.

———. (1969). Decision procedures for real and p-adic fields. *Communications on Pure and Applied Mathematics*, **22**, 131–51.

Dales, H. G. (2001). *Banach Algebras and Automatic Continuity*. London Mathematical Society Monographs 24. Oxford: Clarendon Press.

Dales, H. G., and Woodin, W. H. (1987). *An Introduction to Independence Results for Analysts.* London Mathematical Society Lecture Note Series 115. Cambridge: Cambridge University Press.

Davis, M. (1973). Hilbert's tenth problem is unsolvable. *American Mathematical Monthly,* **80**, 233–69. [Repr. with corrections in (1973). *Computability and Unsolvability.* Dover.]

Dawson, J. (2005). *Logical Dilemmas: The Life and Work of Kurt Gödel.* Wellesley, MA: A K Peters.

Debs, G., and Saint Raymond, J. (1996). Compact covering and game determinacy. *Topology Applications,* **68**, 153–85.

———. (1999). Cofinal $\Sigma 11$ and $\Pi 11$ subsets of NN. *Fundamenta Mathematicae,* **159**, 161–93.

———. (2001). Compact covering mappings and cofinal families of compact subsets of a Borel set. *Fundamenta Mathematicae,* **167**, 213–49.

———. (2004). Compact covering mappings between Borel sets and the size of constructible reals. *Transactions of the American Mathematical Society,* **356**, 73–117.

———. (2007). *Borel Liftings of Borel Sets: Some Decidable and Undecidable Statements.* American Mathematical Society Memoirs.

Dimitracopoulus, C., and Gaifman, H. (1982). Fragments of Peano's arithmetic and the MRDP theorem. In *Logic and Algorithmics: An International Symposium Held in Honour of Ernst Specker,* pp. 317–29. L'enseignement Mathématique 30. Université de Genève.

Dragalin, A. (1980). New forms of realizability and Markov's rule (Russian). *Doklady,* **251**, 534–37. [Trans. *SM,* **21**, 461–64.]

Ehrenfeucht, A., and Mostowski, A. (1956). Models of axiomatic theories admitting automorphisms. *Fundamenta Mathematicae,* **43**, 50–68.

Ehrenfeucht, A., and Mycielski, J. (1971). Abbreviating proofs by adding new axioms. *Bulletin of the American Mathematical Society,* **77**, 366–67.

Feferman, S. (1960). Arithmetization of mathematics in a general setting. *Fundamenta Mathematicae,* **49**, 35–92.

———. (1964). Systems of predicative analysis. I. *Journal of Symbolic Logic,* **29**, 1–30.

———. (1968). Systems of predicative analysis. II. *Journal of Symbolic Logic,* **33**, 193–220.

———. (1998). *In the Light of Logic: Logic and Computation in Philosophy.* Oxford: Oxford University Press.

———. (2004). Tarski's conceptual analysis of semantical notions. In *Sémantique et Épistémologie,* ed. A. Benmakhlouf, pp. 79–108. Casablanca: Le Fennec.

Folina, J. (1992). *Poincaré and the Philosophy of Mathematics.* London: Macmillan.

Friedman, H. (1971). Higher set theory and mathematical practice. *Annals of Mathematical Logic,* **2**, 325–57.

———. (1973). The consistency of classical set theory relative to a set theory with intuitionistic logic. *Journal of Symbolic Logic,* **38**, 314–19.

———. (1978). Classical and intuitionistically provably recursive functions. In *Higher Set Theory,* ed. G. H. Müller and D. S. Scott, pp. 21–27. Lecture Notes in Mathematics 669. Berlin: Springer.

———. (1981). On the necessary use of abstract set theory. *Advances in Mathematics,* **41**, 209–80.

———. (1998). Finite functions and the necessary use of large cardinals. *Annals of Mathematics,* **148**, 803–93.

———. (1999). A complete theory of everything: Satisfiability in the universal domain. http://www.math.ohio-state.edu/%7Efriedman/.

———. (2001). Subtle cardinals and linear orderings. *Annals of Pure and Applied Logic,* **107**, 1–34.

———. (2002a). Internal finite tree embeddings. In *Reflections on the Foundations of Mathematics: Essays in Honor of Solomon Feferman,* ed. W. Sieg, R. Sommer, and C. Talcott, pp. 62–93. Lecture Notes in Logic 15. Providence, RI: Association for Symbolic Logic.

———. (2002b). Philosophical problems in logic. http://www.math.ohio-state.edu/%7Efriedman/.

———. (2003a). Primitive independence results. *Journal of Mathematical Logic*, **3**, 67–83.

———. (2003b). Three quantifier sentences. *Fundamenta Mathematicae*, **177**, 213–40.

———. (2005). Selection for Borel relations. In *Logic Colloquium '01, Lecture Notes in Logic*, vol. 20, ed. J. Krajicek, pp. 151–69. ASL.

———. (2006a). 271: Clarification of Smith article. Foundations of Mathematics Archives. http://www.cs.nyu.edu/pipermail/fom/2006-March/010244.html.

———. (2006b). 272: Sigma01/optimal. Foundations of Mathematics Archives. http://www.cs.nyu.edu/pipermail/fom/2006-March/010260.html.

———. (2006c). 273: Sigma01/optimal/size. Foundations of Mathematics Archives. http://www.cs.nyu.edu/pipermail/fom/2006-March/010279.html.

———. (2006d). n(3) < Graham's number < n(4) < TREE[3]. Foundations of Mathematics Archives. http://www.cs.nyu.edu/pipermail/fom/2006-March/010290.html.

———. (2006e). 274: Subcubic graph numbers. Foundations of Mathematics Archives. http://www.cs.nyu.edu/pipermail/fom/2006-April/010305.html.

———. (2006f). 279: Subcubic graph numbers/restated. Foundations of Mathematics Archives. http://www.cs.nyu.edu/pipermail/fom/2006-April/010362.html.

———. (2006g). Integer thresholds in FFF. Foundations of Mathematics Archives. http://www.cs.nyu.edu/pipermail/fom/2006-June/010627.html.

———. (2007a). Formal statements of Gödel's second incompleteness theorem (abstract). http://www.math.ohio-state.edu/%7Efriedman/manuscripts.html.

———. (2007b). New Borel independence results. http://www.math.ohio-state.edu/%7Efriedman/manuscripts.html.

———. (2009). Upper shift fixed points and large cardinals, upper shift fixed points and large cardinals/correction. http://www.cs.nyu.edu/pipermail/fom/2009-October/014119.html; http://www.cs.nyu.edu/pipermail/fom/2009-October/014123.html.

———. (2011). *Boolean Relation Theory and Incompleteness*. Lecture Notes in Logic, Association for Symbolic Logic, to appear. http://www.math.ohio-state.edu/%7Efriedman/manuscripts.html.

Friedman, H., Robertson, N., and Seymour, P. (1987). The metamathematics of the graph minor theorem. In *Logic and Combinatorics*, ed. S. Simpson, pp. 229–61. Contemporary Mathematics Series 65.

Friedman, H., Montalban, A., and Weiermann, A. (2011). Logical analysis of forms of Kruskal's theorem. Manuscript in preparation.

Gentzen, G. (1969). *The Collected Papers of Gerhard Gentzen*. Edited and translated by M. E. Szabo. Amsterdam, Netherlands: North-Holland.

Gödel, K. (1929). On the completeness of the calculus of logic. PhD diss. In *Collected Works*, vol. 1 (1986), pp. 61–101.

———. (1930). Die Vollständigkeit der Axiome des logischen Funktionenkalküls. *Monatshefte für Mathematik und Physik*, **37**, 349–60. [Published PhD diss. Also in *Collected Works*, vol. 1 (1986), pp. 102–23.]

———. (1931). Über formal unentscheidbare Sätze der *Principia Mathematica* und verwandter Systeme I. *Monatshefte für Mathematik und Physik*, **38**, 173–98. [English trans. J. van Heijenoort, ed. (1967). *From Frege to Gödel: A Source Book on Mathematical Logic*. Cambridge, MA: Harvard University Press, pp. 596–616. Repr. with facing English trans. On formally undecidable propositions of *Principia Mathematica* and Related Systems. I. In *Collected Works*, vol. 1 (1986), pp. 145–95.]

———. (1932a). On completeness and consistency. In *Collected Works*, vol. 1 (1986), pp. 235–37.

———. (1932b). On the intuitionistic propositional calculus. In *Collected Works*, vol. 1 (1986), pp. 223–25.

————. (1933a). An interpretation of the intuitionistic propositional calculus. In *Collected Works*, vol. 1 (1986), pp. 301–3.

————. (1933b). On intuitionistic arithmetic and number theory. In *Collected Works*, vol. 1 (1986), pp. 287–95.

————. (1934). On undecidable propositions of formal mathematical systems. In *Collected Works*, vol. 1 (1986), pp. 346–71.

————. (1936). On the lengths of proofs. In *Collected Works*, vol. 1 (1986), pp. 397–99.

————. (1938). The consistency of the axiom of choice and of the generalized continuum-hypothesis. *Proceedings of the National Academy of Sciences of the United States of America*, **24**, 556–57. [Also in *Collected Works*, vol. 2 (1990), pp. 26–27.]

————. (1939a). The consistency of the generalized continuum hypothesis. In *Collected Works*, vol. 2 (1990), p. 27.

————. (1939b). Consistency-proof for the generalized continuum hypothesis. *Proceedings of the National Academy of Sciences of the United States of America*, **25**, 220–24. [Also in *Collected Works*, vol. 2 (1990), pp. 27–32.]

————. (1940). *The Consistency of the Axiom of Choice and of the Generalized Continuum-Hypothesis with the Axioms of Set Theory*. Annals of Mathematics Studies 3. Princeton, NJ: Princeton University Press. [Rev. ed. 1953. Also in *Collected Works*, vol. 2 (1990), pp. 33–101.]

————. (1944). Russell's mathematical logic. In *Collected Works*, vol. 2 (1990), pp. 119–41.

————. (1946). Remarks before the Princeton Bicentennial Conference on problems in mathematics. [First published M. Davis, ed. (1965). *The Undecidable: Basic Papers on Undecidable Propositions, Unsolvable Problems, and Computable Functions*. Hewlett, NY: Raven Press. Also in *Collected Works*, vol. 2 (1990), pp. 150–53.]

————. (1958). Über eine bisher noch nicht benüzte Erweiterung des finiten Standpunktes. *Dialectica*, **12**, 280–87. [Repr. in English trans. On a hitherto unutilized extension of the finitary standpoint. In *Collected Works*, vol. 2 (1990), pp. 241–51.]

————. (1964 [1947]). What is Cantor's continuum problem? *American Mathematical Monthly*, 54, 515–25. [Rev. version P. Benacerraf and H. Putnam (1984). *Philosophy of Mathematics*. Englewood Cliffs, NJ: Prentice Hall, p. 483. Also in *Collected Works*, vol. 2 (1990), pp. 176–87 (1947 version); pp. 254–70 (1964 version).]

————. (1972). On an extension of finitary mathematics which has not yet been used. In *Collected Works*, vol. 2 (1990), pp. 271–80.

————. (1986). *Collected Works*. Vol. 1. *Publications 1929–1936*. New York: Oxford University Press.

————. (1990). *Collected Works*. Vol. 2. *Publications 1938–1974*. New York: Oxford University Press.

————. (1995). *Collected Works*. Vol. 3. *Unpublished Essays and Lectures*. New York: Oxford University Press.

————. (2003a). *Collected Works*. Vol. 4. *Correspondence A–G*. Oxford: Clarendon Press.

————. (2003b). *Collected Works*. Vol. 5. *Correspondence H–Z*. Oxford: Clarendon Press.

Gogol, D. (1979). Sentences with three quantifiers are decidable in set theory. *Fundamenta Mathematicae*, **CII**, 1–8.

Goodstein, R. L. (1944). On the restricted ordinal theorem. *Journal of Symbolic Logic*, **9**, 33–41.

Grunewald, F. J., and Segal, D. (1981). How to solve a quadratic equation in integers. *Mathematical Proceedings of the Cambridge Philosophical Society*, **89**, 1–5.

Hajek, P., and Pudlak, P. (1993). *Metamathematics of First-Order Arithmetic*. Perspectives in Mathematical Logic. Berlin: Springer.

Hilbert, H. (1902). Mathematical problems. Lecture delivered before the International Congress of Mathematicians at Paris in 1900. *Bulletin of the American Mathematical Society*, **8**, 437–79.

Hilbert, D., and Bernays, P. (1934). *Grundlagen der Mathematik*. Vol. 1. Berlin: Springer. [Written by Bernays alone; 2nd ed. 1968.]

———. (1939). *Grundlagen der Mathematik*. Vol. 2. Berlin: Springer. [Written by Bernays alone; 2nd ed. 1970.]

Hosoi, T., and Ono, H. (1973). Intermediate propositional logics (a survey). *Journal of Tsuda College*, **5**, 67–82.

Jech, T. (2006). *Set Theory*. 3rd ed. Springer Monographs in Mathematics. Berlin: Springer.

Jerosolow, R. G. (1973). Redundancies in the Hilbert-Bernays derivability conditions for Gödel's second incompleteness theorem. *Journal of Symbolic Logic*, **38**, 359–67.

Jones, J. P. (1982). Universal Diophantine equation. *Journal of Symbolic Logic*, **47**, 549–71.

Kechris, A. S. (1994). *Classical Descriptive Set Theory*. Graduate Texts in Mathematics. Berlin: Springer.

Kohlenbach, U. (2005). Some logical metatheorems with applications in functional analysis. *Transactions of the American Mathematical Society*, **357**, 89–128.

———. (2008). Effective bounds from proofs in abstract functional analysis. In *New Computational Paradigms: Changing Conceptions of What Is Computable*, ed. B. Cooper, B. Loewe, and A. Sorbi, pp. 223–58. Berlin: Springer.

Kohlenbach, U., and Gerhardy, P. (2008). General logical metatheorems for functional analysis. *Transactions of the American Mathematical Society*, **360**, 2615–60.

Kohlenbach, U., and Oliva, P. (2003). Proof mining: A systematic way of analyzing proofs in mathematics. *Proceedings of the Steklov Institute of Mathematics*, **242**, 136–64.

Kreisel, G. (1968a). Functions, ordinals, species. In *Logic, Methodology, and Philosophy of Sciences*, vol. 3, ed. B. van Rootselaar and J. F. Staal, pp. 143–58. Amsterdam, Netherlands: North-Holland.

———. (1968b). A survey of proof theory. *Journal of Symbolic Logic*, **33**, 321–88.

Kruskal, J. B. (1960). Well-quasi-ordering, the tree theorem, and Vazsonyi's conjecture. *Transactions of the American Mathematical Society*, **95**, 210–25.

Leivant, D. (1985). Syntactic translations and provably recursive functions. *Journal of Symbolic Logic*, **50**, 682–88.

Maes, K. (2007). A 5-quantifier $(\in, =)$-expression ZF-equivalent to the axiom of choice. Preprint.

Martin, D. A. (1975). Borel determinacy. *Annals of Mathematics*, **102**, 363–71.

———. (1985). A purely inductive proof of Borel determinacy. *Recursion Theory, Proceedings of Symposia in Pure Mathematics*, **42**, 303–8.

Martin, D. A., and Steel, J. R. (1989). A proof of projective determinacy. *Journal of the American Mathematical Society*, **2**, 71–125.

Masser, D. W. (1998). How to solve a quadratic equation in rationals. *Bulletin of the London Mathematical Society*, **30**, 24–28.

Matiyasevich, Y. (1993). *Hilbert's Tenth Problem*. Cambridge, MA: MIT Press.

McLarty, Colin (2010). What does it take to prove Femat's Last Theorem? Grothendieck and the logic of number theory. *The Bulletin of Symbolic Logic*, **16**(1), 359–377.

Minari, P. (1983). Intermediate logics: A historical outline and a guided bibliography. *Rapporto matematico*, **79**, 1–71.

Mostowski, A. (1939). Über die Unabhängigkeit des Wohlordnungssatzes vom Ordnungsprinzip. *Fundamenta Mathematicae*, **32**, 201–52.

———. (1952). *Sentences Undecidable in Formalized Arithmetic: An Exposition of the Theory of Kurt Gödel*. Amsterdam, Netherlands: North-Holland.

Myhill, J. (1974). Embedding classical type theory in "intuitionistic" type theory: A correction. In *Axiomatic Set Theory, Proceeding of Symposia in Pure Mathematics*, vol. 13, part 2, ed. T. Jech, pp. 185–88. Providence, RI: American Mathematical Society.

Nash-Williams, C. S. J. A. (1963). On well-quasi-ordering finite trees. *Proceedings of the Cambridge Philosophical Society*, **59**, 833–35.

Parikh, R. (1971). Existence and feasibility in arithmetic. *Journal of Symbolic Logic*, **36**, 494–508.

———. (1973). Some results on the lengths of proofs. *Transactions of the American Mathematical Society*, **177**, 29–36.

Paris, J., and Harrington, L. (1977). A mathematical incompleteness in Peano arithmetic. In *Handbook of Mathematical Logic*, ed. J. Barwise, pp. 1133–42. Amsterdam, Netherlands: North-Holland.

Pudlak, P. (1985). Improved bounds on the lengths of proofs of finitistic consistency statements. In *Contemporary Mathematics, Logic and Combinatorics*, **65**, 309–32.

Quine, W. V. O. (1951). *Mathematical Logic*. Rev. ed. Cambridge, MA: Harvard University Press. [1st ed. 1940.]

Rabin, M. (1977). Decidable theories. In *Handbook of Mathematical Logic*, ed. J. Barwise, pp. 595–629. Amsterdam, Netherlands: North-Holland.

Rathjen, M., and Weiermann, A. (1993). Proof-theoretic investigations on Kruskal's theorem. *Annals of Pure and Applied Logic*, **60**, 49–88.

Rayo, A., and Wiliamson, T. (2003). A completeness theorem for unrestricted first-order languages. In *Liars and Heaps*, ed. J. C. Beall, pp. 331–56. Oxford: Clarendon Press.

Renegar, J. (1992a). On the computational complexity and geometry of the first-order theory of the reals, Part I: Introduction. Preliminaries. The geometry of semi-algebraic sets. The decision problem for the existential theory of the reals. *Journal of Symbolic Computation*, **13**, 255–300.

———. (1992b). On the computational complexity and geometry of the first-order theory of the reals, Part II: The general decision problem. Preliminaries for quantifier elimination. *Journal of Symbolic Computation*, **13**, 301–28.

———. (1992c). On the computational complexity and geometry of the first-order theory of the reals, Part III: Quantifier elimination. *Journal of Symbolic Computation*, **13**, 329–52.

Robertson, N., and Seymour, P. D. (1985). Graph minors – a survey. In *Surveys in Combinatorics 1985*, pp. 153–71. London Mathematical Society Lecture Note Series 103. Cambridge: Cambridge University Press.

———. (2004). Graph minors. XX. Wagner's conjecture. *Journal of Combinatorial Theory, Series B*, **92**, 325–57.

Robinson, R. M. (1952). An essentially undecidable axiom system. In *Proceedings of the International Congress of Mathematicians*, pp. 729–30. Providence, RI: American Mathematical Society.

Rosser, J. B. (1936). Extensions of some theorems of Gödel and Church. *Journal of Symbolic Logic*, **1**, 87–91.

Schütte, K. (1977). Proof theory. *Grundlehren der mathematischen Wissenschaften*, XX, 225.

Scott, D. S. (1961). Measurable cardinals and constructible sets. *Bulletin de l'Academie Polonaise des Sciences*, **9**, 521–24.

Shelah, S. (1969–2007). Shelah's Archive. http://shelah.logic.at/.

Siegel, C. L. (1972). Zur Theorie der quadratischen Formen. Nachricten der Akademie der Wissenschafen. *Göttingen. II. Mathematisch-Physikalische Klasse*, **3**, 21–46.

Simpson, S. G. (1999). *Subsystems of Second Order Arithmetic*. Perspectives in Mathematical Logic. Berlin: Springer.

Skolem, T. (1922). Some remarks on axiomatized set theory. In *From Frege to Gödel: A Source Book in Mathematical Logic, 1879–1931*, ed. J. van Heijenoort, pp. 290–301. Cambridge, MA: Harvard University Press. [Cited as "Skolem 1923a" in Gödel's letter to Wang; see Endnote 2.]

Smith, R. L. (1985). The consistency strengths of some finite forms of the Higman and Kruskal theorems. In *Harvey Friedman's Research on the Foundations of Mathematics*, ed. L. A. Harrington et al., pp. 119–59. New York: Elsevier.

Solovay, R. M. (1970). A model of set theory in which every set of reals is Lebesgue measurable. *Annals of Mathematics*, **92**, 1–56.

Spector, C. (1962). Provably recursive functions of analysis: A consistency proof of analysis by an extension of principles formulated in current intuitionistic mathematics. In *Recursive Function*

Theory, Proceedings of Symposia in Pure Mathematics, vol. 5, ed. J. C. E. Dekker, pp. XX–XX. Providence, RI: American Mathematical Society.

Stanley, L. J. (1985). Borel diagonalization and abstract set theory: Recent results of Harvey Friedman. In *Harvey Friedman's Research on the Foundations of Mathematics*, ed. L. A. Harrington et al., pp. 11–86. New York: Elsevier.

Tarski, A. (1952). Some notions and methods on the borderline of algebra and metamathematics. In *Proceedings of the International Congress of Mathematicians*, vol. 1, pp. 705–20. Providence, RI: American Mathematical Society.

Tarski, A., and McKinsey, J. C. C. (1951). *A Decision Method for Elementary Algebra and Geometry*. 2nd rev. ed. Berkeley: University of California Press.

Tarski, A., and Vaught, R. L. (1957). Arithmetical extensions of relational systems. *Compositio Mathematica*, **13**, 81–102.

Tarski, A., Mostowski, A., and Robinson, R. M. (1953). *Undecidable Theories*. Amsterdam, Netherlands: North-Holland.

Wang, H. (1987). *Reflections on Kurt Gödel*. Cambridge, MA: MIT Press.

———. (1996). *A Logical Journey: From Gödel to Philosophy*. Cambridge, MA: MIT Press.

Weyl, H. (1918). *Das Kontinuum. Kritische Untersuchungen über die Grundlagen der Analysis*. Leipzig, Germany: Velt.

———. (1987). *The Continuum: A Critical Examination of the Foundations of Analysis*. Translated by S. Pollard and T. Bole. Philadelphia: Thomas Jefferson University Press.

Wiedijk, F., ed. (2006). *The Seventeen Provers of the World*. Foreword by Dana S. Scott. New York: Springer.

Williamson, T. (2000). Existence and contingency. *Proceedings of the Aristotelian Society*, **100**, 117–39.

———. (2003). Everything. *Philosophical Perspectives*, **17**, 415–65.

———. (2006). Absolute identity and absolute generality. In *Absolute Generality*, ed. A. Rayo and G. Uzquiano, pp. 369–89. Oxford: Oxford University Press.

The Realm of Set Theory

CHAPTER 19

My Interaction with Kurt Gödel: The Man and His Work

Paul J. Cohen

Editors' Note

Sadly, Paul Cohen, the only Fields Medal Laureate in the field of logic as of this writing, died on March 23, 2007, shortly after completing this chapter. This is his final published work, which was also reprinted in the revised edition of his book *Set Theory and the Continuum Hypothesis* (Dover, 2008; originally published by W. A. Benjamin, 1966).

On the centenary of Kurt Gödel's birth in 2006, it was most appropriate to celebrate the man who, more than any other, guided logic out of its philosophical past to become a vibrant part of present-day mathematics. His contributions are so well known, and the recognition he has received is so plentiful, that to recite them here would be extraneous. Rather I have chosen to relate how first his work, and then my interaction with him, affected me so strongly.

Let me begin by explaining, and perhaps apologizing for, the somewhat personal tone of my remarks. Since the publication of Gödel's (1986, 1990, 1995, 2003a, 2003b) *Collected Works*, with their very rich introductions to the various articles, as well as historical material, I can add little of a purely biographical nature to Gödel's mathematical contributions. Thus I shall have to speak mostly about what I know best: my own development and interest in set theory, how the example of Gödel's life work deeply affected me, the story of my own discoveries, and finally, my personal interaction with Gödel when I presented my work to him.

19.1 Background

It was my great fortune and privilege to be the person who fulfilled the expectations of Gödel in showing that the continuum hypothesis (CH), in addition to other questions in set theory, is independent of the usual set theory axioms. Gödel foresaw this possibility in his article "What Is Cantor's Continuum Problem?" (Gödel, 1964 [1947]). Reading it greatly inspired me to work on the CH problem. Gödel's point of departure in discussing CH questions always seemed to be grounded in philosophical discussions and in the work of previous researchers. In my case, I felt it was necessary to start afresh and to treat the CH as an ordinary problem of mathematics, perhaps similarly to how Cantor, and later Hilbert, regarded it.

Here I would like to sketch briefly some of the evolution of my own ideas about the CH and their relation to Gödel's work as well as his published opinions about my eventual resolution of the problem. None of my work, however, would have been possible without the momentous discovery of the constructible universe, which Gödel achieved around 1938. His approach in discovering the notion of a constructible set seems to have been decidedly influenced by the philosophical discussions of the past, especially the notion of impredicativity, which was emphasized by Poincaré and others. In this sense, I found the tone of his article difficult to understand as it seemed to hover between philosophy and mathematics.

When I began my own attempts at solving the CH problem in 1962, I was strongly motivated by the idea of constructing a model. In this sense, my spiritual ancestor was perhaps Thoralf Skolem more so than Gödel. Of course, I eventually realized that Gödel's syntactical approach, where a constructible set could satisfy a certain property and constructible sets could be collected to form a model, was not very different. After my work was completed, I realized that a prejudice against models was widespread among logicians and even led to serious doubts about the correctness of my work. Nevertheless, as I was trying to construct interesting new models of set theory, I had to start with a given model of it. Thus it seemed most natural to assume that this model was a collection of actual sets, in which the membership relation has its usual meaning. Of course, the Löwenheim-Skolem theorem guarantees the existence of such a model, even a countable one, but this cannot be proved within Zermelo-Fraenkel set theory (ZF). Even so, its truth would intuitively be accepted by everyone, even though, technically, it goes beyond ZF. On the other hand, Skolem specifically mentioned the problem of constructing more interesting models of set theory. In a remarkably prescient remark, he said that adjoining new sets of integers would be difficult but probably of great interest. Although Skolem's remark was not known to me when I started my work, it did seem like a natural way to proceed. Thus, to violate the CH in this way, one would have to adjoin $\aleph_2$ new sets of integers. This seemed an almost obvious way to proceed, so rather soon, I found myself thinking in a manner similar to Skolem.

At first, I had not actually understood Gödel's work on the CH, finding the exposition of his monograph rather difficult. However, as I thought about it more deeply, I realized that all my attempts were intimately allied to his methods, and I eventually mastered them completely. Suddenly, I had a tool with which I could make my own somewhat vague ideas more precise. Thanks to Gödel's discovery, I saw that the new universes of set theory I wished to construct consisted of sets that were constructible from a limited number of new sets, which I would introduce in a very particular way.

I would say that had the idea of constructing models by introducing new sets been established earlier, it would have been a natural development for Gödel to have achieved my own work. That my methods are a kind of synthesis of the work of Gödel and Skolem has not been previously pointed out by authors, but I think it is fitting to honor the intuitions, fulfilled many years later, of these two pioneers.

A third theme of my work, namely, the analysis of so-called truth, is made precise in the definition of *forcing*. To people who read my work for the first time, this may very well strike them as the essential ingredient, rather than the two aspects of constructibility and the introduction of new sets, as mentioned earlier. Although I emphasized the model approach in my earlier remarks, this point of view is more akin to the syntactical approach that I associate with Gödel. At least as I first presented it, the approach depends heavily on an analysis and ordering of statements and, by implication, on an analysis of proof.

Initially, I was not very comfortable with this approach; it seemed to be too close to philosophical discussions, which I felt would not ultimately be fruitful. In a sense, my enormous admiration for Gödel, as well as the fact that so much of his work was based on this kind of philosophical analysis, was unconsciously a driving factor in my persevering in this syntactical analysis. Of course, in the final form, it is very difficult to separate what is model-theoretic from what is syntactical. As I struggled to make these ideas precise, I vacillated between two approaches: the model-theoretic, which I regarded as roughly more mathematical, and the syntactical-forcing, which I thought of as more philosophical. In this way, the body of Gödel's work, rather than particular results, helped me to believe that something could come from a syntactical analysis. I would say that my notion of forcing can also be viewed as arising from the analysis of how models are constructed, something in the spirit of the Löwenheim-Skolem theorem.

If I were to comment on previous work that might have led to an earlier discovery of the independence results, following would be some of the salient points:

1. We have Skolem's emphasis on models. It seems that Gödel hardly discussed models of set theory. In particular, John Shepherdson's result about the impossibility of inner models to solve the problem should have received more attention (Shepherdson, 1951, 1952, 1953). Now, Gödel wrote no textbook on set theory, so we can only speculate as to what he would have had to say about models in general. His emphasis, as mentioned earlier, was syntactical, and thus he was concerned with consistency. That a standard model for set theory (i.e., Skolem's paradox) was a new axiom received scant attention. Of course, it is stronger than mere consistency.

2. It is easy to show that assuming constructibility, no uncountable model that violates the CH can exist. Had this been realized, a key step in forcing – namely, the use of countable models – might have led to more rapid development of a solution.

3. In the spirit of Skolem, the addition of new sets was discussed briefly by András Hajnal (1956) and Azriel Levy (1960). I do not know whether this work intrigued Gödel to any extent.

My perhaps too dogmatic conclusion is that Gödel had no new method for a proof of independence, as he had told some people. Yet I cannot overestimate how the body of his work – his belief that a precise analysis of the axioms and consistency results was possible – played a crucial role in allowing me to go forward. As I shall now describe, these attitudes were reflected in my personal talks with Gödel after my work was completed.

19.2 Meeting Gödel

Soon after proving my first results, I had a great desire to meet Gödel and to do him and myself the great honor of personally explaining the proofs to him. I met him at Princeton with a preprint that gave all the essential details. He very graciously agreed to read it. After just a few days, he pronounced my proof correct. We had a series of intense discussions, and I asked whether he would communicate the paper to the *Proceedings of the National Academy of Sciences of the United States of America*, where his own results were first announced (Gödel, 1938; see also Gödel, 1939a, 1939b). He kindly consented to do this, and we had a lively correspondence about the best way to present it. He suggested several revisions. With this acknowledgment, and with his later overly kind remarks about my discovery, I felt that we shared a great bond in that we had successfully discovered, each in his own way, new fundamental methods in set theory.

I visited Princeton again for several months and had many meetings with Gödel. I brought up the question of whether, as rumor had it, he had proved the independence of the axiom of choice. He replied that he had, evidently by a method related to my own, but he gave me no precise idea or explanation of why his method evidently failed to succeed with the CH. His main interest seemed to lie in discussing the truth or falsity of these questions, not merely their undecidability. He struck me as having an almost unshakable belief in this realist position that I found difficult to share. His ideas were grounded in a deep philosophical belief as to what the human mind could achieve. I greatly admired this faith in the power and beauty of Western culture, as he put it, and would have liked to understand more deeply the sources of his strongly held beliefs. Through our discussions, I came closer to his point of view, although I never shared completely his realist perspective that all questions of set theory were, in the final analysis, either true or false.

Let me turn now to a brief review of my intellectual development in mathematics.

19.3 High School: My Earliest Interest in Logic

In high school, number theory attracted me the most in mathematics, probably because of the simplicity of the statements of the results and the complexity and ingenuity of the proofs. I looked at the famous work of Edmund Landau (1927), *Vorlesungen über Zahlentheorie* [*Lectures on Number Theory*]. The most complicated proof in the book was the famous result of Carl Siegel, generalizing the work of Axel Thue on the approximation of algebraic numbers and applications to Diophantine equations. This proof is one of the rare, truly nonconstructive proofs in mathematics, and I was rather disappointed by this. My basic instinct was to prefer constructive proofs, of course, and so I began to think of exactly what I thought a constructive proof should be.

Thus I was clumsily reconstructing elementary recursion theory, primitive recursive functions, and so on. Simultaneously, results in partitions of integers led me to think how the results about sums of squares found by Jakob Jacobi and others could be generalized. As the early methods used infinite power series, I wondered whether – because we could answer all questions about polynomials by finite constructive methods – it would be possible to find (what I later learned are called) decision procedures for a larger class of problems.

At this time, I did not at all know Gödel's incompleteness theorem, which made this dream impossible. Eventually, I became dimly aware of Gödel's result, although not the precise statement. I felt it was probably primarily a theorem of philosophical interest, having little to do with the concrete questions of number theory that so infatuated me in those years.

19.4 Graduate School: The Continued Pull of Logic

When I started graduate school, for the first time, I met students who had studied logic. I was told that the search for a decision procedure was probably doomed – although I did not clearly say exactly which class of problems concerning infinite power series most interested me. As fate would have it, Professor Stephen Kleene visited the University of Chicago, where I was enrolled, and he questioned me a bit about my "dream." In no uncertain terms, he told me that what I hoped to accomplish was impossible. Therefore I started to read his book on metamathematics (Kleene, 1952). There, in just a few pages, was a complete sketch of the incompleteness theorem. The rest of the book held little interest for me, and I was confused about many basic ideas. I still had a feeling of skepticism about Gödel's work, but it was skepticism mixed with awe and admiration.

I can say my feeling was roughly this: how can someone thinking about logic in almost philosophical terms discover a result that had implications for Diophantine equations? How very different it was from Landau's book. For a brief period, I believed that the famous liar's paradox, which is almost at the very heart of Gödel's reasoning, itself entailed a paradox – and even that his theorem, as I wanted to apply it, was false. However, after a few days, I closed the book and tried to rediscover the proof, which I still feel is the best way to understand things. I totally capitulated. The incompleteness theorem was true, and Gödel was far superior to me in understanding the nature of mathematics.

Although the proof was basically simple, when stripped to its essentials, I felt that its discoverer was above me and other mere mortals in his ability to understand what mathematics – and even human thought, for that matter – really was. From that moment on, my regard for Gödel was so high that I almost felt it would be beyond my wildest dreams to meet him and discover for myself how he thought about mathematics and the fount from which his deep intuition flowed. I could imagine myself as a clever mathematician solving difficult problems, but how could I emulate a result of the magnitude of the incompleteness theorem? There it stood, in splendid isolation and majesty, not allowing any kind of completion or addition because it answered the basic questions with such finality. I returned to the study of pure mathematics, having tasted a bit of the depth of logic. However, the rest of Kleene's book seemed rather philosophical, and by temperament, even to the present day, I could not really be involved in philosophical controversy. So much seemed to depend on personalities and how one was able to express oneself with wit and elegance. As fate would have it, I was wrong in several regards.

Part of the standard curriculum at the university was a not-very-technical introduction to set theory, mostly the axiom of choice, facts about cardinality, and so on. I saw these results as interesting but merely as pieces of clever reasoning, inferior in that regard to the great body of mathematics that was attracting me – analysis, algebra, and so on. As I have said, a career in logic seemed not very appealing because, in some sense, Gödel had preempted the entire field.

Thus logic seemed to me to have little of the element of combinatorial thinking – or, more crudely put, cleverness – that I found in analysis and number theory. There was one exception, however. A small group of students were very interested in Emil Post's problem about maximal degree of unsolvability. I did dally with the thought of working on it, but in the end, I did not. Suddenly, one day, a letter arrived containing a sketch of the solution by Richard Friedberg (1957), and it was brought to my office. Amid a certain degree of skepticism, I checked the proof and could find nothing wrong. It was exactly the kind of thing I would like to have done. I mentally resolved that I would not let an opportunity like that pass again. However, there seemed no way for me to find an entry into the kind of deep philosophical thinking that I so admired in Gödel and translate that into concrete mathematics.

19.5 The Beginnings of My Research Career

I wrote a thesis on harmonic analysis, on topics similar to those that led Georg Cantor to his discovery of set theory. Cantor's set theory impetus was well known to me, but I saw no real connection with logic, at least not enough for me to begin to think about set theory as a possible research field. Thus I essentially floated around the periphery of logic. Kleene's book contained essentially no mention of Gödel's result on the consistency of the CH, and I still had never actually seen a precise statement of the axioms of Zermelo and Fraenkel. I was, however, aware of Skolem's paradox concerning the existence of countable models of set theory. I was able to sketch a proof for myself, which seemed deep but not difficult. I might also mention that one of my office mates was Michael Morley, who wrote a very good thesis in model theory, but I knew nothing about it.

I then spent two years in Princeton at the Institute for Advanced Study and saw Gödel occasionally in the common room, but I had no contact with him. Princeton was an awe-inspiring place because of the famous people there, and Gödel was certainly one of them.

19.6 The Quest for Consistency

I then went to Stanford. At a departmental lunch, the conversation turned to the consistency of mathematics, mostly revolving around the ideas of Solomon Feferman (who is represented in this volume). At that point, something jelled, and I thought I saw a way to prove the consistency of mathematics by some kind of construction. I gave a few lectures to a small audience, but I became discouraged in thinking that

I was engaged in a futile enterprise – Gödel had shown that a consistency proof along the lines I had envisaged was impossible.

At some point, my attention turned to set theory and the CH. In Gödel's Princeton monograph, I read the precise statement of the axioms for the first time (Gödel, 1940). Essentially, the only weapon in my armory was the Löwenheim-Skolem theorem, but I was convinced that the key question was constructing models of set theory. The Skolem construction offered little help. I did not see at first how it could be adapted to controlling the truth or falsity of given statements in the new model. It seemed that I had no real tools with which to work – set theory was too big to handle, and Gödel's syntactical approach seemed more appealing for a while.

I read Gödel's article on Cantor's continuum problem (Gödel, 1964 [1947]) before I had even considered attacking the problem of the CH. It was just of general interest for me, and to a large extent, it was not even comprehensible to me. For one thing, it assumed a certain philosophical point of view and the philosophical ramifications of various attitudes toward it. I had never been attracted to philosophy, as I have explained, and thus I was not really aware of the various attitudes the mathematicians of earlier years had held.

Having said this, it is necessary to add that, of course, I had intuitive ideas of what constituted a correct proof in mathematics. In particular, in number theory, I had begun to think about certain applications in which nonconstructive methods were used and was quite aware that these methods raised certain problems. Of course, these were very far from the fundamental questions of set theory.

Did Gödel have unpublished methods for the CH? This is a tantalizing question. Let me state some incontrovertible facts. First, much effort was spent analyzing Gödel's notes and papers, and no idea has emerged about what kinds of methods he might have used. Second, I did ask him point-blank whether he had proved the independence of CH, and he said no, but that he had had success with the axiom of choice. I asked him what his methods were, and he said only that they resembled my own; he seemed extremely reluctant to give any further information.

My conclusion is that Gödel did not complete any serious work on this topic that he thought was correct. In our discussions, the word *model* almost never occurred. Therefore I assume that he was looking for a syntactical analysis that was in the spirit of his definition of constructibility. His total lack of interest in a model-theoretic approach quite astounded me. Thus, when I mentioned to him my discovery of the minimal model also found by John Shepherdson, he indicated that this was clear and, indirectly, that he knew of it. However, he did not mention the implication that no purely inner model could be found. Given that I also believe he was strongly wedded to the syntactical approach, this would have been of great interest. My conclusion, perhaps uncharitable, is that he totally ignored questions of models and was perhaps only subconsciously aware of the minimal model.

The history of Gödel's involvement with the independence results, to be sure, has a somewhat sad conclusion. I think, without concrete evidence, of course, that it became interwoven in his mind with larger questions of the reality of set theory. His unshakable belief that all questions of set theory are decidable in some sense led him to think about the independence question in a somewhat confused way and eventually led to a belief that he had discovered a method that would resolve the issue. Otherwise, I find no plausible reason why he would so adamantly refuse to discuss it with me. Of course, his general mental health was deteriorating in the last years, which no doubt was a contributing factor. It is not clear from biographical remarks of people who knew him at the time whether his breakdown in the early 1940s was a result of his frustration at not being able to solve the independence of CH.

Clearly Gödel's personal life was almost tragic in this period, and as far as the history of mathematics is concerned, I think no new information will arise.

References

Friedberg, R. M. (1957). Two recursively enumerable sets of incomparable degrees of unsolvability (solution of Post's problem). *Proceedings of the National Academy of Sciences of the United States of America*, **43**, 236–38.

Gödel, K. (1938). The consistency of the axiom of choice and of the generalized continuum-hypothesis. *Proceedings of the National Academy of Sciences of the United States of America,* **24**, 556–57. [Also in *Collected Works*, vol. 2 (1990), pp. 26–27.]

———. (1939a). The consistency of the generalized continuum hypothesis. In *Collected Works*, vol. 2 (1990), p. 27.

———. (1939b). Consistency-proof for the generalized continuum hypothesis. *Proceedings of the National Academy of Sciences of the United States of America*, **25**, 220–24. [Also in *Collected Works*, vol. 2 (1990), pp. 27–32.]

———. (1940). *The Consistency of the Axiom of Choice and of the Generalized Continuum-Hypothesis with the Axioms of Set Theory.* Annals of Mathematics Studies 3. Princeton, NJ: Princeton University Press. [Rev. ed. 1953. Also in *Collected Works*, vol. 2 (1990), pp. 33–101.]

———. (1964 [1947]). What is Cantor's continuum problem? *American Mathematical Monthly*, 54, 515–25. [Rev. version P. Benacerraf and H. Putnam, eds. (1984 [1964]). *Philosophy of Mathematics.* Englewood Cliffs, NJ: Prentice Hall, p. 483. Also in *Collected Works*, vol. 2 (1990), pp. 176–87 (1947 version); pp. 254–70 (1964 version).]

———. (1986). *Collected Works.* Vol. 1. *Publications 1929–1936.* New York: Oxford University Press.

———. (1990). *Collected Works.* Vol. 2. *Publications 1938–1974.* New York: Oxford University Press.

———. (1995). *Collected Works.* Vol. 3. *Unpublished Essays and Lectures.* New York: Oxford University Press.

———. (2003a). *Collected Works.* Vol. 4. *Correspondence A–G.* Oxford: Clarendon Press.

———. (2003b). *Collected Works.* Vol. 5. *Correspondence H–Z.* Oxford: Clarendon Press.

Hajnal, A. (1956). On a consistency theorem connected with the generalized continuum problem. *Zeitschrift für mathematische Logik und Grundlagen der Mathematik*, **2**, 131–36.

Kleene, S. C. (1952). *Introduction to Metamathematics.* Amsterdam, Netherlands: North-Holland.

Landau, E. (1927). *Vorlesungen über Zahlentheorie [Lectures on Number Theory].* Leipzig, Germany: Hirzel.

Levy, A. (1960). A generalization of Gödel's notion of constructibility. *Journal of Symbolic Logic,* **25**, 147–55.

Shepherdson, J. C. (1951). Inner models of set theory. *Journal of Symbolic Logic*, **16**, 161–90.

———. (1952). Inner models of set theory. Part II. *Journal of Symbolic Logic*, **17**, 225–37.

———. (1953). Inner models of set theory. Part III. *Journal of Symbolic Logic*, **18**, 146–67.

Additional Reading

The following bibliography was added after Paul Cohen's death to reflect sources of information that may be helpful to readers.

The following two references contain additional information on the discovery and reception of forcing as well as further information on the author's intellectual development obtained from personal interviews by the writers; the second reference incorporates some material from the first.

Moore, C. H. (1988). The origins of forcing. In *Logic Colloquium '86*, ed. F. R. Drake and J. K. Truss, pp. 143–73. Amsterdam, Netherlands: North-Holland.

Yandell, B. H. (2001). *The Honors Class: Hilbert's Problems and Their Solvers*. Wellesley, MA: A K Peters.

The following references contain details about works relevant to this essay.

Church, A. (1968). Paul J. Cohen and the continuum problem. In *Proceedings of the International Congress of Mathematicians*, pp. 15–20. Moscow: International Congress of Mathematicians.

Cohen, P. J. (1963a). The independence of the continuum hypothesis. I. *Proceedings of the National Academy of Sciences of the United States of America*, **50**, 1143–48.

———. (1963b). A minimal mode for set theory. *Bulletin of the American Mathematical Society*, **69**, 537–40.

———. (1964). The independence of the continuum hypothesis. *II. Proceedings of the National Academy of Sciences of the United States of America*, **51**, 105–10.

———. (1965). Independence results in set theory. In *The Theory of Models: Proceedings of the 1963 International Symposium at Berkeley*, pp. 39–54. Amsterdam, Netherlands: North-Holland.

———. (1966). *Set Theory and the Continuum Hypothesis*. New York: W. A. Benjamin. [Repr. 2008 Mineola, NY: Dover.]

Davis, M. (1965). *The Undecidable: Basic Papers on Undecidable Propositions, Unsolvable Problems, and Computable Functions*. Hewlett, NY: Raven Press.

Dawson, J. W., Jr. (1997). New light on the continuum problem. In *Logical Dilemmas: The Life and Work of Kurt Gödel*, pp. 215–28. Wellesley, MA: A K Peters.

Feferman, S. (1965). Some applications of the notions of forcing and generic sets. *Fundamenta Mathematicae*, **56**, 325–45.

Feferman, S., and Levy, A. (1963). Independence results in set theory by Cohen's method. II (abstract). *Notices of the American Mathematical Society*, **10**, 593.

Fraenkel, A. A. (1919). *Einleitung in die Mengenlehre*. Berlin: Springer.

———. (1922a). Zu den Grundlagen der Cantor-Zermeloschen Mengenlehre. *Mathematische Annalen*, **86**, 230–37.

———. (1922b). Der Begriff "definit" und die Unabhängigkeit des Auswahl-axioms. In *Sitzungsberichte der Preussischen Akademie der Wissenschaften, Physikalisch-mathematische Klasse* [*The Concept "Definite" and the Independence of the Auswahl Axioms*], pp. 253–57. [English trans. J. van Heijenoort, ed. (1967). *From Frege to Gödel: A Source Book in Mathematical Logic, 1879–1931*. Cambridge, MA: Harvard University Press, pp. 284–89.]

———. (1922c). Zu den Grundlagen der Mengenlehre. *Jahresbericht der Deutschen Mathematiker-Vereinigung (Angelegenheiten)*, **31**, 101–2.

———. (1927). *Zehn Vorlesungen über die Grundlegung der Mengenlehre, gehalten in Kiel auf Einladung der Kant-Gesellschaft, Ortsgruppe Kiel, vom 8–12, Juni 1925*. Leipzig, Germany: B. G. Teubner.

Fraenkel, A. A., Bar-Hillel, Y., and Levy, A. (1973). *Foundations of Set Theory*. 2nd rev. ed. Amsterdam, Netherlands: North-Holland.

Gödel, K. (1946). Remarks before the Princeton Bicentennial Conference on problems in mathematics. [First published in M. Davis (1965). *The Undecidable: Basic Papers on Undecidable Propositions, Unsolvable Problems, and Computable Functions.* Hewlett, NY: Raven Press. Also in *Collected Works*, vol. 2 (1990), pp. 150–53.]

Hajnal, A. (1961). On a consistency theorem connected with the generalized continuum problem. *Acta Mathematica Academiae Scientiarum Hungaricae*, **12**, 321–76.

Hallett, M. (1984). *Cantorian Set Theory and Limitation of Size.* New York: Oxford University Press.

Hilbert, D. (1926). Über das Unendliche. *Mathematische Annalen*, **95**, 161–90. [English trans. J. van Heijenoort, ed. (1967). *From Frege to Gödel: A Source Book in Mathematical Logic, 1879–1931.* Cambridge, MA: Harvard University Press, pp. 367–92.]

———. (1928). Die Grundlagen der Mathematik. *Abhandlungen aus dem mathematischen Seminar der Hamburgischen Universität*, **6**, 65–85. [English trans. J. van Heijenoort, ed. (1967). *From Frege to Gödel: A Source Book in Mathematical Logic, 1879–1931.* Cambridge, MA: Harvard University Press, pp. 464–79.]

Levy, A. (1963). Independence results in set theory by Cohen's method. I, III, IV (abstract). *Notices of the American Mathematical Society*, **10**, 592–93.

———. (1965). Definability in axiomatic set theory. I. Logic, methodology, and philosophy of science. In *Proceedings of the 1964 International Congress*, ed. Y. Bar-Hillel, pp. 127–51. Amsterdam, Netherlands: North-Holland.

Moore, G. H. (1982). *Zermelo's Axiom of Choice: Its Origins, Development, and Influence.* New York: Springer.

Mostowski, A. (1939). Über die Unabhängigkeit des Wohlordnungssatzes vom Ordnungsprinzip. *Fundamenta Mathematicae*, **32**, 201–52.

———. (1950). Some impredicative definitions in the axiomatic set theory. *Fundamenta Mathematicae*, **37**, 111–24.

———. (1952). On models of axiomatic systems. *Fundamenta Mathematicae*, **39**, 133–58.

Skolem, T. (1923). Einige Bemerkungen zur axiomatischen Begründung der Mengenlehre. In *Matematikerkongressen i Helsinfors den 4–7 Juli 1922, Den femte skandinaviska Matematikerkongressen, Redogörelse*, pp. 217–32. Helsinki: Akademiska Bokhandeln. [English trans. J. van Heijenoort, ed. (1967). *From Frege to Gödel: A Source Book in Mathematical Logic, 1879–1931.* Cambridge, MA: Harvard University Press, pp. 290–301.]

———. (1930). Einige Bemerkungen zu der Abhandlung von E. Zermelo: Über die Definitheit in der Axiomatik. *Fundamenta Mathematicae*, **25**, 37–41.

———. (1934). Über die Nicht-charakterisierbarkeit der Zahlenreihe mittels endlich oder abzählbar unendlich vieler Aussagen mit ausschließlich Zahlenvariablen. *Fundamenta Mathematicae*, **23**, 150–61.

Specker, E. (1957). Zur Axiomatik der Mengenlehre (Fundierungs- und Auswahl-axiom). *Zeitschrift für mathematische Logik und Grundlagen der Mathematik*, **3**, 173–210.

Van Heijenoort, J., ed. (1967). *From Frege to Gödel: A Source Book in Mathematical Logic, 1879–1931.* Cambridge, MA: Harvard University Press.

Von Neumann, J. (1923). Zur Einführung der transfiniten Zahlen. *Acta litterarum ac scientiarum Regiae Universitatis Hungaricae Francisco-Josephinae, Sectio scientiarum mathematicarum*, **1**, 199–208. [English trans. J. van Heijenoort, ed. (1967). *From Frege to Gödel: A Source Book in Mathematical Logic, 1879–1931.* Cambridge, MA: Harvard University Press, pp. 346–54.]

———. (1925). Eine Axiomatisierung der Mengenlehre. *Journal für die reine und angewandte Mathematik*, **154**, 219–40. [English trans. J. van Heijenoort, ed. (1967). *From Frege to Gödel: A Source Book in Mathematical Logic, 1879–1931.* Cambridge, MA: Harvard University Press, pp. 393–413.]

_____. (1929). Über eine Widerspruchsfreiheitsfrage in der axiomatischen Mengenlehre. *Journal für die reine und angewandte Mathematik*, **160**, 227–41.

Zermelo, E. (1908). Untersuchungen über die Grundlagen der Mengenlehre. I. *Mathematische Annalen*, **65**, 261–81.

_____. (1929). Über den Begriff der Definitheit in der Axiomatik. *Fundamenta Mathematicae*, **14**, 339–44.

_____. (1930). Über Grenzzahlen und Mengenbereiche: Neue Untersuchungen über die Grundlagen der Mengenlehre. *Fundamenta Mathematicae*, **16**, 29–47.

_____. (1932). Über Stufen der Quantifikation und die Logik des Unendlichen. *Jahresbericht der Deutschen Mathematiker-Vereinigung (Angelegenheiten)*, **41**, 85–88.

Gödel and the Higher Infinite

CHAPTER 20

The Transfinite Universe

W. Hugh Woodin

The twentieth-century choice for the axioms[1] of set theory are the Zermelo-Frankel axioms together with the axiom of choice; these are the ZFC axioms. This particular choice has led to a twenty-first-century problem:

The ZFC Dilemma: *Many of the fundamental questions of set theory are formally unsolvable from the ZFC axioms.*

Perhaps the most famous example is given by the problem of the continuum hypothesis: suppose X is an infinite set of real numbers; must it be the case that either X is countable or that the set X has cardinality equal to the cardinality of the set of all real numbers?
   One interpretation of this development is as follows:

Skeptic's Attack: The continuum hypothesis is neither true nor false because the entire conception of the universe of sets is a complete fiction. Furthermore, all the theorems of set theory are merely finitistic truths, a reflection of the mathematician and not of any genuine mathematical "reality."

Here and in what follows, the "Skeptic" simply refers to the metamathematical position that denies any genuine meaning to a conception of uncountable sets. The counterview is that of the "Set Theorist":

The Set Theorist's Response: The development of set theory, after Cohen, has led to the realization that formally unsolvable problems have *degrees of unsolvability* that can be calibrated by *large cardinal axioms*.

Elaborating further, as a consequence of this calibration, it has been discovered that in many cases, very different lines of investigation have led to problems whose degree of unsolvability is the same. Thus the hierarchy of large cardinal axioms emerges as an intrinsic, fundamental conception within set theory. To illustrate this, I discuss two examples.

---

[1] This chapter is dedicated to the memory of Paul J. Cohen.

An excellent reference for both the historical development and the background material for much of what I will discuss is the book by Kanamori (1994). The present chapter is not intended to be a survey: my intent is to discuss some very recent results that I think have the potential to be relevant to the concept of the universe of sets. The danger, of course, is that this invariably involves speculation, and this is compounded whenever such speculation is based on research in progress (as in Woodin, 2009).

20.1 The Examples

20.1.1 The First Example: Infinitary Combinatorics

A natural class of objects for study comprises the subsets of ω_1, which is the least uncountable ordinal. Recall that ω_1 is the set of all countable ordinals, and so the collection of all subsets of ω_1 is exactly the collection of all sets of countable ordinals. I shall be concerned with two varieties of subsets of ω_1, which I define below.

Definition 20.1

1. A set $C \subseteq \omega_1$ is *closed* if, for all $\alpha < \omega_1$, if $C \cap \alpha$ is cofinal in α, then $\alpha \in C$.
2. A set $S \subseteq \omega_1$ is *stationary* if $S \cap C \neq \emptyset$ for all closed, cofinal sets $C \subseteq \omega_1$. □

The sets, $S \subseteq \omega_1$, which are stationary and costationary, are in many respects the simplest manifestation of the axiom of choice. For example, one can show, without appealing to the axiom of choice, that if there exists a well ordering of the real numbers, then such a set S must exist. The converse is not true as the existence of such a set S does not imply the existence of a well ordering of the real numbers. Recall that a well-ordering of the real numbers is the total order of the real numbers relative to which every nonempty set of real numbers has a least element.

Therefore it is natural to ask how complicated the structure of the stationary subsets of ω_1 (modulo nonstationary subsets of ω_1) is or even if there can exist a small generating family for these sets. Consider the following stationary basis hypothesis (SBH):

There exists ω_1 many stationary sets, $\langle S_\alpha : \alpha < \omega_1 \rangle$, such that for *every* stationary set $S \subseteq \omega_1$, there exists $\alpha < \omega_1$ such that $S_\alpha \subseteq S$ modulo a nonstationary set.

The assertion that $S_\alpha \subseteq S$ modulo a nonstationary set is simply the assertion that the set

$$S_\alpha \backslash S = \{\beta < \omega_1 \mid \beta \in S_\alpha \text{ and } \beta \notin S\}$$

is not stationary. Such a sequence $\langle S_\alpha : \alpha < \omega_1 \rangle$ of stationary subsets of ω_1 would give, in a natural sense, a *basis* for the stationary subsets of ω_1, which is of cardinality $\aleph_1$.

There is a remarkable theorem of Shelah (1986):

Theorem 20.2 (Shelah, 1986) *The hypothesis* SBH *implies that* CH *is false.* □

This theorem, in conjunction with the subsequent consistency results of Woodin (1999), suggests the following intriguing conjecture: *the hypothesis* SBH *implies that* $2^{\aleph_0} = \aleph_2$.

20.1.2 The Second Example: Infinite Games

Suppose $A \subseteq \mathcal{P}(\mathbb{N})$, where $\mathcal{P}(\mathbb{N})$ denotes the set of all sets $\sigma \subseteq \mathbb{N}$ and $\mathbb{N} = \{1, 2, \ldots, k, \ldots\}$ is the set of all natural numbers. Associated with the set A is an infinite game involving two players, player I and player II. The players alternate, declaring at stage k whether $k \in \sigma$ or $k \notin \sigma$, as follows:

Stage 1: Player I declares $1 \in \sigma$ or declares $1 \notin \sigma$
Stage 2: Player II declares $2 \in \sigma$ or declares $2 \notin \sigma$
Stage 3: Player I declares $3 \in \sigma$ or declares $3 \notin \sigma \ldots$

After infinitely many stages, a set $\sigma \subseteq \mathbb{N}$ is specified. Player I wins this run of the game if $\sigma \in A$; otherwise, player II wins. (Note that player I has control of which odd numbers are in σ, and player II has control of which even numbers are in σ.)

A *strategy* is simply a function that provides moves for the players given just the current state of the game. More formally, a strategy is a function

$$\tau : [\mathbb{N}]^{<\omega} \times \mathbb{N} \to \{0, 1\},$$

where $[\mathbb{N}]^{<\omega}$ denotes the set of all finite subsets of $\mathbb{N}$. At each stage k of the game, the relevant player can choose to follow τ by declaring $k \in \sigma$ if $\tau(a, k) = 1$ and by declaring $k \notin \sigma$ if $\tau(a, k) = 0$, where

$$a = \{i < k \mid i \in \sigma \text{ was declared at stage } i\} \, .$$

The strategy τ is a *winning strategy* for player I if, by following the strategy at each stage k where it is player I's turn to play (i.e., for all odd k), player I wins the game *no matter how player II plays*. Similarly, τ is a *winning strategy* for player II if, by following the strategy at each stage k where it is player II's turn to play (i.e., for all even k), player II wins the game *no matter how player I plays*. The game is *determined* if there is a *winning strategy* for one of the players. Clearly it is impossible for there to be winning strategies for both players.

It is easy to specify sets $A \subseteq \mathcal{P}(\mathbb{N})$ for which the corresponding game is determined. For example, if $A = \mathcal{P}(\mathbb{N})$, then any strategy is a winning strategy for player I. On the other hand, if A is countable, then one can fairly easily show that there exists a strategy that is a winning strategy for player II.

The problem of specifying a set $A \subseteq \mathcal{P}(\mathbb{N})$ for which the corresponding game is not determined turns out to be quite a bit more difficult. The axiom of determinacy (AD), is the axiom that asserts that for all sets $A \subseteq \mathcal{P}(\mathbb{N})$, the game given by A, as described earlier, is determined. This axiom was first proposed by Mycielski and Steinhaus (1962) and contradicts the axiom of choice. So the problem here is whether the axiom of choice is necessary to construct a set $A \subseteq \mathcal{P}(\mathbb{N})$ for which the corresponding game is not determined. Clearly, if the axiom of choice is necessary, then the existence of such a set A is quite a subtle fact.

20.1.3 Three Problems and Three Formal Theories

I now add a third problem to the list and specify formally a list of three problems. As indicated, the first and third problems are within ZFC, and the second problem is within just the theory ZF (this is the theory given by the axioms ZFC without the axiom of choice):

Problem 1: (ZFC) Does SBH hold?
Problem 2: (ZF) Does AD hold?
Problem 3: (ZFC) Do there exist infinitely many Woodin cardinals?

I shall not give the formal definition of a Woodin cardinal here as it is a large cardinal notion whose definition is a bit technical; see Kanamori (1994) for one definition.

The first problem, problem 1, is *formally unsolvable* if, assuming the axioms ZFC, one can neither prove nor refute the SBH. Similarly, problem 2 is formally unsolvable if, assuming the axioms ZF, one can neither prove nor refute AD. Finally, problem 3 is formally unsolvable if, assuming the axioms ZFC, one can neither prove nor refute the existence of infinitely many Woodin cardinals. In each case, the assertion of formal unsolvability is simply a statement of number theory. The remarkable fact is that these three assertions of number theory are equivalent, and this is a theorem of number theory from the classical (Peano) axioms for number theory. Thus two completely different lines of investigation have resulted in problems whose degree of formal unsolvability is the same, and this is exactly calibrated by a large cardinal axiom.

Assuming that the axioms ZFC are formally consistent, then, for the three problems indicated here, the only possible formal solutions are as follows: no for the first and third problems and yes for the second problem. Therefore it is more natural to rephrase these assertions of formal unsolvability as assertions that particular theories are formally consistent. I have implicitly defined three theories, and the assertions of unsolvability discussed correspond to the assertions that these theories are each formally consistent:

Theory 1: ZFC + SBH
Theory 2: ZF + AD
Theory 3: ZFC + "There exist infinitely many Woodin cardinals"

The following theorem is the theorem that implies that the degree of unsolvability of the three problems that I have listed is the same; see Kanamori (1994) for a discussion of this theorem.

Theorem 20.3   *The three theories, theory 1, theory 2, and theory 3, are equiconsistent.*   □

20.2 A Prediction and a Challenge for the Skeptic

Are the three theories I have defined really formally consistent? The claim that they are consistent is a prediction that can be refuted by finite evidence (a formal contradiction).

Just knowing that the first two theories are equiconsistent does not justify this prediction at all. I claim the following:

> It is through the calibration by a large cardinal axiom in conjunction with our understanding of the hierarchy of such axioms as true axioms about the universe of sets that this prediction is justified.

As a consequence of my belief in this claim, I make a prediction:

> In the next ten thousand years, there will be no discovery of an inconsistency in these theories.

This is a specific and unambiguous prediction about the physical universe. Furthermore, it is a prediction that does not arise by a reduction to a previously held truth (e.g., as is the case for the prediction that no counterexample to Fermat's last theorem will be discovered).

This is a genuinely new prediction that I make based on the development of set theory over the last fifty years and on my belief that the conception of the transfinite universe of sets is meaningful. Finally, I make this prediction independently of all speculation of what computational devices might be developed in the next ten thousand years that will increase the effectiveness of research in mathematics – it is a prediction based on mathematics and not on consideration of the mathematician.

Now the Skeptic might object that this prediction is not interesting or natural because the formal theories are not interesting or natural, but such objections are not allowed in physics: the ultimate physical theory should explain all (physical) aspects of the physical universe, not just those that we regard as natural. How can we apply a lesser standard for the ultimate mathematical theory? In fact, I make a stronger prediction:

> There will be no discovery, ever, of an inconsistency in these theories.

One can arguably claim that if this stronger prediction is true, then it is a physical law.

> **Skeptic's Retreat:** OK, I accept the challenge, noting that I only have to explain the predictions of formal consistency given by the large cardinal axioms. The formal theory of set theory as given by the axioms ZFC is so "incomplete" that either any large cardinal axiom, in the natural formulation of such axioms, is consistent with the axioms of set theory or there is an elementary proof that the axiom cannot hold.

We shall see that this is a very shrewd counterattack, even framed within the specific context of the current list of large cardinal axioms, where it is a much more plausible position. I shall need to review some elementary concepts from set theory. This is necessary to specify the basic template for large cardinal axioms.

20.3 The Cumulative Hierarchy of Sets

As is customary in modern set theory, V denotes the universe of sets. The purpose of this notation is to facilitate the (mathematical) discussion of set theory; it does not presuppose any meaning to the concept of the universe of sets.

The ordinals calibrate V through the definition of the cumulative hierarchy of sets (zermelo, 1930). The relevant definition is given as follows.

Definition 20.4   Define for each ordinal α a set V_α by induction on α:

1. $V_0 = \emptyset$
2. $V_{\alpha+1} = \mathcal{P}(V_\alpha) = \{X \mid X \subseteq V_\alpha\}$

1. If β is a limit ordinal, then $V_\alpha = \bigcup\{V_\beta \mid \beta < \alpha\}$.   □

It is a consequence of the ZF axioms that for every set a, there must exist an ordinal α, such that $a \in V_\alpha$.

A set N is transitive if every element of N is a subset of N. Transitive sets are fragments of V that are analogous to initial segments. For each ordinal α, the set V_α is a transitive set.

Every ordinal is a transitive set; in fact, the ordinals are precisely those transitive sets X with the property that for all $a, b \in X$, if $a \neq b$, then either $a \in b$ or $b \in a$. Thus, if X is an ordinal and if $Y \in X$, then necessarily, Y is an ordinal. This defines a natural order on the ordinals. If α and β are ordinals, then $\alpha < \beta$ if $\alpha \in \beta$. Thus every ordinal is simply the set of all ordinals that are smaller than the given ordinal, relative to this order.

The simplest (proper) class is the class of all ordinals. This class is a transitive class, and more generally, a class $M \subseteq V$ is a transitive class if every element of M is a subset of M. The basic template for large cardinal axioms is as follows:

There is a transitive class M and an elementary embedding

$$j : V \to M$$

that is not the identity.

With the exception of the definition of a Reinhardt cardinal, which I shall discuss later, one can always assume that the classes, M and j, are classes that are logically definable from parameters by formulas of a fixed bounded level of complexity (Σ_2-formulas). Moreover, the assertion that j is an elementary embedding is the follwing assertion:

For all formulas $\phi(x)$ and for all sets a, $V \models \phi[a]$ if and only if $M \models \phi[j(a)]$;

this is equivalent to the assertion that

for all formulas $\phi(x)$, for all ordinals α, and for all sets $a \in V_\alpha$, $V_\alpha \models \phi[a]$ if and only if $j(V_\alpha) \models \phi[j(a)]$.

Therefore this template makes no essential use of the notion of a class. It is simply for convenience that I refer to classes (and this is the usual practice in set theory).

Suppose that M is a transitive class and that $j : V \to M$ is an elementary embedding that is not the identity. Suppose that $j(\alpha) = \alpha$ for all ordinals α. Then one can show by transfinite induction that for all ordinals α, the embedding j is the identity on V_α. Therefore, because j is not the identity, there must exist an ordinal α such that $j(\alpha) \neq \alpha$, and because j is order preserving on the ordinals, this is equivalent to the requirement that $\alpha < j(\alpha)$. The least such ordinal is called the "critical point" of j, and it can be shown that this must be a cardinal. The critical point of j is the large cardinal specified, and the existence of the transitive class M and the elementary embedding j are the witnesses for this.

A cardinal κ is a *measurable cardinal* if there exists a transitive class M and an elementary embedding $j : V \to M$ such that κ is the critical point of j. It is by requiring M to be closer to V that one can define large cardinal axioms far beyond the axiom that "there is a measurable cardinal." In general, the closer one requires M to be to V, the stronger the large cardinal axiom. In Reinhardt (1970), the natural maximum axiom was proposed ($M = V$). The associated large cardinal axiom is that of a Reinhardt cardinal.

Definition 20.5   A cardinal κ is a *Reinhardt cardinal* if there is an elementary embedding $j : V \to V$ such that κ is the critical point of j.   □

The definition of a Reinhardt cardinal makes essential use of classes, but the following variation does not and is more useful for this discussion. The definition requires a logical notion. Suppose that α and β are ordinals such that $\alpha < \beta$. Then we write $V_\alpha \prec V_\beta$ to mean that for all formulas $\phi(x)$ and for all $a \in V_\alpha$, $V_\alpha \models \phi[a]$ if and only if $V_\beta \models \phi[a]$.

Definition 20.6   A cardinal κ is a *weak Reinhardt cardinal* if there exist $\gamma > \lambda > \kappa$ such that

1. $V_\kappa \prec V_\lambda \prec V_\gamma$
2. there exists an elementary embedding $j : V_{\lambda+2} \to V_{\lambda+2}$ such that κ is the critical point of j □

The definition of a weak Reinhardt cardinal only involves sets. The relationship between Reinhardt cardinals and weak Reinhardt cardinals is unclear; however, given the original motivation for the definition of a Reinhardt cardinal, one would conjecture that, at least in terms of consistency strength, Reinhardt cardinals are stronger than weak Reinhardt cardinals, hence my choice of terminology. In any case, the concept of a weak Reinhardt cardinal is better suited to illustrate the key points I am trying to make.

The following theorem is an immediate corollary of the fundamental inconsistency results of Kunen (1971).

Theorem 20.7 (Kunen, 1971)   *There are no weak Reinhardt cardinals.*   □

The proof is elementary, so this does not refute the Skeptic's Retreat. But Kunen's proof makes essential use of the axiom of choice. The problem is open without this assumption, and this is not just an issue for weak Reinhardt cardinals, which is just a notion of large cardinals defined in this chapter. There really is no known interesting example of a strengthening of the definition of a Reinhardt cardinal that yields a large cardinal axiom that can be refuted without using the axiom of choice. The difficulty is that without the axiom of choice, it is extraordinarily difficult to prove anything about sets.

Kunen's proof leaves open the possibility that the following large cardinal axiom might be consistent with the axiom of choice. This therefore is essentially the strongest large cardinal axiom not known to be refuted by the axiom of choice; see Kanamori (1994) for more on this as well as for the actual statement of Kunen's theorem.

Definition 20.8 A cardinal κ is an *ω-huge cardinal* if there exists $\lambda > \kappa$ and an elementary embedding $j : V_{\lambda+1} \to V_{\lambda+1}$ such that κ is the critical point of j. □

One could strengthen this axiom still further by requiring in addition that for some $\gamma > \lambda$, we have $V_\kappa \prec V_\lambda \prec V_\gamma$ and so match in form the definition of a weak Reinhardt cardinal, the only modification being that $\lambda + 2$ is replaced by $\lambda + 1$. This change would not affect any of the following claims concerning ω-huge cardinals.

The issue of whether the existence of a weak Reinhardt cardinal is consistent with the axioms ZF is an important issue for the Set Theorist because, by the results of Woodin (2009), the theory

$$\text{ZF} + \text{``There is a weak Reinhardt cardinal''}$$

proves the formal consistency of the theory

$$\text{ZFC} + \text{``There is a proper class of } \omega\text{-huge cardinals.''}$$

This number-theoretic statement is a theorem of number theory. However, as indicated earlier, the notion of an ω-huge cardinal is essentially the strongest large cardinal notion that is not known to be refuted by the axiom of choice.

Therefore the number-theoretic assertion that the theory

$$\text{ZF} + \text{``There is a weak Reinhardt cardinal''}$$

is consistent is a stronger assertion than the number-theoretic assertion that the theory

$$\text{ZFC} + \text{``There is a proper class of } \omega\text{-huge cardinals''}$$

is consistent. More precisely, the former assertion implies, but is not implied by, the latter assertion, unless, of course, the theory

$$\text{ZFC} + \text{``There is a proper class of } \omega\text{-huge cardinals''}$$

is formally inconsistent. This raises an interesting question:

> How could the Set Theorist ever be able to argue for the prediction that the existence of weak Reinhardt cardinals is consistent with axioms of set theory without the axiom of choice?

Moreover, this one prediction implies all the predictions (of formal consistency) the Set Theorist can currently make based on the entire large cardinal hierarchy as presently conceived (in the context of a universe of sets that satisfies the axiom of choice). My point is that by appealing to the Skeptic's Retreat, one could reasonably claim that the theory

$$\text{ZF} + \text{``There is a weak Reinhardt cardinal''}$$

is formally consistent – and in making this single claim, one would subsume all the claims of consistency that the set theorist can make based on our current understanding of the universe of sets (without abandoning the axiom of choice).

The only tools currently available seem powerless to resolve this issue. Reinterpreting the number-theoretic statement that the theory

$$ZF + \text{"There is a weak Reinhardt cardinal"}$$

is formally consistent, in a way that allows the Set Theorist to argue for the truth of this statement, seems equally hopeless. Finally, unlike the axiom AD, there is no candidate presently known for a fragment of V for which the existence of weak Reinhardt cardinals is the correct (or even a possible) axiom.

20.4 Probing the Universe of Sets: The Inner Model Program

The *inner model program* is the detailed study of large cardinal axioms. The first construction of an inner model is from Gödel (1938, 1940). This construction founded the inner model program, and the transitive class constructed is denoted by L. This is the minimum possible universe of sets containing all ordinals.

If X is a transitive set, then $\text{Def}(X)$ denotes the set of all $A \subseteq X$ such that A is logically definable in the structure $(X, \in)$ from parameters. The definition of L is simply given by replacing the operation $\mathcal{P}(X)$ in the definition of $V_{\alpha+1}$ with the operation $\text{Def}(X)$. More precisely, we have the following:

Definition 20.9

1. Define L_α by induction on the ordinal α:
   a. $L_0 = \emptyset$ and $L_{\alpha+1} = \text{Def}(L_\alpha)$
   b. If α is a limit ordinal, then $L_\alpha = \bigcup \{L_\beta \mid \beta < \alpha\}$.
2. L is the class of all sets a such that $a \in L_\alpha$ for some ordinal α. □

It is perhaps important to note that though there must exist a proper class of ordinals α such that

$$L_\alpha = L \cap V_\alpha,$$

this is not true for all ordinals α.

The question of whether $V = L$ is an important question for set theory. The answer has profound implications for the conception of the universe of sets.

Theorem 20.10 (Scott, 1961) *Suppose there is a measurable cardinal. Then* $V \neq L$. □

The *axiom of constructibility* is the axiom that asserts $V = L$; more precisely, this is the axiom that asserts that for each set a, there exists an ordinal α such that $a \in L_\alpha$. Scott's theorem provided the first indication that the axiom of constructibility is independent of the ZFC axioms. At the time, there was no compelling reason to believe that the existence of a measurable cardinal was consistent with the ZFC axioms, so one could not make the claim that Scott's theorem established the formal independence of the axiom of constructibility from the ZFC axioms. Of course, it is an immediate corollary of Cohen's results that the axiom of constructibility is formally independent of the ZFC axioms. The modern significance of Scott's theorem is more profound:

Scott's theorem establishes that the axiom of constructibility is false. This claim (that $V \neq L$) is not universally accepted, but in my view, no one has come up with a credible argument against it.

The inner model program seeks generalizations of L for the large cardinal axioms; in brief, it seeks generalizations of the axiom of constructibility that are compatible with large cardinal axioms (such as the axioms for measurable cardinals and beyond). It has been a very successful program, and its successes have led to the realization that the large cardinal hierarchy is a very robust notion. The results that have been obtained provide some of our deepest glimpses into the universe of sets, and its successes have led to a metaprediction:

> **A Set Theorist's Cosmological Principle:** The large cardinal axioms for which there is an inner model theory are consistent; the corresponding predictions of unsolvability are true because the axioms are true.

Despite its rather formidable merits, as indicated earlier, there is a fundamental difficulty with the prospect of using the inner model program to counter the Skeptic's Retreat. The problem is in the basic methodology of the inner model program, but to explain this, I must give a (brief) description of the (technical) template for inner models.

The inner models that are the goal and focus of the inner model program are defined layer by layer, working up through the hierarchy of large cardinal axioms, which in turn is naturally revealed by the construction of these inner models. Each layer provides the foundation for the next, and L is the first layer. Roughly (and in practice), in constructing the inner model for a specific large cardinal axiom, one obtains an exhaustive analysis of all weaker large cardinal axioms. There can be surprises here in that seemingly different notions of large cardinals can coincide in the inner model. Finally, as one ascends through the hierarchy of large cardinal axioms, the construction generally becomes more and more difficult.

20.5 The Building Blocks for Inner Models: Extenders

Suppose that M is a transitive class and that $j : V \to M$ is an elementary embedding with critical point κ. As with the basic template for large cardinal axioms I discussed earlier, here and subsequently, one can restrict onself to the classes that are definable classes (by Σ_2-formulas), for example, so that no essential use of classes is involved.

It is immediate from the definitions that for all ordinals γ, $j(\gamma)$ is an ordinal, and moreover, $j(\gamma) \geq \gamma$. Suppose that $\kappa < \gamma < j(\kappa)$ and that $\mathcal{P}(\gamma) \subseteq M$, where $\mathcal{P}(\gamma) = \{A \mid A \subseteq \gamma\}$. The function

$$E(A) = j(A) \cap \gamma$$

with domain $\mathcal{P}(\gamma)$ is the *extender* E of length γ defined from j. Note that because $\gamma > \kappa$, necessarily, E is not the identity function. Extenders are nontrivial fragments of the elementary embedding j. (The concept of an extender is from Mitchell.) The definition that I have given is really that of a strong extender because of the assumption that $\mathcal{P}(\gamma) \subseteq M$. This I do for expository reasons. In the case that $\gamma < j(\kappa)$, which is the present case, one could drop this requirement without affecting much of the discussion.

The importance of the concept of an extender is as follows: suppose that E is an extender of length γ derived from an elementary embedding $j : V \to M$ and that N is a transitive class such that $N \models \text{ZFC}$. Suppose that $E \cap N \in N$ and that $\gamma = \kappa + 1$, where κ is the critical point of j. Then there exists a transitive class $M_E \subseteq N$ and an elementary embedding

$$j_E : N \to M_E$$

such that $E \cap N$ is the extender of length γ derived from j_E. The point here, of course, is that the assumption is only that $E \cap N \in N$, as opposed to the much stronger assumption that $E \in N$. Both M_E and j_E can be chosen to be definable classes of N (by Σ_2-formulas) using just the parameter $E \cap N$.

Without the assumption that $\gamma = \kappa + 1$, which is a very special case, these claims still hold, provided that one drops the requirement $\mathcal{P}(\gamma) \subseteq M$ in the definition of an extender that I have given.

These remarks suggest that one should seek, as generalizations of L, transitive classes N such that N contains enough extenders of the form $E \cap N$ for some extender $E \in V$ to witness that the targeted large cardinal axiom holds in N. One can then regard such transitive classes as refinements of V that are constructed to preserve certain extenders of V. The complication is in specifying just which extenders are to be preserved.

For each set A, one can naturally define a class $L[A]$ that is L relativized to the set A, as follows:

Definition 20.11

1. Define $L_\alpha[A]$ by induction on the ordinal α:
   a. $L_0 = \emptyset$ and $L_{\alpha+1}[A] = \text{Def}(X) \cap \mathcal{P}(L_\alpha[A])$,   where $X = L_\alpha[A] \cup \{L_\alpha[A] \cap A\}$.
   b. If α is a limit ordinal, then $L_\alpha[A] = \bigcup\{L_\beta[A] \mid \beta < \alpha\}$.
2. $L[A]$ is the class of all sets a such that $a \in L_\alpha[A]$ for some ordinal α.   □

If F is a function, then $L[F]$ is defined to be $L[A]$, where $A = F$. Thus, if the domain of F is disjoint from L, then $L[F] = L$.

Constructing from a single extender E yields $L[E]$, which is a true generalization of L and solves the inner model problem for the large cardinal axiom that "there is a measurable cardinal." This claim follows from the results and methods of Kunen (1970) and is illustrated in part by the following theorem.

There is a feature of inner models for large cardinals that is implicit in this example:

> For a specific large cardinal axiom, there is, in general, no unique inner model for that axiom but rather a family of inner models. However, all these inner models are equivalent in a natural (but technical) sense.

One illustration of this is given by the following theorem, which is a modern formulation of the fundamental results of Kunen (1970) on the inner model problem for one measurable cardinal. For each extender E, let κ_E denote the least ordinal α such that $E(\alpha) \neq \alpha$. This coincides with the critical point of the elementary embedding $j : V \to M$ from which E is derived.

Theorem 20.12   *Suppose that E and F are extenders.*

1. *If $\kappa_E = \kappa_F$, then $L[E] = L[F]$.*
2. *If $\kappa_E < \kappa_F$, then $L[F] \subset L[E]$, and there is an elementary embedding*

$$j : L[E] \to L[F].$$ $\square$

For the generalizations of $L[E]$ that one must consider to solve the inner model problem for large cardinals beyond the level of measurable cardinals, this ambiguity is much more subtle and lies at the core of the difficulty in even defining the inner models.

By the preceding theorem, one cannot use a single extender to build an inner model for essentially any large cardinal axiom beyond the level of a single measurable cardinal. For example, suppose that E is an extender. Combining elements of Gödel's basic analysis of L, generalized to an analysis of $L[E]$ with Theorem 20.12, it follows that the inner model $L[E]$ will fail to satisfy the large cardinal axiom that "there are two measurable cardinals." There is an obvious remedy: to reach inner models for stronger large cardinal axioms, one should use sequences

$$\tilde{E} = \langle E_\alpha : \alpha < \theta \rangle,$$

where each E_α is the extender derived from some elementary embedding, as earlier. The complication is in how actually to define the sequence; in fact, one must ultimately allow the sequence to contain partial extenders, which creates still further complications.

A partial extender of length γ is the extender of length γ derived from a Σ_0-elementary embedding $j : N \to M$, where N and M are transitive sets that are only assumed to be closed under the Gödel operations, and instead of requiring $\mathcal{P}(\gamma) \subseteq M$, one requires that

$$\mathcal{P}(\gamma) \cap N = \mathcal{P}(\gamma) \cap M.$$

The requirement that j be a Σ_0-elementary embedding is the requirement that for a very restricted collection of formulas, $\phi(x_0)$, and for all $a \in N$, $N \models \phi[a]$ if and only if $M \models \phi[j(a)]$. The relevant formulas are the Σ_0-formulas.

The difficulty mentioned is in determining exactly when such partial extenders are acceptable. In fact, things get so complicated that unlike the situation with measurable cardinals, one can only define the inner model by simultaneously developing the detailed analysis of the inner model in an elaborate induction.

The current state of the art is found in the inner models defined by Mitchell and Steel (1994). The definition of these inner models is the culmination of a nearly twenty-year program of developing the theory of inner models. The Mitchell-Steel inner models can accommodate large cardinals up to the level of superstrong cardinals, but existence has only been proved – from the relevant large cardinal axioms – at the level of a Woodin cardinal, which is a limit of Woodin cardinals. In this program of establishing existence of Mitchell-Steel inner models, the best results to date are from Neeman (2002).

The distinction between developing the theory of the inner models and proving the existence of the inner models is perhaps a confusing one at first glance. The precise explanation requires details of the Mitchell-Steel theory, which are beyond the scope of the present discussion. Roughly, the Mitchell-Steel theory reduces the problem of the existence of the generalization of L for the large cardinal axiom under consideration to a specific combinatorial hypothesis, provided that the large cardinal axiom is at the level of a superstrong cardinal or below. This combinatorial (iteration) hypothesis can be specified without any reference to the Mitchell-Steel theory and, more generally, without reference to inner model theory at all. There is, of course, the possibility that this is symptomatic of a far more serious problem and that by answering one of the test questions of the inner model program negatively, one can prove that the inner model program as presently conceived fails for some large cardinal axiom below the level of a superstrong cardinal.

20.6 The Inner Model Program, the Core Model Program, and the Skeptic's Retreat

As I have claimed, there is a fundamental problem with appealing to the inner model program to counter the Skeptic's Retreat. The precise nature of the problem is subtle, so I shall begin by describing what might seem to be a plausible version of the problem. I will then briefly try to describe the actual problem. This will involve the core model program, which is a variant of the inner model program.

Suppose (for example) that a hypothetical large cardinal axiom Ω provides a counterexample to the Skeptic's Retreat, and this is accomplished by the inner model program:

> To use the inner model program to refute the existence of an Ω-cardinal, one first must be able to successfully construct the inner models for all smaller large cardinals, and this hierarchy would be fully revealed by the construction.

Perhaps this could happen, but it can only happen once – this is the problem. Having refuted the existence of an Ω-cardinal, how could one then refute the existence of any smaller large cardinals, for one would have solved the inner model problem for these smaller large cardinals? This would refute the Set Theorist's cosmological principle, so the fundamental problem is as follows:

> The inner model program seems inherently unable, by virtue of its inductive nature, to provide a framework for an evolving understanding of the boundary between the possible and the impossible (large cardinal axioms).

On close inspection, it is perhaps not entirely convincing that there is a problem here. Arguably, there is the potential for a problem, but the specific details of how the inner model program might succeed in countering the Skeptic's Retreat are clearly critical in determining whether there really is a problem.

Although the idea that the inner model program could ever yield an inconsistency result has always seemed unlikely, there is another way that the inner model program might succeed in establishing inconsistency results in a manner that refutes the Skeptic's Retreat. The *core model program* can be described as follows: suppose that $L[\tilde{E}]$ is an inner model as constructed by the inner model program. In general, would one not expect the inner model $L[\tilde{E}]$ to contain even all the real numbers; for example, if the continuum hypothesis is false in V, then necessarily, there are real numbers that are not in $L[\tilde{E}]$. Therefore every extender on the sequence $\tilde{E}$, when restricted to $L[\tilde{E}]$, is necessarily a partial extender in V. This suggests that one might attempt to construct the inner model $L[\tilde{E}]$ without using extenders at all but rather just partial extenders.

Though this might seem reasonable, there is, of course, a problem. If there are no extenders in V, then there are no measurable cardinals in V, and so one cannot in general expect to be able to build an inner model in which there are measurable cardinals. But suppose one assumes that there is no proper transitive class N in which a particular large cardinal axiom holds. Then a reasonable conjecture is that there is an inner model of the form $L[\tilde{E}]$ that is close to V.

One measure of the closeness of an inner model N to V is a *weak covering principle*. This requires a definition: a cardinal γ is *singular* if there exists a cofinal set $X \subseteq \gamma$ such that $|X| < \gamma$, and γ^+ refers to the least cardinal κ such that $\kappa > \gamma$. Suppose that N is a (proper) transitive class and that $N \models \mathrm{ZFC}$. Then *weak covering* holds for V relative to the inner model N, if for all uncountable singular cardinals γ, if $\gamma = |V_\gamma|$, then

$$(\gamma^+)^N = \gamma^+.$$

Allowing the case that $\tilde{E} = \emptyset$, so that $L[\tilde{E}] = L$, this becomes a very interesting problem. The program to solve this family of problems is the core model program. Both the inner model program and the core model program seek to construct exactly the same form of an inner model – the only difference is in the assumptions from which one starts. The inner model program starts with the assumption that a particular large cardinal axiom holds in V, whereas the core model program starts with the assumption that a particular large cardinal axiom does not hold in any transitive class $N \subseteq V$. It is customary to refer to the transitive classes constructed by the methods of the core model program as core models.

The core model program was inspired by Jensen's covering lemma and began with the results of Dodd and Jensen (1981). The strongest results to date are primarily from Steel, who extended the core model program to the level of Woodin cardinals (Steel, 1996). As with the inner model program, the solutions provided by the Core Model Program increase in complexity as the associated large cardinal axiom is strengthened.

The core model program has been quite successful, and out of it have come a number of deep combinatorial theorems. For example, the methods and constructions of the core model program play an essential role in the proof of Theorem 20.3. One might suspect that the utility of the core model program is limited for proving the kinds of theorems that an inconsistency result would require because of the requisite hypothesis that there be no proper transitive class N in which a specific large cardinal axiom holds. Despite this requirement, though, the core model program has yielded some surprising theorems of set theory.

Recall that the *generalized continuum hypothesis* (GCH) is the assertion that for all infinite cardinals, γ, $2^\gamma = \gamma^+$, where 2^γ is the cardinality of $\mathcal{P}(\gamma) = \{X \mid X \subseteq \gamma\}$. The following theorem is an example of a theorem proved by the methods of the core model program.

Theorem 20.13 (Woodin, 1996) *Suppose that there exists a countable set A such that A is a set of ordinals and $V = L[A]$. Then the GCH holds.* □

On general grounds, to prove the theorem, it suffices from the hypothesis of the theorem just to prove that the continuum hypothesis holds. The special case where $A \subseteq \omega$ follows from Gödel's analysis of L, generalized to the analysis of $L[A]$. For the general case where one does not assume $A \subseteq \omega$, there is no elementary proof of the continuum hypothesis known.

By a theorem of Jensen (Beller et al., 1982) for any sentence ϕ, if the sentence is consistent with the axioms ZFC, then the existence of a proper class N within which the sentence holds is consistent with the hypothesis of Theorem 20.13. If, in addition, the sentence is consistent with the axioms ZFC + GCH, then one can even require that the transitive class N be close to $L[A]$. For example, one can require that N and $L[A]$ have the same cardinals. While this additional consistency assumption may seem like a very restrictive assumption, at least for the current generation of large cardinal axioms, it is not.

Therefore it is perhaps not unreasonable that the core model program might yield that some proposed large cardinal axiom is inconsistent and, in doing so, refute the Skeptic's Retreat. But to accomplish this, though, the core model program would seem to have to produce an ultimate core model corresponding to the ultimate inner model. However, if this is an inner model of the form $L[\tilde{E}]$ for some sequence of (partial) extenders, as is the case for essentially all core models that have been constructed to date, then the nature of the extenders on the sequence $\tilde{E}$ should reveal the entire large cardinal hierarchy – and we again are in a situation where further progress looks unlikely.

Thus it would seem that the Skeptic's Retreat is in fact a powerful counterattack. However, there is something wrong here – some fundamental misconception. The answer lies in understanding large cardinal axioms that are much stronger than those within reach of the Mitchell-Steel hierarchy of inner models. Ironically, one outcome of my proposed resolution to this misconception is that for these large cardinal axioms, the Set Theorist's cosmological principle is either false or useless. I shall discuss these ramifications after Theorem 20.22.

20.7 Supercompact Cardinals and Beyond

We begin with a definition.

Definition 20.14 (Solovay) A cardinal κ is a *supercompact cardinal* if, for each ordinal α, there exists a transitive class M and an elementary embedding $j : V \to M$ such that

1. κ is the critical point of j and $j(\kappa) > \alpha$
2. M contains all functions, $f : \alpha \to M$ □

Still stronger are *extendible cardinals*, *huge cardinals*, and *n-huge cardinals* where $n < \omega$. These I shall not define here. As I have already indicated, the strongest large cardinal axioms not known to be inconsistent with the axiom of choice are the family of axioms asserting the existence of ω-huge cardinals. These axioms have seemed so far beyond any conceivable inner model theory that they simply are not understood.

The possibilities for an inner model theory at the level of supercompact cardinals and beyond have essentially been a complete mystery until recently. The reason lies in the nature of extenders. Suppose that E is an extender of length γ derived from an elementary embedding $j : V \to M$ with critical point κ such that $\mathcal{P}(\gamma) \subseteq M$. If $\gamma \leq j(\kappa)$, then E is a short extender; otherwise, E is a long extender. Up to this point, I have only considered short extenders. The properties of long extenders can be quite subtle, and it is for this reason that I impose the requirement $\mathcal{P}(\gamma) \subseteq M$. Even with this requirement, many subtleties remain. For example, Theorem 20.12, which I implicitly stated for short extenders, is false for long extenders. One can prove the following variation, provided that the extenders are not too long. For expository purposes, let me define an extender E to be a *suitable* extender if E is the extender of length γ derived from an elementary embedding $j : V \to M$ such that $\mathcal{P}(\gamma) \subseteq M$ and such that $\gamma < j(\alpha)$ for some $\alpha < j(\kappa)$, where κ is the critical point of j. Suitable extenders can be long extenders, but they cannot be too long.

Theorem 20.15 *Suppose that E and F are suitable extenders. Then*

$$\mathbb{R} \cap L[E] \subseteq \mathbb{R} \cap L[F] \quad or \quad \mathbb{R} \cap L[F] \subseteq \mathbb{R} \cap L[E]. \qquad \square$$

Without the restriction to suitable extenders, it is not known if this theorem holds.

The following lemma of Magidor, reformulated in terms of suitable extenders, gives a useful reformulation of supercompactness (Woodin, 2009). The statement of the lemma involves the following notation, which I previously defined for short extenders. Suppose that E is an extender of length γ derived from an elementary embedding $j : V \to M$. Then κ_E is the critical point of j. Suppose that $\alpha < \gamma$. Then, by the definition of E, and given that $\alpha = \{\beta \mid \beta < \alpha\} \subseteq \gamma$, we find that α is in the domain of E and either $E(\alpha) = \gamma$ or $E(\alpha) < \gamma$. Moreover, the following hold:

1. $E(\alpha) < \gamma$ if and only if $j(\alpha) < \gamma$ and $E(\alpha) = j(\alpha)$.
2. $E(\alpha) = \gamma$ if and only if $j(\alpha) \geq \gamma$.

Thus κ_E is simply the least α such that $E(\alpha) \neq \alpha$.

Lemma 20.16 *Suppose that δ is a cardinal. Then the following are equivalent:*

1. δ is supercompact.
2. For each ordinal $\gamma > \delta$, there exists a suitable extender E of length γ such that $E(\kappa_E) = \delta$. $\qquad \square$

My convention in what follows is that a class $\mathcal{E}$ of extenders witnesses that δ is a supercompact cardinal if, for each $\gamma > \delta$, there exists an extender $E \in \mathcal{E}$ such that

1. E has length γ
2. $E(\kappa_E) = \delta$ and for some $\alpha < \delta$, $E(\alpha) = \gamma$.

Note that condition 2 implies that E is a suitable extender.

The Mitchell-Steel inner models are constructed from sequences of short extenders. To build inner models at the level of supercompact cardinals and beyond, one must have long extenders on the sequence, and this creates serious obstacles if these extenders are too long. In fact, Steel has isolated a specific obstacle that becomes severe at the level of one supercompact cardinal with a measurable cardinal.

By some fairly recent theorems (Woodin, 2009), something completely unexpected and remarkable happens. Suppose that N is a transitive class for some cardinal δ,

$$N \models \text{"δ is a supercompact cardinal,"}$$

and that this is witnessed by the class of all $E \cap N$ such that $E \cap N \in N$ and such that E is an extender. Then the transitive class N is close to V, and N inherits essentially all large cardinals from V.

For example, suppose that for each n, there is a proper class of n-huge cardinals in V. Then in N, for each n, there is a proper class of n-huge cardinals. The amazing thing is that this must happen no matter how N is constructed. This would seem to undermine my earlier claim that inner models should be constructed as refinements of V that preserve enough extenders from V to witness that the targeted large cardinal axiom holds in the inner model. It does not, and the reason is that by simply requiring that $E \cap N \in N$ for enough suitable extenders from V to witness that the large cardinal axiom, "there is a supercompact cardinal," holds in N, one (and this is the surprise) necessarily must have $E \cap N \in N$ for a much larger class of extenders of V. So the principle that there are enough extenders of N that are of the form $E \cap N$ for some extender $E \in V$, to witness that the targeted large cardinal axiom holds in N, is preserved. The change, in the case that N is constructed from a sequence of extenders, is that these extenders do not have to be on the sequence from which N is constructed. In particular, in the case that the sequence of extenders from which N is constructed contains only suitable extenders, large cardinal axioms can be witnessed to hold in N by the "phantom" extenders (these are extenders of N that are not on the sequence), which cannot be witnessed to hold by any extender on the sequence.

As a consequence of this, one can completely avoid the cited obstacles because

one does not need to have the kinds of long extenders on the sequence that give rise to the obstacles.

Specifically, one can restrict consideration to extender sequences of just suitable extenders, and this is a paradigm shift in the whole conception of inner models.

The analysis yields still more. Suppose that there is a positive solution (in ZFC) to the inner model problem for just one supercompact cardinal. Then, as a corollary, one would obtain a proof of the following conjecture:

Conjecture (ZF) *There are no weak Reinhardt cardinals.* □

It is possible to isolate a specific conjecture that must be true if there is a positive solution to the inner problem for one supercompact cardinal that itself suffices for this inconsistency result. To explain this further, I must give one last definition.

This is the definition of the class HOD that originates in remarks of Gödel at the Princeton Bicentennial Conference in December 1946 (Levy, 1965).

Definition 20.17 (ZF)

1. For each ordinal α, let HOD_α be the set of all sets a such that there exists a set $A \subseteq \alpha$ such that

a. A is definable in V_α from ordinal parameters
b. $a \in L_\alpha[A]$.
2. HOD is the class of all sets a such that $a \in \text{HOD}_\alpha$ for some α.   □

The definition of HOD_α combines features of the definition of L_α and features of the definition of V_α. I caution that, just as is the case for L_α, in general, we have

$$\text{HOD}_\alpha \neq \text{HOD} \cap V_\alpha,$$

although for a proper class of ordinals α, it is true that $\text{HOD}_\alpha = \text{HOD} \cap V_\alpha$.

The class HOD is quite interesting for a number of reasons, one of which is illustrated by the following observation of Gödel, which, as indicated, is stated within just the theory ZF, in other words, without assuming the axiom of choice.

Theorem 20.18 (ZF) $\text{HOD} \models \text{ZFC}$.   □

This theorem gives a completely different approach to showing that if ZF is formally consistent, then so is ZFC.

One difficulty with HOD is that the definition of HOD is not absolute; for example, in general, HOD is not even the same as defined within HOD. As a consequence, almost any set-theoretic question one might naturally ask about HOD is formally unsolvable. Two such immediate questions are whether $V = \text{HOD}$ and, more simply, whether HOD contains all the real numbers. Both of these questions are formally unsolvable but are of evident importance because they specifically address the complexity of the axiom of choice. If $V = \text{HOD}$, then there is no mystery as to why the axiom of choice holds, but of course, one is left with the problem of explaining why $V = \text{HOD}$. If $V = L$, then it is easy to verify that $V = \text{HOD}$. Furthermore, the inner models of Mitchell-Steel (in the situations where existence can be proved) can always be constructed to be contained in HOD, even though the axiom $V = \text{HOD}$ can fail in a Mitchell-Steel inner model.

I now present a key conjecture. This conjecture involves the notion that an uncountable cardinal γ is a *regular cardinal*. This is the property that for all $X \subseteq \gamma$, if $|X| < \gamma$, then X is bounded in γ. Alternatively, referring to notions already defined, an uncountable cardinal γ is a regular cardinal if it is not a singular cardinal.

The HOD Conjecture: (ZFC) *Suppose that κ is a supercompact cardinal. Then there exists a regular cardinal $\gamma > \kappa$ that is not a measurable cardinal in HOD.*   □

If there is a supercompact cardinal in V, then the HOD conjecture implies that HOD is close to V. Assuming that a slightly stronger large cardinal hypothesis holds in V (and that the HOD conjecture also holds in V), one actually obtains that HOD is quite close to V. As evidence for the latter claim, I note the following theorem. Part 3 of this theorem is a very strong closure condition for HOD, and part 2 follows directly from part 3. The analogous closure condition holds for any transitive (proper) class $N \models \text{ZFC}$ such that for some cardinal δ,

$$N \models \text{``}\delta \text{ is a supercompact cardinal,''}$$

and such that this is witnessed by the class of all restrictions $E \cap N$, where E is an extender and $E \cap N \in N$. Referring back to the discussion of the distinction between the core model program and the inner model program, this closure condition establishes that in the context of the existence of just one supercompact cardinal, there is no difference between the primary objectives of these two programs.

Theorem 20.19 (ZFC) *Suppose that there is an extendible cardinal and that the* HOD *conjecture holds. Then the following hold:*

1. *There exists an ordinal α such that for all cardinals $\gamma > \alpha$, if γ is a singular cardinal, then $\gamma^+ = (\gamma^+)^{\mathrm{HOD}}$.*
2. *Suppose for each n that there is a proper class of n-huge cardinals. Then for each n,*

$$\mathrm{HOD} \models \text{``There is a proper class of n-huge cardinals.''}$$

3. *There exists an ordinal α such that for all $\gamma > \alpha$, if*

$$j : \mathrm{HOD} \cap V_{\gamma+1} \to \mathrm{HOD} \cap V_{j(\gamma)+1}$$

   is an elementary embedding with critical point above α, then $j \in \mathrm{HOD}$.   □

If the HOD conjecture is provable in ZFC, then there is a striking corollary in ZF:

Theorem 20.20 (ZF) *Suppose that λ is a limit of supercompact cardinals and that there is an extendible cardinal below λ. Then there is no elementary embedding*

$$j : V_{\lambda+2} \to V_{\lambda+2}$$

that is not the identity.   □

This corollary would prove the preceding conjecture (that there are no weak Reinhardt cardinals) and would give an inconsistency result that more closely matches the version of Theorem 20.7 that Kunen (1970) actually proved (assuming the axiom of choice).

Theorem 20.21 (Kunen, 1970) *Suppose that λ is an ordinal. Then there is no elementary embedding $j : V_{\lambda+2} \to V_{\lambda+2}$ that is not the identity.*   □

The connection between the HOD conjecture and the inner model problem for one supercompact cardinal is illustrated by the next theorem. Arguably, statement 3 of this theorem would follow from any reasonable solution to the inner model problem for one supercompact cardinal.

Theorem 20.22 (ZFC) *Suppose that there is an extendible cardinal. Then the following are equivalent:*

1. *The* HOD *conjecture holds.*
2. *There is a cardinal δ such that*

$$\mathrm{HOD} \models \text{``δ is a supercompact cardinal,''}$$

   and this is witnessed by the class of all $E \cap \mathrm{HOD}$ such that E is an extender and $E \cap \mathrm{HOD} \in \mathrm{HOD}$.
3. *There exists a class $N \subseteq \mathrm{HOD}$, and there exists a cardinal δ such that*

$$N \models \mathrm{ZFC} + \text{``δ is a supercompact cardinal,''}$$

and this is witnessed by the class of all $E \cap N$ such that E is an extender and $E \cap N \in N$.   □

These developments come with a price. For the large cardinal axioms stronger than the axiom that asserts the existence of one supercompact cardinal, the set theorist's cosmological principle must be either abandoned or revised. The reason for this is that the solution to the inner model problem for the specific axiom "there exists one supercompact cardinal" necessarily will solve the inner model problem (as currently defined) for essentially all the known large axioms up to and including the axiom "there is an ω-huge cardinal."

We therefore face a very simple dichotomy of possibilities: the inner model problem for the axiom "there exists one supercompact cardinal" is solvable, or it is not. The other possibility – this is the possibility that this solvability question is itself unsolvable – is not an option based on any reasonable notion of mathematical truth. The simple reason for this is that if the solvability question is itself formally unsolvable, then the inner model problem for the axiom "there exists one supercompact cardinal" is not solvable. The situation here is exactly like the situation for a number (but not all) of the prominent open questions of modern mathematics. For example, if the Riemann hypothesis is formally unsolvable (and there is absolutely no evidence for this), then the Riemann hypothesis is true.

Whatever the outcome to this dichotomy of possibilities, one outcome seems certain: the Set Theorist's cosmological principle cannot be applied to argue for the truth of (any) large cardinal axioms beyond the axiom "there exists one supercompact cardinal." Of course, it could be that the solution to the inner model problem for the axiom "there exists one supercompact cardinal" involves the construction of an inner model that is not of the form $L[\tilde{E}]$, where $\tilde{E}$ is a sequence of partial extenders, for example. But this alone would not suffice to resolve the issue raised by the preceding theorem. The reason is that statement 3 of the theorem makes no assumption that the inner model N is constructed from an extender sequence. Furthermore, the necessity of the closeness of N to V (which is the only issue here) does not require that $N \subseteq \text{HOD}$ but only requires that for some cardinal δ,

$$N \models \text{ZFC} + \text{``}\delta \text{ is a supercompact cardinal."}$$

This is witnessed by the class of all $E \cap N$ such that E is an extender and $E \cap N \in N$. A solution that solves the inner model problem for the axiom "there exists one supercompact cardinal," and yet involves only the construction of inner models for which this fails would be completely unlike all the current solutions to the inner model problem for the various large cardinal axioms where a solution exists.

One can correctly speculate that the difficulty is in the requirement that for the inner model N, the relevant large cardinal axiom is witnessed to hold by extenders of N that are of the form $E \cap N$ for some extender $E \in V$. However, extenders are the witnesses for large axioms, and therefore any genuine construction of an inner model N should arguably satisfy this requirement. Moreover, to avoid trivial solutions, one has to require that the associated inner models satisfy some form of being close to V. More precisely, one has to require that the large cardinals of the inner models N that constitute the solution have some form of ancestry in the large cardinals of V.

There is a silver lining to this dark cloud. Suppose that the inner model problem at the level of one supercompact cardinal can be solved and that the solution does involve defining inner models that are of the form $L[\tilde{E}]$, where $\tilde{E}$ is a sequence of (partial) extenders. Then it is possible to analyze the relationship of the inner models $L[\tilde{E}]$ to V without knowing how the corresponding extender sequences $\tilde{E}$ are actually constructed. The results to date (Woodin, 2009) have greatly clarified the axioms that assert the existence of ω-huge cardinals and have revealed a new hierarchy of such axioms. The emerging structure theory for these axioms could well develop to the point where it serves as a surrogate for the existence of an inner model theory in a revised version of the Set Theorist's cosmological principle. On general grounds, one can argue that if these axioms are consistent, then $L[\tilde{E}]$ must provide a structure theory for these axioms because if the axioms can hold in V, then they can hold in $L[\tilde{E}]$ (by the closeness of $L[\tilde{E}]$ to V). Therefore the required revision of the Set Theorist's cosmological principle may actually not be so severe. The revision would be in what constitutes an inner model theory for those large cardinal axioms beyond the large cardinal axiom that asserts the existence of one supercompact cardinal. My point is that simply requiring that there be a generalization of the axiom of constructibility that is compatible with the large cardinal axiom may (for evident reasons) not be sufficient for these axioms. In addition, one may have to impose much stronger requirements, perhaps more in line with the often-quoted speculation of Gödel (1947):

> There might exist axioms so abundant in their verifiable consequences, shedding so much light upon a whole discipline, and furnishing such powerful methods for solving given problems (and even solving them, as far as possible, in a constructivistic way) that quite irrespective of their intrinsic necessity they would have to be assumed at least in the same sense as any established physical theory.

To summarize the point I am attempting to make:

> The foundational basis for asserting that large cardinal axioms beyond the level of one supercompact cardinal are true might lie in the structural consequences for $L[\tilde{E}]$ that their existence implies. Moreover, this claim of truth may require (and reinforce) some version of the claim that $V = L[\tilde{E}]$.

This speculation is grounded in a number of preliminary results (Woodin, 2009).

20.8 Summary

There is now a body of mathematical evidence that if there is a supercompact cardinal, then there is a transcendent version, say, L^{Ω}, of Gödel's inner model L: in brief, there is an ultimate L. This development, if realized, will yield a much deeper understanding of the large cardinal axioms:

1. identifying much more precisely the transition for large cardinal axioms from the possible to the impossible
2. providing a framework for a continuing evolution in the understanding of this transition

The analysis will reveal some very subtle theorems about the nature of sets, which in turn will eliminate essentially all the large cardinal axioms known to contradict the axiom of choice. How, then, could one account for the new predictions of consistency (and formal unsolvability) that will arise? Certainly not by invoking the Skeptic's Retreat.

Finally, we know that Gödel rejected the axiom $V = L$. The current view rejects this axiom primarily because it is a limiting axiom that denies large cardinal axioms. This particular argument would not apply to the axiom $V = L^{\Omega}$. Furthermore, assuming that the analysis can be carried out to construct L^{Ω},

> there is no known candidate for a sentence that is independent from the axiom $V = L^{\Omega}$ and that is not a consequence of some large cardinal axiom.

However, all large cardinal axioms are merely axioms for the height of L^{Ω}, given that no (known) large cardinal axiom can transcend L^{Ω}.

As this point is a key point of the thesis of this chapter, I shall discuss it a bit further. It is well known that large cardinal axioms yield new theorems of number theory; more precisely, assuming the large cardinal axioms to be true, one can infer as true specific statements of number theory that arguably cannot otherwise be proved. The foundational issue raised by this is that the large cardinals do not exist in the universe of number theory and yet their existence generates new truths of that universe. How, then, can the number theorist account for these truths? This, of course, is presented in the previous discussion as the debate between the Skeptic and the Set Theorist.

Why do similar issues not arise for the universe given by L^{Ω}? Because the large cardinals can (and therefore do) exist in this universe. This is the key new feature that L^{Ω} would possess that sets it apart from all the current known generalizations of L. As a consequence of this feature, L^{Ω} would provide an unambiguous conception of the transfinite universe, giving an example of an axiom that achieves this goal and that is arguably compatible with all large cardinal axioms where there is no example currently known.

Would this alone be sufficient to argue that the axiom $V = L^{\Omega}$ is the axiom for V? No more than one could argue that the axiom $V = L$ is the axiom for V should all the large cardinal axioms that imply that $V \neq L$ turn out to be inconsistent (which will not happen). However, the successful construction of L^{Ω} would provide substantial evidence that there is a single axiom for V that yields a conception of the universe of sets that is in fact (for the reasons articulated earlier) more unambiguous than our present conception of number theory. Moreover, the successful construction of L^{Ω} would provide a starting point for discovering that axiom.

This development would be a significant milestone in our understanding of the transfinite universe. I make this claim completely independently of any speculation that there are number-theoretic problems that are orthogonal to all large cardinal axioms such as is the case for the problem of the continuum hypothesis.

References

Beller, A., Jensen, R. B., and Welch, P. (1982). *Coding the Universe*. London Mathematical Society Lecture Note Series 47. Cambridge: Cambridge University Press.

Dodd, A., and Jensen, R. (1981). The core model. *Ann. Math. Logic*, **20**, 43–75.

Gödel, K. (1938). Consistency-proof for the generalized continuum-hypothesis. *Proc. Nat. Acad. Sci. U.S.A.* **25**, 220–24.

———. (1940). *The Consistency of the Continuum Hypothesis*. Annals of Mathematics Studies 3. Princeton, NJ: Princeton University Press.

Kanamori, A. (1994). *The Higher Infinite*. Perspectives in Mathematical Logic. Berlin: Springer.

Kunen, K. (1970). Some applications of iterated ultrapowers in set theory. *Ann. Math. Logic*, **1**, 179–227.

———. (1971). Elementary embeddings and infinitary combinatorics. *J. Symbolic Logic*, **36**, 407–13.

Lévy, A. (1965). Definability in axiomatic set theory. I. In *Logic, Methodology and Philosophy of Science (Proceedings 1964 International Congress)*, pp. 127–51. Amsterdam, Netherlands: North-Holland.

Mitchell, W. J., and Steel, J. R. (1994). *Fine Structure and Iteration Trees*. Berlin: Springer-Verlag.

Mycielski, J., and Steinhaus, H. (1962). A mathematical axiom contradicting the axiom of choice. *Bull. Acad. Polon. Sci. Sér. Sci. Math. Astronom. Phys.*, **10**, 1–3.

Neeman, I. (2002). Inner models in the region of a Woodin limit of Woodin cardinals. *Ann. Pure Appl. Logic*, **116**, 67–155.

Reinhardt, W. N. (1970). Ackermann's set theory equals ZF. *Ann. Math. Logic*, **2**, 189–249.

Scott, D. (1961). Measurable cardinals and constructible sets. *Bull. Acad. Polon. Sci. Sér. Sci. Math. Astronom. Phys.*, **9**, 521–24.

Shelah, S. (1986). *Around Classification Theory of Models*. Lecture Notes in Mathematics 1182. Berlin: Springer.

Steel, J. R. (1996). *The Core Model Iterability Problem*. Berlin: Springer.

Woodin, W. H. (1996). The universe constructed from a sequence of ordinals. *Arch. Math. Logic*, **35**, 371–83.

———. (1999). *The Axiom of Determinacy, Forcing Axioms, and the Nonstationary Ideal*. Berlin: Walter de Gruyter.

———. (2010). Suitable extender sequences. To appear in *Journal of Mathematical Logic*, pp. 1–676.

Zermelo, E. (1930). Über Grenzzahlen und Mengenbereiche: Neue Untersuchungenuberdie Grundlagen der Mengenlehre. *Fundamenta Math.*, **16**, 29–47.

Gödel and Computer Science

CHAPTER 21

The Gödel Phenomenon in Mathematics: A Modern View

Avi Wigderson

What are the limits of mathematical knowledge? The purpose of this chapter is to introduce the main concepts from *computational complexity theory* that are relevant to algorithmic accessibility of mathematical understanding. In particular, I'll discuss the $\mathcal{P}$ versus $\mathcal{NP}$ problem, its possible impact on research in mathematics, and how interested Gödel himself was in this computational viewpoint.

Much of the technical material will be necessarily sketchy. The interested reader is referred to the standard texts on computational complexity theory, primarily Arora and Barak (2009), Goldreich (2008), Papadimitriou (1994a), and Sipser (1997).

21.1 Overview

Hilbert believed that all mathematical truths are *knowable*, and he set the threshold for mathematical knowledge at the ability to devise a "mechanical procedure." This dream was shattered by Gödel and Turing. Gödel's incompleteness theorem exhibited true statements that can never be proved. Turing formalized Hilbert's notion of computation and of finite algorithms (thereby initiating the computer revolution) and proved that some problems are *undecidable* – they have no such algorithms.

Though the first examples of such unknowables seemed somewhat unnatural, more and more natural examples of unprovable or undecidable problems were found in different areas of mathematics. The independence of the continuum hypothesis and the undecidability of Diophantine equations are famous early examples. This became known as the *Gödel phenomenon*, and its effect on the practice of mathematics has been debated since. Many argued that though some of the inaccessible truths above are natural, they are far from what is really of interest to most working mathematicians. Indeed, it would seem that in the seventy-five years since the incompleteness theorem, mathematics has continued thriving, with remarkable achievements such as the recent settlement of Fermat's last "theorem" by Wiles and the Poincaré conjecture by Perelman. Are there *interesting* mathematical truths that are *unknowable*?

The main point of this chapter is that when *knowability* is interpreted by modern standards, namely, via *computational complexity*, the Gödel phenomenon is very much with us. We argue that to understand a mathematical structure, having a *decision procedure* is but a first approximation; a real understanding requires an *efficient* algorithm. Remarkably, Gödel was the first to propose this modern view in a letter to von Neumann in 1956, which was discovered only in the 1990s.

Meanwhile, from the mid-1960s on, the field of theoretical computer science has made formal Gödel's challenge and has created a theory that enables quantification of the difficulty of *computational* problems. In particular, a reasonable way to capture *knowable* problems (which we can efficiently solve) is the class $\mathcal{P}$, and a reasonable way to capture *interesting* problems (which we would like to solve) is the class $\mathcal{NP}$. Moreover, assuming the widely believed $\mathcal{P} \neq \mathcal{NP}$ conjecture, the class $\mathcal{NP}$-complete captures interesting *unknowable* problems.

In this chapter, I define these complexity classes, explain their intuitive meaning, and show how Gödel foresaw some of these definitions and connections. I relate computational difficulty to mathematical difficulty and argue how such notions, developed in computational complexity, may explain the difficulty mathematicians have in accessing structural, noncomputational information. I also survey proof complexity: the study of lengths of proofs in different proof systems.

Finally, I point out that this modern view of the limits of mathematical knowledge is adopted by a growing number of mathematicians working on a diverse set of interesting structures in different areas of mathematics. This activity increases interaction with computer scientists and benefits both fields. Results are of two types, as in standard computer science problems. On one hand, there is a growing effort to go beyond characterizations of mathematical structures, and attempts are made to provide efficient recognition algorithms. On the other, there is a growing number of $\mathcal{NP}$-completeness results, providing the stamp of difficulty for achieving useful characterizations and algorithms. This second phenomenon, now ubiquitous in science and mathematics, may perhaps better capture Gödel's legacy on what is unknowable in mathematics.

21.1.1 Decision Problems and Finite Algorithms

Which mathematical structures can we hope to understand? Let us focus on the most basic mathematical task of *classification*. We are interested in a particular class of objects and a particular property. We seek to *understand* which of the objects have the property and which do not. Let us consider the following examples:

1. Which valid sentences in first-order predicate logic are provable?
2. Which Diophantine equations have solutions?
3. Which knots are unknotted?
4. Which elementary statements about the reals are true?

It is clear that each object from the preceding families has a finite representation. Hilbert's challenge was about the possibility to find, for each such family, a finite procedure that would solve the decision problem *in finite time* for every object in the family. In his seminal paper, Turing (1936, 1937) formulated the *Turing machine*, a mathematical definition of an *algorithm*, capturing such finite mechanical procedures. This allowed a mathematical study of Hilbert's challenge.

Hilbert's *Entscheidungsproblem* – problem 1 – was the first to be resolved, in the same paper by Turing. Turing showed that problem 1 is undecidable, namely, that there is no algorithm that distinguishes provable from unprovable statements of first-order predicate logic. Hilbert's tenth problem – problem 2 – was shown to be undecidable as well, in a series of works by Davis, Putnam, Robinson, and Matiasevich (Matiasevich, 1993). Again, no algorithm can distinguish solvable from unsolvable Diophantine equations.

The crucial ingredient in those (and all other) undecidability results is showing that each of these mathematical structures can *encode computation*. This is known today to hold for many different structures in algebra, topology, geometry, analysis, logic, and more, even though a priori the structures studied seem to be completely unrelated to computation. It is as though much of contemporary work in mathematics is pursuing, unwittingly and subconsciously, an agenda with essential computational components. I shall return to refined versions of this idea later.

The notion of a decision procedure as a minimal requirement for understanding of a mathematical problem has also led to direct positive results. It suggests that we look for a decision procedure as *a means*, or as *a first step*, for understanding a problem. For problems 3 and 4, this was successful. Haken (1961) showed how knots can be so understood, with his decision procedure for problem 3, and Tarski (1951) showed that real-closed fields can be understood with a decision procedure for problem 4. Naturally, significant *mathematical, structural* understanding was needed to develop these algorithms. Haken developed the theory of *normal surfaces*, and Tarski invented *quantifier elimination* for their algorithms, both cornerstones of the respective fields.

This reveals only the obvious: mathematical and algorithmic understanding are related and often go hand in hand. There are many ancient examples. The earliest is probably Euclid's greatest common divisor (GCD) algorithm. Abel's proof that roots of real polynomials of degree at least 5 have no *formula* with radicals is the first impossibility result (of certain classes of algorithms). Of course, Newton's *algorithm* for approximating such roots is a satisfactory practical alternative. What was true in previous centuries is truer in this one: the language of algorithms is slowly becoming competitive with the language of equations and formulas (which are special cases of algorithms) for explaining complex mathematical structures.

Back to our four problems. The undecidability of problems 1 and 2 certainly suggests that these structures cannot be mathematically understood in general. This led researchers to consider special cases of problems that are decidable. But does decidability, for example, that of problems 3 and 4, mean that we completely understand these structures? Far from it. Both algorithms are extremely complex and time consuming. Even structurally speaking, there are no known complete knot invariants or characterizations of real varieties. How can one explain such mathematical difficulties for decidable problems?

It is natural to go beyond decidability and try to quantify the level of understanding. We will use a computational yardstick for it. We argue that better mathematical understanding goes hand in hand with better algorithms for "obtaining" that understanding from the given structures. To formalize this, we will introduce the computational terms that are central to the theory of computational complexity. There is no better way to start this than with Gödel's letter to von Neumann.

21.2 Gödel's Letter to von Neumann

In Gödel's letter, which I include in its entirety for completeness, complexity is discussed only in the second paragraph – it is remarkable how much this paragraph contains! It defines *asymptotic* and *worst-case* time complexity, suggests the study of the *satisfiability* problem, discusses the mathematical significance of having a fast algorithm for it, and even speculates on the possibility of the existence of such an algorithm. In the section that follows the letter, I identify the letter's contributions and insights in modern language.

21.2.1 The Letter

First, a bit on the context. The letter was written when von Neumann was in the hospital, already terminally ill with cancer (he died a year later). It was written in German, and here we reproduce a translation. The surveys by Sipser (1992) and Hartmanis (1989) provide the original letter as well as more details on the translation.

As far as we know, the discussion Gödel was trying to initiate never continued. Moreover, we have no evidence of Gödel's thinking more about the subject. Again, we will refer later only to the second paragraph, which addresses the computational complexity issue.

```
Princeton, 20 March 1956

Dear Mr. von Neumann:

With the greatest sorrow I have learned of your
illness. The news came to me as quite unexpected.
Morgenstern already last summer told me of a bout of
weakness you once had, but at that time he thought
that this was not of any greater significance. As I
hear, in the last months you have undergone a radical
treatment and I am happy that this treatment was
successful as desired, and that you are now doing
better. I hope and wish for you that your condition
will soon improve even more and that the newest
medical discoveries, if possible, will lead to a
complete recovery.

Since you now, as I hear, are feeling stronger, I
would like to allow myself to write you about a
mathematical problem, of which your opinion would
```

very much interest me: One can obviously easily construct a Turing machine, which for every formula F in first order predicate logic and every natural number n, allows one to decide if there is a proof of F of length n (length = number of symbols). Let $\psi(F, n)$ be the number of steps the machine requires for this and let $\phi(n) = \max_F \psi(F, n)$. The question is how fast $\phi(n)$ grows for an optimal machine. One can show that $\phi(n) \geq k \cdot n$. If there really were a machine with $\phi(n) \approx k \cdot n$ (or even $\phi(n) \approx k \cdot n^2$), this would have consequences of the greatest importance. Namely, it would obviously mean that in spite of the undecidability of the Entscheidungsproblem, the mental work of a mathematician concerning Yes-or-No questions (apart from the postulation of axioms) could be completely replaced by a machine. After all, one would simply have to choose the natural number n so large that when the machine does not deliver a result, it makes no sense to think more about the problem. Now it seems to me, however, to be completely within the realm of possibility that $\phi(n)$ grows that slowly. Since 1.) it seems that $\phi(n) \geq k \cdot n$ is the only estimation which one can obtain by a generalization of the proof of the undecidability of the Entscheidungsproblem and 2.) after all $\phi(n) \approx k \cdot n$ (or $\phi(n) \approx k \cdot n^2$) only means that the number of steps as opposed to trial and error can be reduced from N to $\log N$ (or $(\log N)^2$). However, such strong reductions appear in other finite problems, for example in the computation of the quadratic residue symbol using repeated application of the law of reciprocity. It would be interesting to know, for instance, the situation concerning the determination of primality of a number and how strongly in general the number of steps in finite combinatorial problems can be reduced with respect to simple exhaustive search.

I do not know if you have heard that "Post's problem," whether there are degrees of unsolvability among problems of the form $(\exists y)\phi(y, x)$, where ϕ is recursive, has been solved in the positive sense by a very young man by the name of Richard Friedberg. The solution is very elegant. Unfortunately, Friedberg does not intend to study mathematics, but rather medicine (apparently under the influence of his father). By the way, what do you think of the

```
attempts to build the foundations of analysis on
ramified type theory, which have recently gained
momentum? You are probably aware that Paul Lorenzen
has pushed ahead with this approach to the theory of
Lebesgue measure. However, I believe that in
important parts of analysis non-eliminable
impredicative proof methods do appear.

I would be very happy to hear something from you
personally. Please let me know if there is something
that I can do for you. With my best greetings and
wishes, as well to your wife,

Sincerely yours,

Kurt Gödel

P.S. I heartily congratulate you on the award that
the American government has given to you.
```

21.2.2 Time Complexity and Gödel's Foresight

In this section, I go through the main ingredients of Gödel's letter, most of which were independently[1] identified and developed in the evolution of computational complexity in the 1960s and 1970s. These include the basic model, input representation (asymptotic, worst-case), time complexity,[2] brute-force algorithms and the possibility of beating them, proving lower bounds, and last but not least, Gödel's choice of a focus problem and its significance. I comment on the remarkable foresight of some of Gödel's choices. When introducing some of these notions, I use common, modern notation.

- **Computability versus complexity:** All problems we will discuss here are computable, that is, they have a finite algorithm. Gödel states that for his problem, "one can...easily construct a Turing machine." We take for granted that the given problem is decidable and ask for the complexity of the best or optimal algorithm. Gödel is the first on record to suggest shifting focus from computability to complexity.
- **The model:** Gödel's computational model of choice is the Turing machine. Time will be measured as the number of elementary steps on a Turing machine. We note that this is a nontrivial choice for a discussion of time complexity back in 1956. The Church-Turing thesis of the equivalence of all feasible computational models was around, and all known algorithmic models were known to simulate each other. However, that thesis

---

[1] Recall that the letter was discovered only in the 1990s.
[2] I shall focus on time as the primary resource of algorithms when studying their efficiency. Other resources, such as memory, parallelism, and more, are studied in computational complexity, but I will not treat them here.

did not include time bounds, and it seems that Gödel takes for granted (or knows) that all known models can simulate each other *efficiently* (something that is now well established).

- **The problem:** Gödel focuses on one problem to define complexity, a finite version of the Entscheidungsproblem, in which the task is to determine if a formula F has a proof of length n. This is by no means an arbitrary choice, and Gödel is well aware of it, as will be discussed later. Put differently, Gödel's problem asks if we can *satisfy* a first-order logic verifier that F is provable using only n symbols. We shall later meet its cousin, the problem SAT (abbreviating "satisfiability"), in which the task is to determine if a *propositional* formula has a satisfying assignment. It is not hard to see that SAT captures Gödel's problem,[3] and so we will call his problem SAT as well.

- **Input representation:** Every finite object (formulas, integers, graphs, etc.) can be represented as a binary string. We let Im stand for the set of all finite binary strings. Let f be a decision problem ("Yes-or-No question" in Gödel's letter), like those in Section 21.1.1. Then $f : I \rightarrow \{0, 1\}$ is what we are trying to compute. Thus we consider Turing machines that for every input x halt with the answer $f(x)$. In Gödel's problem, the input is the pair (F, n), and the task is computing whether the first-order formula F has a proof of length at most n.

- **Input length:** This is always taken in computational complexity to be the binary length of the input. Clearly Gödel makes the same choice: when he talks about the complexity of testing primality of an integer N, he takes the input length to be $\log N$, the number of bits needed to describe N. Of course, finite objects may have many representations that differ in length, and one must pick "reasonable" encodings.[4]

- **Time complexity:** Complexity is measured as a function of *input length*. Adapting Gödel's notation, we fix any Turing machine M computing f. We let $\phi(n)$ denote the maximum,[5] over all inputs x of length n, of the number of steps M takes when computing $f(x)$. We then consider the *growth* of $\phi(n)$ for the "optimal machine"[6] M. It is quite remarkable that Gödel selects both the asymptotic viewpoint and worst-case complexity – neither is an obvious choice but both proved extremely fruitful in the development of computational complexity.

- **Brute-force algorithms:** All problems mentioned by Gödel – SAT, primality testing, and computing quadratic residue – have what Gödel calls "trial and error" and "simple exhaustive search" algorithms. All these problems happen to be in the class $\mathcal{NP}$, and these obvious trivial algorithms require exponential time (in the input length).

---

[3] Simply assign propositional variables to the n possible proof symbols and encode the verification process so that indeed they form a proof of F as a Boolean formula.

[4] For example, in Gödel's problem, it is possible to encode the input using $|F| + \log n$ bits, where $|F|$ is the encoding length (in bits) of the given first-order formula. However, as we will see later, it is more natural to take it to be $|F| + n$, allowing for n variables to encode the possible values of the potential proof symbols. For this problem, Gödel indeed measures complexity as a function of n.

[5] This worst-case analysis is the most common in complexity theory, though of course, other measures are important and well studied. In particular, average-case analysis (typical behavior of algorithms under natural input distributions) is central both for algorithm design and cryptography.

[6] Strictly speaking, this is not well defined.

- **Efficient algorithms:** Gödel specifically suggests that time complexity[7] $O(n)$ or $O(n^2)$ is efficient (or tractable) enough for practical purposes. This did not change for fifty years, despite enormous changes in computer speed, and is a gold standard of efficiency. I'll try to explain why computational complexity theory chose to call (any) polynomial-time algorithms efficient and why the class $\mathcal{P}$ includes all problems with such algorithms.

- **The complexity of SAT:** Gödel is completely aware of what it would mean if the problem SAT were to have an efficient algorithm: "the mental work of a mathematician concerning Yes-or-No questions (apart from the postulation of axioms) could be completely replaced by a machine." So he realizes that solving this problem will solve numerous others in mathematics. While not stating precisely that it is $\mathcal{NP}$-complete ("captures" all problems in $\mathcal{NP}$), Gödel does ask "how strongly in general the number of steps in finite combinatorial problems can be reduced with respect to simple exhaustive search," a set of problems we naturally identify now with $\mathcal{NP}$.

- **Gödel's speculation:** The letter states, "Now it seems to me...to be completely within the realm of possibility that $\phi(n)$ grows that slowly," namely, that $SAT \in \mathcal{P}$, which, as we shall, see implies $\mathcal{P} = \mathcal{NP}$. Gödel suggests this possibility based on the existence of highly nontrivial efficient algorithms that improve exponentially over brute-force search: "such strong reductions appear in other finite problems, for example in the computation of the quadratic residue symbol using repeated application of the law of reciprocity" (the trivial exponential time algorithm would rely on factoring, whereas Gauss's reciprocity allows a GCD-like polynomial-time algorithm[8]). This lesson, that there are sometimes ingenious ways of cutting the exponential search space for many important problems, was taught again and again in the last half century.

- **Lower bounds:** Gödel tried proving that the complexity of SAT, $\phi(n)$, cannot grow too slowly, but he could only prove a linear lower bound[9] $\phi(n) = \Omega(n)$. For Turing machines, not much has changed in this half century – we still have no superlinear lower bounds. Proving such lower bounds, namely, proving that computational problems are intrinsically difficult, is a central task of complexity. I shall discuss partial success as well as the difficulty of this task later.

The next sections elaborate on these and other notions and issues and provide some historical background on their development in computational complexity theory. I first deal with the complexity of computation and then move on to discuss the complexity of proofs.

---

[7] We use standard asymptotic notation: $g(n) = O(h(n))$ if, for some constant k and for all n, we have $g(n) \leq kh(n)$. In this case, we also say that $h(n) = \Omega(g(n))$.

[8] In any reasonable implementation Gödel had in mind, this would take time n^3, which shows that Gödel may have indeed considered the class $\mathcal{P}$ for efficiently solved problems.

[9] Note that though this lower bound is trivial in the modern formulation of SAT, in which the propositional formula has length at least n, it is far from trivial in Gödel's formulation. Gödel does not supply a proof, and Buss provides one in Buss (1995).

21.3 Complexity Classes, Reductions, and Completeness

In this section, I define efficient computations, efficient reductions between problems, efficient verification, and the classes $\mathcal{P}$, $\mathcal{NP}$, co$\mathcal{NP}$, and $\mathcal{NP}$-complete. I will keep referring back to Gödel's letter.

21.3.1 Efficient Computation and the Class $\mathcal{P}$

In all that follows, I focus on asymptotic complexity and analyze time spent on a problem as a function of input length. The asymptotic viewpoint is inherent to computational complexity theory. We shall see in this chapter that it reveals structure that would probably be obscured by finite, precise analysis.

Efficient computation (for a given problem) will be taken to be one whose run time on any input of length n is bounded by a *polynomial* function in n. Let I_n denote all binary sequences in I of length n.

Definition 21.3.1 (the class $\mathcal{P}$)   *A function $f : I \rightarrow I$ is in the class $\mathcal{P}$ if there is an algorithm computing f and positive constants A, c such that for every n and every $x \in I_n$, the algorithm computes $f(x)$ in at most An^c steps.*

Note that the definition applies in particular to Boolean functions (whose output is $\{0, 1\}$), which capture classification problems. For convenience, we will sometimes think of $\mathcal{P}$ as the class containing only these classification problems. Observe that a function with a long output can be viewed as a sequence of Boolean functions, one for each output bit.

This definition was suggested by Cobham (1965), Edmonds (1965b), and Rabin (1967), all attempting to formally delineate *efficient* from just *finite* (in their cases, exponential time) algorithms. Of course, nontrivial polynomial-time algorithms were discovered earlier, long before the computer age. Many were discovered by mathematicians, who needed efficient methods to calculate (by hand). The most ancient and famous example is Euclid's GCD algorithm, which bypasses the need to factor the inputs when computing their greatest common factor.

21.3.1.1 Why Polynomial?

The choice of polynomial time to represent efficient computation seems arbitrary, and different possible choices can be made.[10] However, this particular choice was extremely useful and has justified itself over time from many points of view. We list some important ones.

Polynomials typify slowly growing functions. The closure of polynomials under addition, multiplication, and composition preserves the notion of efficiency under natural programming practices such as using two programs in sequence or using one as a subroutine of another. This choice removes the necessity to describe the computational model precisely (e.g., it does not matter if we allow arithmetic operations only on single

---

[10] And indeed such choices are studied in computational complexity.

digits or on arbitrary integers because long addition, subtraction, multiplication, and division have simple polynomial-time algorithms taught in grade school). Similarly, we need not worry about data representation: one can efficiently translate between essentially any two natural representations of a set of finite objects.

From a practical viewpoint, while a running time of, say, n^2 is far more desirable than n^{100}, very few known efficient algorithms for natural problems have exponents above 3 or 4. On the other hand, many important natural problems that so far resist efficient algorithms cannot at present be solved faster than in *exponential* time. Thus reducing their complexity to (any) polynomial is a huge conceptual improvement (and when this is achieved, often, further reductions of the exponent are found).

The importance of understanding the class $\mathcal{P}$ is obvious. There are numerous computational problems that arise (in theory and practice) that demand efficient solutions. Many algorithmic techniques were developed in the past four decades and enable solving many of these problems (see, e.g., the textbook by Cormen et al. (2001)). These drive the ultrafast home computer applications we now take for granted, such as Web searching, spell checking, data processing, computer game graphics, and fast arithmetic, as well as heavier-duty programs used across industry, business, math, and science. However, many more problems (some of which we shall meet soon), perhaps of higher practical and theoretical value, remain elusive. The challenge of *characterizing* this fundamental mathematical object – the class $\mathcal{P}$ of efficiently solvable problems – is far beyond us at this point.

I end this section with a few examples of nontrivial problems in $\mathcal{P}$ of mathematical significance. In each, the interplay of mathematical and computational understanding needed for the development of these algorithms is evident.

- **Primality testing:** Given an integer, determine if it is prime. Gauss challenged the mathematical community to find an efficient algorithm, but it took two centuries to solve. The story of this recent achievement of Agrawal et al. (2004) and its history are beautifully recounted in Granville (2005).
- **Linear programming:** Given a set of linear inequalities in many variables, determine if they are mutually consistent. This problem and its optimization version capture numerous others (finding optimal strategies of a zero-sum game is one), and the convex optimization techniques used to give the efficient algorithms (Khachian, 1979; Karmarkar, 1984) for it do much more (see, e.g., Schrijver, 2003).
- **Factoring polynomials:** Given a multivariate polynomial with *rational* coefficients, find its irreducible factors over $\mathbb{Q}$. Again, the tools developed in Lenstra et al. (1982) (mainly regarding short bases in lattices in Re^n) have numerous other applications.
- **Hereditary graph properties:** Given a finite graph, test if it can be embedded on a fixed surface (like the plane or the torus). A vastly more general result is known, namely, testing *any* hereditary property (one that is closed under vertex removal and edge contraction). It follows the monumental structure theory (Robertson and Seymour, 1983–1995) of such properties, including a *finite basis theorem* and its algorithmic versions.[11]

---

[11] To be fair, while running in polynomial time, the algorithms here have huge exponents. We expect that further, deeper structural understanding will be needed for more efficient algorithms.

- **Hyperbolic word problem:** Given any presentation of a hyperbolic group by generators and relations, and a word w in the generators, determine if w represents the identity element. The techniques give isoperimetric bounds on the Cayley graphs of such groups and more (Gromov, 1987).

21.3.2 Efficient Verification and the Class $\mathcal{NP}$

Now we change our view of classification problems from classifying functions to classifying sets. We shall see that some subtle (and highly relevant mathematical) aspects of complexity come to life when we focus on classifying sets. Thus a classification problem will be any subset $C \subset I$. It is convenient for this section to view C as defining a property; $x \in C$ are objects that have the property, and $x \notin C$ are objects that do not. While this is formally equivalent to having a Boolean function $f : I \to \{0, 1\}$ that gives opposite values on C and $I\backslash C$, this view allows us to distinguish between the complexity of these two sets.

We are given an input $x \in I$ (describing a mathematical object) and are supposed to determine if $x \in C$. If we had an efficient algorithm for C, we could simply apply it to x, but if we don't, what is the next best thing? One answer is a *convincing proof* that $x \in C$. Before defining it formally, let us see a motivating example.

As Gödel points out, the working mathematician meets such examples daily, reading a typical math journal paper. In it, we typically find a (claimed) theorem, followed by an (alleged) proof. Thus we are verifying claims of the type $x \in THEOREMS$, where *THEOREMS* is the set of all provable statements in, say, set theory. It is taken for granted that the written proof is *short* (page limit) and *easily verifiable* (otherwise the referee and editor would demand clarification), regardless of how long it took to discover.

The class $\mathcal{NP}$ contains all properties C for which membership (namely, statements of the form $x \in C$) have *short, efficiently verifiable* proofs. As before, we use polynomials to define both terms. A candidate proof y for the claim $x \in C$ must have a length at most polynomial in the length of x, and the verification that y indeed proves this claim must be checkable in polynomial time. Finally, if $x \notin C$, no such y should exist.

Definition 21.3.2 (the class $\mathcal{NP}$)   The set C is in the class $\mathcal{NP}$ if there is a function $V_C \in \mathcal{P}$ and a constant k such that

- if $x \in C$, then $\exists y$ with $|y| \leq |x|^k$ and $V_C(x, y) = 1$
- if $x \notin C$, then $\forall y$ we have $V_C(x, y) = 0$

Thus each set C in $\mathcal{NP}$ may be viewed as a set of theorems in the complete and sound proof system defined by the verification process V_C.

A sequence y that convinces V_C that $x \in C$ is often called a *witness* or *certificate* for the membership of x in C. Again, we stress that the definition of $\mathcal{NP}$ is not concerned with how difficult it is to come up with a witness y. Indeed, the acronym $\mathcal{NP}$ stands for "nondeterministic polynomial time," where the nondeterminism captures the ability of a *hypothetical* "nondeterministic" machine to "guess" a witness y (if one exists) and then verify it deterministically.

Nonetheless, the complexity of finding a witness is, of course, important, as it captures the *search problem* associated with $\mathcal{NP}$ sets. Every decision problem C (indeed every verifier V_C for C) in $\mathcal{NP}$ comes with a natural search problem associated with it: given $x \in C$, find a short witness y that convinces V_C. A correct solution to this search problem can be easily verified by V_C.

Though it is usually the search problems that occupy us, from a computational standpoint, it is often more convenient to study the decision versions. Both versions are almost always equivalent.[12]

These definitions of $\mathcal{NP}$ were first given (independently and in slightly different forms) by Cook (1971) and Levin (1973), although Edmonds discusses "good characterization" that already captures the essence of efficient verification (Edmonds, 1965a). There is much more to these seminal papers than this definition, and we shall discuss this later at length.

It is evident that decision problems in $\mathcal{P}$ are also in $\mathcal{NP}$. The verifier V_C is simply taken to be the efficient algorithm for C, and the witness y can be the empty sequence.

Corollary 21.3.3 $\mathcal{P} \subseteq \mathcal{NP}$.

A final comment is that problems in $\mathcal{NP}$ have trivial *exponential time* algorithms. Such algorithms search through all possible short witnesses and try to verify each. Now we can make Gödel's challenge more concrete. Can we always speed up this brute-force algorithm?

21.3.3 The $\mathcal{P}$ versus $\mathcal{NP}$ Question: Its Meaning and Importance

The class $\mathcal{NP}$ is extremely rich (we shall see examples of this later). There are literally thousands of $\mathcal{NP}$ problems in mathematics, optimization, artificial intelligence, biology, physics, economics, industry, and other applications that arise naturally out of different necessities and whose efficient solutions will benefit us in numerous ways. They beg for efficient algorithms, but decades of effort (and sometimes more) has succeeded for only a few. Is it possible that all sets in $\mathcal{NP}$ possess efficient algorithms, and these simply have not been discovered yet? This is the celebrated $\mathcal{P}$ versus $\mathcal{NP}$ question, which appeared explicitly first in the aforementioned papers of Cook and Levin and was foreshadowed by Gödel.

Open Problem 21.3.4   Is $\mathcal{P} = \mathcal{NP}$?

What explains the abundance of so many natural, important problems in the class $\mathcal{NP}$? Probing the intuitive meaning of the definition of $\mathcal{NP}$, we see that it captures many tasks of human endeavor *for which a successful completion can be easily recognized*. Consider the following professions and the typical tasks they are facing (this will be extremely superficial but nevertheless instructive):

- **Mathematician:** Given a mathematical claim, come up with a proof for it.
- **Scientist:** Given a collection of data on some phenomena, find a theory explaining it.

---

[12] A notable possible exception is the set *COMPOSITES* (with a verification procedure that accepts as witness a nontrivial factor). Note that though *COMPOSITES* $\in \mathcal{P}$ is a decision problem, the related search problem is equivalent to integer factorization, which is not known to have an efficient algorithm.

- **Engineer:** Given a set of constraints (on cost, physical laws, etc.), come up with a design (of an engine, bridge, laptop, etc.) that meets these constraints.
- **Detective:** Given the crime scene, find "who done it."

What is common to all these tasks is that we can typically tell a good solution when we see one (or we at least think we can). In various cases, "we" may be the academic community, the customers, or the jury, but we expect the solution to be *short* and *efficiently verifiable*, just as in the definition of $\mathcal{NP}$.

The richness of $\mathcal{NP}$ follows from the simple fact that such tasks abound, and their mathematical formulation is indeed an $\mathcal{NP}$-problem. For all these tasks, efficiency is paramount, and so the importance of the $\mathcal{P}$ versus $\mathcal{NP}$ problem is evident. The colossal implications of the possibility that $\mathcal{P} = \mathcal{NP}$ are evident as well: every instance of these tasks can be solved, optimally and efficiently.

One (psychological) reason people feel that $\mathcal{P} = \mathcal{NP}$ is unlikely is that tasks such as those described earlier often seem to require a degree of *creativity* that we do not expect a simple computer program to have. We admire Wiles' proof of Fermat's last theorem; the scientific theories of Newton, Einstein, Darwin, and Watson and Crick; the design of the Golden Gate Bridge and the pyramids; and sometimes even Hercule Poirot's and Miss Marple's analysis of a murder precisely because they seem to require a leap that cannot be made by everyone, let alone by a simple mechanical device. My own view is that when we finally understand the algorithmic processes of the brain, we may indeed be able to automate the discovery of these specific achievements, and perhaps many others. However, can we automate them all? Is it possible that for every task for which verification is easy, finding a solution is not much harder? If $\mathcal{P} = \mathcal{NP}$, the answer is positive, and creativity (of this abundant, verifiable kind) can be completely automated. Most computer scientists believe that this is not the case.

Conjecture 21.3.5 Is $\mathcal{P} \neq \mathcal{NP}$?

Back to mathematics! Given the preceding discussion, one may wonder why it is so hard to prove that indeed $\mathcal{P} \neq \mathcal{NP}$ – it seems completely obvious. We shall discuss attempts and difficulties soon, developing a methodology that will enable us to identify the *hardest* problems in $\mathcal{NP}$. Before that, though, we discuss a related question with a strong relation to mathematics: the $\mathcal{NP}$ versus co$\mathcal{NP}$ question.

21.3.4 The $\mathcal{NP}$ versus co$\mathcal{NP}$ Question: Its Meaning and Importance

Fix a property $C \subseteq I$. We already have the interpretations

- $C \in \mathcal{P}$ if it is easy to check that object x has property C
- $C \in \mathcal{NP}$ if it is easy to certify that object x has property C

to which we now add

- $C \in$ co$\mathcal{NP}$ if it is easy to certify that object x *does not have* property C,

where we formally define the following:

Definition 21.3.6 (the class co$\mathcal{NP}$) A set C is in the class co$\mathcal{NP}$ if and only if its complement $\bar{C} = I \setminus C$ is in $\mathcal{NP}$.

While the definition of the class $\mathcal{P}$ is symmetric,[13] the definition of the class $\mathcal{NP}$ is asymmetric. Having nice certificates that a given object has property C does not automatically entail having nice certificates that a given object does not have it.

Indeed, when we can do both, we are achieving a mathematics holy grail of understanding structure, namely, *necessary and sufficient* conditions, sometimes phrased as a *duality theorem*. As we know well, such results are rare. When we insist (as we shall do) that the given certificates are *short, efficiently verifiable* ones, they are even rarer. This leads to the following conjecture:

Conjecture 21.3.7 $\mathcal{NP} \neq \mathrm{co}\mathcal{NP}$.

First note that if $\mathcal{P} = \mathcal{NP}$, then $\mathcal{P} = \mathrm{co}\mathcal{NP}$ as well, and hence this conjecture implies $\mathcal{P} \neq \mathcal{NP}$. We shall discuss at length refinements of this conjecture in Section 21.5, on proof complexity.

Despite the shortage of such efficient complete characterizations, namely, properties that are simultaneously in $\mathcal{NP} \cap \mathrm{co}\mathcal{NP}$, they nontrivially exist. Following is a list of some exemplary ones

- **Linear programming:** Systems of consistent linear inequalities[14]
- **Zero-sum games:**[15] Finite zero-sum games in which one player can gain at least (some given value) v:
- **Graph connectivity:** The set of graphs in which *every* pair of vertices is connected by (a given number) k disjoint paths
- **Partial order width:** Finite partial orders whose largest antichain has at most (a given number) w elements
- **Primes:** Prime numbers

These examples of problems in $\mathcal{NP} \cap \mathrm{co}\mathcal{NP}$ were chosen to make a point. At the time of their discovery (by Farkas, von Neumann, Menger, Dilworth, and Pratt, respectively), the mathematicians working on them were seemingly interested only in characterizing these structures. It is not known if they attempted to find efficient algorithms for these problems. However, all these problems turned out to be in $\mathcal{P}$, with some solutions entering the pantheon of efficient algorithms, for example, the Ellipsoid method of Khachian (1979) and the interior-point method of Karmarkar (1984), both for linear programming, and the recent breakthrough of Agrawal et al. (2004) for primes.[16]

Is there a moral to this story? Only that sometimes, when we have an efficient characterization of structure, we can hope for more: efficient algorithms. Conversely, a natural stepping-stone toward an elusive, efficient algorithm may be first to get an efficient characterization.

---

[13] Having a fast algorithm to determine if an object has a property C is obviously equivalent to having a fast algorithm for the complementary property $\bar{C}$.

[14] Indeed, this generalizes to other convex bodies given by more general constraints such as *semidefinite* programming.

[15] This problem was later discovered to be equivalent to linear programming.

[16] It is interesting that a simple polynomial-time algorithm, whose correctness and efficiency rely on the (unproven) extended Riemann hypothesis, was given thirty years earlier by Miller (1976).

Can we expect this magic always to happen? Is $\mathcal{NP} \cap co\mathcal{NP} = \mathcal{P}$? There are several natural problems in $\mathcal{NP} \cap co\mathcal{NP}$ that have resisted efficient algorithms for decades, and for some (e.g., factoring integers, computing discrete logarithms), humanity literally banks on their difficulty for electronic commerce security. Indeed, the following is generally believed:

Conjecture 21.3.8 $\mathcal{NP} \cap co\mathcal{NP} \neq \mathcal{P}$.

Note again that conjecture 21.3.8 implies $\mathcal{P} \neq \mathcal{NP}$ but that it is independent of Conjecture 21.3.7.

We now return to develop the main mechanism that will help us study such questions: efficient reductions.

21.3.5 Reductions: A Partial Order of Computational Difficulty

In this section, I deal with relating the computational difficulty of problems for which we have no efficient solutions (yet).

Recall that we can regard any classification problem (on finitely described objects) as a subset of our set of inputs I. Efficient reductions provide a natural partial order on such problems that capture their relative difficulty.

Definition 21.3.9 (efficient reductions)   Let $C, D \subset I$ be two classification problems; $f : I \to I$ is an efficient reduction from C to D if $f \in \mathcal{P}$, and for every $x \in I$, we have $x \in C$ if and only if $f(x) \in D$. In this case, we call f an *efficient reduction* from C to D. We write $C \leq D$ if there is an efficient reduction from C to D.

The definition of efficient computation allows two immediate observations on the usefulness of efficient reductions: first, that indeed, $\leq$ is transitive and thus defines a partial order, and second, that if $C \leq D$ and $D \in \mathcal{P}$, then also $C \in \mathcal{P}$.

Intuitively, $C \leq D$ means that solving the classification problem C is *computationally* not much harder than solving D. In some cases, one can replace *computationally* by the (vague) term *mathematically*. Often such usefulness in mathematical understanding requires more properties of the reduction f than merely being efficiently computable (e.g., we may want it to be represented as a linear transformation or a low-dimension polynomial map), and indeed, in some cases, this is possible. When such a connection between two classification problems (which look unrelated) can be proved, it can mean the portability of techniques from one area to another.

The power of efficient reductions to relate seemingly unrelated notions will unfold in later sections. We shall see that they can relate not only classification problems but such diverse concepts as hardness to randomness; average-case to worst case difficulty; proof length to computation time; the relative power of geometric, algebraic, and logical proof systems; and last but not least, the security of electronic transactions to the difficulty of factoring integers. In a sense, *efficient reductions are the backbone of computational complexity*. Indeed, given that polynomial-time reductions can do all these wonders, it may not be surprising that we have a hard time characterizing the class $\mathcal{P}$!

21.3.6 Completeness

We now return to classification problems. The partial order of their difficulty, provided by efficient reductions, allows us to define the *hardest* problems in a given class. Let $\mathcal{C}$ be any collection of classification problems (namely, every element of $\mathcal{C}$ is a subset of I). Of course, here we shall mainly care about the class $\mathcal{C} = \mathcal{NP}$.

Definition 21.3.10 (hardness and completeness) A problem D is called $\mathcal{C}$-*hard* if, for every $C \in \mathcal{C}$, we will have $C \leq D$. If we further have that $D \in \mathcal{C}$, then D is called $\mathcal{C}$-*complete*.

In other words, if D is $\mathcal{C}$-complete, it is a hardest problem in the class $\mathcal{C}$: if we manage to solve D efficiently, we will have done so for all other problems in $\mathcal{C}$. It is not a priori clear that a given class has any complete problems! On the other hand, a given class may have many complete problems, and by definition, they all have essentially the same complexity. If we manage to prove that *any* of them cannot be efficiently solved, then we automatically have done so for *all* of them.

It is trivial, and uninteresting, that every problem in the class $\mathcal{P}$ is in fact $\mathcal{P}$-complete under our definition. It becomes interesting when we find such universal problems in classes of problems for which we do not have efficient algorithms. By far, the most important of all such classes is $\mathcal{NP}$.

21.3.7 $\mathcal{NP}$-Completeness

As mentioned earlier, the seminal papers of Cook (1971) and Levin (1973) defined $\mathcal{NP}$, efficient reducibilities, and completeness, but the crown of their achievement was the discovery of a *natural* $\mathcal{NP}$-complete problem.

Definition 21.3.11 (the problem *SAT*) A Boolean formula is a logical expression over Boolean variables (that can take values in $\{0, 1\}$) with connectives $\wedge, \vee, \neg$, for example, $(x_1 \vee x_2) \wedge (\neg x_3)$. Let *SAT* denote the set of all satisfiable Boolean formulas (namely, those formulas for which there is a Boolean assignment to the variables that gives it the value 1).

Theorem 21.3.12 (Cook, 1971; Levin 1973) *SAT is $\mathcal{NP}$-complete.*

It is a simple exercise to show that Gödel's original problem, as stated in his letter, is $\mathcal{NP}$-complete as well. Moreover, he clearly understood, at least intuitively, its universal nature, captured formally by $\mathcal{NP}$-completeness, a concept discovered fifteen years later.

We recall again the meaning of this theorem. For *every* set $C \in \mathcal{NP}$, there is an efficient reduction $f : I \rightarrow I$ such that $x \in C$ if and only if the formula $f(x)$ is satisfiable! Furthermore, the proof gives an extra bonus that turns out to be extremely useful: given any witness y that $x \in C$ (via some verifier V_C), the same reduction converts the witness y to a Boolean assignment satisfying the formula $f(x)$. In other words, this reduction translates not only between the decision problems but also between the associated search problems.

You might (justly) wonder how one can prove a theorem like that. Certainly the proof cannot afford to look at all problems $C \in \mathcal{NP}$ separately. The gist of the proof is a generic transformation, taking a description of the verifier V_C for C and emulating its computation on input x and hypothetical witness y to create a Boolean formula $f(x)$ (whose variables are the bits of y). This formula simply tests the validity of the computation of V_C on (x, y) and ensures that this computation outputs 1. Here the locality of algorithms (say, described as Turing machines) plays a central role as checking the consistency of each step of the computation of V_C amounts simply to a constant size formula on a few bits. To summarize, *SAT* captures the difficulty of the whole class $\mathcal{NP}$. In particular, the $\mathcal{P}$ versus $\mathcal{NP}$ problem can now be phrased as a question about the complexity of one problem instead of infinitely many.

Corollary 21.3.13 $\mathcal{P} = \mathcal{NP}$ *if and only if SAT* $\in \mathcal{P}$.

A great advantage of having one complete problem at hand (like *SAT*) is that now, to prove that another problem (say, $D \in \mathcal{NP}$) is $\mathcal{NP}$-complete, we only need to design a reduction from *SAT* to D (namely, to prove *SAT* $\leq D$). We already know that for every $C \in \mathcal{NP}$, we have $C \leq$ *SAT*, and the transitivity of $\leq$ takes care of the rest.

This idea was used powerfully in Karp's (1972) seminal paper. There he listed twenty-one problems from logic, graph theory, scheduling, and geometry that are $\mathcal{NP}$-complete. This was the first demonstration of the wide spectrum of $\mathcal{NP}$-complete problems and initiated an industry of finding more. A few years later, Garey and Johnson (1979) published their book on $\mathcal{NP}$-completeness, which contains hundreds of such problems from diverse branches of science, engineering, and mathematics. Today, thousands are known, in a remarkably diverse set of scientific disciplines.

21.3.8 The Nature and Impact of $\mathcal{NP}$-Completeness

It is hard to do justice to this notion in a couple of paragraphs, but I shall try. More can be found in, for example, Papadimitriou (1997).

$\mathcal{NP}$-completeness is a unique scientific discovery – there seems to be no parallel scientific notion that pervaded so many fields of science and technology. It became a standard for hardness for problems whose difficulty we have yet no means of proving. It has been used both technically and allegorically to illustrate a difficulty or failure to understand natural objects and phenomena. Consequently, it has been used as a justification for channeling efforts to less ambitious (but more productive) directions. We elaborate on this effect within mathematics.

$\mathcal{NP}$-completeness has been an extremely flexible and extensible notion, allowing numerous variants that enabled capturing universality in other (mainly computational, but not only) contexts. It led to the ability of defining whole classes of problems by single, universal ones, with the benefits mentioned earlier. Much of the whole evolution of computational complexity, the theory of algorithms, and most other areas in theoretical computer science have been guided by the powerful approach of reduction and completeness.

It would be extremely interesting to explain the ubiquity of $\mathcal{NP}$-completeness. Being highly speculative for a moment, we can make the following analogies of this mystery with similar mysteries in physics. The existence of $\mathcal{NP}$-completeness in such diverse fields of inquiry may be likened to the existence of the same building blocks of matter in remote galaxies, begging for a common explanation of the same nature as the big bang theory. On the other hand, the near lack of natural objects in the (seemingly huge) void of problems in $\mathcal{NP}$ that are neither in $\mathcal{P}$ nor $\mathcal{NP}$-complete raises questions about possible "dark matter," which we have not developed the means of observing yet.

21.3.9 Some $\mathcal{NP}$-Complete Problems in Mathematics

Again, I note that all $\mathcal{NP}$-complete problems are equivalent in a very strong sense. Any algorithm solving one can be simply translated into an equally efficient algorithm solving any other. Conversely, if one problem is difficult, then all of them are. I'll explore the meaning of these insights for a number of mathematical classification problems in diverse areas that are NP-complete.

Why did Gödel's incompleteness theorem seemingly have such small effect on working mathematicians? A common explanation is that the unprovable and undecidable problems are far too general compared to those actively being studied. As we shall see, this argument will not apply to the $\mathcal{NP}$-complete problems we discuss. Indeed, many of these $\mathcal{NP}$-completeness results were proven by mathematicians!

We exemplify this point with two classification problems: *2DIO*, of quadratic Diophantine equations, and *KNOT*, of knots on three-dimensional manifolds.

2DIO: Consider the set of all equations of the form $Ax^2 + By + C = 0$ with integer coefficients A, B, C. Given such a triple, does the corresponding polynomial have a positive integer root (x, y)? Let *2DIO* denote the subset of triples for which the answer is yes. Note that this is a very restricted subproblem of the undecidable Hilbert's tenth problem, problem 2 from the introduction; indeed, it is simpler than even elliptic curves. Nevertheless:

Theorem 21.3.14 (Adleman and Manders, 1975) *The set 2DIO is $\mathcal{NP}$-complete.*

KNOT: Consider the set of all triples (M, K, G), representing,[17] respectively, a three-dimensional manifold M, a knot K embedded on it, and an integer G. Given a triple (M, K, G), does the surface that K bounds have genus at most G? Let *KNOT* denote the subset for which the answer is yes.

Theorem 21.3.15 (Agol et al., 2006) *The set KNOT is $\mathcal{NP}$-complete.*

Recall that to prove $\mathcal{NP}$-completeness of a set, one has to prove two things: that it is in $\mathcal{NP}$ and that it is $\mathcal{NP}$-hard. In almost all $\mathcal{NP}$-complete problems, membership in $\mathcal{NP}$ (namely, the existence of short certificates) is easy to prove. For example, for *2DIO*, one can easily see that if there is a positive integer solution r to the equation $Ax^2 + By + C = 0$, then indeed, there is one whose length (in bits) is polynomial in

---

[17] A finite representation can describe M by a triangulation (finite collection of tetrahedra and their adjacencies), and the knot K will be described as a link (closed path) along edges of the given tetrahedra.

the lengths of A, B, C, and given such r, it is easy to verify that it is indeed a root of the equation. In short, it is very easy to see that $2DIO \in \mathcal{NP}$. However, *KNOT* is an exception, and proving $KNOT \in \mathcal{NP}$ is highly nontrivial. The short witnesses that a given knot has a small genus requires Haken's algorithmic theory of normal surfaces, considerably enhanced even short certificates for unknottedness in Re^3 are hard to obtain; see Hass et al., 1999.

Let us discuss what these $\mathcal{NP}$-completeness results mean, first about the relationship between the two and then about each individually.

The two preceding theorems and the meaning of $\mathcal{NP}$-completeness together imply that there are *simple* translations (in both directions) between solving $2DIO$ and problem *KNOT*. More precisely, it provides efficiently computable functions $f, h: \text{I} \to \text{I}$ performing these translations:

$(A, B, C) \in 2DIO$ if and only if $f(A, B, C) \in KNOT$

$(M, K, G) \in KNOT$ if and only if $h(M, K, G) \in 2DIO$.

If we have gained enough understanding of topology to solve, for example, the knot genus problem, it means that we automatically have gained enough number-theoretic understanding for solving these quadratic Diophantine problems (and vice versa).

The proofs that these problems are complete both follow by reduction from (variants of) *SAT*. The combinatorial nature of these reductions may cast doubt on the possibility that the computational equivalence of these two problems implies the ability of real "technology transfer" between topology and number theory. Nevertheless, now that we know of the equivalence, perhaps simpler and more direct reductions can be found between these problems. Moreover, we stress again that for any instance, say, $(M, K, G) \in KNOT$, if we translate it using this reduction to an instance $(A, B, C) \in 2DIO$ and happen (either by sheer luck or special structure of that equation) to find an integer root, the same reduction will translate that root back to a description of a genus G manifold that bounds the knot K. Today, many such $\mathcal{NP}$-complete problems are known throughout mathematics, and for some pairs, the equivalence can be mathematically meaningful and useful (as it is between some pairs of computational problems).

But regardless of the meaning of the connection between these two problems, there is no doubt what their individual $\mathcal{NP}$-completeness means. Both are mathematically "nasty," as both embed in them the full power of $\mathcal{NP}$. If $\mathcal{P} \neq \mathcal{NP}$, there are no efficient algorithms to describe the objects at hand. Moreover, assuming the stronger $\mathcal{NP} \neq \text{co}\mathcal{NP}$, we should not even expect complete characterization (e.g., we should not expect short certificates that a given quadratic equation *does not* have a positive integer root).

In short, $\mathcal{NP}$-completeness suggests that we lower our expectations of fully understanding these properties and study perhaps important special cases, variants, and so on. Note that such reaction of mathematicians may follow the frustration of unsuccessful attempts at general understanding. However, the stamp of $\mathcal{NP}$-completeness *may* serve as moral justification for this reaction. I stress the word *may* as the judges for accepting such a stamp can only be the mathematicians working on the problem and how well the associated $\mathcal{NP}$-completeness result captures the structure they try to reveal. I merely point out the usefulness of a formal stamp of difficulty (as opposed to a general feeling) and its algorithmic meaning.

The two preceding examples come from number theory and topology. I list in the following more $\mathcal{NP}$-complete problems from algebra, geometry, optimization, and graph theory. There are numerous other such problems. It is hoped that this will demonstrate how wide this *modern Gödel phenomena* on is in mathematics. The preceding discussion is relevant to them all.

- **Quadratic equations:** Given a system of multivariate polynomial equations of degree at most 2, over a finite field (say, GF(2)), do they have a common root?
- **Knapsack:** Given a sequence of integers $a_1, \ldots, a_n$ and b, decide if there exists a subset J such that $\sum_{i \in J} a_i = b$.
- **Integer programming:** Given a polytope in Re^n (by its bounding hyperplanes), does it contain an integer point?
- **Shortest lattice vector:** Given a lattice L in Re^n and an integer k, is the shortest nonzero vector of L of (Euclidean) length $\leq k$?
- **3Color:** Given a graph, can its vertices be colored from {Red, Green, Blue} with no adjacent vertices having the same color?
- **Clique:** Given a graph and an integer k, are there k vertices with all pairs mutually adjacent?

21.4 Lower Bounds and Attacks on $\mathcal{P}$ versus $\mathcal{NP}$

To prove that $\mathcal{P} \neq \mathcal{NP}$, we must show that for a given problem in $\mathcal{NP}$, no efficient algorithm exists. A result of this type is called a *lower bound* (limiting from below the computational complexity of the problem). Several powerful techniques for proving lower bounds have emerged in the past decades. They apply in two (very different) settings. Subsequently, I describe both and try to explain our understanding of why they seem to stop short of proving $\mathcal{P} \neq \mathcal{NP}$. I mention the first, diagonalization, only very briefly and concentrate on the second, Boolean circuits.

21.4.1 Diagonalization and Relativization

The diagonalization technique goes back to Cantor and his argument that there are more real numbers than algebraic numbers. It was used by Gödel in his incompleteness theorem and by Turing in his undecidability results and then was refined to prove computational complexity lower bounds. A typical theorem in this area is that more time buys more computational power, for example, there are functions computable in time n^3, say, that are not computable in time n^2. The heart of such arguments is the existence of a universal algorithm that can simulate every other algorithm with only a small loss in efficiency.

Can such arguments be used to separate $\mathcal{P}$ from $\mathcal{NP}$? This depends on what we mean by "such arguments." The paper by Baker et al. (1975) suggests a formal definition by exhibiting a feature shared by many similar complexity results, called *relativization*. For example, the result mentioned previously separating Turing machines running in time n^3 from those with runnning time n^2 would work perfectly well if all machines in questions were supplied with an "oracle" for any fixed function f that would answer queries of the form x by $f(x)$ in unit time. Therefore relativizing lower bounds hold in a "relativized world" in which any fixed such f is an easy function for the algorithms we consider and so should hold in *all* such worlds. The Baker et al. (1975) paper then proceeded to show that relativizing arguments do not suffice to resolve the $\mathcal{P}$ versus $\mathcal{NP}$ question. This is done by showing that equipping the machines with different functions f gives different answers to the $\mathcal{P}$ versus $\mathcal{NP}$ question in the respective "relativized worlds." In the three decades since that paper, complexity theory grew far more sophisticated, but nevertheless almost all new results obtained do relativize (one of the few exceptions is in Vinodchandran, 2004). More on this subject can be found in chapter 14.3 of Papadimitriou (1994a), chapter 9.2 of Sipser (1997), and in Fortnow (1994).

21.4.2 Boolean Circuits

A Boolean circuit may be viewed as the hardware analogue of an algorithm (software). Computation on the binary input sequence proceeds by a sequence of Boolean operations (called *gates*) from the set $\{\wedge, \vee, \neg\}$ (logical AND, OR, and NEGATION) to compute the output(s). We assume that $\wedge, \vee$ are applied to two arguments. We note that whereas an algorithm can handle inputs of any length, a circuit can handle only one input length (the number of input "wires" it has). A circuit is commonly represented as a (directed, acyclic) graph, with the assignments of gates to its internal vertices. We note that a Boolean formula is simply a circuit whose graph structure is a tree.

Recall that I denotes the set of all binary sequences and that I_k is the set of sequences of length exactly k. If a circuit has n inputs and m outputs, it is clear that it computes a function $f: I_n \rightarrow I_m$. The efficiency of a circuit is measured by its size, which is the analogue of time in algorithms.

Definition 21.4.1 (circuit size)   Denote by $S(f)$ the size of the smallest Boolean circuit computing f.

As we care about asymptotic behavior, we shall be interested in sequences of functions $f = \{f_n\}$, where f_n is a function on n input bits. We shall study the complexity $S(f_n)$ asymptotically as a function of n and denote it $S(f)$. For example, let PAR be the parity function, computing if the number of 1s in a binary string is even or odd. Then PAR_n is its restriction to n-bit inputs, and $S(PAR) = O(n)$.

It is not hard to see that an algorithm (say, a Turing machine) for a function f that runs in time T gives rise to a circuit family for the functions f_n of sizes (respectively) $(T(n))^2$, and so efficiency is preserved when moving from algorithms to circuits. Thus proving lower bounds for circuits implies lower bounds for algorithms, and we can try to attack the $\mathcal{P}$ versus $\mathcal{NP}$ question this way.

Definition 21.4.2 (the class $\mathcal{P}/\text{poly}$)   Let $\mathcal{P}/\text{poly}$ denote the set of all functions computable by a family of polynomial-sized circuits.

Conjecture 21.4.3 $\mathcal{NP} \not\subseteq \mathcal{P}/\text{poly}$.

Is this a reasonable conjecture? As mentioned earlier, $\mathcal{P} \subseteq \mathcal{P}/\text{poly}$. Does the reverse inclusion hold? It actually fails badly! There exist undecidable functions f (which cannot be computed by Turing machines at all, regardless of their running time) that have linear-sized circuits. This extra power comes from the fact that circuits for different input lengths share no common description (and thus this model is sometimes called "nonuniform").

One might expect, therefore, that proving circuit lower bounds is a much harder task than proving $\mathcal{P} \neq \mathcal{NP}$. However, there is a strong sentiment that the extra power provided by nonuniformity is irrelevant to $\mathcal{P}$ versus. $\mathcal{NP}$. This sentiment comes from a result of Karp and Lipton (1982), proving that $\mathcal{NP} \subseteq \mathcal{P}/\text{poly}$ implies a surprising uniform "collapse" – not quite $\mathcal{NP} = \text{co}\mathcal{NP}$ but another (somewhat similar but weaker) unlikely collapse of complexity classes.

Still, what motivates replacing the Turing machine by the stronger circuit model when seeking lower bounds? The hope is that focusing on a finite model will allow the use of combinatorial techniques to analyze the power and limitations of efficient algorithms. This hope has been realized in the study of interesting though restricted classes of circuits. The resulting lower bounds for such restricted circuits fall short of resolving the big questions, but nevertheless have had important applications to computational learning theory, pseudorandomness, proof complexity, and more.

21.4.2.1 Basic Results and Questions

We have already mentioned several basic facts about Boolean circuits, in particular that they can efficiently simulate Turing machines. The next basic fact, first observed by Shannon (1947), is that *most Boolean functions require exponential sized circuits.*

This lower bound follows from the gap between the number of functions and the number of small circuits. Fix the number of input bits n. The number of possible functions on n bits is precisely 2^{2^n}. On the other hand, the number of circuits of size s is (via a crude estimate of the number of graphs of that size) at most 2^{s^2}. Because every circuit computes one function, we must have $s \gg 2^{n/3}$ for most functions.

Theorem 21.4.4 (Shannon, 1949)   *For almost every function $f : I_n \to \{0, 1\}$, $S(f) \geq 2^{n/3}$.*

Therefore hard functions for circuits (and hence for Turing machines) abound. However, the hardness is proved via a counting argument and thus supplies no way of putting a finger on one hard function. I shall return to the nonconstructive nature of this problem in Section 21.5. So far, we cannot prove such hardness for any *explicit* function f (e.g., for an $\mathcal{NP}$-complete function like SAT).

Conjecture 21.4.5 $S(SAT) = 2^{\Omega(n)}$.

The situation is even worse: no *nontrivial* lower bound is known for any explicit function.[18] Note that for any function f on n bits (which depends on all its inputs), we

---

[18] The notion "explicit," which we repeatedly use here, is a bit elusive and context dependent at times. However, in all cases we seek natural functions, usually of independent mathematical, scientific, or technological interest, rather than functions whose existence is shown by some counting or simulation arguments.

trivially must have $S(f) \geq n$ just to read the inputs. The main open problem of circuit complexity is beating this trivial bound.

Open Problem 21.4.6   Find an explicit function $f : I_n \to I_n$ for which $S(f) \neq O(n)$.

Here natural explicit candidate functions may be the multiplication of two $(n/2)$-bit integers or of two $\sqrt{n} \times \sqrt{n}$ matrixes over $GF(2)$. But for any function in $\mathcal{NP}$, such a lower bound would be a breakthrough. Unable to prove any nontrivial lower bound, we now turn to restricted models. There has been some remarkable success in developing techniques for proving strong lower bounds for natural restricted classes of circuits. I discuss in some detail only one such model.

The 1980s saw a flurry of new techniques for proving circuit lower bounds on natural, restricted classes of circuits. Razborov developed the *approximation method*, which allowed proving exponential circuit lower bounds for *monotone* circuits, for such natural problems as *CLIQUE*. The *random restriction* method, initiated by Furst et al. (1984) and Ajtai (1983), was used to prove exponential lower bounds on constant depth circuits for such natural problems as *PARITY*. The communication complexity method of Karchmer and Wigderson (1990) was used to prove lower bounds on monotone formulas, for example, for the *PERFECT MATCHING* problem. See the survey (Boppana and Sipser, 1990) for these and more. However, these all fall short of obtaining any nontrivial lower bounds for general circuits and in particular of proving that $\mathcal{P} \neq \mathcal{NP}$.

21.4.2.2 Why Is It Hard to Prove Circuit Lower Bounds?

Is there a fundamental reason for this failure? The same may be asked about any long-standing mathematical problem (e.g., the Riemann hypothesis). A natural (vague) answer would be that, probably, the current arsenal of tools and ideas (which may well have been successful at attacking related, easier problems) does not suffice.

Remarkably, complexity theory can make this vague statement into a theorem. Thus we have a formal excuse for our failure so far: we can classify a general set of ideas and tools that are responsible for virtually all restricted lower bounds known yet must necessarily fail for proving general ones. This introspective result, developed by Razborov and Rudich (1997), suggests a framework called *natural proofs*. Very briefly, a lower-bound proof is natural if it applies to a large, easily recognizable set of functions. The authors first show that this framework encapsulates all known lower bounds, then they show that natural proofs of general circuit lower bounds are unlikely in the following sense: any natural proof of a lower bound surprisingly implies (as a by-product) subexponential algorithms for inverting every candidate one-way function. Again, I stress this irony – natural lower bounds lead to efficient algorithms for the type of problem we want to prove hard!

Specifically, a *natural* (in this formal sense) lower bound would imply subexponential algorithms for such functions as integer factoring and discrete logarithm, generally believed to be difficult (to the extent that the security of electronic commerce worldwide relies on such assumptions). This connection strongly uses *pseudorandomness*, which will be discussed later. A simple corollary is that no natural proof exists to show that integer factoring requires circuits of size $2^{n^{1/100}}$ (the best current upper bound is $2^{n^{1/3}}$).

One interpretation of the aforementioned result is an independence result of general circuit lower bounds from a certain natural fragment of Peano arithmetic. This may suggest that the $\mathcal{P}$ versus $\mathcal{NP}$ problem may be independent from Peano arithmetic, or even set theory, which is certainly a possibility.

Finally, note that it has been over ten years since the publication of the natural proof paper. The challenge it raised, *prove a nonnatural lower bound*, has not yet been met.

21.5 Proof Complexity

Gödel's letter focuses on lengths of proofs. This section highlights some of the research developments within complexity theory on the many facets of this issue in the last couple decades. For extensive surveys and different takes on this material and its relation to proof theory, independence results, and Gödel's letter, see Beame and Pitassi (1998), Buss (1995), Pudlak (1995, 1996a, 1996b, 1998), and Rudich and Wigderson (2000).

The concept of *proof* is what distinguishes the study of mathematics from all other fields of human inquiry. Mathematicians have gathered millennia of experience to attribute such adjectives to proofs as "insightful," "original," "deep," and, most notably, "difficult." Can one quantify mathematically the difficulty of proving various theorems? This is exactly the task undertaken in proof complexity. It seeks to classify theorems according to the difficulty of proving them, much like circuit complexity seeks to classify functions according to the difficulty of computing them. In proofs, just like in computation, there will be a number of models, called *proof systems*, capturing the power of reasoning allowed to the prover.

Proof systems abound in all areas of mathematics (and not just in logic). Let us see some examples:

1. Hilbert's *Nullstellensatz* is a (sound and complete) proof system in which *theorems* are inconsistent sets of polynomial equations. A *proof* expresses the constant 1 as a linear combination of the given polynomials.
2. Each finitely presented group can be viewed as a proof system in which *theorems* are words that reduce to the identity element. A *proof* is the sequence of substituting relations to generate the identity.
3. Reidemeister moves are a proof system in which *theorems* are trivial, unknotted knots. A *proof* is the sequences of moves reducing the given plane diagram of the knot into one with no crossings.
4. Von Neumann's Minimax theorem gives a proof system for every zero-sum game. A *theorem* is an optimal strategy for White, and its *proof* is a strategy for Black with the same value.

In these and many other examples, the length of the proof plays a key role, and the quality of the proof system is often related to the shortness of the proofs it provides:

1. In the Nullstellensatz (over fields of characteristic 0), length (of the coefficient polynomials, measured usually by their degree and height) usually plays a crucial role in the efficiency of commutative algebra software (e.g., Gröbner basis algorithms).

2. The word problem, in general, is undecidable. For hyperbolic groups, Gromov's polynomial upper bound on proof length has many uses, of which perhaps the most recent is in his own construction of finitely presented groups with no uniform embedding into Hilbert spac (Gromove, 2003).

3. Reidemeister moves are convenient combinatorially, but the best upper bounds on the length of such proofs (namely, on the number of moves) that a given knot is unknotted, are exponential in the description of the knot (Hass and Lagarias, 2001). Whether one can improve this upper bound to a polynomial is an open question. We note that stronger proof systems were developed to give polynomial upper bounds for proving unknottedness (Hass et al., 1999).

4. In zero-sum games, happily, all proofs are of linear size.

I stress that the asymptotic viewpoint, considering *families* of "theorems" and measuring their proof length as a function of the description length of the theorems, is natural and prevalent. As for computation, this asymptotic viewpoint reveals structure of the underlying mathematical objects, and economy (or efficiency) of proof length often means a better understanding. Though this viewpoint is appropriate for a large chunk of mathematical work, you may object that it cannot help explain the difficulty of *single* problems such as the Riemann hypothesis or $\mathcal{P}$ versus $\mathcal{NP}$. This is, of course, a valid complaint. We note, however, that even such theorems (or conjectures) may be viewed asymptotically (though not always with improved illumination). The Riemann hypothesis has equivalent formulations as a sequence of finite statements (e.g., about cancellations in the Möbius function). More interestingly, we shall see later a formulation of the $\mathcal{P}/poly$ versus $\mathcal{NP}$ problem as a sequence of finite statements that are strongly related to the natural proofs paradigm mentioned earlier.

All theorems that will concern us in this section are *universal* statements (e.g., an inconsistent set of polynomial equations is the statement that *every* assignment to the variables fails to satisfy them). A short proof for a universal statement constitutes an equivalent formulation that is *existential* – the existence of the proof itself (e.g., the existence of the coefficient polynomials in Nullstellensatz that implies this inconsistency). The mathematical motivation for this focus is clear: the ability to describe a property both universally and existentially constitutes *necessary and sufficient* conditions – the aforementioned holy grail of mathematical understanding. Here we shall be picky and quantify that understanding according to our usual computational yardstick: the *length* of the existential certificate.

We shall restrict ourselves to *propositional* tautologies. This will automatically give an exponential (thus a known, finite) upper bound on the proof length and will restrict the ballpark (as with $\mathcal{P}$ vs. $\mathcal{NP}$) to the range between polynomial and exponential. The type of statements, theorems, and proofs with which we shall deal is best illustrated by the following example.

21.5.1 The Pigeonhole Principle: A Motivating Example

Consider the well-known pigeonhole principle, stating that there is no injective mapping from a finite set to a smaller one. Though trivial, this principle was essential for the counting argument proving the *existence* of exponentially hard functions

(Theorem 21.4.4) – this partially explains our interest in its proof complexity. More generally, this principle epitomizes *nonconstructive* arguments in mathematics such as Minkowski's theorem that a centrally symmetric convex body of sufficient volume must contain a lattice point. In both results, the proof does not provide any information about the object proved to exist. Other natural tautologies capture the combinatorial essence of topological proofs (e.g., Brauer's fixed-point theorem, the Borsuk-Ulam theorem, Nash's equilibrium, see Papadimitriou (1994b) for more).

Let us formulate it and discuss the complexity of proving it. First, we turn it into a sequence of finite statements. Fix $m > n$. Let PHP^m_n stand for the statement *there is no one-to-one mapping of m pigeons to n holes*. To formulate it mathematically, imagine an $m \times n$ matrix of Boolean variables x_{ij} describing a hypothetical mapping (with the interpretation that $x_{ij} = 1$ means that the ith pigeon is mapped to the jth hole).[19]

Definition 21.5.1 (the pigeonhole principle)   The pigeonhole principle PHP^m_n now states that

- either pigeon i is not mapped anywhere (namely, *all* x_{ij} for a fixed i are zeros)
- some two are mapped to the same hole (namely, for some different i, i', and some j, we have $x_{ij} = x_{i'j} = 1$).

These conditions are easily expressible as a formula in the variables x_{ij} (called a *propositional formula*), and the pigeonhole principle is the statement that this formula is a *tautology* (namely, satisfied by every truth assignment to the variables).

Even more conveniently, the negation of this tautology (which is a *contradiction*) can be captured by a collection of constraints on these Boolean variables that are mutually contradictory. These constraints can easily be written in different languages:

- **Algebraic:** as a set of constant degree polynomials over GF(2)
- **Geometric:** as a set of linear inequalities with integer coefficients (to which we seek a {0, 1} solution)
- **Logical:** as a set of Boolean formulas

We shall see soon that each setting naturally suggests (several) reasoning tools, such as variants of the Nullstellensatz in the algebraic setting, of Frege systems in the logical setting and integer programming heuristics in the geometric setting. All these can be formalized as proof systems that suffice to prove this (and any other) tautology. Our main concern will be in the efficiency of each of these proof systems and their relative power, measured in *proof length*. Before turning to some of these specific systems, we discuss this concept in full generality.

21.5.2 Propositional Proof Systems and $\mathcal{NP}$ versus co$\mathcal{NP}$

Most definitions and results in this section come from the paper by Cook and Reckhow (1979) that initiated this research direction. I define proof systems and the complexity

---

[19] Note that we do not rule out the possibility that some pigeon is mapped to more than one hole – this condition can be added, but the truth of the principle remains valid without it.

measure of proof length for each and then relate these to complexity questions we have already met.

All theorems we shall consider will be propositional tautologies. Here are the salient features that we expect[20] from any proof system:

- **Completeness:** Every true statement has a proof.
- **Soundness:** No false statement has a proof.
- **Verification efficiency:** Given a mathematical statement T and a purported proof π for it, it can be easily checked if indeed π proves T in the system. Note that here efficiency of the verification procedure refers to its running time measured in terms of the *total length of the alleged theorem and proof.*

Remark 21.5.2 Note that we dropped the requirement used in the definition of $\mathcal{NP}$, limiting the proof to be short (polynomial in the length of the claim). The reason is, of course, that proof length is our measure of complexity.

All these conditions are concisely captured, for propositional statements, by the following definition.

Definition 21.5.3 (proof systems (Cook and Reckhow, 1979)) A (propositional) proof system is a polynomial-time Turing machine M with the property that T is a tautology if and only if there exists a ("*proof*") π such that $M(\pi, T) = 1.$[21]

As a simple example, consider the following "truth-table" proof system M_{TT}. Basically, this machine will declare a formula T a theorem if evaluating it on every possible input makes T true. A bit more formally, for any formula T on n variables, the machine M_{TT} accepts (π, T) if π is a list of all binary strings of length n, and for each such string σ, $T(\sigma) = 1$.

Note that M_{TT} runs in polynomial time in its input length, which is the combined length of formula and proof. However, in the system M_{TT}, proofs are (typically) of exponential length in the size of the given formula. This leads us to the definition of the efficiency (or complexity) of a general propositional proof system M – how short is the shortest proof of each tautology.

Definition 21.5.4 (proof length (Cook and Reckhow, 1979)) For each tautology T, let $S_M(T)$ denote the size of the shortest proof of T in M (i.e., the length of the shortest string π such that M accepts (π, T)). Let $S_M(n)$ denote the maximum of $S_M(T)$ over all tautologies T of length n. Finally, we call the proof system M *polynomially bounded* if and only if, for all n, we have $S_M(n) = n^{O(1)}$.

Is there a polynomially bounded proof system (namely, one that has polynomial-sized proofs for all tautologies)? The following theorem provides a basic connection of this question with computational complexity and the major question of Section 21.3.4.

---

[20] Actually, even the first two requirements are too much to expect from strong proof systems, as Gödel famously proved in his incompleteness theorem. However, for propositional statements that have finite proofs, there are such systems.

[21] In agreement with standard formalisms (see later), the proof is seen as coming before the theorem.

Its proof follows quite straightforwardly from the $\mathcal{NP}$-completeness of *SAT*, the problem of satisfying propositional formulas (and the fact that a formula is unsatisfiable if and only if its negation is a tautology).

Theorem 21.5.5 (Cook and Reckhow, 1979)   *There exists a polynomially bounded proof system if and only if $\mathcal{NP} = \text{co}\mathcal{NP}$.*

In the next section, I focus on natural restricted proof systems. A notion of reduction between proof systems, called "polynomial simulation," was introduced in Cook and Reckhow (1979) and allows us to create a partial order of the relative power of some systems. This is but one example of the usefulness of the methodology developed within complexity theory after the success of $\mathcal{NP}$-completeness.

21.5.3 Concrete Proof Systems

All proof systems in this section are familiar to every mathematician, ever since *The Elements* of Euclid, who formulated a deductive system for plane geometry. In all deductive systems, one starts with a list of formulas and, using simple (and sound) derivation rules, infers new ones (each formula is called a "line" in the proof).

Normally the initial formulas are taken to be (self-evident) *axioms*, and the final formula derived is the desired theorem. However, here it will be useful to reverse this view and consider *refutation systems*. In the refutation systems following, I start with a contradictory set of formulas and derive a basic contradiction (e.g., $\neg x \wedge x$, $1 = 0$, $1 < 0$), depending on the setting. This serves as proof of the theorem that the initial formulas are mutually inconsistent. I highlight some results and open problems on the proof length of basic tautologies in algebraic, geometric, and logical systems.

21.5.3.1 Algebraic Proof Systems

We restrict ourselves to the field GF(2). Here a natural representation of a Boolean contradiction is a set of polynomials with no common root. We always add to such a collection the polynomials $x^2 - x$ (for all variables x) that ensure Boolean values (and so we can imagine that we are working over the algebraic closure).

Hilbert's Nullstellensatz suggests a proof system. If $f_1, f_2, \ldots, f_n$ (with any number of variables) have no common root, there must exist polynomials $g_1, g_2, \ldots, g_n$ such that $\sum_i f_i g_i \equiv 1$. The g_is constitute a proof, and we may ask how short its description is.

A related but far more efficient system (intuitively based on computations of Gröbner bases) is polynomial calculus (PC), which was introduced in Clegg et al., (1996). The *lines* in this system are polynomials (represented explicitly by all coefficients), and it has two *deduction rules*, capturing the definition of an ideal: for any two polynomials g, h and variable x_i, we can use g, h to derive $g + h$, and we can use g and x_i to derive $x_i g$. It is not hard to see (using linear algebra) that if this system has a proof of length s for some tautology, then this proof can be found in time polynomial in s. Recalling our discussion of $\mathcal{P}$ versus $\mathcal{NP}$, we do not expect such a property from really strong proof systems.

The PC is known to be exponentially stronger than Nullstellensatz. More precisely, there are tautologies that require exponential-length Nullstellensatz proofs but only polynomial PC proofs. However, strong size lower bounds (obtained from degree lower bounds) are known for PC systems as well. Indeed, the pigeonhole principle is hard for this system. For its natural encoding as a contradictory set of quadratic polynomials, Razborov (1998) proved the following:

Theorem 21.5.6 (Razborov 1998) *For every* n *and every* $m > n$, $S_{PC}(PHP_n^m) \geq 2^{n/2}$ *over every field.*

21.5.3.2 Geometric Proof Systems

Yet another natural way to represent Boolean contradictions is by a set of regions in space containing no integer points. A wide source of interesting contradictions comprises integer programs from combinatorial optimization. Here the constraints are (affine) linear inequalities with integer coefficients (so the regions are subsets of the Boolean cube carved out by half spaces). A proof system infers new inequalities from old ones in a way that does not eliminate integer points.

The most basic system is called cutting planes (CP), introduced by Chvátal (1973). Its *lines* are linear inequalities with integer coefficients. Its *deduction rules* are (the obvious) addition of inequalities and the (less obvious) dividing of the coefficients by a constant (and rounding, taking advantage of the integrality of the solution space).[22]

Let us look at the pigeonhole principle PHP_n^m again. It is easy to express it as a set of contradictory linear inequalities: for every pigeon, the sum of its variables should be *at least* 1. For every hole, the sum of its variables should be *at most* 1. Thus adding up all variables in these two ways implies $m \leq n$, a contradiction. Thus the pigeonhole principle has polynomial-sized CP proofs.

Though PHP_n^m is easy in this system, exponential lower bounds were proved for other tautologies. Consider the tautology $CLIQUE_n^k$: no graph on n nodes can simultaneously have a k-clique and a legal $k - 1$-coloring. It is easy to formulate this technology as a propositional formula. Notice that it somehow encodes *many* instances of the pigeonhole principle, one for every k-subset of the vertices.

Theorem 21.5.7 (Pudlak, 1997) $S_{CP}\left(CLIQUE_n^{\sqrt{n}}\right) \geq 2^{n^{1/10}}$.

The proof of this theorem by Pudlak (1997) is quite remarkable. It *reduces* this proof complexity lower bound to a circuit complexity lower bound. In other words, he shows that any short CP proof of tautologies of certain structure yields a small circuit computing a related Boolean function. You probably guessed that for the tautology at hand, the function is indeed the *CLIQUE* function mentioned earlier. Moreover, the circuits obtained are *monotone* but of the following, very strong form. Rather than allowing only $\wedge$, $\vee$ as basic gates, they allow any monotone binary operation on real numbers. Pudlak then goes on to generalize Razborov's approximation method for such circuits and proves an exponential lower bound on the size they require to compute *CLIQUE*.

---

[22] For example, from the inequality $2x + 4y \geq 1$, we may infer $x + 2y \geq \frac{1}{2}$, and by integrality, $x + 2y \geq 1$.

21.5.3.3 Logical Proof Systems

The proof systems in this section will all have lines that are Boolean formulas, and the differences between them will be in the structural limits imposed on these formulas. We introduce the most important ones: Frege, capturing polynomial-time reasoning, and Resolution, the most useful system used in automated theorem provers.

The most basic proof system, called the Frege system, puts no restriction on the formulas manipulated by the proof. It has one *derivation rule*, called the *cut rule*: from the two formulas $A \vee C$, $B \vee \neg C$, we may infer the formula $A \vee B$. Every basic book in logic has a slightly different way of describing the Frege system – one convenient outcome of the computational approach, especially the notion of efficient reductions between proof systems, is a proof (in Cook and Reckhow, 1979) that they are all equivalent in the sense that the shortest proofs (up to polynomial factors) are independent of which variant you pick.

The Frege system can polynomially simulate *both* the PC and the CP systems. In particular, the counting proof described earlier for the pigeonhole principle can be carried out efficiently in the Frege system (not quite trivially), yielding the following:

Theorem 21.5.8 (Buss, 1987) $S_{Frege}(PHP_n^{n+1}) = n^{O(1)}$.

Frege systems are basic in the sense that they are the most common in logic and in the sense that polynomial-length proofs in these systems naturally correspond to polynomial-time reasoning about feasible objects. In short, this is the proof analogue of the computational class $\mathcal{P}$. The major open problem in proof complexity is to find any tautology (as usual, we mean a family of tautologies) that has no polynomial-size proof in the Frege system.

Open Problem 21.5.9   Prove superpolynomial lower bounds for the Frege system.

As lower bounds for Frege are hard, we turn to subsystems of Frege that are interesting and natural. The most widely studied system is Resolution. Its importance stems from its use by most propositional (as well as first-order) *automated theorem provers*, often called "Davis–Putnam" or (DLL) procedures (Davis et al., 1962). This family of algorithms is designed to find proofs of Boolean tautologies, arising in diverse applications from testing computer chips or communication protocols to basic number theory results.

The lines in Resolution refutations are clauses, namely, disjunctions of literals (like $x_1 \vee x_2 \vee \neg x_3$). The *inference cut rule* simplifies to the *resolution rule*: for two clauses A, B and variable x, we can use $A \vee x$ and $B \vee \neg x$ to derive the clause $A \vee B$.

Historically, the first major result of proof complexity was Haken's[23] (Haken, 1985) exponential lower bound on Resolution proofs for the pigeonhole principle:

Theorem 21.5.10 (Haken, 1985) $S_{Resolution}(PHP_n^{n+1}) = 2^{\Omega(n)}$.

To prove it, Haken developed the *bottleneck method*, which is related to both the random restriction and approximation methods mentioned in the circuit complexity chapter. This lower bound was extended to *random tautologies* (under a natural

---

[23] Armin Haken, the son of Wolfgang Haken, cited earlier for his work on knots.

distribution) in Chvátal and Szemerédi (1988). The *width method* of Ben-Sasson and Wigderson (1999) much simpler proofs for both results.

21.5.4 Proof Complexity versus Circuit Complexity

These two areas look like very different beasts, despite the syntactic similarity between the local evolution of computation and proof. To begin with, the number of objects they care about differs drastically. There are doubly exponentially number of functions (on n bits) but only exponentially many tautologies of length n. Thus a counting argument shows that some functions (albeit nonexplicit) require exponential circuit lower bounds (Theorem 21.4.4), but no similar argument can exist to show that some tautologies require exponential-sized proofs. So though we prefer lower bounds for natural, explicit tautologies, *existence* results of hard tautologies for strong systems are interesting in this setting as well.

Despite the different nature of the two areas, there are deep connections between them. Quite a few of the techniques used in circuit complexity, most notably *random restrictions*, were useful for proof complexity as well. The lower bound we saw in the previous section is extremely intriguing: a monotone circuit lower bound directly implies a (nonmonotone) proof system lower bound. This particular type of reduction, known as the *interpolation method*, was successfully used for other, weak proof systems, such as Resolution. This leads to the question, can reductions of a similar nature be used to obtain lower bounds for a strong system (like Frege) from (yet unproven) circuit lower bounds?

Open Problem 21.5.11   Does $\mathcal{NP} \not\subseteq \mathcal{P}/\text{poly}$ imply superpolynomial Frege lower bounds?

Why are Frege lower bounds hard? The truth is that we do not know. The Frege system (and its relative, extended Frege) capture *polynomial-time reasoning* as the basic objects appearing in the proof are polynomial-time computable. Thus superpolynomial lower bounds for these systems are the proof complexity analogue of proving superpolynomial lower bounds in circuit complexity. As we saw, for circuits, we at least understand to some extent the limits of existing techniques via natural proofs. However, there is no known analogue of this framework for proof complexity.

I conclude with a tautology capturing the $\mathcal{P}/\text{poly}$ versus $\mathcal{NP}$ question, thus using proof complexity to try to show that proving circuit lower bounds is difficult.

This tautology, suggested by Razborov, simply encodes the statement $\mathcal{NP} \not\subseteq \mathcal{P}/\text{poly}$, namely, that *SAT* does not have small circuits. More precisely, fix n, an input size to *SAT*, and s, the circuit size lower bound we attempt to prove.[24] The variables of our lower bound formula LB_n^s encode a circuit C of size s, and the formula simply checks that C disagrees with *SAT* in at least one instance ϕ of length n (namely, that either $\phi \in SAT$ and $C(\phi) = 0$ or $\phi \notin SAT$ and $C(\phi) = 1$). Note that LB_n^s has size $N = 2^{O(n)}$, so we seek a superpolynomial in N lower bound on its proof length.[25]

---

[24] For example, we may choose $s = n^{\log \log n}$ for a superpolynomial bound or $s = 2^{n/1000}$ for an exponential bound.

[25] Of course, if $\mathcal{NP} \subseteq \mathcal{P}/\text{poly}$, then this formula is not a tautology, and there is no proof at all.

Proving that LB_n^s is hard for Frege will in some sense give another explanation for the difficulty of proving circuit lower bounds. Such a result would be analogous to the one provided by natural proofs, only without relying on the existence of one-way functions. But paradoxically, the same inability to prove circuit lower bounds seems to prevent us from proving this proof complexity lower bound. Even proving that LB_n^s is hard for Resolution has been extremely difficult. It involves proving hardness of a weak pigeonhole principle[26] – one with exponentially more pigeons than holes. It was finally achieved with the tour de force of Raz (2004)and with further strengthening of Razborov (2004).

Acknowledgments

I am grateful to Scott Aaronson and Christos Papadimitriou for carefully reading and commenting on an earlier version this chapter. I acknowledge support from NSF grant CCR-0324906. Some parts of this chapter are revisions of material taken from my ICM 2006 paper "$\mathcal{P}, \mathcal{NP}$ and Mathematics: A Computational Complexity View."

References

Adleman, L., and Manders, K. (1975). Computational complexity of decision problems for polynomials. *Proceedings of 16th IEEE Symposium on Foundations of Computer Science*, pp. 169–77. Los Alamitos, CA: IEEE Comput. Soc. Press.

Agol, I., Hass, J., and Thurston, W. P. (2006). The computational complexity of knot genus and spanning area. *Trans. Amer. Math. Sci.*, **358**, 3821–50.

Agrawal, M., Kayal, N., and Saxena, N. (2004). Primes is in $\mathcal{P}$. *Ann. Math.*, **160**, 781–93.

Ajtai., M. (1983). Σ_1-formulae on finite structures. *Ann. Pure Appl. Logic*, **24**, 1–48.

Arora, S., and Barak, B. (2009). *Computational Complexity: A Modern Approach*. New York: Cambridge University Press.

Baker, T., Gill, J., and Solovay, R. (1975). Relativizations of the P = ?NP question. *SIAM J. Comput.*, **4**, 431–42.

Beame P., and Pitassi, T. (1998). Propositional proof complexity: Past, present, and future. *Bull. EATCS*, **65**, 66–89.

Ben-Sasson, E., and Wigderson., A. (1999). Short proofs are narrow – resolution made simple. In *Proceedings of the 31st Annual ACM Symposium on Theory of Computing*, pp. 517–26. New York: ACM Press.

Boppana, R., and Sipser, M. (1990). The complexity of finite functions. In *Handbook of Theoretical Computer Science, Volume A, Algorithms and Complexity*, ed. J. van Leeuwen, pp. 757–804. Amsterdam, Netherlands: Elsevier Science.

Buss. S. (1987). Polynomial size proofs of the propositional pigeonhole principle. *J. Symbolic Logic*, **52**, 916–27.

Buss. S. (1995). On Gödel's theorems on lengths of proofs II: Lower bounds for recognizing k symbol provability. In *Feasible Mathematics II*, ed. P. Clote and J. Remmel, pp. 57–90. Boston: Birkhauser.

Chvátal, V. (1973). Edmonds polytopes and a hierarchy of combinatorial problems. *Discrete Math.*, **4**, 305–37.

---

[26] This explicates the connection mentioned between the pigeonhole principle and the counting argument proving existence of hard functions.

Chvátal, V. and Szemerédi, E. (1998). Many hard examples for resolution. *J. ACM*, **35**, 759–68.

Clegg, M., Edmonds, J., and Impagliazzo, R. (1996). Using the Groebner basis algorithm to find proofs of unsatisfiability. In *Proceedings of the 28th Annual ACM Symposium on Theory of Computing*, pp. 174–83. New York: ACM Press.

Cobham, A. (1965). The intrinsic computational difficulty of functions. In *Logic, Methodology, and Philosophy of Science*. Amsterdam Netherlands: North-Holland.

Cook, S. A. (1971). The complexity of theorem-proving procedures. In *Proceedings of the 3rd Annual ACM Symposium on Theory of Computing*, pp. 151–8. New York: ACM Press.

Cook, S. A., and Reckhow, R. A. (1979). The relative efficiency of propositional proof systems. *J. Symbolic Logic*, **44**, 36–50.

Cormen, T. H., Leiserson, C., and Rivest, R. (2001). *Introduction to Algorithms*. 2nd ed. Cambridge, MA: MIT Press.

Davis, M., Logemann, G., and Loveland., D. (1962). A machine program for theorem proving. *J. ACM*, **5**, 394–97.

Edmonds, J. (1965a). Minimum partition of a matroid into independent sets. *J. Res. Nat. Bur. Standards (B)*, **69**, 67–72.

Edmonds, J. (1965b). Paths, trees, and flowers. *Can. J. Math.*, **17**, 449–67.

Fortnow. L. (1994). The role of relativization in complexity theory. *Bull. EATCS*, **52**, 229–44.

Furst, M., Saxe, J., and Sipser, M. (1984). Parity, circuits and the polynomial time hierarchy. *Math. Systems Theory*, **17**, 13–27.

Garey, M. R., and Johnson, D. S. (1979). *Computers and Intractability: A Guide to the Theory of NP-Completeness*. New York: W. H. Freeman.

Goldreich, O. (2008). *Computational Complexity: A Conceptual Perspective*. New York: Cambridge University Press.

Granville, A. (2005). It is easy to determine whether a given integer is prime. *Bull. Amer. Math. Soc.*, **42**, 3–38.

Gromov, M. (1987). Hyperbolic groups. In *Essays in Group Theory*, ed. S. M. Gersten, pp. 75–264. New York: Springer.

Gromov, M. (2003). Random walk in random groups. *Geom. Funct. Anal.*, **13**, 73–146.

Haken, W. (1961). Theorie der Normalflächen: ein Isotopiekriterium für den Kreisknoten. *Acta Math.*, **105**, 245–375.

Haken, A. (1985). The intractability of resolution. *Theor. Comput. Sci.*, **39**, 297–308.

Hartmanis, J. (1989). Gödel, von Neumann and the P = ?NP problem. *Bull. EATCS*, **38**, 101–7.

Hass, J., and Lagarias, J. C. (2001). The number of Reidemeister moves needed for unknotting. *J. Amer. Math. Soc.*, **14**, 399–428.

Hass, J., Lagarias, J. C., and Pippenger, N. (1999). The computational complexity of knot and link problems. *J. ACM*, **46**, 185–211.

Karchmer, M., and Wigderson, A. (1990). Monotone circuits for connectivity require super-logarithmic depth. *SIAM J. Discrete Math.*, **3**, 255–65.

Karmarkar, N. (1984). A new polynomial-time algorithm for linear programming. *Combinatorica*, **4**, 373–94.

Karp, R. (1972). Reducibility among combinatorial problems. In *Complexity of Computer Computations*, ed. R. E. Miller and J. W. Thatcher, pp. 85–103. New York: Plenum Press.

Karp, R., and Lipton, R. J. (1982). Turing machines that take advice. *Enseign. Math.*, **28**, 191–209.

Khachian, L. (1979). A polynomial time algorithm for linear programming. *Soviet Math. Doklady*, **10**, 191–94.

Lenstra, A. K., Lenstra, H. W., Jr., and Lovász, L. (1982). Factoring polynomials with rational coefficients. *Math. Ann.*, **261**, 515–34.

Levin, L. A. (1973). Universal search problems. *Probl. Peredaci Inform*, **9**, 115–16; [English trans. *Probl. Inf. Transm.*, **9** (1973), 265–66.]

Matiasevich, Y. V. (1993). *Hilbert's Tenth Problem*. Cambridge, MA: MIT Press.

Miller, G. L. (1976). Riemann's hypothesis and tests for primality. *J. Comput. System Sci.*, **13**, 300–17.

Papadimitriou, C. H. (1994a). *Computational Complexity*. Reading, MA: Addison Wesley.

Papadimitriou, C. H. (1994b). On the complexity of the parity argument and other inefficient proofs of existence. *J. Comput. System Sci.*, **48**, 498–532.

Papadimitriou, C. H. (1997). NP-completeness: A retrospective. In *Automata, Languages and Programming* (ICALP 1997), Lecture Notes in Computer Science 1256, ed. P. Degano, R. Gorrieri, and A. Marchetti-Spaccamela, pp. 2–6. Berlin: Springer.

Pudlak, P. (1995). Logic and complexity: Independence results and the complexity of propositional calculus. In *Proceedings of International Congress of Mathematicians 1994*, ed. S. D. Chatterji, pp. 288–97. Basel, Switzerland: Birkhäuser.

Pudlak, P. (1996a). A bottom-up approach to foundations of mathematics. In *Gödel '96: Logical Foundations of Mathematics, Computer Science and Physics – Kurt Godel's Legacy* (Proceedings), Lecture Notes in Logic 6, ed. P. Hajek, pp. 81–97. Berlin: Springer.

Pudlak, P. (1996b). On the lengths of proofs of consistency. *Collegium Logicum, Ann. Kurt-Godel-Society*, **2**, 65–86.

Pudlak, P. (1997). Lower bounds for resolution and cutting planes proofs and monotone computations. *J. Symbolic Logic*, **62**, 981–98.

Pudlak, P. (1998). The lengths of proofs. In *Handbook of Proof Theory*, ed. S. R. Buss, pp. 547–637, Amsterdam, Netherlands: Elsevier.

Rabin, M. (1967). Mathematical theory of automata. In *Mathematical Aspects of Computer Science*, Proceedings of Symposia in Applied Mathematics 19. Providence, RI: American Mathematical Society.

Raz, R. (2004). Resolution lower bounds for the weak pigeonhole principle. *J. ACM*, **51**, 115–38.

Razborov, A. A. (1998). Lower bounds for the polynomial calculus. *Comput. Complexity*, **7**, 291–324.

Razborov, A. A. (2004). Resolution lower bounds for perfect matching principles. *J. Comput. System Sci.*, **69**, 3–27.

Razborov, A. A., and Rudich, S. (1997). Natural proofs. *J. Comput. System Sci.*, **55**, 24–35.

Robertson, N., and Seymour, P. (1983–1995). Graph Minors I–XIII. *J. Combin. Theory B*.

Rudich, S., and Wigderson, A. eds. (2000). *Computational Complexity Theory*. IAS/Park City Mathematics Series 10. American Mathematical Society/Institute for Advanced Study.

Schrijver, A. (2003). *Combinatorial Optimization: Polyhedra and Efficiency*. Algorithms and Combinatorics 24. Berlin: Springer.

Shannon, C. E. (1949). The synthesis of two-terminal switching circuits. *Bell Systems Technical J.*, **28**, 59–98.

Sipser, M. (1992). The history and status of the P versus NP question. In *Proceedings of the 24th Annual ACM Symposium on Theory of Computing*, pp. 603–18. New York: ACM Press.

Sipser, M. (1997). *Introduction to the Theory of Computation*. Boston: PWS.

Tarski, A. (1951). *A Decision Method for Elementary Algebra and Geometry*. University of California Press.

Turing, A. M. (1936). On computable numbers, with an application to the Entscheidungsproblem. *Proc. London Math. Soc., Ser. 2*, **42**, 230–65.

Turing, A. M. (1937). On computable numbers, with an application to the Entscheidungsproblem: A correction. *Proc. London Math. Soc., Ser. 2*, **43**, 544–46.

Vinodchandran, N. V. (2004). $AM_{exp} \not\subseteq (NP \cap coNP)/poly$. *Inform. Process. Lett.*, **89**, 43–47.

Index

Printed in the United States
By Bookmasters